Bioethics of Displacement and Its Implications

Manuel Lozano Rodríguez
American University of Sovereign Nations, Spain

IGI Global
PUBLISHER of TIMELY KNOWLEDGE

A volume in the Advances in Medical Education, Research, and Ethics (AMERE) Book Series

Published in the United States of America by
IGI Global
Medical Information Science Reference (an imprint of IGI Global)
701 E. Chocolate Avenue
Hershey PA, USA 17033
Tel: 717-533-8845
Fax: 717-533-8661
E-mail: cust@igi-global.com
Web site: http://www.igi-global.com

Library of Congress Cataloging-in-Publication Data

Names: Lozano Rodriguez, Manuel, 1978- author.
Title: Bioethics of displacement and its implications / by Manuel Lozano Rodriguez.
Description: Hershey, PA : Medical Information Science Reference, [2023] | Includes bibliographical references and index. | Summary: "This book deals with the topic that has been understudied of what Bioethics of Displacement is and what are their implications, subverting the anthropocentric moral view for the resulting loving, knowing and experiencing from the displaced ones, those who have lost their center and offering a case study from the Salvadorian humanitarian crises that will shed a new light on how political decision-making works"-- Provided by publisher.
Identifiers: LCCN 2022022393 (print) | LCCN 2022022394 (ebook) | ISBN 9781668448083 (hardcover) | ISBN 9781668448090 (ebook)
Subjects: LCSH: Bioethics. | Displacement (Psychology)--Philosophy.
Classification: LCC QH332 .L69 2022 (print) | LCC QH332 (ebook) | DDC 174.2--dc23/eng/20220603
LC record available at https://lccn.loc.gov/2022022393
LC ebook record available at https://lccn.loc.gov/2022022394

This book is published in the IGI Global book series Advances in Medical Education, Research, and Ethics (AMERE) (ISSN: 2475-6601; eISSN: 2475-661X)

British Cataloguing in Publication Data
A Cataloguing in Publication record for this book is available from the British Library.

For electronic access to this publication, please contact: eresources@igi-global.com.

Advances in Medical Education, Research, and Ethics (AMERE) Book Series

ISSN:2475-6601
EISSN:2475-661X

Mission

Humans are living longer now than ever as a result of advances in the medical field. Having the tools available to train knowledgeable and ethical future generations of doctors and medical researchers is essential to continuing to advance our understanding of the human body and develop new ways of treating and curing sickness and disease.

The **Advances in Medical Education, Research, and Ethics (AMERE)** book series highlights publications pertaining to advancements in pedagogical practice for developing future healthcare professionals, research methods, and advancements in the medical field, as well as moral behavior and practice of healthcare professionals, students, and researchers. Featuring research-based book publications that are highly relevant to the healthcare community, this series is ideally designed for library inclusion at medical universities and research institutions as well as personal use by medical professionals, researchers, and upper-level students entering the field.

Coverage

- Clinical Research
- Conflicts of Interest
- Ethics in Medicine
- Healthcare Pedagogy
- Medical Curricula
- Medical Simulation
- Patient Data
- Professional Development
- Research Methods
- Scientific Misconduct

IGI Global is currently accepting manuscripts for publication within this series. To submit a proposal for a volume in this series, please contact our Acquisition Editors at acquisitions@igi-global.com or visit: https://www.igi-global.com/publish/.

The Advances in Medical Education, Research, and Ethics (AMERE) Book Series (ISSN 2475-6601) is published by IGI Global, 701 E. Chocolate Avenue, Hershey, PA 17033-1240, USA, www.igi-global.com. This series is composed of titles available for purchase individually; each title is edited to be contextually exclusive from any other title within the series. For pricing and ordering information please visit https://www.igi-global.com/book-series/advances-medical-education-research-ethics/132365. Postmaster: Send all address changes to above address.

Titles in this Series

For a list of additional titles in this series, please visit: www.igi-global.com/book-series/advances-medical-education-research-ethics/132365

Cases on Teaching Pharmacology to Complementary and Alternative Medicine Students
Muhammad Shahzad Aslam (School of Traditional Chinese Medicine, Xiamen University Malaysia, Malaysia)
Medical Information Science Reference • © 2023 • 209pp • H/C (ISBN: 9781668478288) • US $295.00

Handbook of Research on Instructional Technologies in Health Education and Allied Disciplines
Manuel B. Garcia (FEU Institute of Technology, Philippines) Mildred Vanessa Lopez Cabrera (Tecnológico de Monterrey, Mexico) and Rui Pedro Pereira de Almeida (University of Algarve, Portugal)
Medical Information Science Reference • © 2023 • 439pp • H/C (ISBN: 9781668471647) • US $425.00

Cases on Diversity, Equity, and Inclusion for the Health Professions Educator
Chasity Beth O'Malley (Wright State University Boonshoft School of Medicine, USA) Arkene Levy (Nova Southeastern University Dr. Kiran C. Patel College of Allopathic Medicine, USA) Amanda Chase (Nova Southeastern University Dr. Kiran C. Patel College of Allopathic Medicine, USA) and Samiksha Prasad (Nova Southeastern University Dr. Kiran C. Patel College of Allopathic Medicine, USA)
Medical Information Science Reference • © 2023 • 369pp • H/C (ISBN: 9781668454930) • US $295.00

Contemporary Issues in Global Medicine and Moving Toward International Healthcare Equity
Nick Comninellis (University of Missouri-Kansas City, USA) and Steven D. Waldman (University of Missouri-Kansas City School of Medicine, USA)
Medical Information Science Reference • © 2022 • 463pp • H/C (ISBN: 9781799884903) • US $325.00

Simulation and Game-Based Learning for the Health Professions
Rachel Umoren (University of Washington, USA)
Medical Information Science Reference • © 2022 • 275pp • H/C (ISBN: 9781799843788) • US $270.00

Handbook of Research on Developing Competencies for Pre-Health Professional Students, Advisors, and Programs
Rohini Ganjoo (The George Washington University, USA) and Lisa S. Schwartz (The George Washington University, USA)
Medical Information Science Reference • © 2022 • 492pp • H/C (ISBN: 9781668459690) • US $470.00

Handbook of Research on Improving Allied Health Professions Education Advancing Clinical Training and Interdisciplinary Translational Research
Rui Pedro Pereira Almeida (University of Algarve, Portugal)
Medical Information Science Reference • © 2022 • 409pp • H/C (ISBN: 9781799895787) • US $435.00

IGI Global
PUBLISHER of TIMELY KNOWLEDGE

701 East Chocolate Avenue, Hershey, PA 17033, USA
Tel: 717-533-8845 x100 • Fax: 717-533-8661
E-Mail: cust@igi-global.com • www.igi-global.com

To Elsa and for her, too.

To those few who stood by me against all odds.

Table of Contents

Preface viii

Acknowledgment xx

Introduction xxi

Chapter 1
Bioethics From Science to Dogma and the Intersections With Religion, Politics, and Gender 1

Chapter 2
Preemptive Bioethics and Syncretist Approach for Ethical Challenges in AI and Robotics: Good v Evil I 24

Chapter 3
Psychology, Moral Principia, and Gehenna: Good v Evil II 35

Chapter 4
Transcendental Bioethics of Displacement: Good v Evil III 49

Chapter 5
On the Misery of Bioethics: Good v Evil IV 58

Chapter 6
Bioethics of Displacement Across Time and Eras: Genesis v Gehenna I 72

Chapter 7
Examining Biopolitics and Bioethics Through Abusive Pain Management: Genesis v Gehenna II 92

Chapter 8
Critiquing the Psychological Inflexibility and the Politics of Pain in Woke Politics and Neoliberalism's Moral Governance: Genesis v Gehenna III 104

Chapter 9
Individual Mindset Change and Civil Disobedience in Shaping Societal Re-Evolution: Praxis I 123

Chapter 10
Dignity and Engagement for Making Bioethics of Displacement Real: Praxis II 132

Chapter 11
The Paradigmatic Displacement of Migration: Belonging v Set Apart I 144

Chapter 12
Displaced, Aging, and Cyborgs' Migration Through Borders and Beyond Borders: Belonging v Set Apart II 164

Chapter 13
Exploring Human Identity From Social, Cultural, Philosophical, and Biopolitical Perspectives: Free v Confined I 181

Chapter 14
An In-Depth Analysis of the Bioethical Implications of Stigma: Free v Confined II 199

Chapter 15
Examining Confinement Beyond Bars From Philosophical and Practical Perspectives: Free v Confined III 217

Chapter 16
Meditations on the Intrinsic Value of Life and the Present and Future Nature of Death and Dying: Worthy v Lost I 239

Chapter 17
The Ethics Behind Neuroethics, Neurobioethics, Sensory Enhancement, and Surrogate Realities: Worthy v Lost II 256

Chapter 18
Extraterrestrial Bioethics: Worthy v Lost III 271

Chapter 19
The Cosmos of Bioethics of Displacement.. 282

Conclusion 294

Appendix... 300

Compilation of References 314

About the Author 382

Index.. 383

Preface

THINKING ABOUT DISPLACEMENT AS NEVER BEFORE

This book is not your typical bioethics book, and readers looking for medical bioethics and forcibly-displaced persons as core topics should look elsewhere. Instead, it explores the concept of belonging and the feeling of being accepted, valued, and connected to a community or group of individuals. Belonging is an important aspect of human existence, providing a sense of identity, purpose, and contributing to overall well-being and mental health. This chapter challenges dominant narratives and assumptions about what it means to be human, alive, or worthy, and explores alternative possibilities for belonging and freedom. It also examines the political and ethical implications of different ways of thinking about and valuing life, including an ethical view on disability, transhumanism, and other areas of study. The chapter encourages critical thinking, inspires communal action for belonging, freedom, and human worth, and explores new possibilities for human understanding and discovery.

DON'T READ THIS BOOK IF YOU LOOK FOR...

Before anything else, I want you to know I appreciate your time. It won't be back and the world is distracting enough. Since life is very inelastic and the library has plenty of books, don't read this one if you are looking for medical bioethics and forcibly-displaced persons as core topics. Both topics will appear (and soon) but as secondary actors. Here, displacement is more about being pushed away from a meaningful life. So, I aim for a bigger picture and resorted more often to social sciences, politics and history than medicine. Even if bioethics won't be explained as usual, you will find rigorous work. Bet on it! But, on the other hand, nothing here is shoehorned in academicism. Indeed, you won't find only academic content and over 5-syllable words across these pages. There is also anomia, strangeness and fear gliding like fall leaves. On the other hand, plenty of music, movies and stories worth to be told and remembered await. Warning: bittersweet irony, pop culture references and even (OMG!) slang awaits you ahead.

Bioethics of Displacement (Bodi to its friends) is a bioethics for politically suspicious minds. Take it whichever way you want but, after so many generations of swindlers, only with our cognitive biases there is enough to keep the causal arrow pointing from elites to people. And it happens regardless of democratic whoppers. As a result, the human brain, which should be a grey daring frontier, is closer to a grey dumping site. There are billions of pounds of plastic in the Ocean awaiting removal and about the same weight of brains in dry land casted aside when it comes to choosing the future of our planet. We are muffled. We are aside. We trust. But while we trust, those who suffer the most from displacement walk covered in ashes of practical invisibility. In 2010, there were 40 million displaced persons suffering from a disabling condition, many of them with their needs falling outside the social minimum of what humanitarian healthcare should be. But don't read this book if you want to know more about the bioethical dilemmas in the healthcare of forcibly displaced populations. All those biomedical issues are well-covered by mainstream bioethics. Here, you'll find how those lives have been dragged away from a true human life and turned into a digestible human mass by the ones who talk on behalf of human life. Meanwhile, most people pay attention to all those who can raise their political voice louder hoping their success leaks on them. Till the whip changes hands, we await. We obey. We trust. But while we trust a concealed slippery slope may be under our feet.

In this book, you'll find the word "elite" many times so, just like happens in many similar books, the chances for tinfoil hats wearing are real. For this reason, I think that we should talk about whip holders now. Most times, examples of elites are Davos, Bilderberg Group and credit rating agencies along with many unglued examples such as reptilians and alike. To begin with, I don't think that there is a global elite ruling the world over. Not as such, at least. As I see it, those visible global elites as Bilderberg Group are nothing but another node in a larger network of individuals who have a say in local oligopolies of violence and debt. The main difference is the visibility of the node rather than its relative relevance, which is hard to say. Global elites' network is built on membership fluidity, privileged information, and a shared goal of human herding. But a shared goal doesn't mean a common goal. Thus, hardly there is an apex of apexes -unless you believe in a puppeteer Satan, alien Masters race, Zion protocols or any other rigmarole. Therefore, don't read this book looking for satisfying your confirmation bias on conspiracy theories.[1]

What makes me wonder if there is a conspiracy is how lightly the bioethics debate on obedience is taken. If you look for "bioethics of obedience" in Google Scholar there are no results. There it was about time! Human life and new choices on human life take place along a structural framework of rules. Needless to say, rules are not the same for all. Just like happens in Japanese, the same word can have a different meaning (or become meaningless) if you read it top down or bottom up. Rules that will determine if you live or die are written by people away whose closer common good hardly overlaps with yours and who give a damn about your existence. Besides the musty due obedience, displacing people's flesh and views to make them fall into the line is done in many ways we'll address later. Back to the rule o'er life, it is significant that this fits into the interests of those who can sway individuals, groups and peoples till they get empty and without anything to share. It doesn't always happen in lively and self-evident ways, but underpowered demographics are, individually and collectively, turned into an Other. A displace-able quasi subject for politics but a worthy subject of research for bioethics. Bioethics is pre-human but written down bioethics by this name is mostly *postmo* and has grown with the ethics industry. In its mainstream form, bioethics appeared in the US peakest, when its presidents' hubris in Vietnam casted shadows of technological wariness. It was the time of the Tuskegee experiments where low-income Black people were used as guinea pigs. Both the Vietnam War and Tuskegee experiments ended a few years later, but as technology empowered asymmetrical warfare abroad and Earth degraded, all this moral mess sprouted in more and more countries rather than end. These pressures excite a present-day relentless political will to tell apart the lives of those worthy of being guarded. Cherry-picking those who are unlikely to suffer from biomedical derangement, life-sustaining environment decline and profiteering from the Others. If you want to know more on this telling apart life from life, wellness from wellness, human from human, even beyond Earth and living bodies, then you *must* read this book.

WHAT IS THIS BOOK ABOUT?

It's a book about belonging versus being casted aside because a brightspark suit bloody says so. Belonging as a human being refers to the feeling of being accepted, valued, and connected to a community or group of individuals. It involves a sense of fitting in and being understood, as well as being able to contribute and make a meaningful impact. This feeling of belonging can arise from shared experiences, beliefs, or cultural background, among other things. It is an important aspect of human existence, as it helps to provide a sense of identity and purpose, and contributes to overall well-being and mental health. To belong you need to be alive and able to love. It means being a subject for bioethics. The diagram below illustrates the interdependence of the main topics addressed in this book. At the centre of the diagram is the concept of belonging, which is closely linked to the idea of being human and alive. Belonging also involves navigating boundaries, whether they are self-imposed or externally imposed, as well as the concept of otherness, which can lead to feelings of alienation and confinement.

There is a despair and loneliness that is an inherent aspect of the human condition. We are doomed to confront it. It is necessary to mature and come to terms with the sticky absurdity that stains existence so many times. But we won't talk about these unavoidable deep, blue, bad streaks that reach the most. Rather than this we'll look into the crafted despair oozing from the above in the form of politics. Just like "elites", "politics" is also a common word in this book and, therefore, an important concept for Bodi. We'll define politics as a network for human herding and getting the means for doing so. Just like happens with most Neolithic game changes such as agriculture, nowadays politics has a family resemblance anywhere: a very few people at the helm, an extended human livestock and, unlike big apes, a bureaucracy in the middle. Almost any political analysis you find will aim at this or that individual elite, generally presenting it as the most global among the global elites. Rather than this, we'll focus on the evolution of the global elite network as a whole.

To belong you need others but not Others. I mean that you need other people you can share with. But if a moral agent is so far away from your-self that you can't share, then it's an Other. And you can't belong to the Others. It's not bad that Others exist. The river crabs are, overall, my Others but I wish no evil to them. Without the Others everything would be too psychedelic for our own good. The problems come when Otherness is fueled by an impersonal powerthirst wrapping the life and future of Others. The only way of belonging that people with this condition accept is hierarchical. Thanks to how decisions spread across bureaucratic organisations, this evil can be done in heartrending banal ways. As History shows (sic).

We'll extend far and thoroughly in how this evil is done. But there will be room for world repairing. Bodi emphasises the importance of recognizing and respecting individual differences and the importance of building strong and supportive communities committed to the values of justice and humanity.

WHAT IS THIS BOOK FOR?

The main goal of this Bioethics of Displacement *is exposing tyranny by other names and how it works against love of belonging.*

Other goals are:

- Challenging dominant narratives and assumptions about what it means to be human, alive, or worthy, and exploring alternative possibilities for belonging and freedom.
- Examining the political and ethical implications of different ways of thinking about and valuing life, including the intersections of bioethics, disability, transhumanism, and other areas of study.
- Exploring the possibilities and limitations of emerging technologies such as cybernetics, exobiology, and spiritual AIs, and the ways in which these technologies might challenge or reinforce existing power structures.
- Going beyond the woke's echo chamber when examining the historical and ongoing effects of systems of oppression and discrimination against marginalised groups, and the ways in which these systems shape our understandings of belonging, otherness, and worthiness.
- Encourage critical thinking and challenge beliefs and assumptions, as well as to question the political, social, and cultural interests that perpetuate injustice and inequality.
- Inspire communal action for belonging, freedom, and human worth by highlighting the experiences of those who have overcome marginalisation and discrimination.
- Bring attention to the important social issues around shunning, that are often overlooked or ignored.
- Explore the unknown by merging topics like exobiology, transhumanism, and spiritual AI and examining the opening up of new possibilities for human understanding and discovery.
- Inspiring creative critical thinking by incorporating elements of pop culture, futurism, cybernetics, and exobiology.

Besides philosophers, scientists, researchers and activists (whether armchair or not) this book can also be useful for the general public, who can find hope. By exploring topics like near-death experiences and the potential for collective freedom, the book can offer readers hope for a better future and a more just and compassionate society. Furthermore, by exploring the role of serendipity and liminal life, the book can help readers to appreciate the unexpected and to find meaning in the most challenging experiences. Bodi can encourage readers to reflect on their own experiences and beliefs and consider how they might expand their perspectives.

WHAT DOES THIS BOOK CONTAIN AND HOW DOES IT FIT IN THE REAL WORLD?

Structure

Now we'll see what's up with these topic families.

Call Me Bodi

This section begins by introducing the concept of Displacement and its impact on our daily lives. It highlights that displacement is not limited to physical movement but can be experienced in various situations, such as identity struggles and a sense of aimlessness. The unspeakable nature of displacement is highlighted, acknowledging that certain experiences are difficult to articulate and fully comprehend.

Understanding this unspeakability can help us navigate the complexities of our inner lives and the world around us.

Table 1. Book chapters by topic families and sections

Family or Section	Chapters
Call Me Bodi	Introduction, 1
Good v Evil	2, 3, 4 and, 5
Genesis v Gehenna	6, 7 and 8
Praxis v Lazy	9 and 10
Belonging v Set Apart	11 and 12
Free v Confined	13, 14 and 15
Worthy v Lost	16, 17 and 18
Outro	19, Recommendations, Conclusion

Continuing the exploration, Chapter 1 delves into the rise of bioethics as a relevant field of study in today's world. Bioethics is defined as a discipline that addresses ethical issues related to life, health, and medicine. The chapter stresses the significance of bioethics in navigating intricate ethical dilemmas arising from scientific advancements, medical practices, and social policies. It also explores how bioethics intersects with religion, cultural diversity, and environmental sustainability in the realms of healthcare and research. Furthermore, it addresses the relationship between bioethics and biopolitics, which involves the use of political power to regulate biological processes.

In addition, the chapter discusses the influence of traditional religions and highlights the importance of different faiths within the field of bioethics. It acknowledges the challenges involved in achieving cultural competency and underscores the need for a broader vision that goes beyond limiting oneself to particular traditions. The politicisation of bioethics is also explored, considering both traditional religions and secular ideologies associated with partisan politics. The chapter emphasizes the burden of political feminism, highlighting the importance of situational and contextual considerations.

Overall, the Introduction and Chapter 1 of "Call Me Bodi" present a continuum of exploration into the concept of Displacement and its communion with Bioethics. The Introduction establishes the groundwork by discussing the unspeakable nature of displacement, while Chapter 1 delves into the rise of bioethics, traditional and secular religions and politics influences, and the necessity of adopting a comprehensive and open-minded approach to ethical decision-making in diverse contexts. Together, these chapters contribute to a cohesive examination of Displacement and its implications in the realm of bioethics.

Good v Evil

This family has as a common thread grappling with ethical challenges and seeking morally desirable outcomes. This overarching theme of ethical decision-making and moral considerations is where the concept of Good v Evil emerges. The second chapter introduces the ethical issues associated with AI and robotics, including bias, discrimination, and safety concerns. These challenges can be seen as manifestations of potential "evils" that need to be addressed in order to ensure the "good" outcomes and ethical practices. The focus on preemptive bioethics highlights the proactive approach to anticipate and mitigate these potential negative impacts, thereby emphasising the importance of aligning with what is considered good. In the third chapter, psychological flexibility is discussed as a key requirement to navigate complex ethical issues related to displacement. The emphasis on reflective decision-making and mindful communication suggests a conscious effort to distinguish between the morally right and wrong actions. This reflects the struggle between good and evil, as individuals grapple with making ethical choices that align with what is considered morally good. The fourth chapter explores the definition of "good bioethics" by emphasising genuine communication, consideration of future generations, and the Transcendental Bioethics of Displacement. The inclusion of the term "Transcendental" implies a search for higher ethical principles that can guide decision-making in bioethical issues. By delving into the relationship between energy and displacement, this family of topics seeks to understand the experiences of individuals with certain conditions and promote what is considered good in terms of their well-being and quality of life. Finally, the fifth chapter addresses the downstream burden of illness and highlights the need for solidarity, justice, and equality for vulnerable populations. The discussion of crony capitalism, left-wing populism, and intersectionality reflects an awareness of the potential negative influences that can perpetuate inequality and injustice, opposing what is generally considered good. The focus on collective freedom, belonging, and social justice aligns with the broader struggle of promoting the good and overcoming the forces of evil. Together, this family of topics contribute to a collective discussion of ethics, where the concept of Good v Evil is implicit in the exploration of ethical challenges, the search for proactive approaches, the emphasis on reflective decision-making, the consideration of well-being and vulnerability, and the pursuit of justice and equality. By addressing these topics, the abstracts collectively engage with the fundamental question of how to navigate the complexities of bioethics and social issues in a way that promotes the good and minimises or counters the influence of evil.

One real-life example that can exemplify the application of Bioethics of Displacement is the displacement of indigenous communities caused by the construction of a large hydroelectric dam. Let's consider the case of the Belo Monte Dam in Brazil, which has led to the forced displacement of several indigenous populations, including the Juruna and Xikrin communities. In this case, the moral philosophy of Bioethics of Displacement can be employed to address the ethical issues related to the displacement. The framework of Preemptive bioethics can be utilised to anticipate and address potential challenges arising from the construction of the dam, such as the impact on the communities' physical and psychological well-being, their cultural heritage, and their rights to land and resources. Applying the Syncretist approach to ethical decision-making, guidelines and principles for the dam's design and operation can be established. This approach would take into account the perspectives, values, and needs of the affected indigenous communities. It would involve meaningful participation and consultation with the communities throughout the decision-making process, respecting their rights to self-determination and considering potential bias, discrimination, and safety concerns associated with the project. The case also highlights the importance of psychological flexibility when dealing with the ethical issues related to displacement. The affected communities experience profound existential and psychological impacts as they are uprooted from their ancestral lands. It is crucial to recognize and address their emotional well-being and cultural identity throughout the process, fostering resilience and ensuring their dignity and rights are upheld. Furthermore, the downstream burden of illness resulting from the displacement can be addressed through a bottom-up approach to solidarity and governance. This involves fairness, justice, and equality for the vulnerable populations affected by the dam's construction. Inclusive participation, transparent communication, and fair compensation for the communities are essential aspects of this approach. By analysing the complexities and consequences of this case through the lens of Bioethics of Displacement, we can stimulate critical thinking and contribute to a more ethical approach. This example demonstrates the importance of considering the vulnerabilities, perspectives, and rights of real populations facing displacement and the need for policies and actions that prioritise justice, equality, and collective well-being.

Genesis v Gehenna

Starting with the exploration of displacement across different periods of time in the sixth chapter, the historical perspective emphasises the long-term consequences of our actions on future generations and the environment. This sets the stage for understanding the shifts in power dynamics, resource access, and technology that occurred in each era, from the extinction of dinosaurs to the era of data and computing. These historical patterns provide valuable insights into the present global landscape and the challenges we face. Moving on to the politics of pain, in chapter seventh, the examples and stories illustrate the intersection of biopolitics and bioethics. Animal experimentation, the Tuskegee experiment, and the Guatemala study highlight how decision-makers have disregarded the well-being of vulnerable populations in the pursuit of scientific progress. The psychological inflexibility of individuals in high-income societies contributes to inadequate pain management practices, perpetuating a society of winners and losers where material progress and the politics of pain intersect. Finally, in the eighth chapter, exploring the concept of psychological inflexibility, the discussion extends beyond political ideologies, encompassing far-right, far-left, woke politics, and neoliberalism. This inflexible commitment to ideas that prioritise political or economic gain over ethical considerations perpetuates inequalities and injustices. It critiques the abuse of intersectionality theory and the creation of hierarchies that devalue certain groups. The conclusion warns against a narrow definition of morality and the dangers of moral governance led by technological and financial elites, advocating for the involvement of lesser social and economic stakeholders to shape a more equitable future. The Genesis v Gehenna discussion serves as a unifying thread that connects the different elements presented in this family of topics. By adopting a historical perspective and examining various epochs, we gain a deeper understanding of the long-term consequences of displacement on future generations and the environment. This historical analysis reveals the power dynamics driven by resource access, energy, and technology throughout different periods, underscoring the need to consider the ethical implications of our actions. The exploration of the politics of pain further reinforces this discussion by highlighting the ways in which decision-makers have exploited vulnerable populations and perpetuated inequalities in the pursuit of scientific progress. The concept of psychological inflexibility, encompassing far-right, far-left, woke politics, and neoliberalism, emphasises the dangers of rigid ideologies that prioritise political or economic gain over ethical considerations. It also critiques the abuse of intersectionality theory, urging for a more balanced approach to social justice. The family of topics concludes by advocating for a broader definition of morality and the importance of lesser social and economic stakeholders taking the lead in shaping a more equitable and ethical governance framework. In this way, the Genesis v Gehenna discussion ties together the complex interplay between displacement, power dynamics, ethics, and the need for a more responsible approach to decision-making.

In the context of displacement and its long-term consequences, one well-studied case that exemplifies the intersection of power dynamics and the politics of pain is the Tuskegee experiment in the United States. The Tuskegee study, conducted from 1932 to 1972, involved withholding treatment for syphilis from African American men without their informed consent. This unethical research disregarded the well-being of a vulnerable population, perpetuating inequalities and injustices. The psychological inflexibility of decision-makers prioritised scientific purposes and political gain over ethical considerations, contributing to the perpetuation of a society of winners and losers. By examining this historical example, we gain insights into the harmful consequences that arise when vulnerable populations are used for scientific purposes without regard for their well-being, underscoring the importance of ethical decision-making and the need to challenge narrow definitions of morality in shaping a more equitable future. The experiment's disregard for the long-term consequences and well-being of the participants aligns with the Gehenna perspective, emphasising the negative consequences that arise when ethical considerations are overshadowed by the willful ignorance of others' humanity. This case serves as a poignant reminder of the ethical challenges and power dynamics involved in displacement and highlights the need for a balanced approach prioritising the well-being and rights of individuals and communities over partisan ideas of progress.

Praxis v Lazy

The praxis of Bioethics of Displacement serves as the common thread throughout ninth and tenth chapters related to social injustice. By adopting a stance of civil disobedience and challenging caste systems, Bodi aims to address the injustices perpetuated by social structures. Quite the opposite, the role of patient rebels create a sense of belonging and draw attention to these injustices. Furthermore, the need for a global response to injustices and threats is highlighted, emphasising the moral agency of individuals in rebelling against injustice. The praxis of Bioethics of Displacement centres around moral legitimacy, emphasising the importance of rebellion rooted in principles rather than opportunism. Additionally, the discussion explores the attack on dignity in various forms, from forced organ donation to the marginalised populations of the Global South. It sheds light on how responsibility is displaced onto the disadvantaged themselves and how political dignity is compromised when revealing disadvantage becomes a requirement for compensation. However, the eleventh chapter also showcases successful challenges to the concept of dignity, such as the recognition of extensive geographical regions in New Zealand as legal persons, merging personhood and nonhuman dignity. Alongside dignity, the examination of engagement underscores the significance of recognizing and respecting the humanity, agency, and knowledge of individuals and communities affected by displacement. In essence, the praxis of Bioethics of Displacement connects these abstracts by providing a framework for addressing social injustices, protecting dignity, and promoting active engagement in creating a more just society.

One well-studied case that exemplifies the Bioethics of Displacement and its implications for real populations is the Nepali case of bonded labourers, also known as Kamaiyas. The author sheds light on the atrocities arising from the social structure that perpetuates caste-based inequality and the dual global pyramids. The Kamaiyas, belonging to the lowest caste, were subjected to bonded labour, where they were forced to work without wages and endured harsh living conditions. By naming the Kamaiya population and their struggle, the author highlights the need for patient rebels to create a sense of belonging and draw attention to the injustices faced by marginalised communities.

Belonging v Set Apart

Across chapters eleventh and twelfth, Belonging v Set Apart runs through the exploration of migration and displacement in various contexts. Whether it is the refugee crisis or transitions within life stages, the underlying theme revolves around the human desire for belonging and the challenges individuals face in their quest to be part of a community. This family of topics highlights the struggle to overcome being set apart from stable networks, infrastructure, and prosperous communities. It underscores the need for understanding, empathy, and a more inclusive response to address the barriers and challenges faced by migrants and displaced populations. The thread also delves into the impact of social, political, and cultural dimensions on identity, emphasising the importance of recognizing and respecting individuals' humanity, agency, and knowledge. Ultimately, the exploration of Belonging v Set Apart aims to shed light on the complexities of the different migrations, urging societies to foster a sense of belonging and solidarity while addressing the structural inequalities that perpetuate separation and marginalisation.

A well-studied case that exemplifies the complex issues involving Belonging v Set Apart dichotomy is the Syrian refugee crisis. Millions of Syrians have been forced to flee their homes due to the ongoing civil war in their country. They face significant challenges and structural barriers as they seek safety and a new life in other countries. The lack of economic diversity, destruction of local human networks and infrastructure, and discriminatory policies all contribute to their displacement and struggles. The public health implications are also evident, as refugees often face limited access to healthcare and increased vulnerability to diseases. This real-life example highlights the urgent need for a more humane response to the refugee crisis, with policies aimed at reducing discrimination and ensuring solidarity with displaced populations. It underscores the importance of understanding the human experience of migration and addressing the true impact of migration on global social change.

Free v Confined

The common thread of Free v Confined is explored through various dimensions of human identity and confinement at chapters thirteenth, fourteenth and fifteenth. This family of topics begins by examining social identity and its influence on self-concept and body-image, highlighting the ways in which technological progress and social structures shape our understanding of identity. The ethical dilemmas arising from the merging of body and information technologies are also discussed, including concerns about income inequality and genetic editing. The philosophical grounds of human identity are examined, emphasising the narrative nature of selfhood and its significance in bioethics. Additionally, the impact of public health and biopolitics on identity is explored, including the creation of fictive ethnicities and the challenges associated with displacement. About surrogate realities, we'll analyze the potential risks and concerns for identity related to the centralization of the metaverse, including the manipulation of mental health, privacy issues, distortion of reality, cyberbullying, and toxic social dynamics. Away from technology and closer to traditional philosophy, the concept of confinement is examined from historical and practical perspectives. The development of confinement as a colonial institution designed for punishment and isolation is traced, and the broader implications of confinement on individuals' identity are analysed. This family of topics also delves into the biopolitical and phenomenological aspects of confinement, exploring its ethical implications in relation to memory, disability, caretaking, and trauma. Ultimately, it calls for a more humane and thoughtful approach to various forms of confinement, one that recognizes the impact of technology on social and cultural norms and values.

In the context of exploring human identity and confinement, a well-studied case that exemplifies the themes discussed is the experience of Japanese Americans during World War II. The confinement of Japanese Americans in internment camps serves as a real-life example of how confinement affects people's identity. The historical development of confinement can be traced back to the establishment of these camps, where thousands of Japanese Americans were forcibly removed from their homes and confined under the pretext of national security. This case illustrates the profound impact of confinement on individuals' sense of self, their cultural identity, and their relationships with the broader society. The ethical implications of this confinement are evident, as it involved the infringement of civil liberties and the violation of personal freedoms based on racial profiling. Furthermore, the experience of confinement had long-lasting effects on the Japanese American community, including trauma, stigma, and the disruption of social and familial ties. Reflecting on the lessons learned from this case, it becomes evident the importance of preserving individual identity, safeguarding civil liberties, and addressing the ethical complexities that arise from such practices.

Worthy v Lost

The common thread of Worthy v Lost family of topics is centred around the exploration of human worth, value, and the potential for loss or devaluation. It highlights the ethical considerations and challenges associated with recognizing and preserving human dignity in various contexts. In the sixteenth chapter, the discussion revolves around end-of-life care and the creation of conscious machines. This raises questions about what makes an individual worthy and the potential loss of human value when merging body and information technologies. It prompts reflection on the complex issues surrounding death and the creation of conscious machines, emphasising the need to differentiate between the dead and the alive to ensure a better future. The seventeenth chapter deals with the impact of surrogate realities on human perception and quality of life. It raises ethical concerns about the potential risks and benefits of these technological devices and how they may affect human worth. By discussing the convergence of philosophy, science, and technology, the abstract underscores the importance of maintaining a balance between enhancing human experiences and preserving human dignity. It highlights the potential for loss or devaluation of human worth when delving into the realm of surrogate realities. The eighteenth and last chapter in this family focus on off-Earth bioethics, xenobioethics, and the ethical considerations surrounding the disposal of a cosmonaut's corpse. This abstract delves into the challenges of encountering extraterrestrial life and establishing a non-speciesist standard of personhood. It raises questions about the worth and value of life beyond traditional human boundaries. It emphasises the ethical dilemmas associated with the disposal of a cosmonaut's corpse and underscores the importance of preserving human dignity even in extreme circumstances.

For example, the Native American tribes in North America faced displacement, cultural disruption, and forced assimilation by European colonists, along with the genocide perpetrated by the US. It highlights the interrelation of individuals in their social, cultural, and historical contexts. Within this case study, the psychological factors that allowed colonists to participate in atrocities can be explored. Cognitive dissonance, psychological inflexibility, a desire for revenge, a sense of moral superiority, and obedience to authority were all influential factors that contributed to the mistreatment of indigenous populations. Leaders with psychopathic traits and a tendency towards oppressive politics further exacerbated the situation. Reflecting on this complex issue raises important ethical questions about the devaluation of indigenous lives and the consequences of displacing and reducing populations to mere objects or resources. It prompts us to consider the nature of death and the need to recognize the interconnectedness of all individuals, irrespective of their cultural backgrounds. This example showcases how the concepts discussed in the text, such as the ethical implications of technology and the examination of psychological factors, can be applied to a real-life scenario. It emphasises the need to understand the historical, social, and cultural contexts of populations and the impact of power dynamics on bioethical considerations.

Outro

This section is the end of this song about Belonging. It shares a common theme centred around the Bioethics of Displacement and its impact on humanity.

Chapter underscores the rapid advancements of society, particularly in Big Data empires, and the potential consequences that come with these changes. Using a metaphorical calendar, the chapter compresses significant historical events into a one-year timeline to stress the speed at which transformations occur. The text warns of the dangers of not adapting to these changes and the negative effects they can have on humanity. It encourages the adoption of values like self-kindness, human warmth, solidarity, and a sense of belonging to navigate displacement.

The Conclusion section further explores the longing for belonging and the ethical challenges associated with displacement. It argues for the development of a moral philosophy known as the Bioethics of Displacement (Bodi) to address these issues. Bodi puts the vulnerable first, considers the bigger picture, promotes proactive approaches, and maintains a healthy dose of scepticism. The section also examines concepts like off-Earth bioethics, xenobioethics, dignity, engagement, and the impact of the COVID-19 pandemic. It calls for a holistic approach to wellness, responsible governance, and recognizes the humanity and agency of displaced individuals and communities.

The cohesion between these chapters lies in their shared focus on the ethical implications of displacement and the need for thoughtful approaches to address them. Chapter 19 underscores the rapid changes and potential consequences of displacement, urging adaptation and the embrace of essential values. The Conclusion expands on these ideas, reflecting on the moral philosophy of the Bioethics of Displacement. It offers recommendations to navigate the complex ethical landscape, acknowledging the challenges faced by displaced individuals and communities. Together, both chapters contribute to the overarching theme of the book, shedding light on the ethical dimensions of displacement and advocating for inclusive approaches to tackle these issues.

Fitting in the Real World

A book with the contents mentioned above fits in the real world for good because it sheds light on the challenges faced by individuals and communities and emphasises the need for a more sensible response to such situations. The book also delves into the ethical implications of displacement and the complexities of decision-making in challenging circumstances. It introduces concepts like preemptive bioethics and syncretist ethical decision-making, which provide frameworks for considering and addressing potential ethical issues before they escalate and for approaching ethical dilemmas in a comprehensive and context-dependent manner. Furthermore, the book promotes empathy, understanding, and a broader perspective on human experiences and identity. It highlights the interconnectedness of individuals and the impact of power dynamics on bioethical considerations. By recognizing the intrinsic value and dignity of all human beings and prioritising the well-being of vulnerable populations, the book advocates for a more compassionate society but also a more sceptical one. Shortly, this book helps us understand the real-world problems faced by people and encourages us to stand by ourselves.

Bioethics of Displacement brings timely knowledge today because it exposes the complex issues surrounding Displacement and the ethical implications it entails. This book stresses the urgent need for a more creative response to Displacement, calling for policies that address the true impact of migration on global social change. It exposes the inauthenticity and necrophilic tendencies of the biopolitical Gehenna side, where individuals are treated as failures rather than acknowledging the failures of society. These pages promote a bioethics that prioritises the intrinsic value and dignity of all human beings, emphasising vulnerability, big-picture thinking, proactivity, and scepticism in ethical decision-making. Therefore, Bodi broadens prevailing views in bioethics and highlights the importance of a future-oriented approach that considers cultural, contextual, and communicational factors. It advocates for genuine, open, and well-grounded personal communication that welcomes all concerned voices and considers the interests of future generations. Overall, the book's critical and sceptical approach exposes the flaws in political power and the corrupt governance prevalent in Gehenna politics and puts emphasis on the value of loyalty. By shedding light on the psychological factors that contribute to human evil, this book serves as a warning against blindly following immoral ideologies and encourages readers to uphold our humanity in the face of adversity and wickedness.

Manuel Lozano Rodriguez
American University of Sovereign Nations, Spain

Acknowledgment

In Huang Fan's short story Dianti (The elevator), the protagonist, an overworked and confined night student, pushes his granny telephone number in the office elevator and it sends him back to home. When in the future I feel somewhat displaced I will dial the American University of Sovereign Nations (AUSN) number in the nearest elevator!

I feel grateful beyond words to the person who first trusted me when I was an academic dropout Prof. Darryl Macer and, extensively, to the AUSN to trust in me twice for a full grant. Their trust also wrote my PhD. thesis, from which this research originates from. I want to also give special thanks to all my colleagues, including Drs. Sheriff Dhastagir, Suma Parahakaran, Philippo Petros, Osama Rajhkan, Mihaela Serbulea, Sukran Sevimli, Ryo Takahashi, Anke Weisheit and Maria Yasuoka for the time and effort devoted to making the AUSN a great and greater educational project.

Special thanks to my unelected families and many teachers and classmates before joining the AUSN since their mocking, mistrust, hate, bizarre compassion and disregard towards my scientific vocation have turned me into a bold and bolder thinker throughout my life. I only point this out to cheer any disabled person, especially from a disadvantaged background, to think that there's also very good people around. May she spot this fleeting chance that enables us to go further in many human enterprises and that a bunch of self-confident mediocrities and overvalued individuals don't pay attention to.

Hopefully, just after sending this book to IGI, I will stop hearing my wife's bitter complaints about my long study hours (no way, sweet!). Then I will give back to her, as well as with my beloved daughter Elsa and with my lifelong friends, the amusement time that I owe to them. To deal with an ironic guy who is usually lost in thoughts, jabbering and digressing while still swayed by the burst of creativity he felt last late night is very hard (such a lie!). Thus I thank my closest circle for practising patience with me daily: don't give up and you'll see how practice leads to mastery.

Introduction

WHY IS DISPLACEMENT SO DIFFERENT HERE?

As I warned, displacement, as used here, doesn't necessarily involve movement or a measurable change. To taste Displacement we don't need to migrate or start a new life, it can be felt in daily life. Even if it's the difference between a slap and a bullet when compared with some of the cases of study on extreme disgraces referred here, even the luckiest can feel displaced today. Every time we are requested about a role or label that identifies us strongly we can feel it. For example, when we are asked for our documentation at a border. Or every time we feel a sense of aimlessness and need to reevaluate if we should keep in or get out of a place. Should a distinguished gentleman like me drink his rosé reserve in this new lesbian grunge pub after it was served at room temperature? Displacement. Are you navigating a social event with plenty of in-laws and ex-things while the implicit rules we set for ourselves are shaken? Annoying displacement.

There is something in the human body's beauty that makes some string within us tremble, especially when brief and sudden. So, to feel a bit of displacement let's consider seeing nudity in movies. I'm not talking about porn but neither about classic nudity as you might see in Phidias' work. I mean mild disturbing scenes like the butter one in *Last Tango in Paris*, the orgies in the *I, Claudius* TV series or Emilia Clarke's nudity in *Game of Thrones*.

If your modesty is so great, the experiment can go on thanks to films, stories and music exposing us to life, death and annihilation, for example, classic films like *The Virgin Spring*. Maybe this kind of experience would come easier to you when dancing rather than with images. Perhaps with the dreadful spontaneity of Stephen Crane's writings (Crane, 1968).

Unsurprisingly, music also deserves special attention. Schumann and Beethoven's music reveals how the displacement of climax doesn't change the metre (McCune, 1985, p. 14; Richards, 2011, pp. 70–76).[2]

Music is not only a way to challenge the everyday nature grappling us. It's an example of a displacement not so dependent on movement. Musical displacement refers to how music changes from one rhythm or melody to another. It's like when you're walking and then suddenly you start running, or when you're jumping and then you start spinning. In music, this change can create a sense of surprise or excitement, and it can make the listener feel like the music is going on a journey. In the example, the Papillons Op.2 score, it can be heard in the way that the different sections of the piece are introduced and then changed. For example, in the first section (No. 1), the left hand plays a steady, walking bass line while the right hand plays a simple melody. But as the piece progresses, the melody in the right hand becomes more complex and starts to move around the keyboard more, creating a sense of musical displacement.

Figure 1. Still from the movie The Virgin Spring
Source: Bergman (1960)

Figure 2. Papillons Op.2 score
Source: Retrieved from the Robert Schummans' Compositionen Für Das Pianoforte (Kühner, 1900, p. 12)

Musical movement, in turn, is about section changes, like when a song goes from a chorus to a verse. This change is more like when you move from one room to another in your house. It's a natural progression from one place to another and can create a sense of flow and structure in the music. Papillons Op.2 exemplifies the way that the different sections of the piece flow into each other. For example, in the second section (No. 2), the left hand plays a chord progression while the right hand plays a fast, swirling melody. When the third section (No. 3) begins, the left hand changes to a more stately, march-like rhythm, but the right hand continues to play a similar swirling melody. This creates a sense of musical movement as the listener is taken from one section to the next.

Finally, musical divergence points to two or more melodies or rhythms that were originally the same start to go in different directions. It's like when you and your friends are walking together, but one of you decides to take a different path. In music, this can create a sense of tension or conflict, and it can make the listener feel like they're waiting to see how the music will resolve itself. Thus, the displacement is a mismeasure of musical *movement*. That's not the case when it comes to the binomial dissonance / divergence (Krebs 1987). Musical divergence can be heard in Papillons Op.2 in the way that different melodies in the piece start to move in different directions. For example, in the fourth section (No. 4), the left hand plays a simple, repeating pattern while the right hand plays a melody that starts out simple but then becomes more complex and varied. As the section goes on, the left hand stays the same, but the right hand continues to evolve and change, creating a sense of musical divergence.

In Bioethics as it is in Music, divergence is a better way to measure displacement than using movement because it focuses on the changes in the Self, rather than just the movement from one section to another. The movements a person does can give us an idea of how the different parts of her life fit together, but are the conflicts that face what helps us understand how she is changing and evolving. Or being displaced.

Another interesting property of musical displacement is that it goes both forward and backward (Krebs, 1999, p. 35). Just like a person walking forward and then walking backward. It brings hope since, regardless of how many bad things happened, a person can find a place to belong again as long as she keeps *beating*.

All this little turmoil within us fed by films, songs, poems or any other little mess, hints to bigger questions about good and evil in displacement. Those bigger questions are about how conflicts and changes in one's life can impact our sense of belonging and understanding of right and wrong. For example, a person may have to make difficult decisions that challenge their moral beliefs, and this can create feelings of displacement and uncertainty. But this emotion drops by the slope of perplexity towards the intellectual problems' box. I mean this feeling of confusion and tentativeness that arises when trying to understand complex intellectual concepts. For example, when we encounter difficult philosophical or ethical questions, we may feel like we are sliding down a slippery slope towards a dungeon of doubt we can get out of. But as soon as hesitation appears, we can't help sliding down. Of course, it is something that happens internally, within our own minds and feelings. It is not something that can be observed by others from the outside. And then we become like a child who has lost track of time wondering about why the sky is blue. Just like with this child, displacement can make us feel like we are disconnected from the normal flow of our lives. We may feel like we have lost our sense of time or direction. It is the shortest way to taste a little sip of *displacement*.

The philosophy itself might have been born as an act of displacement. Plato and Aristotle agreed in setting its beginnings with wondering (Plato, 2004, sec. 155d; Aristotle, 2013, sec. 982b). Unlike them, Descartes in *Passions of the Soul* expresses negative views on wondering (Descartes, 1989, p. 58). The wondering or astonishment (thaumazein, θαυμάζειν), like the musical displacement points in two opposite directions at the same time.

Philosophy is an epistemic displacement. But beware! Displacing moral issues to intellectual problems allows us to use the meaning both as a defence and as a pretext. About this, now is the time to open our encyclopaedia to Ludwig Wittgenstein's page. In the *Philosophical Investigations* he stated: "*the meaning of a word is its use in the language*" (Wittgenstein, 2010a, p. lxxiii). It leads us to an ontological circle. I mean that a good chunk of the language has resulted (and is resulting) from pushing away, from displacing, moral concerns. Ludwig Wittgenstein was a Jewish man with anti-Semitic tantrums mixed with a deep sense of belonging (Szabados, 1999). An intellectual giant who grew in a hell of an opulent family. Such a horrible family that they bet a big wave in the career of an old classmate that Ludwig hated. It doesn't seem a big deal if it weren't for the fact that the old classmate was Adolf Hitler

(Waugh, 2013; Cornish, 1998). Therefore, we can learn a lot from Wittgenstein about *the meaning* of turning moral issues in scholarly lectures. We'll keep his track for a couple of pages, but before then, I want to highlight that this metaphysical goings-on between meaning and use is not a monopoly of the tormented ones. Although Wittgenstein had a knack for it.

The flow between meaning and use is an intrinsic psyche's feature that everybody has. This flow allows us to connect what we perceive with our understanding of what those perceptions mean, and then use that understanding to inform our actions and decisions. It's the way we make sense of the world around us. Indeed, this nexus between the ethics and the limits of language is nothing new. It was formulated (among others) by Wittgenstein himself almost a century ago in the seventh and last basic proposition of the *Tractatus Logico-Philosophicus* (Wittgenstein, 2010b, p. 108). The Austrian philosopher further extended himself in the ethics and language nexus through Wittgenstein's *Lecture on Ethics* (Wittgenstein, 1965).[3] Besides considerations on language, Wittgenstein also equated ethics and aesthetics (Wittgenstein, 2010b, sec. 6.421). It has been a controversial statement for a difficult topic (Collinson, 1985). To realise its scope is enough with putting it in perspective: "*From the psychologist's perspective, 2500 years of philosophical debate failed to yield agreement in the nature of aesthetic objects or aesthetics processes*" (Cupchik & Winston, 1996, p. 61). Anyways, twenty-five centuries is not so much time from an eternal perspective used first by Spinoza and then by Wittgenstein (Tilghman, 1991). Indeed, despite displacement's alignment with positive sciences, it doesn't lack spiritual or even religious symbolism. In Roman Palestine, seventeen centuries before Wittgenstein's most famous quote: "*Whereof one cannot speak, thereof one must be silent*" (Wittgenstein, 2010b, p. 27) the first Kabbalists held that the 22 letters of the Hebrew alphabet are the 22 divine life-giving ways (Scholem, 1991, pp. 28–29).[4] It fits like a glove in the Wittgenstein "*meaning as use*" approach to language. By the same token, after the Covenant, Abram's name was changed to Abraham, afterwards he turned nomad and moved away from the nowadays Iraq (Krupnick, 1983, p. 9; Jewish Publication Society Version, 2010, pt. Gen 17:1-5).[5] Regardless of Abraham's adventures for gaining a new life meaning that somewhat endure today, the idea of displacement is essentially closer to the Self shaping produced by leaving the house of his father Terah and changing his name after the Covenant than to the hundreds of miles travelled.

But better if we focus now on the plain fact that the displacement is more about divergences than convergences. A divergence is a complex and intimate feeling that implies an individual who sees how her agency blends. However, we can see a *family resemblance* with Einstein's spatial reasoning insight: "*The main reason lies in the fact that it is not so easy to free oneself from the idea that coordinates must have an immediate metrical meaning*." (Schilpp, 1949, p. 67) Bluntly, a divergence is a change of rules in the *game* of life.[6] A real life example of a divergence without metrical meaning could be the decision of a person to quit their high-paying corporate job to pursue a career in art, despite the significant decrease in income. From a purely metric standpoint, the decision may not make sense, as the person is giving up financial stability and security for a career path that is often uncertain and unpredictable. However, the individual may see the opportunity for personal fulfilment and self-expression as outweighing the monetary benefits of their previous job, leading to a divergence from the expected path based on conventional metrics. An even more real example is a blue collar worker being kicked out after a factory relocation who must learn the rules of unemployment and, perhaps, the ones of divorce, eviction and beneficence.

So, could the displacement be an outcome of competition? It might be the case in the struggle for resources. It's generally assumed that the shift in the combination of an ecosystem's components and benefits from reduced competition. However, the divergence happens regardless of the level of between-resource competition or whether the direct or indirect interaction between the consumers (Abrams & Cortez, 2015). Divergence happens.

Let's descend from the heightened human displacement brought from classic erudition to the gnarled oink of the practical totality of politics. That's the kind of displacement you don't want to taste. It's a hell of lack of accountability, corruption, and the subversion of collective freedom. A hell fueled by manipulation of public opinion, so much so, that from the state's eyes, a social turning point is poison. When the state faces truth, the gap between political and non-political spheres disappears and its power gets lost (Kouvelakis, 2018, p. 310). Like a musical climax, the social self-criticism appears distant to the rhythmical rolling of the revolving doors. We'll see more on politics in its own next chapter.

Just like in the Beach Boys' song *Good Vibrations* (The Beach Boys, 2016), the eerie theremin is splitted and renewed with discordant harmonies and unpredictable rhythm muddling the listener up. The sound of a society ready for change becomes so intense and dissonant that it feels like it's no longer following the original rhythm of politics. This self-criticism by the citizens creates a sense of tension and even discomfort for the ones who abhor changes in the top down flow. So the required strong emotional response and commitment is constantly hindered through any available stimuli. When the very few used to speak on behalf of the most fail to create their collective delusions, their fake representation gets unveiled. Then, social movements upholding collective freedom appear bearing a new divergence. Actually, they are *social displacements* disclosing the political narratives' veil. When it happens the facts cannot be turned into refugees; they are not forcibly displaced realities while the discordance endures. A people is free only to the extent it can unease their rulers and avoid punishment. One historical example of this is the Civil Rights Movement in the United States during the 1950s and 1960s. African Americans along with European Americans and other minorities that were invisibilized later in favour of a Black agenda, protested and challenged the US racist laws of its time (Sokol, 2008). Segregation laws and policies denied racial minorities (mostly but not only to Black people) basic human rights, such as the right to vote and equal access to education and public accommodations (Tischauser, 2012, p. 17). These protests often faced violent repression from the state, including police brutality and arrests. However, the movement persisted and gained momentum, with people from all walks of life joining the cause. Eventually, the movement's efforts led to significant legal and social changes, including the Civil Rights Act of 1964 and the Voting Rights Act of 1965 (Senate and House of Representatives of the United States of America, 1964; Christopher, 1965). The movement's success demonstrates that when people are committed to change and are willing to work together and even to risk punishment from the ruling class, they can make significant progress in advancing their collective rights and freedoms.

Divergence is deep and *groundbreaking* when it comes to our lives. So, it's enlightening that we can understand seismic activity as displacements rather than forces (Kowalsky et al., 1995). As a good tellurian phenomena, the unwrapping of many major events, both individual and collective, fits in this equation. But Displacement as used in Bodi is not the same as an usual displacement on the plane since it can't always be cancelled. Many times, wrongdoings cannot be fixed. In spite of this, Displacement as used in Bodi shares an interesting nexus with Poincaré's geometrical displacement. The French mathematician's main conclusion is that the set of displacement classes are a group in the mathematical sense and hence, "*pre-exist in our minds*" (Poincare, 2016). Further, Poincaré's intuition about the link between geometry and muscular sensations has also a translation in the pre-human *bios* of this *Bioethics of Displacement* (Poincare, 2012). It has roots in the earliest animals on Earth. In other words, the idea of displacement is not just a mathematical or intellectual concept, but has a basis in the very biology of living organisms. An example of this could be the way that many animals, including humans, respond to environmental changes or disruptions. For instance, when a natural disaster such as an earthquake or hurricane occurs, animals may need to quickly adapt or "displace" themselves in order to survive. This ability to respond and adapt to displacement is not just a learned behaviour, but is also deeply rooted in the biology and evolution of these organisms. Displacement and the idea of the interplay between meaning and use is as part of a larger philosophical inquiry into the meaning of life.

Bodi's displacement is the amount of divergence that a life bears while losing its meaning.

If we bring this definition to the furthest simplicity (zero divergence), an interesting displacement's property can be unearthed. Let's picture a Turing machine, an abstract computer that, sequentially, runs instructions over symbols in a stripe of tape. For each state and symbol there are new ones after the head displacement. So, a displacement happens just before any new bit of code. But this particular or abstract head displacement, in turn, is a fundamental property of computation that cannot be fully captured or expressed in language or any other symbolic system. It is a phenomenon that occurs prior to the establishment of meaning or symbol systems, just like the calculator on the table before the result on the display. If we consider along this that the most fundamental insight into computation is that one machine can run any program whatsoever, we can conclude that:

Displacement precedes the meaning, ergo it's unspeakable.

This can affect our daily lives as there are many experiences that we have that may be difficult to put into words or fully understand. For example, we may have a gut feeling or intuition about something, or we may experience a sense of unease or discomfort without being able to pinpoint exactly why. These experiences may be related to the unspeakable nature of displacement, in that they are rooted in deeper, non-verbal aspects of our experience that are difficult to fully articulate. Understanding Displacement's unspeakability may therefore help us to better appreciate and navigate the complexities of our own inner lives and the world around us.

BIOETHICS GLOBAL TRENDS AS A POLITICAL MIRROR

The main pending subject of Bioethics is diversity, but not in the woke way. It doesn't mean that bioethics had not been following the yankee orthodoxy for most of its modern and postmodern history, as we'll see. Besides lack of diversity, the bioethics plurality presents other worrying traits. There is a trend in bioethics to bifurcate between ethical theory and practice that chases away both the keen rhetoric and the classic humanism (Carter, 1998). The new types of postmodern family, especially when displacing the traditional one, can turn themselves into a problematic concept in bioethics (Florea & Puiescu, 2018). On the other hand, there are constants. The arguments by analogy, mainly in the US, greatly influence ethical reasoning itself detrimental to Hippocratic and phenomenological traditions (Sprinkle, 2011). This displacement from phenomenology is related with adversarialism and loss of trust in the public domain as a result of technologization of ethics and dominance of utilitarianism (Barilan, 2002). In the opposite sense, so much time ago like the seventies, the need to positively describe every connection between social changes and scientific intellectual development has been evident (Mulkay, 1974). In 2023, the answer to how new ideas appear in science has not been solved yet. Thus, the quest for bioethics patterns might be complex and not have a single solution. For this reason, I will focus on global trends coming from sluggish politics rather than overloading the crystal ball.

Even such a bragged topic in bioethics, just like dignity is, can be tracked down to tailored Western political needs. After WW2 the identification of political leaders with biopolitics had lost the last ounce of appeal. Rather than this, the Foucaldian pastoral power became trendy: a good leader was expected to care and promote caring attitudes among her flock members. Regardless of the pastoral power advocates' self-interest it was very good news. But this do-gooding wasn't going to last forever and the old-school biopolitics were unyielding unearthed by the winds of change from stem-cells research (Bennett 2018). Ironically, as happens during car crashes, the crying in pain people commonly present the less worrisome injuries. Those who are in the throat of death hardly can cry. Likewise, the US unlikely is the worst case in brown-nosing politicisation of bioethics, even if the one with more chances of spilling-over due to its preeminence both in academia and biotechnologies. On the other hand, considering the chances for ethical behaviour that free speech and free press offer, the US bioethical sycophants, to get ahead, surely are even more insidious than abroad.

For good and evil, Western-centrism in bioethics started to decline after 2011. From those years we can find regrets such as: "*Bioethical concerns are global, bioethics is Western*" (Chattopadhyay & De Vries, 2008). Or a 2011 study by Kristen Perry on the websites of 32 universities' Institutional Review Boards (IRBs), two issues were found (Perry, 2011):

- University IRBs offer little, if any, guidance about conducting research with participants whose first language is not English
- There is great variation in the definition of vulnerable populations, as well as in the requirements for including, or excluding, their participation in research

Thus, the kind of ethical research to carry out with the most non-English speaking impoverished countries is too malleable as presented by the many online IRBs.

In 2008, Chattopadhyay and De Vries said that, while the West has the money, their delegates will set the agenda in the international bioethics roundtables and the rush for solving riddles on rights will brim the mainstream bioethics (Chattopadhyay and De Vries 2008). But now the world's money is not predominantly in the West. And even back then in 2008, the West was not so united. In the last years of the 20th century, the mainstream Anglo-American Bioethics was displaced by a genuine European bioethics that claimed its origins in Fritz Jahr 1920s works (most focused in "imperatives") rather than in the Van Rensselaer Potter ideas (Muzur, 2017).

In this point, I want to highlight the term Asian Bioethics and advocate for precaution when dealing with it. Asian Bioethics as a relevant scholar topic (not as an idea) comes from the "*Asian values*" rise. Indeed, they started to be popular from the Bangkok Declaration of Human Rights in 1993 (Kam-por, 2002).[7] Let's delve into it. Beside sound observers like NGOs, supranational organisms and other non-Asian states, the 1993 Bangkok Declaration was signed by bureaucrats from Bahrain, Bangladesh, Bhutan, Brunei Darussalam, China, Cyprus, Democratic People's Republic of Korea, Fiji, India, Indonesia, Iran, Iraq, Japan, Kiribati, Kuwait, Lao People's Democratic Republic, Malaysia, Maldives, Mongolia, Myanmar, Nepal, Oman, Pakistan, Papua New Guinea, Philippines, Republic of Korea, Samoa, Singapore, Solomon Islands, Sri Lanka, Syrian Arab Republic, Thailand, United Arab Emirates, Viet Nam. If we look to the 1993 Amnesty International Report we obtain that, with the exception of Japan and Fiji, the most of the aforementioned countries made use of torture or arbitrary imprisonment (Amnesty International, 1993).[8] Namely, torture was extensively used in: Bangladesh, China, India, Indonesia, Iran, Iraq, Myanmar, Pakistan, Papua New Guinea, Sri Lanka, Thailand and (very likely) in North Korea. Bluntly, those governments were so used to *playing politics with the life of others* that their only concern about human rights surely was how to get rid of them.

Anyways, back in my present year 2023, the greatest danger in Asian / Western bioethics tagging is to forget that it's another conventionalism. Pointing to 'Western' and 'Eastern' principles as not the same thing divides the world in two types of persons, as if we were not all human beings (Widdows, 2011). First, it is important don't miss Asian bioethics with the Asian bioethicists' publication belittlement, especially in the beginning of our discipline globalisation. Bioethics written in non-Western languages, like Japanese or Thai, has more difficulty entering the international stage (Grätzel & Guhe, 2017, pp. 196–197). The Thai Buddhist bioethics of Pinit Ratanakul, for example, is the exception rather than the norm (Ratanakul, 1986). However, Asian bioethics is not about vindicating a fair recognition for Asian bioethicists. Their recognition has arrived in an amoral way as Asia's GDP boomed. Actually, Asian bioethics is more about most Asian leaders' interests in getting rid of social stakeholders who may benefit if their countries become closer to Western free speech. Some people have written that Asian Bioethics echoes the de-historicization and re-asianization geopolitical strategies (Robertson, 2005) It is important that we don't miss genesis and validity in Bioethics (ten Have & Gordijn, 2011). Not by chance the persons behind the Asian values movement were the then prime ministers of Singapore and Malaysia, Lee Kuan Yew and Mahathir Mohamad respectively (Widdows 2011). Unsurprisingly, during their tenure there happened *so rightful and humanitarian practices* like: tight press control, political opponent arrests, abuses on detainees and the use of colonial tactics against their own people like setting racial quotas in the university and imposing ethnicity cards among others (Vasagar, 2018; Sleeper, 2015). Both Mahathir Mohamad and Lee Kuan Yew shared strong attacks against the judiciary and both of them were praised after publishing their cringing racist ideas. Mahathir Mohamad's *The Malay Dilemma* (Mohamad, 1970) makes extensive use of stereotypes across the book: the Malays are depicted as graceful, the Chinese as intelligent but greedy, and the British as proud and efficient.[9] In 2018, finally, the Malay PM climbed up the highest peak of originality calling Jews 'hook-nosed' (Bachner & Staff, 2018). On his side, Lee Kuan Yew quotes are worthy of the most soaked in Jack outlawed biker across the route 66:

Not that all the children of gardeners are duds. Occasionally two grey horses produce a white horse but very few. If you have two white horses, the chances are you breed white horses. It's seldom spoken publicly because those who are NOT white horses say, 'You're degrading me'. (Lee & Han, 2011)

Now if democracy will not work for the Russians, a white Christian people, can we assume that it will naturally work with Asians? (Kwang et al., 2015)

Against this moral snobbism and horse breeding, it can be argued that the so-called Asian values are not *so Asian* (Sen, 1997). To put an example, major social shifts like going from a paternalistic society to an individualistic (Westerner) one, like in the Japanese case, are coherent with the observed convergence in the end of life bioethics (Macer, 2005). Some Asian (and not so Asian) authors would agree with: "*Humanization and Harmonization are the two main emphasis of Asian bioethics*" (Tai, 2016). Despite my pitiful level in Asian languages, I believe that the word 'harmony' is a tough bone and some meaning is lost in translation. More than this, considering how the contents are handled in the West, there are real chances of 'harmony' being turned into a catchword that pleases almost everybody. On the other hand, the humanization emphasis of Asian bioethics seems to me a muddled concept. I mean, in which way the emphasis in humanization across the most Asian countries is away from the spectrum of Western alternatives? And what about African bioethics, for example? Is the Asian bioethics emphasis in humanization different because it opposes the Western one or it's said to be unique? On the other hand, statements like: "*in many Asian cultures, patient autonomy is not an agenda*" are less prone to error or at least, are more falsifiable (Bowman & Hui, 2000). Some authors' work seems to be oriented for an Asian-West wide comparison but others hold that the cultural influence is meaningless out of local and well-contextualised cases, for example, on death and dying (Phua et al., 2016; Koh & Hwee, 2015; Hern et al., 1998). Anyhow, it would be simplistic making monolithic claims, disregarding the economic differences between countries or ignoring the critical care medicine's ins and outs and its very own subculture (Ho & Tsai, 2016; Curtis & Rubenfeld, 2005; Cohen & Ezer, 2013). For example, the attitudes of seniors towards end-of-life in New Zealand and Taiwan in 2006 were similar but distinctive views on natural death by ageing appeared in a context-sensitive analysis (Hsin & Macer, 2006).

In addition to the core differences and quicksands between Asian and Western Bioethics, there are some gaps in the way we set differences. For example, in countries with a religion sponsored by the state, it could be that a person who classifies herself as belonging to a religious denomination limits herself to paying lip services to avoid bullying. On the other hand, the South Asian religious adherence influence in bioethics leads to new insights on culture. It should be considered how the decision-making moments that involve patients, families and health care providers create micro-level transient neocultures (Hillman, 2019; Ortiz, 1987, pp. 91–97). Thus, modelling a patient's welfare in a transparent and evidence-based fashion can be a bridge between Western and Asian bioethics' representatives or, in a wider sense, patient-centred and family-centred societies (Sim et al., 2019). Despite that the communitarian approach to bioethics is more streamlined with the Asian institutions (including the family), it gets pulverised against the ground by authoritarianism. Sadly, it spoils important lessons on responsive communitarian bioethics everyday coming from the wealth of cultures that live in Asia.

From a Bioethics of Displacement perspective, the most important lesson from the Western / Asian dichotomy is recognizing the need to find ways to make sure that people from different backgrounds get the care they need in a way that respects their beliefs and ways of doing things and it must be done resorting to solutions that are specific to that situation. Shortly, Eastern or Western, no one size fits all. Only Asianism should be considered in Bodi is the one built upon space and time shaping collective features. For example, the traditional cultural values of Asian Indian migrants of first and second generations in the US, greatly influence end-of-life care preferences and retain importance across generations (Sharma et al., 2012). And the same goes for any other continentalism such as Africanism. For example, in some African bioethicists' eyes, the autonomy and justice principles are a Western bias that buries its cultural heritage and should be surrogate by traditional concepts like 'igwebuike' (strength through unity) (Kanu, 2017). Sounds good, but it will take a critical effort to prevent 'igwebuike' turning into a sort of African *harmony*. Learning about 'igwebuike' across the first and second generation of African diasporas, for example, would be a hot topic for Bioethics of Displacement.

As I said at the beginning of this section, the main pending subject of Bioethics is diversity, but diversity of voices beyond these funded with money constantly flowing from taxpayers' pockets and the underserved people's quality of life merciless devaluation. And it's a global trend that doesn't cease.

Even if Asian Bioethics have been roaming for decades, more than half of papers published on this topic belong to the 2018-2023 period, the time in which China's GDP (perhaps) has surpassed the US one, the relations between both hegemons are in a whirl of blames and mockery and bioethics might result useful for the Party during pandemics. Just like Asian Bioethics and due to the rise of China, too, African Bioethics has only started to be a thing in academia after African leaders started to throw back in the French and German diplomats' face the vanity they have seen for years -even if first papers on African Bioethics were published in 1996. A few weeks before these lines, the Namibian President humiliated a German diplomat lecturing him about China.[10] Actually, it is only master swapping (sic) but this *Radfahrer* got what he deserved and I enjoyed it a lot! Likewise, on December 3, 2022, Burkina Faso suspended the broadcast of France's RFI radio and, on the same day, Mali banned the NGOs funded by France (Reuters, 2022). Unsurprisingly, only in the first four months of 2023 were published about so many papers on "African Bioethics" as on "European Bioethics" or "Bioethics in the United States" -and, to cap it all, most of the Western papers are academic whining.

These global tides of anti-global-ist bioethics may feel exciting and even be funny but the fun won't last. Are tides that flow through fractious problematic policies such as controversial and divisive arguments around GMOs. In addition to this, these streams won't help us to keep adrift on the muddy implications of international collaborations in scientific research. For example the US arch-rival, China, have had no qualm in resorting to a Massachusetts company and a prominent Yale researcher as they cast a DNA-based enormous surveillance and control net upon the Uighur minority -for now (Wee, 2019). While adrift on muddy waters, we'll meet new chances to ask ourselves again and again: "*How much of bioethics should be ethics?*" (Pellegrino, 1997). Take it or leave it, but we have not seen the end of bioethics' simple politicisation rope yet. If it weren't enough with the ethical sump of patents over life and glorious leaders' tailored-bioethics, the nature of scientific discoveries has plenty of political uses and misuses that cannot be predicted (Nathan Emmerich 2018). For example, what about an alleged harmless eugenics through epigenetic social work? (Juengst et al., 2014; Combs-Orme, 2013) Indeed, there are so many mistakes ahead in the road that we must ensure not repeating the many blunders and wrongdoings enumerated since 1962 till present 2023. Let's know more about how Bodi can help.

BIOETHICS OF DISPLACEMENT

Bioethics of Displacement means love of belonging.

It's an easy definition, a legacy of my intellectual father Darryl Macer: "Bioethics is love of life" (Macer, 1998). However, it's an easy definition that hiddens a big deal of complexity, too. To disentangle it, let's redefine bioethics and summarise what we have learned till now. In addition to a field of study, from my personal worldview, Bioethics is a neverending exercise to find what we all have in common that is so authentic we can't discard or falsify it: an unwavering truthfulness quest. Summarising, Bioethics of Displacement emphasises the importance of belonging in the context of ethics and the value of human life. Belonging refers to the sense of being part of a group or community as a fundamental aspect of being human. Displacement, in this context, refers to the loss of meaning or divergence that life can experience. This definition suggests that the loss of belonging can result in a loss of meaning in life, which is considered unspeakable. Given that, Bodi's unique perspective relies upon the centrality of loss (and regain) of meaning in life as the root, as the *arché*, of bioethics. Bodi embraces the complexity and multidimensionality of ethical issues in healthcare and the life sciences. In practice, this embracing means that we shouldn't understand the short-reply to 'what is the Bioethics of Displacement' as a door to its meaning, but as counterfactual causation. If it were not for these times, the definition wouldn't have existed. In short, the Bioethics of Displacement exists because its moment has come. For this reason, the causality as a chain of events would be paradoxical here. We have needed a sexual revolution and a ton of LGTB pride to conceive the *intellectual cross-dressing* of bioethics without getting scandalised (Morreim, 1997). We have needed the postmodern economical crisis shaking the mainstream clinical bioethics and putting the for-profit health plans over physicians and patients (HITC, 2015). If it was not for so many people gaining their own place and many others being forcibly expelled from public healthcare, there wouldn't be a Bioethics of Displacement.

Researchers in Bioethics of Displacement may adopt various interdisciplinary approaches and methodologies to comprehensively analyse the multifaceted nature of displacement. The analytic side and its methodologies include:

1. **Ethical Analysis:** Ethical frameworks and principles to critically examine the ethical implications of displacement. Researchers may analyse the moral dilemmas and ethical challenges associated with forced displacement, such as issues of justice, fairness, autonomy, and human rights. For instance, researchers may use principles of bioethics, such as beneficence, non-maleficence, autonomy, and justice, to evaluate the ethical implications of policies and practices related to displacement.
2. **Sociocultural Analysis:** Sociocultural approaches to explore the social and cultural dimensions of displacement. Research examines the impact of displacement on the identity, culture, and sense of belonging of displaced individuals and communities. For example, researchers may use qualitative methods, such as interviews, focus groups, and ethnographic research, to understand the social and cultural dynamics of displacement, including the challenges and opportunities of rebuilding communities and preserving cultural heritage.
3. **Psychological Analysis:** Psychology's tools to assess the well-being of displaced individuals and communities. The main goal is to explore the psychological impact of displacement, such as trauma, stress, and resilience, and evaluate the effectiveness of psychological interventions for addressing mental health issues related to displacement. Case in point, researchers may use quantitative methods, such as psychological assessments and surveys, to measure the psychological outcomes of displacement and evaluate the effectiveness of mental health interventions.
4. **Environmental Analysis:** Assessing the ecological impact of displacement on ecosystems, natural resources, and environmental sustainability. To illustrate, researchers may use ecological methods, such as remote sensing, mapping, and modelling, to assess the ecological changes caused by displacement, such as deforestation, pollution, and loss of biodiversity.
5. **Policy Analysis:** Evaluating the policies, laws, and regulations related to displacement. To assess the effectiveness, fairness, and ethical implications of policies and practices related to forced displacement, including issues of access to healthcare, education, and social services. E.g., researchers may use policy analysis frameworks, such as cost-benefit analysis or equity analysis, to evaluate the impact of displacement policies on different stakeholders and identify areas for improvement.

On the other hand, the specific approaches and methodologies employed may vary depending on the research questions, context, and disciplinary backgrounds of the researchers and will use any tool for shedding light on the *right to belong*, including music and cinema. The reason for using sources far away from canonical literature (like popular movies) is to display the *ordinariness* of narrative and the foreseeing of art. Not only do the existentialists distort the time in their narratives but also it happens in the *Matrix* movie, for example. Unsurprisingly, some universal features of humans like identity, space and time are underserved in bioethics (McKenny, 2006; Kelly, 2003). Till Bodi. The cases of study are the bricks of bioethics. The bioethical discussions are concrete, substantive, practical and are around *cases*. A case is an admixture of genres where the clinical facts meet the personal (Nelson, 2014, pp. 5–7). As Martha Nussbaum pointed, the narrative style is embedded in the topic, thus there are ways to expose that are more suitable for some contents (Nussbaum, 1992, p. 19). Metaphors play a role in bioethics, they shape our views, influence us. The metaphors have a normative value (Pascalev, 2018). Many bioethical discussions are, actually, discussions about metaphors, e.g. depicting some end-of-life practices as torture. Paradoxically, bioethicists have not considered their literary value. It's wrong. One of the main and underestimated challenges of bioethics today is finding the best style for its widening multidisciplinary discourse. For this reason, the bricks of Bioethics of Displacement *are stories* rather than cases.

Bioethics of Displacement's main subject of research are the stories about how the meaning of life is top down arbitrarily dictated and its implications for the present and future of humankind.

Besides its main subject and non-analytical toolset, Bodi differs from mainstream bioethics, which may primarily focus on issues such as autonomy, beneficence, and justice in the healthcare context. In spite of the bioethicists' efforts to raise multidisciplinarity, the truth is that a narrow medical bioethics focused on health keeps being the mainstream (Beever & Whitehouse, 2017). But Bodi is not unbound from healthcare decision-making at all. Thankfully, the ambiguity is just being appreciated now in medicine (Domen, 2016). This fuzziness, this bearable chaos, gives room for a new relationship that features a culture of respect and psychological flexibility where Bodi can grow. (We'll delve into psychological flexibility in the next section).

Finally, Bioethics of Displacement is distinct from other fields of study by its holistic focus on the ethical implications of displacement, the interconnectedness of alive and non-alive entities, and the recognition of collective freedom and human worthiness first. It emphasises the need to consider the holistic impact of displacement on diverse aspects of life and challenges the homogenization and stigmatisation by other names of individuals and collectives.

Overall, Bodi stresses the importance of self-awareness in ethical decision-making, and encourages critical thinking about the implications of different approaches in different settings. It is not aimed only to raise the number of collectives under the bioethics' umbrella and improve its interdisciplinarity. No, the real thing is to introduce ideological build-ups grounded on phenomenology such as the concept of lived experience as a central tenet in ethical decision-making. Shortly, Bodi wants to look below the carpet of principles that shape our moral life.

REFERENCES

Abrams, P. A., & Cortez, M. H. (2015). Is competition needed for ecological character displacement? Does displacement decrease competition? *Evolution; International Journal of Organic Evolution, 69*(12), 3039–3053. doi:10.1111/evo.12816 PMID:26548922

Amnesty International. (1993, January 1). *Amnesty International Annual Report 1993*. Amnesty International. https://www.amnesty.org/en/documents/pol10/0001/1993/en/

Aristotle. (2013). *The Metaphysics*. Courier Corporation.

Bachner, M., & Staff, T. (2018, October 2). *Anti-Semitic Malaysian PM insists Jews are 'hook-nosed' to BBC*. https://www.timesofisrael.com/anti-semitic-malaysian-pm-insists-jews-are-hook-nosed-to-bbc/

Barilan, M. Y. (2002). On the technologizing and technocratic trends in bioethics. *Eubios Journal of Asian and International Bioethics; EJAIB*, *12*, 176.

Beever, J., & Whitehouse, P. J. (2017). The Ecosystem of Bioethics: Building Bridges to Public Health. *JAHR*, *8*(2), 227–243. doi:10.21860/j.8.2.5

Bergman, I. (Director). (1960). *The Virgin Spring*. https://www.imdb.com/title/tt0053976/

Bowman, K. W., & Hui, E. C. (2000). Bioethics for clinicians: 20. Chinese bioethics. *Canadian Medical Association Journal*, *163*(11), 1481–1485. PMID:11192658

Carter, M. A. (1998). Synthetic Model of Bioethical Inquiry. *Twentieth World Congress of Philosophy*.

Chattopadhyay, S., & De Vries, R. (2008). Bioethical concerns are global, bioethics is Western. *Eubios Journal of Asian and International Bioethics; EJAIB*, *18*(4), 106–109. PMID:19593391

Christopher, W. M. (1965). The Constitutionality of the Voting Rights Act of 1965. *Stanford Law Review*, *18*(1), 1. doi:10.2307/1227379

Cohen, J., & Ezer, T. (2013, December 10). Human rights in patient care: A theoretical and practical framework. *Health and Human Rights Journal*. https://www.hhrjournal.org/2013/12/human-rights-in-patient-care-a-theoretical-and-practical-framework/

Collinson, D. (1985). Ethics and aesthetics are one. *British Journal of Aesthetics*, *25*(3), 266–272. doi:10.1093/bjaesthetics/25.3.266

Combs-Orme, T. (2013). Epigenetics and the Social Work Imperative. *Social Work*, *58*(1), 23–30. doi:10.1093wws052 PMID:23409337

Cornish, K. (1998). *The Jew of Linz: Wittgenstein, Hitler and Their Secret Battle for the Mind*. Century.

Crane, S. (1968). *The Open Boat and Other Stories*. Prabhat Prakashan.

Cupchik, G. C., & Winston, A. S. (1996). Confluence and Divergence in Empirical Aesthetics, Philosophy and Mainstream Psychology. In M. P. Friedman & E. C. Carterette (Eds.), *Cognitive Ecology*. Elsevier. doi:10.1016/B978-012161966-4/50005-0

Curtis, J. R., & Rubenfeld, G. D. (2005). Improving Palliative Care For Patients In The Intensive Care Unit. *Journal of Palliative Medicine*, *8*(4), 840–854. doi:10.1089/jpm.2005.8.840 PMID:16128659

Descartes, R. (1989). *Passions of the Soul*. Hackett Publishing.

Domen, R. E. (2016). The Ethics of Ambiguity. *Academic Pathology*, *3*, 2374289516654712. Advance online publication. doi:10.1177/2374289516654712 PMID:28725771

Florea, Ş., & Puiescu, M. (2018). Contemporary bioethical perspectives regarding the family. *International Journal of Theology. Philosophy of Science*, *2*(2), 34–43.

Grätzel, S., & Guhe, E. (2017). *Life, Body, Person and Self: A Reconsideration of Core Concepts in Bioethics from an Intercultural Perspective*. Verlag Herder GmbH. doi:10.5771/9783495818121

Hern, H. E., Koenig, B. A., Moore, L. J., & Marshall, P. A. (1998). The Difference That Culture Can Make in End-of-Life Decisionmaking. *Cambridge Quarterly of Healthcare Ethics*, *7*(1), 27–40. doi:10.1017/S0963180198701045 PMID:9523039

Hillman, S. (2019). *Religion, Medicine, Bioethics, and the Law in End-of-life Care: South Asian Religious Adherent Perspectives* [Thesis]. https://tspace.library.utoronto.ca/handle/1807/97488

HITC. (2015, March 5). *Has For-Profit Healthcare Destroyed The Soul of Medicine?* https://hitconsultant.net/2015/03/05/entry-of-for-profit-healthcare-destroyed-soul-of-medicine/

Ho, A., & Tsai, D. F.-C. (2016). Making good death more accessible: End-of-life care in the intensive care unit. *Intensive Care Medicine*, *42*(8), 1258–1260. doi:10.100700134-016-4396-2 PMID:27260257

Hsin, D. H.-C., & Macer, D. R. J. (2006). Comparisons of life images and end-of-life attitudes between the elderly in Taiwan and New Zealand. *The Journal of Nursing Research*, *14*(3), 198–208. doi:10.1097/01.JNR.0000387578.58974.42 PMID:16967402

Jewish Publication Society. (2010). *The Torah: The Five Books of Moses*. Jewish Publication Society.

Juengst, E. T., Fishman, J. R., McGowan, M. L., & Settersten, R. A. Jr. (2014). Serving epigenetics before its time. *Trends in Genetics*, *30*(10), 427–429. doi:10.1016/j.tig.2014.08.001 PMID:25242336

Kam-por, Y. (2002). *The alleged Asian values and their implications for bioethics fourth Asian bioethics conference: Asian bioethics in the 21st century*. EUBIOS.

Kanu, I. A. (2017). Igwebuike as the consummate foundation of African bioethical principles. *Nnadiebube Journal of Philosophy, 1*(1). https://journals.ezenwaohaetorc.org/index.php/NJP/article/view/245

Kelly, S. E. (2003). Bioethics and rural health: Theorizing place, space, and subjects. *Social Science & Medicine*, *56*(11), 2277–2288. doi:10.1016/S0277-9536(02)00227-7 PMID:12719181

Koh, M., & Hwee, P. C. (2015). End-of-Life Care in the Intensive Care Unit: How Asia Differs From the West. *JAMA Internal Medicine*, *175*(3), 371–372. doi:10.1001/jamainternmed.2014.7397 PMID:25581079

Kouvelakis, S. (2018). *Philosophy and Revolution: From Kant to Marx*. Verso Books.

Kowalsky, M. J., Priestley, M. J. N., & MacRae, G. A. (1995). Displacement-based design of RC bridge columns in seismic regions. *Earthquake Engineering & Structural Dynamics*, *24*(12), 1623–1643. doi:10.1002/eqe.4290241206

Krebs, H. (1999). *Fantasy pieces: Metrical dissonance in the music of Robert Schumann*. Oxford University Press on Demand. doi:10.1093/acprof:oso/9780195116236.001.0001

Krupnick, M. (1983). *Displacement: Derrida and After*. Indiana University Press.

Kühner, C. (1900). Robert Schummans' compositionen für das Pianoforte. *Henry Litolff's Verlag*, *1704*, 4–20.

Kwang, H. F., Fernandez, W., & Tan, S. (2015). Universality or Democracy? In Lee Kuan Yew: The Man and His Ideas. Marshall Cavendish International Asia Pte Ltd.

Lee, K. Y., & Han, F. K. (2011). On Elitism. In *Lee Kuan Yew: Hard Truths to Keep Singapore Going*. Straits Times Press.

Macer, D. R. J. (1998). *Bioethics is love of life: An alternative textbook*. Eubios Ethics Institute Christchurch.

Macer, D. R. J. (2005). End-of-Life in Japan. In R. H. Blank, J. C. Merrick, & A. L. Caplan (Eds.), *End-of-life Decision Making: A Cross-national Study*. MIT Press.

McCune, M. (1985). Theoria. *Digital Library, 1*. https://digital.library.unt.edu/ark:/67531/metadc287882/citation/

McKenny, G. P. (2006). An anthropological bioethics: Hermeneutical or Critical? Commentary on Elliot. In R. A. Carson & C. R. Burns (Eds.), *Philosophy of Medicine and Bioethics: A Twenty-Year Retrospective and Critical Appraisal*. Springer Science & Business Media.

Mohamad, M. bin. (1970). The Malay dilemma. D. Moore for Asia Pacific Press.

Morreim, E. H. (1997). At the intersection of medicine, law, economics, and ethics: Bioethics and the art of intellectual cross-dressing. In *Philosophy of Medicine and Bioethics* (pp. 299–325). Springer. doi:10.1007/0-306-48133-2_22

Mulkay, M. (1974). Conceptual Displacement and Migration in Science: A Prefatory Paper. *Science Studies*, *4*(3), 205–234. doi:10.1177/030631277400400301

Muzur, A. (2017). European bioethics: A new history guaranteeing a new future. *Socijalna Ekologija, 26*.

Nelson, H. L. (2014). *Stories and Their Limits: Narrative Approaches to Bioethics*. Routledge. doi:10.4324/9781315822167

Nussbaum, M. C. (1992). *Love's Knowledge: Essays on Philosophy and Literature*. Oxford University Press.

Ortiz, F. (1987). *Contrapunteo cubano del tabaco y el azúcar*. Fundacion Biblioteca Ayacuch.

Pascalev, A. (2018). Is Death the Enemy? The Normative Power of Metaphor in Bioethics. *Ethics of Emerging Biotechnologies*, 87–106.

Pellegrino, E. D. (1997). Bioethics as an Interdisciplinary Enterprise: Where Does Ethics Fit in the Mosaic of Disciplines? In R. A. Carson & C. R. Burns (Eds.), *Philosophy of Medicine and Bioethics: A Twenty-Year Retrospective and Critical Appraisal* (pp. 1–23). Springer Netherlands. doi:10.1007/0-306-48133-2_1

Perry, K. H. (2011). Ethics, Vulnerability, and Speakers of Other Languages: How University IRBs (Do Not) Speak to Research Involving Refugee Participants. *Qualitative Inquiry*, *17*(10), 899–912. doi:10.1177/1077800411425006

Phua, J., Joynt, G. M., Nishimura, M., Deng, Y., Myatra, S. N., Chan, Y. H., Binh, N. G., Tan, C. C., Faruq, M. O., Arabi, Y. M., Wahjuprajitno, B., Liu, S.-F., Hashemian, S. M. R., Kashif, W., Staworn, D., Palo, J. E., Koh, Y., Gomersall, C. D., & Du, B.ACME Study Investigators and the Asian Critical Care Clinical Trials Group. (2016). Withholding and withdrawal of life-sustaining treatments in low-middle-income versus high-income Asian countries and regions. *Intensive Care Medicine*, *42*(7), 1118–1127. doi:10.100700134-016-4347-y PMID:27071388

Plato. (2004). Theaetetus. Penguin UK.

Poincare, H. (2012). V. Space and Empiricism. In The Value of Science: Essential Writings of Henri Poincare. Random House Publishing Group.

Poincare, H. (2016). Space and Geometry. In *Science and Hypothesis*. Read Books Ltd.

Ratanakul, P. (1986). *Bio Ethics: An Introduction to the Ethics of Medicine and Life Sciences*. Mahidol University.

Reuters. (2022, December 4). Burkina Faso military government suspends broadcast of France's RFI radio. *Reuters*. https://www.reuters.com/world/burkina-faso-military-government-suspends-broadcast-frances-rfi-radio-2022-12-03/

Richards, M. C. (2011). *Analyzing Tension and Drama in Beethoven's First-Movement Sonata Forms* [PhD Thesis].

Robertson, J. (2005). Dehistoricizing History. *Critical Asian Studies*, *37*(2), 233–250. doi:10.1080/14672710500106309

Schilpp, P. A. (1949). *Albert Einstein, philosopher-scientist* (1st ed.). Library of Living Philosophers.

Scholem, G. G. (1991). *Origins of the Kabbalah*. Princeton University Press. doi:10.1515/9781400820429

Sen, A. (1997). *Human rights and Asian values*. Carnegie Council on Ethics and International Affairs New York.

Senate and House of Representatives of the United States of America. (1964). *Civil rights act of 1964*. https://blowthetrumpet.org/documents/CivilRightsActLBJ_000.pdf

Sharma, R. K., Khosla, N., Tulsky, J. A., & Carrese, J. A. (2012). Traditional Expectations Versus US Realities: First- and Second-Generation Asian Indian Perspectives on End-of-Life Care. *Journal of General Internal Medicine*, *27*(3), 311–317. doi:10.100711606-011-1890-7 PMID:21948206

Sim, S. W., Soh, T. L. G. B., & Radha Krishna, L. K. (2019). Multi-dimensional approach to end-of-life care: The Welfare Model. *Nursing Ethics*, *26*(7–8), 1955–1967. doi:10.1177/0969733018806705 PMID:30318993

Sleeper, J. (2015, April 2). *Lee Kuan Yew's hard truths*. OpenDemocracy. https://www.opendemocracy.net/en/lee-kuan-yews-hard-truths/

Sokol, J. (2008). *There Goes My Everything: White Southerners in the Age of Civil Rights, 1945-1975*. Knopf Doubleday Publishing Group.

Sprinkle, R. H. (2011). Bioethics Without Analogy. In O. P. Wiggins & A. C. Allen (Eds.), *Clinical Ethics and the Necessity of Stories: Essays in Honor of Richard M. Zaner* (pp. 71–85). Springer Netherlands., doi:10.1007/978-90-481-9190-1_6

Szabados, B. (1999). Was Wittgenstein An Anti-Semite? The Significance of Anti-Semitism for Wittgenstein's Philosophy. *Canadian Journal of Philosophy*, *29*(1), 1–27. doi:10.1080/00455091.1999.10717503

Tai, M. C. (2016). End-of-Life Care in Taiwan from the Perspectives of Asian Bioethics. *Asia Pacific Journal of Health Law & Ethics*, *10*, 69.

ten Have, H., & Gordijn, B. (2011). Travelling bioethics. *Medicine, Health Care, and Philosophy*, *14*(1), 1–3. doi:10.100711019-010-9300-1 PMID:21086052

The Beach Boys (Director). (2016, October 10). *The Beach Boys—Good Vibrations* (Official Music Video). https://www.youtube.com/watch?v=apBWI6xrbLY

Tilghman, B. R. (1991). Ethics and Aesthetics in the Tractatus. In *Wittgenstein, Ethics, and Aesthetics: The View from Eternity*. SUNY Press. doi:10.1007/978-1-349-21174-6_3

Tischauser, L. V. (2012). *Jim Crow Laws*. ABC-CLIO.

Vasagar, J. (2018, May 11). Mahathir Mohamad is back. Malaysians' smiles may be brief. *The Guardian*. https://www.theguardian.com/commentisfree/2018/may/11/mahathir-mohamad-malaysia-prime-minister-reelection

Waugh, A. (2013). *The House of Wittgenstein: A Family At War*. Bloomsbury Publishing.

Wee, S.-L. (2019, February 21). China Uses DNA to Track Its People, With the Help of American Expertise. *The New York Times*. https://www.nytimes.com/2019/02/21/business/china-xinjiang-uighur-dna-thermo-fisher.html

Widdows, H. (2011). Western and Eastern Principles and Globalised Bioethics. *Asian Bioethics Review*, *3*(1), 14–22.

Wittgenstein, L. (1965). I: A lecture on ethics. *The Philosophical Review*, *74*(1), 3–12. doi:10.2307/2183526

Wittgenstein, L. (2010a). *Philosophical Investigations*. John Wiley & Sons.

Wittgenstein, L. (2010b). *Tractatus Logico-Philosophicus*. Cosimo, Inc.

ENDNOTES

1. If after reading the line before you have found a hidden message saying "Satan Masters Zion" and now you believe that this book or my humble person is the Antichrist, then you should call the loony bin as well. And get away from Beatles' *Abbey Road* cover!
2. For listening to the displacement in Beethoven, I recommend the Piano Sonata No.11 op.22 - 2nd mvt performed by Arthur Schnabel (Schnabel 1933) in this link: https://www.youtube.com/watch?v=ZYA0kE0kYMs

3. The Wittgenstein's Lecture on Ethics slipped out of Cambridge's radars untitled for almost thirty years until The Philosophical Review published it in 1965.
4. Actually, this quote appears twice in the Tractatus, first in the Preface and then in the end of the book as the seventh basic proposition referenced above.
5. Original text with the changing names highlighted:
ךְ מְ ׳ש ה ,י.הְ ו ם .ר בַא ,ךְ מְ ׳ש־תֶ א דוֹ עְ א ,ר .ק .י־אְ ׳לְ ו`
ךי. .ת .ת נְ ם .יוֹג וֹ מֵ ה־באַ י ,כ ם .ה ר בַא
Torah's quotation retrieved (01/22/19) from: https://www.chabad.org/library/bible_cdo/aid/8212/jewish/Chapter-17.htm
6. I use both family resemblance and game in a deliberate comparison with the Wittgenstein's language game and family resemblance descriptions (Wittgenstein, 1968, 2005, p. 62).
7. https://www.ru.nl/publish/pages/688605/bangkok-eng.pdf
8. No data available on: Kiribati, Oman, Samoa, and Solomon Islands.
9. Thankfully, there are no big enough groups of Hispanic people in Malaysia to be tagged as drug pushers by Mahathir Mohamad.
10. https://www.youtube.com/watch?v=5QRi12ldhos&ab_channel=AfricanInsider

Chapter 1
Bioethics From Science to Dogma and the Intersections With Religion, Politics, and Gender

ABSTRACT

This chapter explores how bioethics has grown to become a common place where religious communities and agencies talk with a broader audience. It takes knowledge of how bioethics is related to biopolitics and how political power is used to control and regulate biological processes. Furthermore, the author delves into the feminist critique of bioethics, particularly the differential lived experience of women and the rejection of principlism. The author argues that the scientific circles captured by sex and gender-related secular religions are no less hermetic than the staunch macho ones. To sum up, constructive roles by important communities, including religious ones, are important in major shifts in society.

INTRODUCTION

In the Preface and Introduction, we explored the concept of belonging and its significance in human existence. We challenged dominant narratives and assumptions about what it means to be human, alive, or worthy. We also delved into the political and ethical implications of different ways of valuing life and examined the need for a more sensible response to displacement. Now, in Chapter 1, we will take a closer look at the rise of bioethics and its relevance to the urgent need for belonging.

As we embark on this exploration of bioethics, it is important to understand its significance in shaping our understanding of ethical principles in medicine and life sciences. By inquiring about its rise and interdisciplinarity, we can gain a deeper appreciation for the role of bioethics in navigating the complex moral landscape of our modern world.

DOI: 10.4018/978-1-6684-4808-3.ch001

THE RISE OF BIOETHICS

Bioethics is, commonly defined, as a field of study that focuses on ethical issues related to life, health, and medicine. It examines the moral implications of scientific advancements, medical practices, and social policies that affect human beings, animals, and the environment. Bioethics is important today because it helps us navigate complex ethical dilemmas in areas such as genetic engineering, organ transplantation, end-of-life care, and healthcare allocation. It also addresses issues of social justice, cultural diversity, and environmental sustainability in healthcare and research. Bioethics often intersects with religion, as religious beliefs and values may influence one's perspectives on topics such as abortion, assisted suicide, and reproductive technologies. Additionally, bioethics is closely related to biopolitics, which involves the use of political power to control and regulate biological processes, such as reproductive rights, healthcare policies, and genetic research. In a broader sense, bioethics can provide a framework for critical reflection, ethical deliberation, and responsible decision-making in the face of rapidly advancing biotechnologies and the challenges they pose to individuals, societies, and the environment.

After WW2 even many European intellectuals shifted away from the diaphanous Descartes and returned to the Aristotelian virtue (Carson & Burns, 2006, p. 48). It was a paved road for the birth of the discipline of bioethics. However, the mix of virtue and plurality has never been easy as we'll see. Mainstream Bioethics, or Bioethics as we know it today, traces its origins back to the tumultuous times of the 1960s and 1970s. It was a time when Bob Dylan sang "Times are a-changin'" (Dylan, 1964) and The Crickets sang "Please don't ever change" (The Crickets, 2016), capturing the spirit of a period marked by social divergence, civil rights movements, and medical progress. It was in 1962 when the today "bioethics" first appeared in a Life magazine article by Shana Alexander, the first female staff writer for the publication, titled "They decide who lives, who dies," which featured a non-bioethicists committee tasked with selecting candidates for hemodialysis treatment (Alexander, 1962). This article, spanning 23 pages, was provocative and thought-provoking, raising the issue of "playing God'' and laying the foundation for the field of bioethics to be fraught with controversy from the outset. This time of groundbreaking advancements in theory and clinical practice, were also the time for unprecedented moral dilemmas related to the identity and goals of medicine (Callahan, 2017). These were the times of VietNam, Thalidomide and such publicly relevant cases such as Roe v Wade on abortion, topics that echoes more than 50 years later.

After the 1970s bioethics started to drift towards pragmatism. During the 1980s and 1990s, palliative care and the withdrawal of nutrients were dominant topics in scientific journals, with space for bioethics to contribute to the discourse. However, they were eventually supplanted by discussions on stem cells, germline engineering, and therapeutic cloning (Turner, 2004). In the 1990s, Bioethics' displacement to pragmatism was evident and questions about the relationship between bioethics and science seemed less pressing at the time (Schermer & Keulartz, 2002; Mayes et al., 2015). Bioethics became more and more important till, in 1995, it finally dethroned the old-fashioned moral philosophy of medicine when bioethicists became the more common and influential professionals in such a clinical topic as genetic engineering debate.

Nevertheless, the traditional medical views on moral and philosophy were not displaced only by the bioethics' perks but also by its own flaws. Putting the physicians to balance the kind of care against its cost is squaring the circle out of a public healthcare system. It's unethical, indeed! (Dobken, 2018) Despite their antagonistic positions, both bioethics and medical ethics are moving away from classical

Figure 1. Professions of common (top) and influential (bottom) authors in Human Genetic Engineering debate
Source: Evans (2002, p. 38)

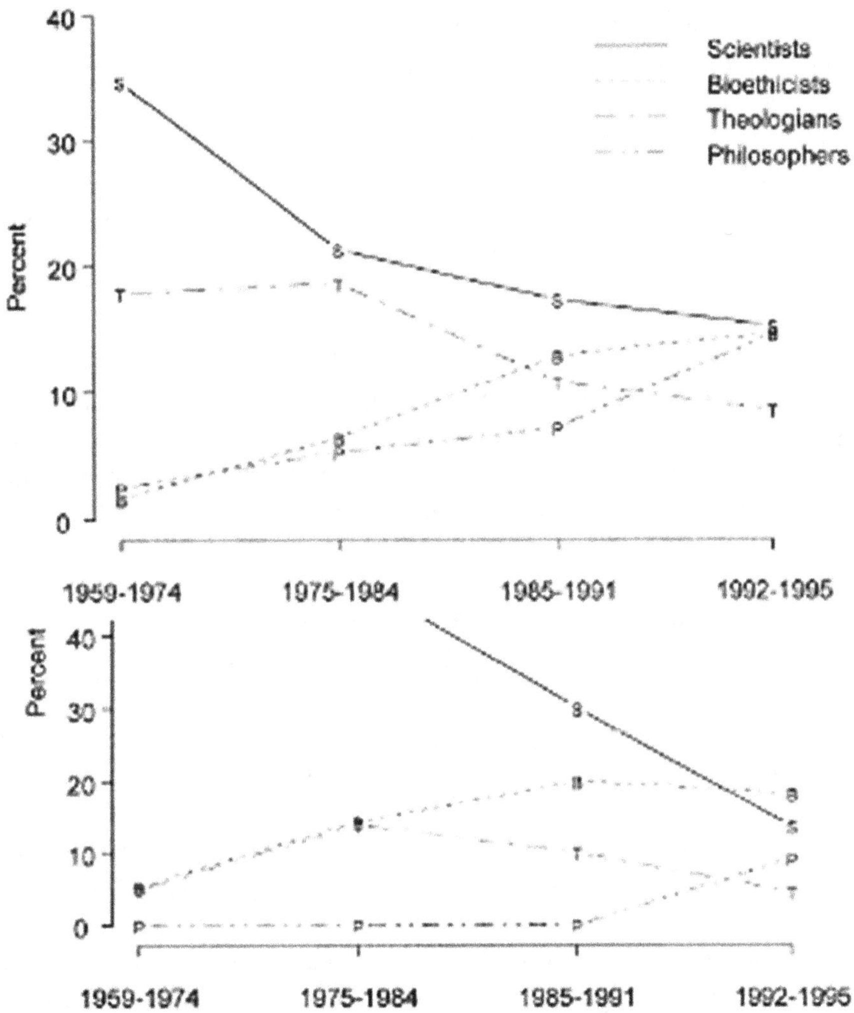

ethics of good and evil, but bioethics has kept enjoying a greater flexibility and adaptability to an ever evolving socio-cultural landscape (Pellegrino, 2013).

On the other hand, since bioethics gained such relevance, from the mid-1990s onwards, it entered its most liquid and global phase.[1] The global bioethics term was coined by Van Rensselaer Potter in 1988, although the term didn't mature quickly. Until HIV/AIDS didn't become pandemic in the middle of 1990s, the bioethics was not *global* (Have, 2016, p. 43). Paradoxically, as bioethics became more global, its principles lost universality. They never were seen to be as universal as before 1995. It was the time of the first pre-Google search engines: WebCrawler, Lycos, Altavista, etc. that was succeeded by the first social media before Facebook: Friendster, Myspace, and Second Life. Short after, in the first years of 21st century those technologies exploded and bioethics increasingly entered the public domain with the advent of information and communication technologies, giving rise to bloggers, influencers, social media campaigns, and advertising that sought answers to questions about children's obesity, transgenics, vaccines, and other topics, while also raising concerns among consumers and patients (Murphy, 2018). The use of Assisted Reproductive Technologies (ART) by the lesbian, gay, bisexual, and transgender (LGBT) communities has also been a subject of controversy and enthusiasm in recent times. The bioethical discourse challenges itself and, as far as it doesn't limit to woke's tears and shark-minded mockery, it's great.

Bioethics has played a crucial role in helping moral science categorise empirical research and evaluate its utility. However, the translation of empirical knowledge into normative content for bioethics has not been entirely successful and, just like happens with mass media cultural landscape, rarely brings about radical new perspectives (Kirkwood 2009; Giddens 2015). But this flaw was quickly prettied up by bioethics' multidisciplinarity and its knack for cultural values. Bioethics has found a place thanks to its capability to discuss ill-defined, broad, speculative issues, especially those able to worry political and religious leaders. Walking transgenic tomatoes? Let's talk about it!

Other factors that influenced the success of bioethics not related with academia were the joint work in advisory commissions and a principle-based ethics as well as the voluntariness and social diversity among the first bioethicists. Those commissions gained jurisdiction of their own while the *principlism* became an efficient system to face the new ethical dilemmas (Evans 2002, 36–38).

TRADITIONAL RELIGIONS AND THE THREADING OF BIOETHICS AS A GLOBAL NETWORK

In the turn of the millennium, with a new rise of religious popularity, bioethics has grown as a common place where the religious communities and agencies talk with a broader audience (Schmidt, 2011). But it didn't have an easy beginning.

The Roman Catholic and Protestant bioethics were bickering when the Beauchamp and Childress' *Principles of Biomedical Ethics* (Beauchamp & Childress, 2001) appeared in 1979. The secular bioethics unjammed the worn out discussions over organ transplantation and assisted reproduction, widening the scope to foetal research and psychosurgery thanks to its canon of research. Nevertheless, the secular bioethics happened not to be less fragmented than the Christian one; so much so that conservative and reformist tags could apply to secular groups in a similar fashion (Carson & Burns, 2006, pp. 90–91). Some said that the secular became the new Christian. Secular principlism has been branded as too reductionist for institutions that heavily rely on spirituality, like hospices (McGrath, 2018). Certainly, it's proven to be hard for bioethics to find a methodology suitable to face the end-of-life issue. In practice, the secular bioethics obtained a Pyrrhic victory that becomes evident each time a public policy is on the table.

To cap it all, the four notes used for creating all that jazz (autonomy, justice, beneficence and, non-maleficence) are usually played off pitch when facing the postmodern *polytheism*. At the end of the day, if it weren't for religions, Bioethics would be a rare discipline out of the West instead of a global network. So we need to assess at least some religious perspectives to better understand Bioethics. Also it will help us to highlight the importance of the beginning and end of life. So, it's a good moment to know more about bioethics and religion, including those from diverse traditions, in shaping global bioethics discussions on topics such as justice, beginning and end of life. To begin with, the many religions of the world bring both diversity and divergence to bioethical issues. These sets of beliefs may guide patients and health care professionals as they seek or provide health care. Doctrinal and theological differences exist even within the same denomination, so the views of patients, family members, and health care providers should be sought.

We should stress here that philosophical and religious traditions raise questions on medical ethics, especially when it comes to justice (Veatch, 2000, pp. xviii–xix). Also the political institutions are a main actor in bioethics. The quasi-religious Hippocratic Oath was replaced by the Soviets by one of its own (Lichterman, 2005). In the seven years before the fall of the USSR, changes happened and a new medical reality appeared (Tichtchenko & Yudin, 1992).

In the West, the religious perspectives on Bioethics are very influenced by its radical scholar tradition. The 13th century model of theology from the European mediaeval universities is connected with the 1970s birth of Bioethics through the Enlightenment. In this line, Engelhardt considers bioethics as a late-Enlightenment surrogate for religious moral authorities and talks even about the "*ordination of bioethicists*" (Engelhardt, 2002b). The idea of philosophy as a servant of theology like in the 13th scholasticism and epitomised by Aquinas has not disappeared. Furthermore, the bioethics has its own non-legalistic Aquinas in Paul Ramsey, one of the most influential ethicists in the 20th century who advocated for the "*love transforming justice*" (Ramsey, 1967, p. 122). In addition to this, the ecumenical agenda of the Christian theology for the 21st century points to an ecological ethics reorientation (Hessel, 2000).

Looking for moral probity with the glass of *systematic* morality is a common trait of secular and Western Christianity. Both mistrust of upholding moral integrity from a subjective or situationist approach. So much so that the Roman Catholic pro-life discourse, for example, is defended by the bioethicists who follow this faith in the same terms that a secular person would do (Cahill, 1990).[2] However, it's not shared at all even by close denominations like the Orthodox faith (Cherry, 2013). In addition to this, the Christianity today compared with the one from the 13-17th centuries has raised both its plurality and disarray, so much so that the secular bioethics' success is partly explained by this quality that religious bioethics can't afford: aseptic neutrality (Engelhardt, 2006). Therefore, the *remaining* great question in the Western religious bioethics is: a plenty relationship with god leads to unique (perhaps revealed) moral insights or it could be attained through discursive moral reasoning anyways? (Engelhardt, 2002a)

The long-lived political hegemony of Western Christian states has led to worldwide sound differences. The concept of human rights, almost entirely based on the groundbreaking socio-political major shifts that happened in the 17th century in Europe and North America, has been a highly disputed canon in contemporary ethical discourse. The main point of disagreement is human dignity. A Jewish-Christian concept of human dignity as *imago dei* is incompatible with non-creationists' set of beliefs as Buddhism (Keown et al., 1998, p. 25). This concept has also been questioned (on non-religious grounds) by important figures like Hyakudai Sakamoto (Sakamoto, 2002) and tagged as 'useless' by a Western author like Ruth Macklin (Macklin, 2003). To cap it all, dignity has besides its religious implications, additional roles as political principle and a moral standard or how to treat the patient as a person -in bioethics, it means a normative framework for biomedical issues (Andorno, 2013). In short, dignity serves as a moral standard for how individuals should be treated as persons, which limits what can be done to them. Therefore, religious views on human dignity will be a source of controversy for a while and the displacement of the geopolitical axis from West to East will raise the bar on the Bioethics discussion about the pertinence of human dignity as a core belief. Let's delve now into those more Eastern traditions.

Buddhism has incorporated itself into the world-wide web of bioethics with a fresh antagonism of the Western idea of dignity and human rights (Tham et al., 2018, pp. 33–34). For example, the concept of human dignity seems to privilege humans against non-humans, which is against the Buddhist teachings.

In the Buddhist point of view, the rights and duties should be balanced in a negative-positive fashion rather than the Western "disparate aggregation" (Keown et al., 1998, pp. 57, 105).

Islam is the most followed religious denomination ever, however the Islamic Bioethics is a quite recent field of study. The Western information services on Islamic Medical and Scientific Ethics (IMSE) started in 2009 and were sponsored by Qatar and partnered with the Kennedy Institute of Ethics and Georgetown University (two Western institutions, by the way). Islamic Bioethics, mirroring the Qur'an, gives special importance to procreation (*The Meaning of the Holy Qur'an*, 2015, v. 13:38, 16:27, 42:50). For example, according to Islam, the procedures of in-vitro fertilisation and embryo transfer are acceptable, although they can be performed only for husband and wife. Thus, unsurprisingly, assisted and reproductive technologies (ART) has been the entrance gate of the Muslim in the global bioethics web. On the other hand, the way that Islamic bioethics deals with new medical technologies is being held back by a specific legal-medical approach that doesn't fully consider ethical beliefs (Padela, 2018).

Besides the centrality of dignity in its discourse, the true touchstone to differentiate bioethical families is the start and the end of life. Curiously, most denominations are quite coherent in how life starts but there is plenty within diversity for when life ends. Major religious groups are active in influencing the public with issues concerning procreation, abortion and infertility therapy. So much so that Abrahamic religions, for example, can distribute themselves along an axis. Judaism, which allows the practice of all techniques of assisted reproduction if the oocyte and spermatozoon originate from the wife and husband respectively, and Islam, which is closer, are in one extreme of the axis; Vatican, in the furthest extreme, holds the most conservative position among major groups while the most Christian denominations lie somewhere in between along this axis (Schenker, 2005).

Quite the opposite from life start, those bioethical issues involving death, just like brain death or organ transplantation from a deceased donor are a way to realise the diversity even within the same denomination and country. For example, among Japanese Buddhists, 9 out of 10 think that there is no unified view on organ transplantation and brain death in their religious denomination (Fujieda, 2018). The medical practice in Western Christian is imbued by the anthropological meaning of death as the experience of vulnerable creatures in relation to God (Ridenour, 2017). However, a zoom in this shows foundational disagreements like the ones we can found in the German Christianity around the euthanasia: both Roman Catholic and Evangelical Lutheran churches have developed advanced directives for treatment choices at the end of life of their own even when admonishing physician-assisted suicide and euthanasia (May, 2003).

Religious denominations with radical views on medicine likely will have a slower incorporation to the global family of bioethics. For example, the Celestial Master movement within the Taoist tradition advocates for a demonological-confessional model of healing where all medical treatments are forbidden but the ingestion of talismans dissolved in water (Camenisch, 2013, p. 197). Hopefully, the personalised treatments, as those being proved in cancer therapy, will help to underpin the personal beliefs in medical choices through a more accurate prognosis (Tan et al., 2018). It would make bioethics more accessible even for the most exotic or cryptic religious traditions. This accessibility surely will offer some interesting turns as the glocal religions, those involving the blending or fusion of global religious expression with local particularity, become more common (Roudometof, 2018, pp. 1–10).

Bioethics' maturity means accepting the particular nature of the many lives (International Association of Bioethics, 1999, pp. 8–9). Those effects seemingly have not faded away. In addition to this, the 2010s has seen a turn in bioethics towards a common good approach from an overwhelming liberal individualism (Gross, 2014). In conclusion, it's important to ground a secular and inter-faith communication organ in order for bioethics to improve as a global discipline. Raising cultural competency in physicians would be helpful here just as has been stressed in the very different context of cosmetic surgery (Bresler & Paskhover, 2018). But also better time management is important. The lion's share of the time spent by Western theologians from the most populated denominations goes to self-referencing their own tradition or reforms, which is delaying a broader and successful vision (Kelsay & Johnson, 1991, p. 25). The same goes for the most small and great religious leaders. Bluntly, the religions (and the interfaith dialogue) have lost the confidence of so many people in the West because important communities didn't see the need for playing a constructive role in nowadays major shifts (Twiss, 2018). It's an example of how divergence can grow not only from competition but also from apathy.

GEOPOLITICAL SECULAR RELIGIONS

In this section on the rise, globalisation and institutionalisation of Bioethics, we cannot limit ourselves to the traditional religions without analysing the secular ones of partisan politics as well. Indeed, from its inception, bioethics' main subject of interest has been the moral issues raised by new technologies, mainly in medicine, but there it has been a gradual politicisation:

Bioethics has always been a biopolitics and the political dimension is only now coming into relief for bioethicists. (Bishop & Jotterand, 2006)

For example, Bill Clinton's National Bioethics Advisory Commission (NBAC) was chaired by Harold T. Shapiro and reflected a certain political agenda and specific moral perspectives. The NBAC was replaced in 2001 by George W. Bush's Council on Bioethics and headed first by L. Kass and later by Edmund D. Pellegrino. Both NBAC and President's Council mirrored the new White House host's sentiments and political opinions in the use of embryonic stem cell research and human cloning.

No other threat on bioethics today is shared everywhere as political lobbies. They are over-representing, swallowing, displacing within, the diversity we need for a critical and substantive reflection on our own goals and guidelines as bioethicists. Bioethics dwells in a high-profile bureaucracy of virtue where everybody but researchers and funders are subjects of research needed for *protection* (Jacob & Riles, 2007). Ironically, this *ethical* institutionalism came as a reaction against the Nazi human subject research (the Nuremberg Code) which soon became "*the institutionalization of over-enthusiastic best practice*" (Strathern, 2006, p. 533). Like in *Animal Farm* the resolution of not to live in the house of the farmer didn't last (Orwell, 2009, p. 43). It began using Guatemalan Amerindian and African American US people as guinea pigs because "All animals are equal but some are more equal than others". After stopping the US Tuskegee-s, the animal farm kept being mismanaged, as in Orwell's tale, till the point in which pigs were drunk in the farmer house while the rest of animals looked through the window. In bioethics, it is the point in which narrow-mindedness and misalignment of the ethical discussions rule on patenting biotechnologies (Forsberg et al., 2018). And where is dignity here?

Talking on perverse sycophancy, we can't ignore the upside down of Western secular religions: authoritarian Asianism in Bioethics. But, first and foremost, it would be good to learn how many Asian nations have solved secularism versus religion-based bioethics in a more efficient way than most of the West. Asia also presented its own Bioethics' bottleneck and adapted in a very different way, so it's worthy to know more about. Let's start with China. Xunzi (313–238 B.C.), a well-known Chinese philosopher, pointed out the relation of universal love with the equal relevance of a good birth as well as a good death (Knoblock 1988, 68–80, 168). Daoism also refers to the rewards in peace and dignity of those who follow nature when "*coming into life and going out of death*" (*chu sheng ru si*) and vice versa (Dong & Wang, 2004). On the other hand, Confucianism, Daoism, and Buddhism have a fatalistic and beyond human power view of life and death and the last is taboo in the Chinese mind (Doring & Chen, 2002). This taboo, even thinking that bad things happen after you say them out loud, is somewhat shared across the most Asian and Pacific Island countries (Cheng et al., 2015; McLaughlin & Braun, 1998). Despite this, a shift in the Chinese views on euthanasia has happened since the 1980s (Shi & Yu, 2004). Likewise, Thailand, a Buddhist country which is a focus of allopathic tech and a universal health-care bastion in Asia, has witnessed how mechanical ventilation became in 2006 the subject of a hot national debate (Stonington & Ratanakul, 2006). As I see it, the main difference in West and Asian bioethics paths to the end of life choices is that most Asian nations have solved their prolonging-life dilemmas through religious adaptation instead of such a heavy emphasis in secularisation like the Western ones. An example of this are the relational autonomy bioethics coming from Singaporean Confucianism or the e-Fatwas for the Sunni Muslim in large communities in the West, e.g. UK (Krishna et al., 2015; Chang et al., 2017; Van den Branden & Broeckaert, 2010, 2011).

To analyse and critique the incoherence of both Western and Asian positions in bioethics, we can turn to the works of a classical scepticism author, Sextus Empiricus. Sextus Empiricus was a Greek philosopher who advocated for scepticism as a philosophical approach that questions the validity and reliability of knowledge and beliefs. His work "Outlines of Pyrrhonism" provides insights that can be applied to the issues raised above on the Western and Asian standards. He said that a proof needs a standard that has been proven in turn, which leads to a vicious circle (Sextus Empiricus, 1933, pp. 263–269). In this case, what *proves* the Western or Asian standards are their hegemonic powers means to project a cultural sphere around them and, in turn, what *proves* their few many ideologies is the docile way they fit in this politico-cultural sphere. Let's see how it has become global.

SEX AND GENDER-RELATED SECULAR RELIGIONS

Now we'll discuss the *religare* (keep tied) around the central dogmas of gender as entirely a social construction and male oppression. To begin with, the feminist critique of bioethics reflects not only the differential lived experience of women, but also adds a very feminist slant (Shildrick, 2015). In general, feminist bioethicists are disappointed with the centrality of traditional moral theories in bioethics given that it is inevitably situated in particular social and historical locations (Tong, 2018). For example, as the Malaysian feminist Sharon Bong states, we must to negotiate the tensions between rights and religions in global and local contexts, doing otherwise, festering positions around universalism and relativism is nothing but "*epistemological violence*" (Bong, 2018). Consequently, a feminist bioethics overlaps with virtue ethics and rejects principlism (Camenisch, 2013, p. 198). The feminist view states that following principles demands another quality distinct from and presupposed by the application of the principle themselves (Carse, 1991). The principlist critique above, extends and echoes feminist authors (Evelyn Fox Keller and Donna Haraway, e.g.) who see in it the result of the dominant 'masculine paradigm' in science and the cause of the lack of parity in Bioethics Committees (Dickenson, 2006). However, the US Presidential Commission for the Study of Bioethical Issues has 50% of women, higher than the percentage of women in medical (40%) and lawyer (37%) professions. More even, the Norwegian National

Research Ethics Committee has 57% of women for a 55% of women in STEM professions. Even Saudi Arabia, which has a very legalist view of Bioethics and includes in its National Committee of Bioethics the Ministry of Defense and a National Guard representative, has 13% of women in this Committee from a 12% of female lawyers.[3,4 5] By the way, US, Norway and Saudi Arabia committees are led by women. Might it happen that you find a bioethics committee stained by misogyny or misandry or even transphobic, but it would be an exception rather than the norm. It would be far more accurate to guess that Bioethics Committee members come from an economically favoured family background, for example.

Indeed, feminist bioethics challenges the non-feminist bioethics holding that the tradition does not view women's rights as human rights, that man has engulfed the human and that it's a source of sickness (Rawlinson, 2001). (Unlike the booming negativity against men that is very healthy). Against this sickness, the feminist movement propose the women's body functions as a meta-language to denounce several types of violence against women, in other words, body talk can easily be applied to better understand issues such as body manipulation, for example, the beauty contests brown-nosing to the politics of indigeneity in Guatemala (Barbosa, 2016). So, if you go to a beauty contest and make a feminist performance, the girl who has her day spoiled and the spotlight robbed should be thankful because her lifelong undeservedness and real sexism will disappear now that she has just heard the *Message* (Martin et al., 2014). And the same goes for any other women who do not fit neatly into a very particular ideological scheme. You are being saved, so shut the fuck up! Considering that when hegemonic feminism wins, women's pain and suffering is publicly elevated above men's and when it loses, suffering is used for self confirmation we could think that, unlike traditional religions, feminism brings no comfort at all. If we resort to the World Values Survey and look for how the feeling of happiness changes among women correlates with their interest in gender politics rise, we'll see that the higher the countries rank, the lower their happiness improvement, even leading to net unhappiness (rs = -0.29785, p (2-tailed) = 0.03766). More even, if we compare the highest and lowest quintiles by cultural groups in a similar fashion to the Inglehart-Welzel World Cultural Map, the result is also evident:[6]

Figure 2. Feeling of happiness among females according attention paid to gender policies

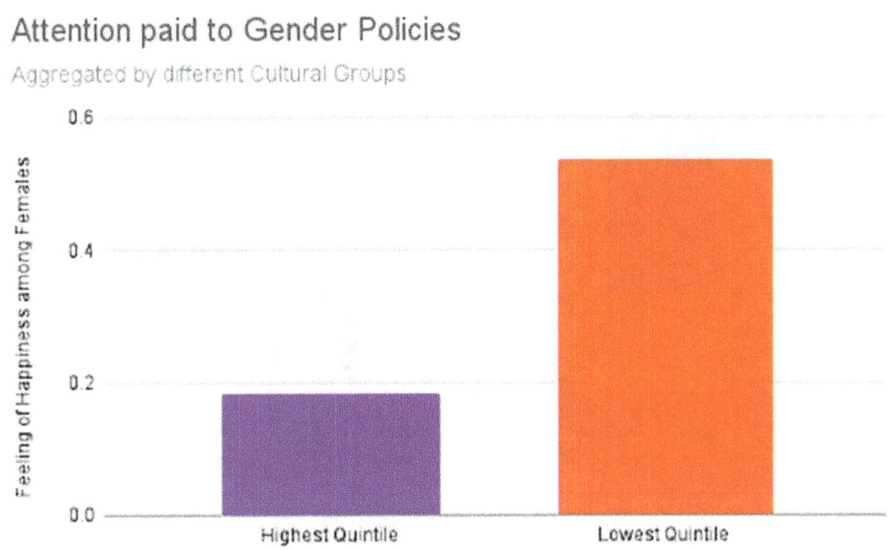

Furthermore, I looked for differences among EU countries with data available plus US, Canada and Argentina, so I divided this list among the countries with political representativity of the fuzzy left in Table 1.

Table 1. Country classification according being ruled either by a party with a clear hegemonic feminist agenda or not

Not Ruled by Parties With a Clear Hegemonic Feminist Agenda	Ruled by Parties With a Clear Hegemonic Feminist Agenda
Hungary Italy Poland Romania	Argentina Canada Finland France Germany Netherlands Spain Sweden United States

And the result when compared how women's happiness has changed and how it has changed regarding men from the period 2005-2009, just before the 4th Wave of feminism, to the 2017-2022 period the result is:

Hegemonic feminism as it's being shovelled down from institutions, actually harms women. So much so that the only point in its favour is that it harms men even more. It doesn't mean that for any culture and any scenario, 4th wave feminism will be a mistake, but very likely it will be a *bad movement.*

Figure 3. Women's happiness and happiness gender gap according country's rule being woke or not

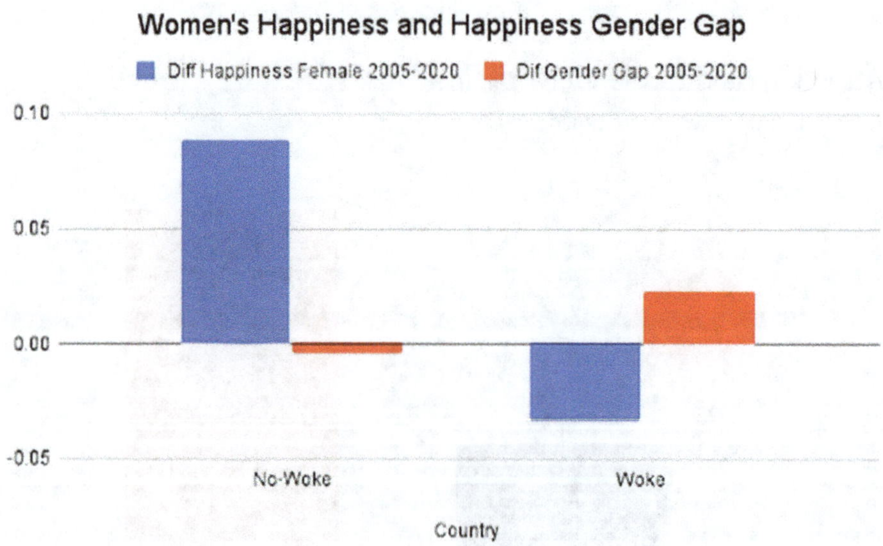

Feminist critique aligns with and extends to bioethics the Zygmunt Bauman modern ethics critique of being only "*about conflict resolution, and about admitting no contradictions except conflicts amenable to, and awaiting resolution*" (Bauman, 1993, p. 8; Harrison et al., 1998). These contradictions relate to conflicts or dilemmas in bioethical decision-making where there are no clear-cut resolutions or where ethical principles may come into conflict with one another. These contradictions can have real-life implications and impact individuals in end-of-life decision-making, reproductive rights or allocation of resources among others. Here are a few examples:

- A patient is terminally ill and in extreme pain: a conflict between the principle of autonomy, which advocates for patient's right to choose to end their suffering through euthanasia or assisted suicide, and the principle of non-maleficence, which requires healthcare providers to "do no harm" and not intentionally cause death.
- Voluntary interruption of pregnancy: a contradiction between the autonomy of the pregnant woman for abortion and the value attributed to the unborn foetus. Or, splitting hairs, the decision to terminate a pregnancy to protect the health or life of the mother conflicting with the moral status attributed to the foetus.
- Limited healthcare resources: deciding how to allocate a life-saving organ transplant may involve considering various factors, such as medical urgency, prognosis, patient's age, social worth, and other ethical considerations, which may sometimes be contradictory and challenging to reconcile.

Even considering the significance for feminism of this postmodern embracing of contradictions, some feminist authors are against it since they see its defence by certain traditionalists as a reaction against the womens' progress as subjects of history (Shildrick, 1997). Indeed, postmodernism's critique of great narratives could also be used as a reactionary response to this one about the progress of women's rights and gender equality. Even if this risk is known for more than a quarter century, the 2012 onwards 4th wave feminism may undermine the gains made by more solid feminism by denying the existence of objective truths, universal values, or shared realities that activism and social change demands. So, ironically, the most postmodern sex and gender-based secular religions are potentially closer to perpetuating gender inequalities and limiting the agency of women as subjects of history than present-day chauvinism.

Seyla Benhabib critique's scope reaches far behind when stating the moral domain theories "... *from Hobbes to Rawls, leads to a privatization of women's experience and to the exclusion of its consideration from a moral point of view*" (Benhabib, 1992, p. 152). However, Benhabib somewhat assumes a linear progression of moral theories across history aiming to commodify women without acknowledging the diverse and evolving nature of moral theories over time. To begin with, the first moral theories that emphasise human rights and provided a framework for advocating for women's rights as human rights appeared in its most bragged form in the period from Hobbes to Rawls -even if they are far older.[7] The social contract idea as it appeared in the works of Hobbes and Locke became the grounds for Carole Pateman and Martha Nussbaum to reinterpret social contract theory in order to address issues of gender inequality (Nussbaum, 2004; Pateman, 2015). The Rawlsian idea of Justice as opposed to gender discrimination is too obvious for deserving further comment, unless you think that Rawls actually was playing a very subtle game. Indeed, there is a popular feminist opinion sustaining that, across history the division of labour has turned the women's voice into "*only marginally moral*" (Hekman, 2014, p. 26). It's as true as unfair that women have not participated from political seats of power as men *with a similar favoured background*. It's very relevant because almost every man and woman in the world, then and now, is a zero on the left of bigger social players. Women, across history, occupied diverse roles in the workforce. For example, during the mediaeval and early modern periods, women in Europe could be involved in trade and commerce, owning and managing their own businesses, and engaging in long-distance trade networks (Grossman, 2004, pp. 274–275; Hilton, 1984). In addition to this, their moral agency was hardly marginal for this. It was not. Women in most societies and eras have always been not only essential but radically socialised individuals able to engage in creative forms of collective resistance such as sex strikes in Ancient Greece, Nigeria, Colombia, Togo, etc (Agbedahin, 2014; Morales, 2013).

An easier explanation to this historical gender gap is that the majority of privileged women had no incentives for changing gender roles since their life was even better than men in some aspects and the underserved big majority of women were pissed along with their men either abusive or caring. In other words, I would be more than surprised if I met a mediaeval European woman from an affluent family who didn't know how to exploit social conveniences in her favour with the same skill and wit if no more than clergy members did. Anyways, the centrality of the workplace in self-realisation is a very postmodern thing (Hancock & Tyler, 2001, pp. 190–209). Hardly success at work was greater than life before dehumanising everyday life for the sake of widespread homogenisation of cognitive (overtime) work before a screen (Berardi, 2003). The wide majority of people didn't think in those terms before the 1990s and feminists fall into presentism when talking about women in the workforce. So, no, during the Victorian era there weren't plenty of young women eagering to work in a lab as scientists but frustrated by the evil patriarchy. It's only a literary and academic cliché that fosters sex reductionism (Burkitt, 1998; Hancock & Tyler, 2001, p. 84). Then and now, Western women have been more present in works that make you invisible but also in highbrow cultural consumption (Lizardo, 2006; Lozano Rodríguez, 2021). Affluent 19th century Western young men and women with intellectual ambitions felt self-realised when proving their worthiness outstanding in (many times inherited) cultural and social circles thanks to their charm, eloquence and erudition. It can be seen in writings of women of this time such as the actress and writer Frances Anne Kemble and the fiction author Mary Russell Mitford (Kemble, 2006; Mitford, 1852). Those youth loved participating in exclusive cultural events such as operas, befriending well-known geniuses or, mostly in the case of men, travelling abroad. Even if some of those intellectual women were asphyxiated by gender roles, most women cope well with their role's vulnerabilities and it's false that men's lives were uniformly better than women's (Szreter & Fisher, 2010, pp. 26–47):

Typical 19th century and early 20th century well-off men enjoyed a degree of freedom above most women, but also above most men. In turn, as the industrial revolution expanded, working-class men paid for their alleged economic *independence* with their own lives.

All that I said above doesn't rule out that discrimination against women in science has always been an issue and, in too many countries, it's still an issue. However, the relevance of the topic as in 2023 is suspicious:

The sex ratio demographic shift refers to the year 2012, when the share of the US population 35 years old and above and four or more years of tertiary education completed shifted from a men's majority towards a women's majority. Till this data the academy seniors in the country that hosts most of the first level world universities were slightly more men than women, although it varies from one discipline to another. Here it is noticeable that the most is published about how sexist against women science is, the most scientifically proven seems that a leader quality is partly, or totally, between her legs. And just like always happens when a human group starts hoarding an insane influence and monopolising narratives, vulnerable people suffer, in this case underserved men but also women who can't or don't want to ride this wave.

Figure 4. Gender gap in life expectancy during the Industrial Revolution in England
Source: Office for National Statistics (2015)

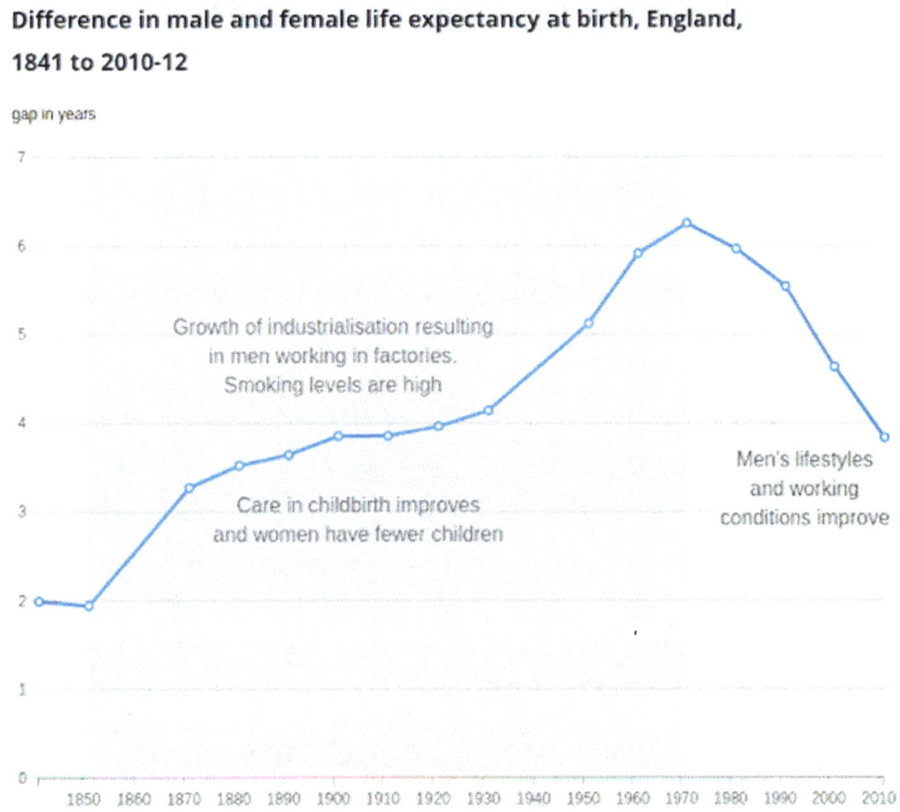

Figure 5. Research papers on "sexism in science" and "women are better leaders"

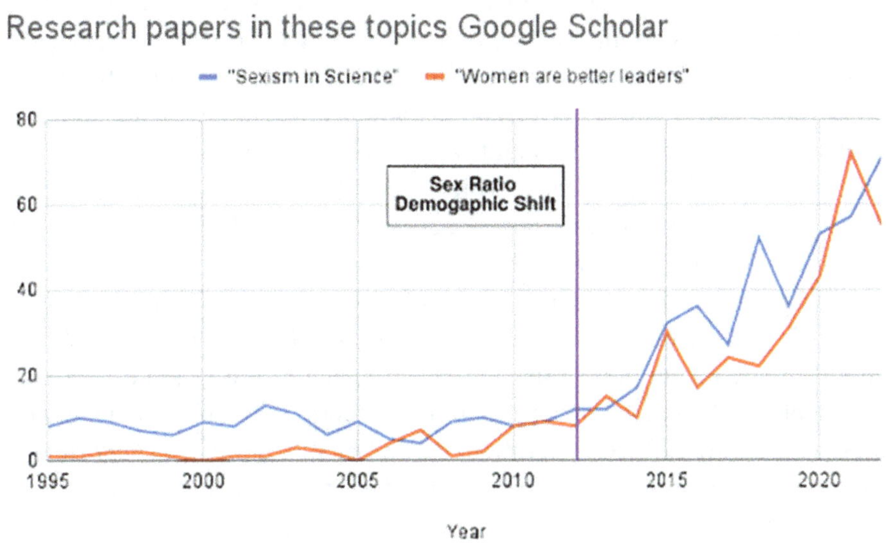

Figure 6. Worldwide Google trends for "women empowerment," "women rights," and "patriarchal"

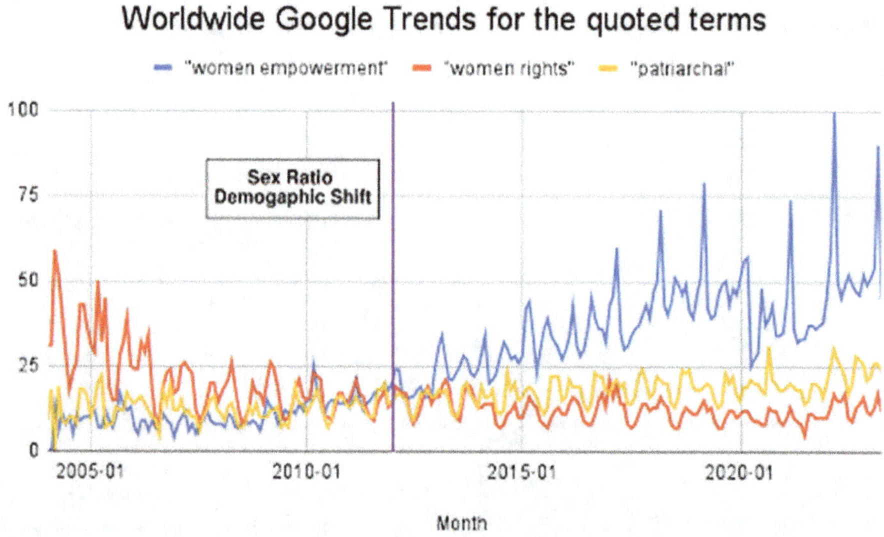

Interest in women (all women!) rights as the human beings they are has been belittled in this woke hegemonic and 4th wave feminism for the sake of women empowerment and heteropatriarchal bores. As the Kenyan Kikuyu people say: "When elephants fight, it is the grass that suffers". We should stop focusing only on between-gender inequality but also pay attention to the within-gender inequality.

On the other hand, a different critique of the feminist critique is also pertinent here. Justice, as the gift of other possibilities, goes beyond sexual duality, involves paying attention to the way in which sexed identity is constituted as an embodied effect (Diprose, 2005). But, gender, by the way, is not a stable identity or locus of agency but a liquid identity grounded on time (Butler, 1988). Justice transcends gender, regardless of its flow. In addition to this, the Western feminist bioethics has still to position itself against the unjust policies from the Global North (IMF, for example) that produces deleterious effects in the population's access to healthcare services, namely the women living in underserved countries (Nnaemeka, 2015). In other words, the feminist identity, even though it is solid, lacks diversity (DeBlaere et al., 2017). The "Woman" doesn't exist: "she" obscures important racial, ethnic, cultural and class differences among women (Tong, 2001). If it weren't enough, most feminist approaches to bioethics fall into one of two clusters: care-focused approaches and power-focused approaches (Tong, 2004). To cap it all, feminist bioethics is hopeless to dissent from political feminism and similar misandry.

Queer theory is a perspective on sexuality aimed to displace heterosexuality from its privileged and institutional status (Bywater & Jones, 2007, p. 33; Richardson, 2000, p. 43). It sounds like the same *drag* as feminism, but there are no definitive rules for Queer bioethics, it opposes categorizations, favours pastiche over principles and, even a thorough analysis will both complement and contradict it (Richie, 2016). Anyways, as Timothy F. Murphy wrote in 2015: "*Bioethics still has important work to do in helping to secure status equality for LGBT people and does not yet have good enough answers in this effort*" (Murphy, 2015).

Indeed, bioethics has not kept up with the growing relevance of queer and trans people in the public space, in spite of the cases of interest (Kimoto, 2018). For example, the trans activist Leslie Feinberg was turned away from a hospital for being trans when suffering an endocarditis and she later died in 2014 due to a chronic tick-borne infection aggravated by her logical mistrust to physicians (Iovannone, 2019; Weber, 2014). The stories like the Leslie's one are quite significant, since the youth queer place and time proficiency proved to be important when "*mapping queer bioethics*" (Marshall, 2016). Namely, the perceived stress related to social disadvantage and marginalisation plays an important role in mental health quality of life disparities among lesbian, gay and bisexual Hispanic midlife and older adults (Kim & Fredriksen-Goldsen, 2017). Indeed, the trans and gender-nonconforming community has a complex relationship with psychiatry. On the one hand, an increasing number of activists refuse the pathologization of their identities through the diagnosis of Gender Dysphoria and some of them reject the whole biomedical model of mental illness. On the other hand, without pathologization the access to publicly funded transition-related medical procedures, counselling services, and disability benefits also get reduced (Smith, 2018, p. ii). On the other hand, turning transition into an open bar is bringing and will bring more harm than good (Canadian Gender Report, 2020; de Blok et al., 2020). The scientific circles captured by sex and gender-related secular religions don't seem less hermetic to me than the macho ones that overlooked Marie Curie and, to cap it all, their theories are even less falsifiable due to subjectivity.

CONCLUSION

The main idea of this chapter is that bioethics plays a crucial role in navigating the complex moral landscape of our modern world. It provides a framework for critical reflection, ethical deliberation, and responsible decision-making in the face of advancing biotechnologies. By exploring the rise of bioethics and its intersections with various domains, we gain a deeper understanding of the ethical challenges we face and the need for a comprehensive and holistic approach to belonging in bioethical discussions and decision-making processes.

REFERENCES

Agbedahin, K. (2014). Interrogating the Togolese historical sex strike. *International Journal on World Peace*, *31*(1), 7.

Alexander, S. (1962, November 9). They decide who lives, who dies: Medical miracle and a moral burden of a small committee. *Life Magazine*, 102–125.

Andorno, R. (2013). The dual role of human dignity in bioethics. *Medicine, Health Care, and Philosophy*, *16*(4), 967–973. doi:10.100711019-011-9373-5 PMID:22173655

Barbosa, M. E. (2016). *Body Talk: Performing Violence Against Women in Contemporary Guatemala*. https://kuscholarworks.ku.edu/handle/1808/25757

Bauman, Z. (1993). *Postmodern ethics*. Blackwell Oxford.

Beauchamp, T. L., & Childress, J. F. (2001). *Principles of Biomedical Ethics*. Oxford University Press.

Benhabib, S. (1992). *Situating the Self: Gender, Community, and Postmodernism in Contemporary Ethics*. Psychology Press.

Berardi, F. (2003). La fábrica de la infelicidad. Nuevas formas de trabajo y movimiento global. Madrid: Traficantes de Sueños, 7-167.

Bishop, J. P., & Jotterand, F. (2006). Bioethics as Biopolitics. *The Journal of Medicine and Philosophy, 31*(3), 205–212. doi:10.1080/03605310600712760 PMID:16760100

Bong, S. A. (2018). Going beyond the Universal-versus-Relativist Rights Discourse and Practice: The Case of Malaysia. In S. Bahun-Radunovic (Ed.), *Violence and Gender in the Globalized World*. doi:10.4324/9781351143363-5

Bresler, A. Y., & Paskhover, B. (2018). Religion and the Plastic Surgeon: An Imam, a Minister, and a Rabbi Walk into a Surgical Centre. *Aesthetic Plastic Surgery, 42*(6), 1699–1703. doi:10.100700266-018-1196-9 PMID:30019243

Burkitt, I. (1998). Sexuality and Gender Identity: From a Discursive to a Relational Analysis. *The Sociological Review, 46*(3), 483–504. doi:10.1111/1467-954X.00128

Butler, J. (1988). Performative Acts and Gender Constitution: An Essay in Phenomenology and Feminist Theory. *Theatre Journal, 40*(4), 519–531. doi:10.2307/3207893

Bywater, J., & Jones, R. (2007). *Sexuality and Social Work*. Learning Matters.

Cahill, L. S. (1990). Can Theology Have a Role in "Public" Bioethical Discourse? *The Hastings Center Report, 20*(4), 10–14. doi:10.2307/3562778

Callahan, D. (2017). Religion and the Secularization of Bioethics. In M. Freeman (Ed.), *Ethics and Medical Decision-Making*. doi:10.4324/9781315209692-5

Camenisch, P. F. (2013). *Religious Methods and Resources in Bioethics*. Springer Science & Business Media.

Canadian Gender Report. (2020, November 12). The Swedish U-Turn on Gender Transitioning for Children. *Canadian Gender Report*. https://genderreport.ca/the-swedish-u-turn-on-gender-transitioning/

Carse, A. L. (1991). The 'voice of care': Implications for bioethical education. *The Journal of Medicine and Philosophy, 16*(1), 5–28. doi:10.1093/jmp/16.1.5 PMID:2010720

Carson, R. A., & Burns, C. R. (2006). *Philosophy of Medicine and Bioethics: A Twenty-Year Retrospective and Critical Appraisal*. Springer Science & Business Media.

Chang, M. K., Lim, M. X., Tay, K., Lee, R. J., Sim, D. S., Menon, S., Kanesvaran, R., Puvanendran, R., & Radha Krishna, L. K. (2017). The Influence of Evolving Confucian Beliefs in the Selection of Proxy Decision-Makers at the End of Life in Singapore. *Asian Bioethics Review, 9*(1), 117–128. doi:10.100741649-017-0003-2

Cheng, S.-Y., Suh, S.-Y., Morita, T., Oyama, Y., Chiu, T.-Y., Koh, S. J., Kim, H. S., Hwang, S.-J., Yoshie, T., & Tsuneto, S. (2015). A Cross-Cultural Study on Behaviors When Death Is Approaching in East Asian Countries. *Medicine, 94*(39), e1573. Advance online publication. doi:10.1097/MD.0000000000001573 PMID:26426631

Cherry, M. (2013). Orthodox Christian Bioethics: Medical Morality in the Mind of the Fathers. In *Religious Perspectives on Bioethics*. Taylor & Francis. doi:10.4324/9781315800394

de Blok, C. J. M., Wiepjes, C. M., Nota, N. M., & den Heijer, M. (2020). Hormone treatment of transgender people: Long-term health effects and safety. *Nederlands Tijdschrift voor Geneeskunde*, *164*, D4481. PMID:32757514

DeBlaere, C., Chadwick, C. N., Zelaya, D. G., Bowie, J.-A., Bass, M. F., & Finzi-Smith, Z. (2017). The Feminist Identity Composite: An Examination of Structural Validity With Sexual Minority Women. *Psychology of Women Quarterly*, *41*(1), 20–31. doi:10.1177/0361684316676046

Dickenson, D. (2006). Gender and Ethics Committees: Where's the 'Different Voice'? *Bioethics*, *20*(3), 115–124. doi:10.1111/j.1467-8519.2006.00485.x PMID:17039630

Diprose, R. (2005). Biomedical ethics and lived, sexed bodies. In *The Bodies of Women: Ethics, Embodiment and Sexual Differences*. Routledge., doi:10.4324/9780203981061-12

Dobken, J. H. (2018). The "New" Medical Morality: Hippocrates or Bioethics? *Journal of American Physicians and Surgeons*, *23*(2).

Dong, P., & Wang, X.-Y. (2004). Life, Death and End-of-Life Care: Taoist Perspective. In R.-Z. Qiu (Ed.), *Bioethics: Asian Perspectives: A Quest for Moral Diversity* (pp. 147–155). Springer Netherlands. doi:10.1007/978-94-017-0419-9_14

Doring, O., & Chen, R. (2002). *Advances in Chinese medical ethics: Chinese and international perspectives*. Institut Fur Asienkunde.

Dylan, B. (1964). *The Times They Are a-Changin* [Vynil 7"].

Engelhardt, H. T. (2002a). Medicine, Philosophy, and Theology: Christian Bioethics Reconsidered. *Christian Bioethics*, *8*(2), 105–117. doi:10.1076/chbi.8.2.105.8750 PMID:12678032

Engelhardt, H. T. (2002b). The ordination of bioethicists as secular moral experts. *Social Philosophy & Policy*, *19*(2), 59–82. doi:10.1017/S026505250219203X PMID:12678082

Engelhardt, H. T. (2006). Bioethics and the Philosophy of Medicine Reconsidered. In R. A. Carson & C. R. Burns (Eds.), *Philosophy of Medicine and Bioethics: A Twenty-Year Retrospective and Critical Appraisal*. Springer Science & Business Media.

Engelhardt, H. T. (2013). Bioethics as a Liberal Roman Catholic Heresy: Critical Reflections on the Founding of Bioethics. In J. R. Garrett, F. Jotterand, & D. C. Ralston (Eds.), *The Development of Bioethics in the United States* (pp. 55–76). Springer Netherlands. doi:10.1007/978-94-007-4011-2_5

Evans, J. H. (2002). *Playing God?: Human Genetic Engineering and the Rationalization of Public Bioethical Debate*. University of Chicago Press.

Forsberg, E.-M., Hanssen, A. B., Nielsen, H. M., & Olesen, I. (2018). Patent Ethics: The Misalignment of Views Between the Patent System and the Wider Society. *Science and Engineering Ethics*, *24*(5), 1551–1576. doi:10.100711948-017-9956-5 PMID:28812235

Fujieda, S. (2018). Between the Secular and the Religious: Japanese Buddhism in the Public Discourse on the Issues of Organ Transplant. *Proceedings of the XXIII World Congress of Philosophy,* 3. 10.5840/wcp232018386

Gross, M. L. (2014). Communitarian Bioethics: Three Case Studies. *Society*, *51*(4), 354–361. doi:10.100712115-014-9792-z

Grossman, A. (2004). *Pious and Rebellious: Jewish Women in Medieval Europe*. UPNE.

Hancock, P., & Tyler, M. (2001). Work, Postmodernism and Organization: A Critical Introduction. *Sage (Atlanta, Ga.)*. Advance online publication. doi:10.4135/9781446218365

Harrison, K., Fahy, K. M., & Thomposn, F. (1998). *Case studies in midwifery ethics: Discourses in knowledge & power*. AAHPSSS.

Hekman, S. J. (2014). *The Feminine Subject*. John Wiley & Sons.

Hessel, D. T. (2000). *Christianity and ecology: Seeking the well-being of earth and humans*. Harvard University Center for the Study of World Religions.

Hilton, R. H. (1984). Women traders in medieval England. *Women's Studies*, *11*(1–2), 139–155. doi:10.1080/00497878.1984.9978607

International Association of Bioethics. (1999). *Embodying Bioethics: Recent Feminist Advances*. Rowman & Littlefield.

Iovannone, J. J. (2019, May 30). *Leslie Feinberg: Transgender Warrior - Queer History For the People*. Medium. https://medium.com/queer-history-for-the-people/leslie-feinberg-transgender-warrior-fcb1bcaf15b2

Jacob, M. A., & Riles, A. (2007). The New Bureaucracies of Virtue: Introduction. *Political and Legal Anthropology Review*, *30*(2), 181–191. doi:10.1525/pol.2007.30.2.181

Kelsay, J., & Johnson, J. T. (1991). *Just War and Jihad: Historical and Theoretical Perspectives on War and Peace in Western and Islamic Traditions*. Greenwood Press.

Kemble, F. A. (2006). *Journal of a Residence on a Georgian Plantation: 1838-1839*. IndyPublish.com.

Keown, D., Prebish, C. S., Husted, W. R., & Husted, W. R. (1998). *Buddhism and Human Rights*. Psychology Press.

Kim, H.-J., & Fredriksen-Goldsen, K. I. (2017). Disparities in Mental Health Quality of Life Between Hispanic and Non-Hispanic White LGB Midlife and Older Adults and the Influence of Lifetime Discrimination, Social Connectedness, Socioeconomic Status, and Perceived Stress. *Research on Aging*, *39*(9), 991–1012. doi:10.1177/0164027516650003 PMID:27193047

Kimoto, T. (2018). The Remedy: Queer and Trans Voices on Health and Health Care edited by Zena Sharman. IJFAB: International Journal of Feminist Approaches to Bioethics. doi:10.3138/ijfab.2018.02.09.1

Krishna, L. K. R., Watkinson, D. S., & Beng, N. L. (2015). Limits to relational autonomy—The Singaporean experience. *Nursing Ethics*, *22*(3), 331–340. doi:10.1177/0969733014533239 PMID:24913544

Lichterman, B. L. (2005). Soviet medical ethics (1917-1991). *Journal International De Bioethique = International Journal of Bioethics*, *16*(3–4), 33–41, 167–168.

Lizardo, O. (2006). The puzzle of women's "highbrow" culture consumption: Integrating gender and work into Bourdieu's class theory of taste. *Poetics*, *34*(1), 1–23. doi:10.1016/j.poetic.2005.09.001

Lozano Rodríguez, M. (2021). Don't Let Kellys Go to Hell: From Welfarism to Misery in the Cleaning Ladies' Sector in Spain. In Handbook of Research on the Impact of COVID-19 on Marginalized Populations and Support for the Future (pp. 284–308). IGI Global.

Macklin, R. (2003). Dignity is a useless concept. *BMJ (Clinical Research Ed.)*, *327*(7429), 1419–1420. doi:10.1136/bmj.327.7429.1419 PMID:14684633

Marshall, D. (2016). Queer Pedagogies Out of Place and Time: Redrawing the Boundaries of Youth, Sexual and Gender Difference, and Education. *Journal of Homosexuality*, *63*(3), 405–415. doi:10.1080/00918369.2016.1124697 PMID:26766505

Martin, P., Washington, J., Sprague, J., & Purkayastha, B. (2014). *Thanks to Tracy Ore for sharing the wonderful pictures from the winter meeting, that you'll see throughout Network News*. https://socwomen.org/wp-content/uploads/2018/04/NN-spring-2014-1.pdf

May, A. T. (2003). Physician-Assisted Suicide, Euthanasia, and Christian Bioethics: Moral Controversy in Germany. *Christian Bioethics*, *9*(2–3), 273–283. doi:10.1076/chbi.9.2.273.30279 PMID:15254994

Mayes, C., Hooker, C., & Kerridge, I. (2015). Bioethics and Epistemic Scientism. *Journal of Bioethical Inquiry*, *12*(4), 565–567. doi:10.100711673-015-9667-z PMID:26646798

McGrath, P. (2018). *A Question of Choice: Bioethical Reflections on a Spiritual Response to the Technological Imperative*. Routledge., doi:10.4324/9780429459948

McLaughlin, L. A., & Braun, K. L. (1998). Asian and Pacific Islander cultural values: Considerations for health care decision making. *Health & Social Work*, *23*(2), 116–126. doi:10.1093/hsw/23.2.116 PMID:9598394

Mitford, M. R. (1852). *Recollections of a Literary Life: Or, Books, Places, and People*. R. Bentley.

Morales, H. (2013). Aristophanes' Lysistrata, the Liberian 'sex strike', and the Politics of Reception. *Greece & Rome, 60*(2), 281–295. doi:10.1017/S0017383513000107

Murphy, T. F. (2015). LGBT People and the Work Ahead in Bioethics. *Bioethics*, *29*(6), ii–v. doi:10.1111/bioe.12168 PMID:26053920

Murphy, T. F. (2018). Bioethics, children, and the environment. *Bioethics*, *32*(1), 3–9. doi:10.1111/bioe.12386 PMID:28873213

Nnaemeka, O. (2015). Feminist bioethics and global responsibility: Exploring health care delivery in Kenya. *International Journal of Feminist Approaches to Bioethics*. Advance online publication. doi:10.3138/ijfab.2.1.71

Nussbaum, M. (2004). Beyond the social contract: Capabilities and global justice. an Olaf Palme lecture, delivered in Oxford on 19 June 2003. *Oxford Development Studies*, *32*(1), 3–18. doi:10.1080/1360081042000184093

Office for National Statistics. (2015). *How has life expectancy changed over time?* Office for National Statistics. https://www.ons.gov.uk/peoplepopulationandcommunity/birthsdeathsandmarriages/lifeexpectancies/articles/howhaslifeexpectancychangedovertime/2015-09-09

Orwell, G. (2009). *Animal Farm: A Fairy Story*. Houghton Mifflin Harcourt.

Padela, A. I. (2018). Conceptualizing the Human Being: Insights from the Genethics Discourse and Implications for Islamic Bioethics. *Islamic Ethics and the Genome Question*, 111–138. doi:10.1163/9789004392137_006

Pateman, C. (2015). Sexual Contract. In *The Wiley Blackwell Encyclopedia of Gender and Sexuality Studies* (pp. 1–3). John Wiley & Sons, Ltd. doi:10.1002/9781118663219.wbegss468

Pellegrino, E. D. (2013). Medical Ethics and Moral Philosophy in an Era of Bioethics. In J. R. Garrett, F. Jotterand, & D. C. Ralston (Eds.), *The Development of Bioethics in the United States* (pp. 241–254). Springer Netherlands. doi:10.1007/978-94-007-4011-2_14

Ramsey, P. (1967). *Deeds and rules in Christian ethics*. Scribner.

Rawlinson, M. C. (2001). The Concept of a Feminist Bioethics. *The Journal of Medicine and Philosophy: A Forum for Bioethics and Philosophy of Medicine, 26*(4), 405–416. 10.1076/jmep.26.4.405.3010

Richardson, H. S. (2000). Specifying, Balancing, and Interpreting Bioethical Principles. *The Journal of Medicine and Philosophy: A Forum for Bioethics and Philosophy of Medicine, 25*(3), 285–307. 10.1076/0360-5310(200006)25:3;1-H;FT285

Richie, C. (2016). Lessons from Queer Bioethics: A Response to Timothy F. Murphy. *Bioethics*, *30*(5), 365–371. doi:10.1111/bioe.12246 PMID:26833492

Ridenour, A. A. (2017). The Meaning of Death and the Goal of Medicine: An Augustinian and Barthian Reassessment. *Christian Bioethics*, *23*(1), 60–76. doi:10.1093/cb/cbw021

Roudometof, V. (2018). *Glocal Religions: An Introduction*. Multidisciplinary Digital Publishing Institute.

Sakamoto, H. (2002). Chapter 2. A Challenge to the Concept of Human Rights and Human Dignity. *Journal International de Bioethique*, *13*(2), 23–30. doi:10.3917/jib.132.0023

Sander, R. (Director). (2016, June 15). *Human Rights and the Cyrus Cylinder: Freedom: A Shared Dream*. https://www.farhang.org/freedom

Sarango, C. (2020). Las Leyes de Burgos y Valladolid un antecedente temprano de una igualdad humana. *Ciencia e Interculturalidad*, *26*(1), 1. Advance online publication. doi:10.5377/rci.v26i01.9887

Schenker, J. G. (2005). Assisted reproduction practice: Religious perspectives. *Reproductive Biomedicine Online*, *10*(3), 310–319. doi:10.1016/S1472-6483(10)61789-0 PMID:15820035

Schermer, M., & Keulartz, J. (2002). How Pragmatic is Bioethics? In J. Keulartz, M. Korthals, M. Schermer, & T. Swierstra (Eds.), *Pragmatist Ethics for a Technological Culture* (pp. 41–68). Springer Netherlands., doi:10.1007/978-94-010-0301-8_4

Schmidt, U. (2011). Church, Public And Bioethics: Religion's Construction Of Public Significance Through The Bioethical Discourse. *The Public Significance of Religion*, 191–213.

Sextus Empiricus. (1933). *Sextus Empiricus: Outlines of Pyrrhonism*. Heinemann.

Shi, D.-P., & Yu, L. (2004). Euthanasia should be Legalized in China: Personal Perspective. In R.-Z. Qiu (Ed.), *Bioethics: Asian Perspectives: A Quest for Moral Diversity* (pp. 157–169). Springer Netherlands. doi:10.1007/978-94-017-0419-9_15

Shildrick, M. (1997). Feminist theory and postmodernism. In *Leaky Bodies and Boundaries*. https://www.taylorfrancis.com/

Shildrick, M. (2015). *Leaky Bodies and Boundaries: Feminism, Postmodernism and (Bio)ethics*. Routledge. doi:10.4324/9781315004952

Smith, S. (2018). *(De)pathologizing Discourse: The Problematization of Trans and Gender-Nonconforming Mental Health in Ontario* [Thesis]. Université d'Ottawa/University of Ottawa. doi:10.20381/ruor-22327

Stonington, S., & Ratanakul, P. (2006). Is There a Global Bioethics? End-of-Life in Thailand and the Case for Local Difference. *PLoS Medicine*, *3*(10), e439. doi:10.1371/journal.pmed.0030439 PMID:17076564

Strathern, M. (2006). Don't eat unwashed lettuce. *American Ethnologist*, *33*(4), 532–534. doi:10.1525/ae.2006.33.4.532

Szreter, S., & Fisher, K. (2010). *Sex Before the Sexual Revolution: Intimate Life in England 1918–1963*. Cambridge University Press. doi:10.1017/CBO9780511778353

Tan, A. C., Kerridge, I. H., & Pavlakis, N. (2018). Refusal of treatment in a young patient with recurrent medulloblastoma. *Journal of Case Reports and Images in Oncology*, *4*, 1–4. doi:10.5348/Z10-2018-48-CR

ten Have, H. (2016). *Global Bioethics: An introduction*. Routledge. doi:10.4324/9781315648378

Tham, J., Durante, C., & Gómez, A. G. (2018). *Religious Perspectives on Social Responsibility in Health: Towards a Dialogical Approach*. Springer. doi:10.1007/978-3-319-71849-1

The Crickets. (2016, July 8). *Please Don't Ever Change 1961-1962 Original Recordings Remastered* [Audio CD]. Jasmine Music.

The Meaning of the Holy Qur'an: Complete Translation with Selected Notes. (2015). Kube Publishing Ltd.

Tichtchenko, P. D., & Yudin, B. G. (1992). Towards a Bioethics in Post-Communist Russia. *Cambridge Quarterly of Healthcare Ethics*, *1*(4), 295–303. doi:10.1017/S0963180100006472 PMID:1342350

Tong, R. (2001). Towards a Feminist Global Bioethics: Addressing Women's Health Concerns Worldwide. *Health Care Analysis*, *9*(2), 229–246. doi:10.1023/A:1011390521518 PMID:11561998

Tong, R. (2004). Feminist Approaches to Bioethics. In G. Khushf (Ed.), *Handbook of Bioethics: Taking Stock of the Field from a Philosophical Perspective* (pp. 143–161). Springer Netherlands., doi:10.1007/1-4020-2127-5_7

Tong, R. (2018). Feminist Reflections on the Role of Theories in a Global Bioethics. In S. Sherwin (Ed.), *Globalizing Feminist Bioethics*., doi:10.4324/9780429500176

Turner, L. (2004). Bioethics needs to rethink its agenda. *BMJ (Clinical Research Ed.), 328*(7432), 175. doi:10.1136/bmj.328.7432.175 PMID:15202475

Twiss, S. B. (2018). Explorations. In *Global Ethics: Comparative Religious Ethics And Interreligious Dialogue*. Routledge. doi:10.4324/9780429500626

Van den Branden, S., & Broeckaert, B. (2010). Necessary Interventions: Muslim Views on Pain and Symptom Control in English Sunni E-Fatwas. *Ethical Perspectives, 17*(4), 626–651.

Van den Branden, S., & Broeckaert, B. (2011). Living in the hands of God. English Sunni e-fatwas on (non-)voluntary euthanasia and assisted suicide. *Medicine, Health Care, and Philosophy, 14*(1), 29–41. doi:10.100711019-010-9280-1 PMID:20809196

Veatch, R. M. (2000). *Cross-cultural Perspectives in Medical Ethics*. Jones & Bartlett Learning.

Weber, B. (2014, November 24). Leslie Feinberg, Writer and Transgender Activist, Dies at 65. *The New York Times*. https://www.nytimes.com/2014/11/25/nyregion/leslie-feinberg-writer-and-transgender-activist-dies-at-65.html

KEY TERMS AND DEFINITIONS

Artificial Intelligence (AI): Technological constellation centred in computer science and concerned with theorising and developing machines able to mimic (and exceed!) traditionally only human tasks related with decision-making, use of language and learning (Lozano Rodríguez & Torres, 2023).

Cyber-Physical: Integration of digital technologies and tangible systems, connecting machines and processes for real-time data exchange.

Data Analytics: Examination of large amounts of data to discover patterns and insights for informed decision-making.

Fourth Industrial Revolution (4IR): The qualitative change in manufacturing and industrial practises powered by disruptive technologies (artificial intelligence, robotics, the Internet of Things, 3D printing, genetic engineering, quantum computing, etc.) that is leading, as in 2023, to the merge of physical, digital, and biological worlds.

Liquid Maintenance: Liquid Maintenance is the Predictive Maintenance (PdM) evolution to thrive in the incoming industrial era. Industry, and subsequently, maintenance and, namely, PdM are unready to face the near-future incoming challenges since they are anchored in an obsolete paradigm alien to the incoming cyber-physical reality and unfit for unbelievable data density (Lozano Rodríguez & Torres, 2023).

Predictive Maintenance (PdM): PdM is a maintenance approach that aims to predict and prevent equipment failures by analyzing real-time data, monitoring performance, and identifying potential issues before they occur. It involves using advanced analytics and machine learning algorithms to make informed decisions about maintenance activities.

Taxonomies: The science and practice of classification and ordering classes according to principles.

ENDNOTES

1. The UNESCO International Bioethics Committee (IBC) was created in 1993; the Universal Declaration on Bioethics and Human Rights was published in 2009.
2. The Second Vatican Council was a Papal political masterstroke to gain a good position in bioethics' discourse (Engelhardt, 2013).
3. US Presidential Commission for the Study of Bioethical Issues: https://bioethicsarchive.georgetown.edu/pcsbi/about.html
4. Norwegian National Research Ethics Committee: https://www.forskningsetikk.no/en/
5. Saudi Arabia National Committee of Bioethics: https://ncbe.kacst.edu.sa/en/about-us/ncbe-members/
6. English-Speaking along with Catholic Europe and Protestant Europe were merged in only one group (n > 5); Philippines was relocated as a West and South Asia member instead of Latin America.
7. Actually, it happened before. The laws from *Leyes de Burgos* published in 1540 yet recognized basic human rights but are ignored in academia because they were not published in an industrial revolution winning country (Sarango, 2020). Likewise, Cyrus the Great published, I mean engraved, his famous cylinder more than 2500 years ago (Sander, 2016). In addition to not being a bloody antiSemite like European royals, he recognized freedom of religion and banned racial discrimination but he and his toy have no political use today so meh!

Chapter 2

Preemptive Bioethics and Syncretist Approach for Ethical Challenges in AI and Robotics:
Good v Evil I

ABSTRACT

This chapter explores the main views of bioethics and their applications in the context of displacement and emerging technologies. It includes ethical implications and outcomes before deploying technologies like AI through preemptive bioethics. The author discusses the four ethical decision-making approaches: legalist, antinomian, situationist, and syncretist. Syncretist approach is related with proactive ethical considerations, integration of perspectives, and collaboration between humans and AI. Emphasizing proactive considerations and collaboration between humans and AI systems, the chapter delves into the complexity of bioethical issues from a mathematical perspective. The chapter concludes by assessing the future impact of bioinformatic privacy and the need for comprehensive frameworks and interdisciplinary approaches.

INTRODUCTION

Last chapter provided an overview of the rise of bioethics and its intersections with various domains, including religion, geopolitics, and sex and gender-related issues. It explored the need for a more sensible response to displacement and the significance of bioethics in navigating complex moral landscapes. This chapter also explores the ethical considerations and adaptive approaches required in the era of autonomous vehicles and postmodern liquid maintenance. By connecting the discussions from past chapters to the new concepts and examples in this, we continue to explore Bodi's practical applications in addressing the complex ethical challenges posed by emerging technologies.

DOI: 10.4018/978-1-6684-4808-3.ch002

MAIN VIEWS OF BIOETHICS

Bioethics of Displacement needs a moral philosophy because it deals with complex ethical issues related to displacement and its implications for human life. Displacement, whether it be physical, psychological, or existential, often involves difficult choices, conflicts, and challenges that have profound moral implications for individuals and communities. A moral philosophy provides a framework for examining and evaluating these issues from an ethical perspective, guiding decision-making and actions in the face of complex and morally challenging situations.

Darryl Macer highlights three main views of bioethics (Macer, 2018, p. 7):

- **Prescriptive Bioethics:** On telling good from evil and ordering ethical principles as well as setting duties and responsibilities
- **Descriptive Bioethics:** On the moral perceptions and interactions when relating with other living beings and ourselves
- **Interactive Bioethics:** On the discussion and debate between social groups and within whole societies about the descriptive and prescriptive bioethics issues.

While streamlined with bringing bioethics beyond space-time boundaries, the clarification and development of Bioethics exposed by Macer can go further even. The Bioethics of Displacement also introduces the Preemptive bioethics. Preemptive bioethics refers to the ethical consideration of possible outcomes in paying more attention to Descriptive, Preemptive or Interactive bioethics given the current trends and the time available. For instance, in the development and use of autonomous robots or AI systems, preemptive bioethics would involve considering the ethical implications and possible outcomes before their widespread deployment. Prescriptive bioethics would provide guidelines and principles for determining what is considered good or evil in the design, programming, and use of these technologies, and setting ethical responsibilities for developers, users, and other stakeholders. Descriptive bioethics would focus on understanding the moral perceptions and interactions that arise in the human-robot relationship, including issues such as privacy, autonomy, and accountability. Interactive bioethics would involve discussions and debates between various social groups and societies about the ethical concerns associated with AI and robotics, involving input from ethicists, technologists, policymakers, and the general public. In turn, Preemptive bioethics would call for anticipating and addressing ethical challenges related to AI and robotics proactively, considering their potential impact on society, culture, economy, and human well-being. This could include addressing concerns about bias, discrimination, job displacement, safety, and transparency, among others. From a preemptive bioethics perspective, it would be essential to ensure that the democratisation and self-representativeness of bioethics and bioethicists are present in major works on bioethics related to AI and robotics, and that forums are highlighted to improve the nature of interaction among stakeholders to foster responsible and ethical development and use of these technologies.

ETHICAL DECISION-MAKING APPROACHES

According to Joseph Fletcher, the ethical decision-making has also three basic approaches (Fletcher, 1997, pp. 17–31):

- **Legalist:** Some moral rules are absolute and inviolable without any further consideration.
- **Antinomian:** There is no rule for the morality to lie upon.
- **Situationist:** There is no absolute moral rules but many helpful moral principles.

By the same token, the Bioethics of Displacement adds a new approach, the Syncretist.

Syncretist ethical decision-making blends the legalist, antinomian, and situationist approaches to create a dynamic and context-dependent approach to ethics. It recognizes that each of these approaches has its strengths and weaknesses and that the best approach may vary depending on the situation. Syncretist decision-making works because it combines and integrates different ethical approaches, allowing for a more comprehensive and nuanced assessment of moral dilemmas. Syncretist decision-making recognizes that decision-making is not a one-size-fits-all process and that rigid adherence to a single ethical approach may be limiting and incomplete. By considering multiple ethical perspectives and blending them in a thoughtful and reflective manner, syncretist decision-making can lead to more informed, balanced, and contextually appropriate ethical decisions. Relying almost solely on one ethical approach, such as legalism or antinomianism, may result in oversimplification, dogmatism, or disregard for relevant contextual factors, leading to suboptimal ethical decisions. Syncretist decision-making, on the other hand, embraces the complexity and diversity of ethical considerations and strives to incorporate different perspectives, making it a more nuanced, dynamic, and robust approach to ethical decision-making. Let's consider a case where a healthcare provider is faced with a decision regarding the treatment of a terminally ill patient who is in severe pain. The legalist approach may dictate that the provider must prioritise the duty to relieve suffering and provide palliative care, adhering strictly to the established medical guidelines and protocols. The antinomian approach, on the other hand, may emphasise the patient's autonomy and personal wishes, even if it means discontinuing medical interventions altogether. The situationist approach may take into account the specific medical condition of the patient, the availability of resources, and the patient's cultural and social context in determining the most appropriate course of action.

A syncretist approach to this ethical dilemma would involve considering all these different approaches and integrating them in a balanced and thoughtful manner. The healthcare provider may prioritise pain relief and palliative care as per the legalist approach, while also respecting the autonomy and personal wishes of the patient as per the antinomian approach. At the same time, the provider may also take into account the patient's individual medical condition, the available resources, and the cultural and social context to determine the most appropriate and compassionate care for the patient, as per the situationist approach. This could involve a collaborative decision-making process involving the patient, their family, and the healthcare team, taking into account all relevant factors and arriving at a decision that is in the best interest of the patient.

In this example, the syncretist ethical decision-making approach allows for a more nuanced and comprehensive assessment of the situation, taking into account the different ethical perspectives and the specific circumstances of the case. It avoids rigid adherence to a single ethical approach and recognizes that ethical decisions are complex and multifaceted, requiring careful consideration of various factors. By blending the legalist, antinomian, and situationist approaches, the healthcare provider can arrive at a decision that is ethically sound, compassionate, and contextually appropriate for the patient's care.

Not by chance, the Darryl Macer and Joseph Fletcher respective classifications form a group that grounds Bioethics of Displacement:

Those approaches are not detrimental to a global and personal view on bioethics grounded in maturity or a mostly practical call to "all to work together for all" as described in the Eubios Declaration for

Table 1. Views of bioethics and ethical decision-making approaches

Main Views of Bioethics	Ethical Decision-Making Approaches
Prescriptive	Legalist
Descriptive	Antinomian
Interactive	Situationist
Preemptive	Syncretist

International Bioethics.1 Rather than this, the Bioethics of Displacement foster them with a bold defence of plurality and individual agency. And vice versa. The Eubios Declaration for International Bioethics emphasises the importance of working together for the common good of all individuals and communities, while also promoting individual agency and pluralism. Therefore, the epistemic maturity fostered by Bioethics of Displacement when valuing diverse perspectives, along with syncretist decision-making, promote Eubios Declaration principles such as the respect of the diversity of human values and beliefs.

ETHICAL CONSIDERATIONS AND ADAPTIVE APPROACHES IN THE ERA OF AUTONOMOUS VEHICLES AND LIQUID MAINTENANCE

One example of the Bioethics of Displacement moral philosophy goes around autonomous vehicles. Preemptive bioethics would focus, as autonomous vehicles become more prevalent, in how ethical concerns arise, such as issues of safety, liability, privacy, and job replacement. Using preemptive bioethics, stakeholders can anticipate and address these ethical challenges before autonomous vehicles are widely deployed. Prescriptive bioethics would provide guidelines and principles for the design, programming, and use of autonomous vehicles, setting ethical responsibilities for manufacturers, users, and other stakeholders. Descriptive bioethics would focus on understanding the moral perceptions and interactions that arise in the human-autonomous vehicle relationship, including issues such as accountability and decision-making. Interactive bioethics would involve discussions and debates between various social groups and societies about the ethical concerns associated with autonomous vehicles, involving input from ethicists, policymakers, and the general public. A syncretist approach to this ethical dilemma would involve considering all these different approaches and integrating them in a balanced and thoughtful manner. For example, a syncretist approach could involve setting guidelines and principles for the design and use of autonomous vehicles (prescriptive bioethics), while also taking into account the perspectives and values of the users and affected communities (descriptive bioethics). It could also involve engaging in discussions and debates with various stakeholders to ensure that the ethical concerns related to autonomous vehicles are addressed in a collaborative and inclusive manner (interactive bioethics). By using preemptive bioethics and a syncretist approach, stakeholders can proactively address ethical challenges associated with autonomous vehicles, consider their potential impact on society, and work towards responsible and ethical development and use of this technology. This example demonstrates how the Bioethics of Displacement and preemptive bioethics can be applied in real-life situations to ensure that emerging technologies are developed and used in an ethical and responsible manner.

Another example is the concept of Liquid Maintenance as a paradigm shift in predictive maintenance (PdM). It recognizes the urgent need for PdM to adapt to the cyber-physical reality of the incoming industrial era and embrace AI as a key enabler. Industry is changing its very core and now hybrid AI-human teams, digital dimensions and sustainable value chains take precedence, necessitating a new approach to maintenance (Lozano Rodríguez & Torres, 2023). More on the importance of accountability and adaptation in the context of booming AI, a paper by Barclay et al. (2018) explores the accountability requirements that extend to the production and utilisation of data and knowledge assets within AI systems, treating them as essential components similar to raw materials in physical goods production (Barclay et al., 2018). Specifically, collective intelligence is made up of data and knowledge generated by crowd-sourced workforces and serves as a core component of AI systems. Furthermore, the researchers propose a supply chain model that tracks the creation and utilisation of crowdsourced collective intelligence assets. The suggested model utilises a decentralised architecture based on blockchain technology, offering validation, accountability, and fairness in the process.

Furthermore, PdM is also challenged by the philosophical hiddenness and taxonomic issues that hinder its progress (Lozano Rodríguez & Torres, 2023). By acknowledging these challenges and embracing flexibility and AI-powered data platforms, PdM professionals can unlock the potential of Liquid Maintenance and navigate the complexities of the new industrial revolution.When it comes to hybrid AI-Human teams, the intersection of Bioethics of Displacement and Liquid Maintenance highlights the significance of ethical decision-making and the integration of different perspectives. In the context of AI systems and predictive maintenance, a syncretic approach to ethical decision-making, as introduced by Bioethics of Displacement, becomes relevant. This approach combines and integrates different ethical approaches, such as legalist, antinomian, and situationist, to ensure a comprehensive and context-dependent assessment of moral dilemmas. It takes into account the potential impact on society, culture, economy, and human well-being. Similarly, Liquid Maintenance recognizes the need for humans and AI systems to collaborate and share experiences and knowledge to achieve optimal maintenance outcomes.

In summary, the intersection of Bioethics of Displacement and Liquid Maintenance with the topic of hybrid AI-Human teams emphasises the importance of proactive ethical considerations, the integration of different ethical perspectives, and the collaboration between humans and AI systems to address the challenges and make informed and responsible decisions in the development and use of emerging technologies.

SENSIBLE MORAL SPACE AND LOCAL INVARIANCE

Before introducing the few most abstract pages in the book, let's warm the brains up reflecting on these questions:

- How far can we predict our future views in bioethics?
- There is a rational limit for a person to discuss bioethics regardless of mental conditions or brain damage? And among many different persons?
- Can we imagine a bioethical issue so complex that understanding it from one point of view would cripple us for understanding it from another one when willingly trying? And to understand it in the future?

Hopefully, the questions above have helped you to think about how predicting the future of bioethics issues is fogged by complexity arising from different perspectives and how we can suffer a bioethical blackout even.

By proposing a different approach based on the concept of a "sensible moral space," Bodi challenges the status quo of ethical decision-making in bioethics and calls for a reevaluation of existing frameworks. As well, Bodi provides a unique perspective on bioethics by integrating mathematical principles into bioethical discourse. We break new ground by bridging the gap between ethics and mathematics, and opening up new avenues for interdisciplinary research and discussion in the field of bioethics. The three main views of bioethics and the ethical decision-making approaches before Bodi form a sensible moral space where ethical decision-making takes place, which is analogous to that of Poincaré, a kind of space that involves no idea of measurement (Poincaré, 1898). For example, we can't say "this abortion seems to me 3.927 times more immoral than this other euthanasia" regardless of how certain we are that one case is more unethical than the other. Furthermore, Bodi's sensible moral space is an invariant group of the prescriptive, descriptive and interactive bioethics in a similar sense than Poincaré naturalistic sensible space is an invariant group, a special set of rules that stay the same no matter how you move or rotate things in space and time, of the Euclid, Lobachevskii and Riemann geometries. Therefore, the Bioethics of Displacement is a sort of manifold, just like a flexible rubber sheet that can bend and stretch, allowing us to describe curved spaces in maths and physics, within the bioethics sensible moral space. It means that Bodi offers a framework that allows for pluralistic and individualistic ethical perspectives, accommodating diverse viewpoints and fostering a bold defence of plurality and individual agency within the broader bioethical landscape. For example, Imagine a patient who has a terminal illness and wishes to have control over their own end-of-life decisions, such as whether to receive palliative care or pursue medical aid-in-dying. However, due to cultural or societal norms, their choices are not respected, and they are facing pressure from family members, healthcare providers, or other external forces to conform to traditional values or beliefs about dying. In this situation, the concept of "sensible moral space" could offer a framework that allows for pluralistic and individualistic ethical perspectives. It could provide a space where the patient's autonomy and agency are respected, and their diverse viewpoint and personal values are taken into consideration. This could mean that the patient's choices are not solely determined by arbitrary top-down decisions, but rather by a thoughtful and inclusive decision-making process that takes into account their unique circumstances, desires, and beliefs. Moreover, by integrating mathematical principles into bioethical discourse, as proposed by Bodi, the concept of "sensible moral space" could provide a structured and systematic approach to ethical decision-making in this complex situation. It could help navigate the ethical complexities and uncertainties involved in end-of-life care, ensuring that the patient's choices are considered with care and respect. In conclusion, the solutions to be promoted are those that raise the sense of meaning and purpose in their final days.

Bioethical issues vanish when localised in the infinitesimal texture of a person mind-flow. We can't see our feelings' gears but we feel emotionally charged, anyways. The unyielding moral astonishment rocks our Prescriptive, Descriptive or Interactive bioethics and leads us to a feeling of displacement. For example, sudden conflicting moral perspectives in bioethics, such as donating organs, can make us feel confused and uncertain. Sometimes conflicts can feel like cradling, sometimes like a full swing. The coherence of the bioethics, its pre-human continuity even, is more easily explained by displaying it on a timeline, but it cannot always be done on the fly. Here, a Preemptive bioethics mindset can be useful. Indeed, Preemptive bioethics allows us to proactively address ethical issues before they arise, providing proactive solutions that Descriptive, Prescriptive, and Interactive bioethics may not.

Along with the "sensible moral space", we'll know about "local invariance", so we will walk this mathematical path a bit further. Hermann Weyl's idea of an affine geometric manifold is also sketched in Bodi. His idea of local invariance in an affine geometric manifold is a mathematical concept where certain properties remain unchanged regardless of changes in perspective or viewpoint. A practical example of local invariance in an affine manifold can be found in the concept of road networks. Consider a road network in a city, with intersections, streets, and bridges. The city can be thought of as an affine manifold, where the road network defines the geometry. Now, imagine a person driving along a particular route in the city, following a specific sequence of streets and intersections. The person starts at a point in the city and navigates along the roads, making turns, crossing bridges, and passing through intersections. As the person drives, they use their GPS device to guide them, which provides directions in terms of distances, angles, and turns to take at each intersection. Despite changes in the person's perspective or viewpoint as they navigate through the city, the geometry of the road network remains invariant locally. In other words, the person's directions and instructions from the GPS remain consistent regardless of their location in the city, as long as they stay within the road network. For example, if the person makes a right turn at an intersection, the GPS instructions would guide them to turn the same direction, regardless of the exact location of the intersection or the orientation of the streets with respect to their perspective. This local invariance in the directions provided by the GPS is a practical example of how an affine manifold, represented by the road network in the city, can exhibit local invariance, where certain properties (i.e., the directions) remain unchanged regardless of changes in perspective or viewpoint. In Bioethics of Displacement, the Legalist, Antinomian, and Situationist ethical decision-making form a 3D map while the Syncretic one is the GPS' arrow. And even if it was not Weyl's gravity theory but Einstein's one the most accurate, both Einstein's Theory of Relativity and Weyl's affine manifold share a similar idea: local invariance (Higa, 1993);(Karananas et al., 2021). According to Einstein, local invariance means that in small regions of space-time, gravity is indistinguishable from acceleration or, in other words, when a lift just starts to go down, you can't tell if everything goes right from an unhinged and free fall short trip. On the one hand, it means that the gravitas of moral issues can appear, change and disappear based on the observer's perspective. Here, disappearing means that the Legalist, Antinomian, and Situationist coordinated axis along with its Syncretic vector are transformed away -displaced from our mind flow to irrelevance. But, on the other hand, a more interesting property can be unearthed, but we won't do it till the end of the section. There is enough differential geometry and tensor calculus… for now.

Bioethics has gained coherence and consistency by incorporating the Syncretic ethical decision-making category alongside Legalist, Antinomian, and Situationist categories. The Syncretic category serves as a guiding principle that helps reconcile conflicting perspectives and viewpoints, and provides a cohesive framework for addressing complex bioethical issues. It helps integrate diverse moral perspectives and promotes a more inclusive and comprehensive approach to ethical decision-making. Back to the previous case about a patient who has a terminal illness and wishes to have control over their own end-of-life decisions, such as whether to receive palliative care or pursue medical aid-in-dying, a Legalist approach may strictly adhere to the laws and regulations surrounding medical aid-in-dying, which may or may not be available in the patient's location. An Antinomian approach may prioritise the patient's autonomy and personal choice above all else, regardless of legal considerations. A Situationist approach may take into account the specific circumstances and context of the patient's situation, such as their cultural or religious beliefs, and make decisions based on those factors. However, a Syncretic approach would consider all of these perspectives and also take into account the local invariance concept, where the effects of moral stress and relief may be intertwined with the effects of being displaced or belonging. This means that the patient's emotional and psychological well-being, as well as their physical comfort, would be taken into consideration in a holistic manner. For example, the Syncretic approach could involve finding alternative ways to address the patient's suffering, such as optimising palliative care and symptom management, and providing emotional and psychological support to both the patient and their family. It could also involve exploring other options for end-of-life care, such as hospice care or spiritual support, that may align with the family's cultural or religious beliefs. By integrating multiple perspectives and considering the local invariance concept, the Syncretic approach could help the patient and their family navigate through the complex and emotionally charged decision-making process, and arrive at a decision that is ethically sound and respectful of the patient's autonomy and well-being.

THE CORE

In the context of AI, the notion of the "non-deterministic polynomial-time hard" complexity class, mentioned in "Liquid Maintenance," highlights the inherent challenges associated with addressing ethical issues in this rapidly evolving field (Lozano Rodríguez and Torres, 2023). The concept of "non-deterministic polynomial-time hard" complexity class implies that certain problems, particularly those related to AI, are difficult to solve efficiently using traditional computational methods or, plainly, their complexity surpasses the capabilities of classical algorithms. By introducing this concept, the text highlights the intricate nature of bioethical issues and the difficulties associated with comprehending and addressing them. In the context of bioethics, complexity arises from the diverse perspectives and viewpoints held by individuals and societies. Bioethical issues are multifaceted, involving ethical, cultural, social, and personal considerations. Attempting to comprehend and resolve these issues solely from one viewpoint can result in a limited understanding and hinder the ability to fully grasp them.

Privacy is not only an ethical issue but also a bioethical issue due to its direct impact on sensitive medical and personal information. In the context of bioethics, privacy becomes particularly complex and challenging because it involves intricate considerations regarding individuals' health, genetics, and medical records. The integration of technology, such as Ambient Intelligence (AmI) systems, further amplifies the complexity (Caire et al., 2016). These systems collect, store, and distribute vast amounts of personal data, necessitating robust privacy preservation measures. The dynamic nature of privacy in bioethics, combined with the rapid advancement of technology and the absence of a universally accepted definition, contributes to its classification as a "non-deterministic polynomial-time hard" issue. This complexity implies that addressing privacy concerns in bioethics requires efficient verification and decision-making processes that exceed traditional computational capabilities. It involves grappling with multifaceted perspectives, evolving norms, and intricate legal and social frameworks.

Bioinformatic privacy can be seen as the intersection of Poincaré's naturalistic sensible space, Weyl's idea of an affine geometric manifold, Einstein's Local Invariance, and Non-deterministic polynomial-time hard problems in a mathematical and bioethical sense. Poincaré's naturalistic sensible space highlights the complexity and intricacy of bioethical issues, including privacy concerns, which arise from the diverse perspectives and factors involved. Weyl's idea of an affine geometric manifold emphasises the concept of local invariance, where certain properties remain unchanged regardless of changes in perspective or viewpoint. In the context of bioinformatic privacy, this means that the protection of personal data and confidentiality should be upheld consistently, regardless of the specific context or technological advancements. Einstein's Local Invariance, rooted in the theory of relativity, suggests that moral issues and ethical dilemmas can appear, change, and disappear based on the observer's perspective. In the case of bioinformatic privacy, different stakeholders, such as individuals, healthcare providers, researchers, and policymakers, may have varying perspectives and interests regarding the use and sharing of genetic and medical data. Finally, the classification of bioinformatic privacy as a non-deterministic polynomial-time hard problem underscores its complexity, indicating that ethical decision-making and privacy-preserving measures in this domain exceed traditional computational capabilities. The intricate nature of bioinformatic privacy, influenced by mathematical concepts and bioethical considerations, highlights the need for comprehensive frameworks, advanced technologies, and interdisciplinary approaches to address the challenges posed by safeguarding sensitive genetic and medical information.

A corrupt government or public administration can devastate its human flock only resorting to bioinformatic privacy abuse.The abuse of the US' Tuskegees or the Stasi's psychologists (Maercker, & Guski-Leinwand, 2018) may fall short in comparison with that information technologies give. As in 2023, we have witnessed biometric data exploitation (facial recognition technology, e.g.) for mass surveillance and control (Feldstein, 2019; Flinkea, 2015; Frederick, 2019). This could involve tracking individuals' movements, identifying them in public spaces, and creating comprehensive profiles of their activities and associations. But in the future we may face political control through blackmailing: the state, in collaboration with technology companies or intelligence agencies, could exploit the private information of individuals, such as their personal relationships, online activities, or sensitive medical records. This information could be used not only for blackmailing political opponents, dissidents, or influential figures and keeping rulers glued to the chair, but to blackmail particular citizens according to electoral districts, opinions on the internet or for turning them on involuntary agents in diverse situations. And it could be done by AI and Big Data.

The future geopolitical impact of bioinformatic privacy is immense and wide-ranging. It entails power dynamics, control, threats to democracy, human rights violations, and significant implications for global relations. The intricate nature of bioethical issues, coupled with technological advancements, underscores the need for comprehensive frameworks and interdisciplinary approaches. As technology progresses, international norms and regulations must adapt to protect individuals' privacy and address the complex ethical and legal challenges. The implications stretch beyond national borders and impact international relations, trade, and intelligence. It is crucial to consider the multifaceted nature of bioinformatic privacy and its potential consequences for the future.

To know how is going to rule the world in 2030 it's enough to know which government or corporation can grab, rob, hoard, deal and mess with more bioinformatic data without fearing reprisals today.

CONCLUSION

Address complex ethical issues related to Displacement and emerging technologies would benefit from preemptive bioethics and proactive approach to anticipate and address ethical challenges. By the same token the syncretist approach to integrate diverse moral viewpoints would be helpful as well. From a stakeholder's perspective, Bodi helps to establish guidelines for using emerging technologies responsibly.

REFERENCES

Barclay, I., Preece, A., & Taylor, I. (2018). *Defining the collective intelligence supply chain*. arXiv preprint arXiv:1809.09444.

Caire, P., Moawad, A., Efthymiou, V., Bikakis, A., & Le Traon, Y. (2016). Privacy challenges in ambient intelligence systems. *Journal of Ambient Intelligence and Smart Environments*, *8*(6), 619–644. doi:10.3233/AIS-160405

Feldstein, S. (2019). *The global expansion of AI surveillance* (Vol. 17). Carnegie Endowment for International Peace.

Finklea, K., Christensen, M. D., Fischer, E. A., Lawrence, S. V., & Theohary, C. A. (2015, July). *Cyber intrusion into US office of personnel management: In brief*. Library of Congress.

Fletcher, J. F. (1997). *Situation Ethics: The New Morality*. Westminster John Knox Press.

Frederick, K. (2019). *Hearing on "How Corporations and Big Tech Leave Our Data Exposed to Criminals, China, and Other Bad Actors"*. Academic Press.

Higa, T. (1993). Weyl manifolds and Einstein-Weyl manifolds. *Rikkyo Daigaku Sugaku Zasshi*, *42*(2), 143–160.

Karananas, G. K., Shaposhnikov, M., Shkerin, A., & Zell, S. (2021). Scale and Weyl invariance in Einstein-Cartan gravity. *Physical Review. D*, *104*(12), 124014. doi:10.1103/PhysRevD.104.124014

Lozano Rodríguez, M., & Torres, C. E. (2023). Evolving From Predictive to Liquid Maintenance in Post-modern Industry. In Encyclopedia of Data Science and Machine Learning (pp. 2182-2198). IGI Global.

Macer, D. (2018). *Application of the essence of bioethics as love of life to some dilemmas of our future*. American University of Sovereign Nations.

Maercker, A., & Guski-Leinwand, S. (2018). Psychologists' involvement in repressive "Stasi" secret police activities in former East Germany. *International Perspectives in Psychology : Research, Practice, Consultation*, *7*(2), 107–119. doi:10.1037/ipp0000085

Poincaré, H. (1898). On the foundations of geometry. *The Monist*, *9*(1), 1–43.

KEY TERMS AND DEFINITIONS

Bioethics: The study of ethical dilemmas, guiding decisions and actions from love for life.

Displacement: The amount of divergence that a life bears while losing its meaning.

Ethical Decision-Making: Weighing moral principles and values to make morally right and appropriate choices.

Local Invariance: In Bodi, refers to the capacity of intertwining or mimicrying of the effects of moral stress and relief with being Displaced or Belonging.

Moral Philosophy: Study of moral principles and values, helping us make ethical decisions based on what's right or wrong.

Preemptive Bioethics: Main view of bioethics focused on the consideration of possible outcomes in paying more attention to Descriptive, Descriptive, or Interactive bioethics given the current trends and the time available.

Syncretist Bioethics: Bioethical decision-making blending different approaches to create a dynamic and context-dependent approach to ethics. It considers multiple perspectives and factors to make balanced and informed decisions.

ENDNOTE

1. https://www.eubios.info/eeidec.htm

Chapter 3
Psychology, Moral Principia, and Gehenna:
Good v Evil II

ABSTRACT

Embark on an enlightening exploration of ethical challenges triggered by displacement. Discover the profound impact of psychological flexibility on navigating complex ethical issues. Unveil the principles of bioethics of displacement: vulnerability first, big picture thinking, proactivity, and skepticism. Explore how psychological inflexibility and moral stress shape ethical decision-making. Embrace diverse perspectives and reflect on the transformative power of open-mindedness. Uncover the importance of mindful communication in addressing ethical dilemmas. Join us on this captivating journey where ethics and displacement intertwine. This chapter explores psychology, moral principia, and Gehenna.

INTRODUCTION

This chapter continues the exploration of Bioethics of Displacement, specifically focusing on psychological flexibility and moral principles. From Chapter 2, this chapter underscores the integration of Preemptive Bioethics and Syncretic ethical decision-making. It also discusses the interconnectedness of psychological inflexibility, moral stress, and ethical decision-making. Additionally, the chapter delves into the issues of alignment, alignment's will, and the dangers of ideological endogamy in bioethics. Also warns against the pitfalls of magic voluntarism and stresses the importance of authenticity and responsible engagement in ethical discourse.

DOI: 10.4018/978-1-6684-4808-3.ch003

PSYCHOLOGICAL FLEXIBILITY AND MORAL PRINCIPIA

Following the previous chapter, let's see another example integrating Preemptive Bioethics and Syncretic ethical decision-making. In the early dawn of postmodern bioethics, philosopher Daniel Callahan was confronted by a feminist pro-abortion who shouted him (Callahan, 1973):

But what right have you to press all of your heavy philosophical questions on people who aren't trained philosophers? What arrogance and cruelty! You should trying to help women nor loading them down with a lot of hard intellectual problems which will you make them suffer all the more. You claim to be interested in ethics. Why don't you practice it!

I don't know if Callahan and his friend met again, but it is important to note that both parties had valid points, but their communication style and approach were not effective in reaching a resolution. On one hand, Callahan was bringing up important ethical questions and trying to engage in philosophical discourse. However, he may have come across as dismissive of the practical concerns of the feminist pro-abortion advocate and insensitive to the lived experiences of women. On the other hand, the feminist pro-abortion advocate had a valid point in that philosophical discourse should not be the only focus, and that practical solutions are needed to help women in their reproductive choices. However, her communication style, which involved shouting and personal attacks, was not effective in facilitating a productive conversation. In this case, preemptive bioethics could have helped by encouraging both parties to engage in ethical reflection and communication *beforehand*, in a respectful and empathetic manner. By doing so, they could have identified common ground and possible solutions that would have addressed the practical concerns of women while also considering the ethical implications of their choices. In addition, syncretic ethical decision-making could have helped by acknowledging the diverse perspectives and values involved in the conflict, and seeking to find a way to integrate them into a coherent ethical framework. By doing so, it would have allowed for a more nuanced and holistic approach to the issue, instead of relying on simplistic black-and-white thinking. In conclusion, conflicts are often complex and nuanced, and it is important to approach them being psychologically flexible. By engaging in preemptive bioethics and syncretic ethical decision-making, we can facilitate productive and ethical dialogue that considers the diverse perspectives and values involved.

The solutions given for the examples mentioned have a common ingredient: psychological flexibility. It involves being open to diverse perspectives, integrating values and beliefs, managing emotions, and communicating effectively. Bioethics of Displacement's moral philosophy and psychological flexibility interact in several ways:

- Being willing to consider multiple viewpoints, and adapting one's beliefs and actions accordingly. By promoting pluralistic and individualistic ethical perspectives, the texts are advocating for psychological flexibility in ethical decision-making.
- Being aware of and adaptable to the context in which one is making decisions, and being willing to consider the unique circumstances and values of individuals. By incorporating contextual considerations, the texts are advocating for psychological flexibility in ethical decision-making.

- Being able to integrate and balance different perspectives, values, and factors in decision-making, and avoiding simplistic black-and-white thinking. By promoting an integrated and holistic approach, the texts are advocating for psychological flexibility in ethical decision-making.
- Being able to communicate in a respectful and empathetic manner, actively listening to others' perspectives, and finding common ground for productive dialogue. By advocating for respectful and empathetic communication, the texts are promoting psychological flexibility in ethical decision-making.

Hopefully, this psychological flexibility can ease a bit of such hard intellectual problems. Indeed, a good chunk of Bodi's complexity lies in how Displacement works like a fuzzy gear between the incommunicable and the reformulation of the Self, connecting things that are hard to express and reshaping how we see ourselves. As a result, the words we choose and how we choose to use them (the paradigmatic and syntagmatic axis of language) unyieldingly tremble as soon as ethical issues appear. The musical tremolo, which is a rapid repetition of a note or chord, can be used to create a sense of tension or unease, and can also be used to evoke a sense of movement or progression. For example, in one of my favourite movies, *The Shawshank Redemption*, tremolo is used outstandingly without falling into scary movies' clichés (Chua, 2011). Similarly, the trembling repetitiveness in the way we choose words and build sentences is a kind of good or bad *vibe* in our communication, which can create a sense of tension or ambiguity when discussing ethical issues. Or a sense of progressing in the topic, if we use them wisely. Bodi's insights encourage us to be more mindful of the words and become more reflective and intentional in our communication.

In a broader sense, the ironclad relation between bioethics and communication is nothing new. As a field of study, bioethics has often been viewed as both a second-order discipline and an integral part of public discourse (Kopelman, 2009). An example of this is the Universal Declaration on Bioethics and Human Rights, which reflects the intersection of bioethics with broader societal debates (Intergovernmental Bioethics Committee, 2005). Sadly, the bioethical discourse is not as unbiased as we may wish. It's not even close. Since Samuel Coleridge-Taylor displaced for the first time the 'natural philosophy' term by the brand-new 'science' in a 24 June 1833 public meeting, its popularity has turned into modern *scientism* (Bryant & la Velle, 2018, pp. 8–10). Those scientific journals that rely upon an internalised scientism, hidden a vigilante subculture that is disclosed as soon as the papers don't come from the West. The scholar journals from low and middle-income countries (LMICs), altogether with the moral diversity they bring, are underrepresented in bioethics which hinders any aspiration to create an authentic global bioethics (Chattopadhyay et al., 2017). This inequality has perpetuated a monocular bioethics as well as a mild discourse on social injustice (Behrens, 2017).

At this stage in the game, we can enunciate the four Bioethics of Displacement's moral principles:

Vulnerable First: Bodi emphasises the intrinsic value and dignity of all human beings, recognizing the importance of respecting and protecting the rights and well-being of those who are displaced first.

Big Picture Thinking: Vulnerable first doesn't mean vulnerable-mania. Even if the emergency criteria should consider risk and vulnerability, we must see the big picture from the first moment and don't lose sight of it. This includes recognizing the interconnectedness of various aspects of displacement, such as physical, psychological, social, and existential dimensions, and integrating insights from different disciplines, including philosophy, medicine, psychology, sociology, anthropology, and others.

Proactivity: Bodi emphasises proactive and preventive approaches, such as preemptive bioethics, that aim to anticipate and address ethical issues before they escalate into conflicts or crises. This includes promoting ethical reflection, dialogue, and collaboration among different stakeholders involved in displacement, such as individuals, communities, healthcare professionals, policymakers, and others.

Scepticism: This includes being open to diverse perspectives, being willing to reconsider and revise ethical decisions in light of new information or changing circumstances, and recognizing the limitations and uncertainties of ethical reasoning in complex situations. Indeed, ethical dilemmas related to displacement often involve conflicting values and interests, and we should mistrust clear-cut solutions -especially those coming from politics.

Before ending this section, we'll go back to the example of the lift that started to go down and nobody can say if the cable has been cut at the very beginning of the descent. Our moral philosophy shovel did "clank!" but this Bodi's relevant learning was temporarily unearthed. Previously, we discussed how the "sensible moral space" and "local invariance" concepts highlight the idea that certain principles or properties remain consistent and invariant, despite changes in perspective or viewpoint. These concepts provide a framework for understanding how ethical decision-making and moral experiences can be influenced by factors such as stress, relief, displacement, belonging, and diverse viewpoints, while still maintaining a sense of stability and consistency within the broader ethical landscape. In other words, ethical decision-making is not fixed or absolute, but rather dependent on various factors such as perspective, context, and individual agency, which can result in a curved and dynamic moral landscape. But, what happens at any point of this plane?

In a small enough lapse of the flow of conscience, one cannot tell apart the effects of moral stress and relief from the effects of being displaced or belonging.

It happens because these experiences can be intertwined and may have similar manifestations in one's thoughts, emotions, and behaviours. For example, when faced with pressing conflicting moral perspectives in bioethics, such as donating organs after a family loss, an individual cannot say if her moral stress at every instant is due to the chances of failing to fulfil self-perceived role on donation or because of the grief. On the other hand, when a proactive approach such as Preemptive bioethics is employed to address ethical issues before they arise, it may result in a sense of moral relief and belonging. Knowing that proactive measures have been taken to address potential ethical conflicts can alleviate stress and uncertainty, and may foster a sense of belonging to a community or group that shares similar values and principles.

Psychological inflexibility and moral stress are related in the sense that they can both influence ethical decision-making and moral experiences. Psychological inflexibility refers to a rigid mindset that is resistant to considering diverse perspectives, adapting beliefs and actions, and managing emotions effectively. Moral stress, on the other hand, arises from conflicting moral perspectives or values, and can result in emotional distress and uncertainty. It means that without considering her personal story, as Bodi's recommends, we can't truly differentiate to what extent a person is slow on the uptake or suffering burn-out. Or, in a Macchiavelian turn, the longest and merciless a person is existentially displaced, the less tools for overcoming she will have. Now, we are talking about power...

CONCUPISCENCE, POWER, AND GEHENNA

Thankfully, bioethics is a relatively young and lively discipline, which is not overburdened with scandals when compared with many others. Thus, most examples of Its Gehenna[1] face don't point to particular authors but describe a hypothetical recurrent scenario.

Bioethicists are usually aligned, which means that they oppose compromised positions that other persons (often other bioethicists) are committed with. To be obviously aligned sometimes is not considered correct since it can undermine an honest epistemology (Devolder & Douglas, 2018). In practice it means a strong ideological bias when choosing sources and methods that leads the bioethicist to consider only the views that match her previous beliefs. Or worst even, the way we discern between a (respectable) opinion and a (justified) working hypothesis might be based on political gain expectancies rather than falsifiable honestly gathered knowledge. When this alignment is the rule, rather than exception, we can talk of revolving doors over science.[2]

Further than alignment, there is the will of alignment. It means that an individual's main goal in bioethics is to be a referent. It's not the usual fame rush, but something more existential described by Kierkegaard, Heidegger and namely by Sartre (Farr, 1998, p. 7):

The books with which I surround myself, the recreation which I enjoy, everything which is mine, all this informs me of my choice -that is, my being. (Sartre, 1992, p. 596)

So, this facet of the Gehenna side is not only about gaining impact in Google Scholar and media no matter how, but very specially about becoming a hermit crab cloistered in an office, in a few acronyms and, foremost, in a toxic relation with the (easy choice) of Utopian sellers. Sartre himself lost his "last good friend", Camus, because he was a difficult choice (Hage, 2013).

Figure 1. Jean-Paul Sartre
Source: Saint-Paul (1955)

One of the greatest works of Gehenna-esque psychology crumbled down just before bioethics as a discipline existed. Denying the animal mind and even animal suffering has been profitable for both the industry and the academy either for exploitation or scientific research (Rollin, 2007). Not by chance, while the animals were useful for working either in the familiar demesnes and cities or cheering the childhood of the future British Empire's intelligentsia, animals had a mind just like us. However, when the great men became passionately modern and urbanites the animals *lost their mind* until the fall of Behaviorism.

Let's see it in a historical context. Classical animal mind advocates, such as the philosophers Lock, Hume and Bentham were children of lawyers and administrators and they weren't brought up in great cities but kept their links with the rural life even during the time they lived in London. B. F. Skinner, the lead behaviourist, was also the son of a lawyer and grew away from the great city before moving in. Nevertheless, Susquehanna Depot (Pennsylvania) was different to the British countryside and the Scottish town markets, it was an industrial region where many modern trains were built. In addition to this, London in 1800s was one of the most advanced cities in Europe but was not close, even in relative terms, to the Skinner's New York, which was the greatest city in the world with seven million habitants in 1925, half a million *fuel-powered* cars in the street and a unique skyline jigsawed by sky-scrapers (Hohenberg & Lees, 2009, pp. 10–11).[3] B. F. Skinner, who was raised in a world where the dependency of animal workforce was dwarfing before the oil, became a frustrated stream-conscience novelist first and the greatest exponent of one of the most Gehenna-like anti-conscience doctrines since ever then (Baars, 2003; Kellert, 1985). Erich Fromm defines neobehaviorism as "*the primacy of egotism and self-interest over all other human passions*" but depicts Skinner as "*a naive rationalist who ignores man's passions*" (Fromm, 2013).

Like in the Skinner youth, we are living in an age in which the dependency of human skills is being belittled by artificial intelligence and robots. Likewise, many psychologies and pseudo-psychologies arise like the misleadingly tagged 'positive thinking' movement and whose greatest exponent is *The Secret* (Byrne, 2011). Put in briefly, this joint is about altering the cosmic behaviour playing to be high, so it's turn behaviourism inside out. Actually, being so bold about cheerfulness ends in the egotism Fromm described as well as to turn oneself in a political twerp on the basis of daily self-blaming (as seen during the 2008 crisis) (Ehrenreich, 2010). Even when the 20th century scientific progresses turned the normative questions in the fertile ground of intellectual legitimacy where the bioethics was born, the behaviourism, among other movements, underplayed the relevance of ethics both in the education and the academia in favour of an inexplicable and aseptic technological ethos (Kiss & Euben, 2010, p. 8). Given the Skinner's doctrine mistrust of ideas like the embodied-self, even the neobehaviorist approach shows itself as a poor way of explaining the purpose of a conduct when facing a cerebral lesion that impairs social decisions just like a bilateral lesion in frontal ventromedial cortex, for example (Hughes & Churchland, 1995; Ettorre, 2017). The behaviourism and neobehaviorism were, to put it bluntly, bad psychology.

And bad politics. B. F. Skinner narrates a Utopian society in his novel *Walden 2* (Skinner, 1966). Even when he tries to depict an ideal society, he actually thinks about a rigmarole to defend a system that incontestably dictates the values and morals of the society (Tuzzeo, 2008). Many times the Walden 2 project has been tried to realise. One of these Waldens, the Walden 7, is 20 minutes away from my home in the Metropolitan Area of Barcelona. In 1975 the Walden 7 was a unique complex of buildings created by the Catalan prestigious architect, Ricardo Bofill. The Walden was created to be a self-sufficient community, but never was close to it and today is somewhat decadent -especially since it was struck by a sort of leper in its façade in the 1990s. However, it still conserves a bit of its old magic inside:

Figure 2. Walden 7
Source: Retrieved from its own web[4]

Some of the first neighbours were hand-picked by the developers of the project (we got off on the wrong foot…) (Iborra, 2016). Many of the original inhabitants were… *la crème de la crème de la gauche divine*. No, I haven't changed this book language to French, but the bourgeois who first moved in were nicknamed collectively as "la gauche divine" (French: the divine left). The point of not choosing the Catalan language was partly Sartre. If you were born with a silver spoon in your mouth and you used to play the cool and progressive know-it-all quoting French leftist intellectuals, the Walden 7 was likely your home (one of them at least). Unlike in Skinner's novel, the idea of maintenance workers from the community system was substituted by hiring half a dozen ones. The Walden 7 progressively became more expensive to maintain: today the monthly service charges are about 300 USD besides any millionaire apportionment for a building that is not so cool now. Therefore, *la gauche divine* gradually moved from the Walden 7, likely to the gentrified Barcelona where you don't need a car to go to the cinema. Nowadays, the Walden 7 is a mixed class dwellers state that destinates the 0.7% of their bulky budget to humanitarian causes.

Skinner apparently didn't understand that a community's sense of belonging cannot be neither the result of scheduling people's life nor putting them on a fancy scenery. In addition to his critique of Skinner, Erich Fromm also describes in *Anatomy of Destructiveness,* biophilia as the passionate love of life.[5] Below, he stresses that there is a necrophilia that opposes biophilia and whose passion is about corpses, dead and inert mechanisms (Fromm 2013). The biophilia idea was developed later as a biological theory (Kellert and Wilson 1995, 426–40). Retroactively, necrophilia can be understood as the biopolitical trait we saw in the rule of Stalin, Hitler and their epigones (Kushma 1974). It's important to consider that the current meaning of bioethics, biopolitics and biophilia were born in the 1970s: forty years after Holodomor (1932-33), thirty years after the Holocaust (1941-1945) and, ten years after the Mao's Great Famine (1959-1961). Indeed, Foucault and Fromm (a gay and a Jewish intellectual, by the way) were able only to give *a posteriori* explanation. In biophilia, every moment counts and "Eros beholding Psyche" is its DNA. On the other hand, necrophilia is predatory, not in the way a jaguar is, but in the same sense as a rapist. Therefore:

Bioethics is the biophilia after trading the passion for a more reflexive judgement.

The Gehenna side is biopolitical necrophilia, the cravings for handling living people as if they were yet ashes or bones.

To my mind, the best explanation of the source of the necrophilia and the Gehenna side was given in the *Enneads* of Plotinus:

In the same way, human beings, when weak on the side of contemplation, find in action their trace of vision and of reason: their spiritual feebleness unfits them for contemplation; they are left with a void, because they cannot adequately seize the vision; yet they long for it; they are hurried into action as their way to the vision which they cannot attain by intellection. They act from the desire of seeing their action, and of making it visible and sensible to others when the result shall prove fairly well equal to the plan. Everywhere, doing and making will be found to be either an attenuation or a complement of vision-attenuation if the doer was aiming only at the thing done; complement if he is to possess something nobler to gaze upon than the mere work produced. (Plotinus, 1952, p. 546)

The source of the Gehenna side is the inauthenticity and the inauthenticity turns viral in a society where a person is usually perceived as a failure rather than perceiving failure as an outcome. Heidegger's philosophy, which also deals with authenticity and the search for meaning, neglected the broader social and political realities that shape human existence and got tainted by his association with the Nazi Party. In contrast, the Bioethics of Displacement framework outlined in the text underscores the importance of understanding and challenging political power dynamics. It highlights the need for transparency, accountability, and independent oversight to mitigate corruption and promote the integrity of civil society initiatives. By incorporating these considerations, Bodi offers a more comprehensive and determined approach to societal change and ethical action than Heidegger's hemorrhagic pedantry (Heidegger, 2010, Chapter 6). He placed a strong emphasis on the individual's authentic existence while ignoring the dangers of arbitrary dictation of the meaning of life from above, either willingly or not (Heidegger, 2010, Chapter 5). Bluntly, Heidegger's understanding of the self as the sole owner of her finite and unique individual life is the perspective of a human animal that didn't realise how domesticated he was. By no means I justify Heidegger's choices but it would be untrue not to recognize, in turn, that we live and he lived in times when Gehenna politics is especially strong but in different ways. So it's not easy to know about our domestication level. At least, I can't determine the one of mine sharply.

Back to authenticity, I've explained the concept with the examples of Sartre who lost his best friend to keep his communist charade and Skinner who preached quite the opposite than he would like to preach and then created a perverted psychology. But they were not evil and peevish: Sartre managed himself to be the love of life of another intellectual celeb and honestly recognized the French crimes during the independence of Algeria. As for Skinner, he defended gender equality, was uprightly involved with his work until the last days of his life. Likewise, he started interesting polemics around Utopian behaviourism and free will that still endures. In addition to this, both of them had a true talent of their own and they were neither craving social climbers nor violent individuals -although pigeons and waiters may differ from me. Given that Sartre and Skinner had very good qualities, they are *good* examples. The true danger today is a simpleton society under the claws of the most vicious Gehenna bringers, the *weakest of the souls* described by Plotinus. Unlike their appearances and speech, the most populists enjoy an intuitive view of politics that is far higher than most political sciences PhDs. The reason is simple: they see opportunities and nothing like having a world beyond the scaffolding of propaganda distracts them from being the first to move. The payback for it is that the expectations they generate in this way are a toxic mirage. Conquering the world, bringing paradise, reducing unemployment, shipping the immigrants back to Africa (?)… Gehenna produces nothing but hoaxes.

Just like we've seen when comparing the birth data of the bio-lexicon (bioethics, biopolitics, biophilia, etc.) with the 20th century's worst crimes against humankind, the global changes driven by Gehenna bringers happened so quickly that the meanings were left behind. The unleashed events to expect in a society that is weak against demagoguery and arrivism are: flat out, shocking, short-lived, meta-violent, unsustainable, biassed, discriminatory and always at the ease of those who are: systemic issues cry-babies, dubious, regretful, tax-haven groupies, boast lovers, pronouns talibans and know-it-all tinpot Atlas cosplayers. A wonderful world for cleaved families, cities and whole societies in favour of a 1% of smoke and mirrors sellers who are the more self-assured the more broken the rest of people are.[6] It's the shadow of Gehenna!

On Trump, Rocky J. Suhayda, founder of the American Nazi Party (sic), held:

We have a wonderful opportunity here folks that may never come again, at the right time Donald Trump's campaign statements, if nothing else, have shown that 'our views' are not so 'unpopular' as the political correctness crowd have told everyone they are! (Heer, 2016, p. 44)

Figure 3. "Make America Grunt Again"
Source: The Mercury News (2016)

But it's not about the raunchiest US Republicans only -consider us lucky if that's all it is! US Democrats and Woke cult in the States and abroad don't lack emotional evocations of violence, even if different from Trump, QAnon, TV preachers and alike. Critical race and gender theories fanboys (bugself / bug-your-self / bug off) show off a graphic imagery of violence, fear, threats and aggression along with a *Message* of hatred and division, extreme emotions, social antagonisms, and international tensions. To begin with, wokism is socially divisive even in its very own terms, those who self-perceive themselves as awaken and in a better position than those who are perceived by them as sleepers. At the same level as Khomeini tagging as Unfaithful the ones with another faith, with another *Message*. By the same token, wokism alienates some other collectives that might benefit from being under the same state umbrella such as mothers in the workplace when, for example, the woke LGBTQ+ charity Stonewall, which operates a Workplace Equality Index in the UK, advised companies to stop using the word "mother". Shaddy quangos there are a lot, but mother there is only one, morons! Indeed, among one of its main targets, the women, wokism has been specially divisive. For example, JK Rowling's jam started when she liked a tweet by Maya Forstater, who lost her job in 2019 after tweeting that "male people [referring to male trans] are not women''. Both Maya and JK faced such disproportionate accusations as killing trans people with her hate. Personally, I disagree with both sides. As I see it, just like an adoptive mother, on her own merit, is finally called mother, a man who ends his transition and overcomes an extremely emotionally punitive condition, even if they never will share the walk of life of a biological woman, deserves to be called woman accordingly. On the other hand, even if the trans collective is completely right on the non-clinical bureaucratic hindrances, the natural born women who express their concerns about non-transitioned men sharing their showers should be heard and respected because, well… you know, women are people. (As a bioethics prof. myself, it's disheartening to me witnessing the kindergarten level of public debate and how education on bioethics keeps away from primary education). Wokism in practice has proven to be hell on Earth. They did a hard social experiment when defunding the state-owned police in twelve US cities but leaving the monopoly of violence on the table because crime is a social construct. Like hell, crime was a social construct! The hard-working and honest people didn't even touch the power police left behind because it happened overnight. No neighbourhood patrols became common and no law-abiding citizen-officers were drafted nor elected because there was no time for this cultural trait to appear. Quite the opposite, the monopoly of violence on the table was hoarded and predated by (oh surprise!) the ones used to it, the ones who, besides police, relentlessly ambitioned it: gangsters and small-time crooks. Further than this, formerly law-abiding citizens who wanted not to become the only dumb that didn't make hay, also indulged in robbery. Of course, no recognition of own flaws, no self-criticism have been raised among woke ranks, not even a miserable apology before all those who have been robbed, hitted, abused and raped.

Gehenna is not less dangerous in the upside down wokism, by the way. The neoliberal idea of leaving the political power on the table for another self-service will only favour those who are used to dealing with power and the powerful, the wealthiest minority. In turn, the non-wealthy majority, just like happened with the non-violent majority, will be abused, again. However, what makes me loathe more wokism than neoliberal-esque stuff is the level of ideological shovelling down. Those who once saw a Jeff Bezos pic in a clumsy office or Elon Musk getting bald and think that being a crypto bro who has overinvested his life in trading and is obsessed with taxes and the right mindset to become millionaire. The same bunch that easily will tag people who mistrust all this drag as parasites programmed for extinction (it includes finding a job) as soon as the state supernanny is utterly destroyed. Yes, those all-in neoliberals, nevertheless, are less prone to get offended. They think that their political opposites are robbing them but they are happier because they believe that they are going to be millionaires and move to a tax-friendly country just like this one-on-a-thousand guru bragging about his success on the Internet. However, wokism doesn't have enough to overlook you, but needs that you scorn yourself if unfit in their identitarian ideals. Neoliberal sharks want you to realise how a loser you are as a result of a process that always starts with her economic epiphany and ends with the others' perennial regret. But wokism is worse even since it demands that you feel inferior and docile from the get-go. It can be seen in its "go woke, get broke" entertainment industry but also in gloomier manifestations: Canadian schools burned almost 5000 books. (In my case, it was the needle that broke the camel's back).

Poisoning the philosophical roots of the public debate on those topics is, of course, the aforementioned magic voluntarism. If, being born as a male, I want to be a woman I only need to wish for it: skirt, heels and ready to go. Of course, it also goes for the Cletus' half: even if the Mexicans have a far lower HIV prevalence than US people, Trump only needed a mic, bigotry and the Internet to turn Mexicans into a sort of all-AIDS nation pumping contagions in the States up.[7] Again, inauthenticity and harmful lies are not even scorned, those fabrications are rewarded. And the more effortless and hateful the more rewarded. We'll get back on politics in the next chapter but now I want to highlight that falling on this Gehenna is not always a point of no return. Just like in the usual crimes and moral messing up, the make-meaning process in which the offender rebuilds her identity and turns back to society also exists (Ferrito et al., 2017). On the other hand, giving up the reinsertion chance for lost in the barren lands of incomprehensibility surely leads to an ultimate nose-diving in the gloomier.

Before ending, I want to stress, in case it is not clear enough yet, that all the criticism in this chapter is not indiscriminate and against any random politically intense person. If you want to spend your whole life moaning for the good ol' times of the cotton and rye in the land of Dixie, it's up to you and I'll hope you meet happiness according to your deeds. And the same goes for those who think that the female concubines and enslavers from the Dahomey tribe were a sort of human perfection. Yes, if you want to litter your flat with plenty of African props and you think that changing your name to Uvuvwevwevwe Onyetenvewve Ugwemubwem Ossas will be the best prop of all, I've nothing to say about it but wishing happiness to you, too. The criticism and puns in this chapter is against proselytism, overreaction, phobia in virtue signalling dress, mass media or violent banal activism or, to put it bluntly, playing the role of a petty politician. Being that ideological endogamy's simp is the most stupid harmful legal activity that a person can do just above Tik Tok challenges.

CONCLUSION

Psychological inflexibility and moral stress have a negative impact on ethical decision-making and moral experiences. Ignoring individuals' personal stories as well as the interplay between different factors contribute to moral stress. Is important considering preemptive bioethics and syncretic ethical decision-making in addressing complex conflicts and aiming to facilitate productive dialogue. Respectful and empathetic communication requires adopting diverse perspectives, and integrating practical solutions with philosophical discourse. Furthermore, alignment and ideological endogamy in bioethics are dangerous and prevent from responsible engagement as well as rejecting inauthenticity and harmful lies.

REFERENCES

Baars, B. (2003). The double life of BF Skinner: Inner conflict, dissociation and the scientific taboo against consciousness. *Journal of Consciousness Studies, 10*(1), 5–25.

Behrens, K. G. (2017). Hearing sub-Saharan African voices in bioethics. *Theoretical Medicine and Bioethics, 38*(2), 95–99. doi:10.100711017-017-9406-z PMID:28361295

Bryant, J. A., & la Velle, L. B. (2018). *Introduction to Bioethics*. John Wiley & Sons.

Byrne, R. (2011). *The Secret*. Simon and Schuster.

Callahan, D. (1973). Bioethics as a Discipline. *The Hastings Center Studies, 1*(1), 66–73. doi:10.2307/3527474

Chattopadhyay, S., Myser, C., Moxham, T., & Vries, R. D. (2017). A Question of Social Justice: How Policies of Profit Negate Engagement of Developing World Bioethicists and Undermine Global Bioethics. *The American Journal of Bioethics, 17*(10), 3–14. doi:10.1080/15265161.2017.1365185 PMID:29020562

Chua, D. K. L. (2011). Listening to the Self: The Shawshank Redemption and the Technology of Music. *19th-Century Music, 34*(3), 341–355. doi:10.1525/ncm.2011.34.3.341

Devolder, K., & Douglas, T. (2018). The epistemic costs of compromise in bioethics. *Bioethics, 32*(2), 111–118. doi:10.1111/bioe.12418 PMID:29280164

Ehrenreich, B. (2010). *Smile Or Die: How Positive Thinking Fooled America and the World*. Granta.

Ettorre, E. (2017). Self, Normativity and Embodiment. In E. Ettorre, E. Annandale, V. M. Hildebrand, A. Porroche-Escudero, & B. K. Rothman (Eds.), Health, Culture and Society: Conceptual Legacies and Contemporary Applications (pp. 37–74). Springer International Publishing. doi:10.1007/978-3-319-60786-3_3

Farr, A. (1998). *Sartre's Radicalism and Oakeshott's Conservatism: The Duplicity of Freedom*. Springer. doi:10.1057/9780230380264

Ferrito, M., Needs, A., & Adshead, G. (2017). Unveiling the shadows of meaning: Meaning-making for perpetrators of homicide. *Aggression and Violent Behavior, 34*, 263–272. doi:10.1016/j.avb.2016.11.009

Fromm, E. (2013). The Relation of Freud's Life and Death Instincts to Biophilia and Necrophilia. In *The Anatomy of Human Destructiveness*. Open Road Media.

Hage, V. (2013, November 6). Philosophical Differences: The Falling-Out of Camus and Sartre. *Spiegel Online*. https://www.spiegel.de/international/zeitgeist/camus-and-sartre-friendship-troubled-by-ideological-feud-a-931969.html

Heidegger, M. (2010). *Being and Time*. SUNY Press.

Hohenberg, P. M., & Lees, L. H. (2009). *The Making of Urban Europe, 1000-1994*. Harvard University Press. doi:10.2307/j.ctvjf9vvp

Hughes, J., & Churchland, P. S. (1995). My behavior made me do it: The uncaused cause of teleological behaviorism. *Behavioral and Brain Sciences*, *18*(1), 130–131. doi:10.1017/S0140525X00037705

Iborra, Y. S. (2016, March 3). *Cuando la puta y el cura entraron al Walden-7*. SomAtents. http://somatents.com/es/magazine-es/cuando-la-puta-y-el-cura-entraron-al-walden-7/

Intergovernmental Bioethics Committee. (2005). *Universal Declaration on Bioethics and Human Rights*. General Conference of the United Nations Educational, Scientific and Cultural Organization.

Kellert, S. R. (1985). Historical trends in perceptions and uses of animals in 20th century America. *Environmental Review*, *9*(1), 19–33. doi:10.2307/3984113

Kiss, E., & Euben, J. P. (2010). *Debating Moral Education: Rethinking the Role of the Modern University*. Duke University Press.

Kopelman, L. M. (2009). Bioethics as Public Discourse and Second-Order Discipline. *The Journal of Medicine and Philosophy: A Forum for Bioethics and Philosophy of Medicine, 34*(3), 261–273. 10.1093/jmp/jhp019

Macer, D. R. J. (1998). *Bioethics is love of life: An alternative textbook*. Eubios Ethics Institute.

Plotinus. (1952). *The six enneads* (S. Mackenna & B. S. Page, Trans.). Encyclopaedia Britannica.

Rollin, B. E. (2007). Animal Mind: Science, Philosophy, and Ethics. *The Journal of Ethics*, *11*(3), 253–274. doi:10.100710892-007-9018-3

Saint-Paul, R. (1955). *0112683—Jean-Paul Sartre (1905-1980)*. Granger. https://www.granger.com/results.asp?image=0112683

Saramo, S. (2017). The meta-violence of Trumpism. *European Journal of American Studies*, *12*(2), 12–2. doi:10.4000/ejas.12129

Sartre, J.-P. (1992). *Being and Nothingness*. Simon and Schuster.

Skinner, B. F. (1966). *Walden Two*. Hackett Publishing.

Sosa Wagner, F. (2004). *El mito de la autonomia universitaria*. Academic Press.

The Mercury News. (2016, May 13). Cartoons: Playing politics with Donald Trump. *The Mercury News*. https://www.mercurynews.com/2016/05/13/cartoons-playing-politics-with-donald-trump-2/

Torres, D. (2018, April 17). *Spain's degree scandal hits at heart of government*. POLITICO. https://www.politico.eu/article/cristina-cifuentes-spain-corruption-mariano-rajoy-pp-degree-scandal-hits-at-heart-of-government/

Tuzzeo, J. (2008). Changing our Minds: Dystopian Psychological Conditioning in Nineteen Eighty-Four, Brave New World, and Walden Two. *English Master's Theses*. https://digitalcommons.brockport.edu/eng_theses/88

KEY TERMS AND DEFINITIONS

Activism: Taking action to create social or political change, through campaigns, protests, and other means.

Communication: Exchange of information and ideas through talking, listening, writing, and fostering understanding.

Concupiscence: The inclination to do evil when moved by a strong desire.

Feminism: Secular religion grounded in the belief of an evil ubiquitous (cis-herero) Patriarchy that oppresses women and, many times, the minority sexual orientations.

Ideology: A set of beliefs and values that shape the way people think and behave, especially within a specific group.

Philosophical Discourse: Thoughtful discussions about fundamental questions, drawing upon theories and concepts.

Syncretic Ethical Decision-Making: A method that seeks to integrate diverse perspectives and values in ethical decision-making.

ENDNOTES

1. Etymologically, Gehenna comes from the Hebrew גֵיא בֶן־הִנֹּם (Valley of the Son of Hinnon) that is a real place 10 minutes away by car from Jerusalem. According to Jeremiah 19:5-7, some Judah Kings burn their children as a sacrifice to Baal:
   And have built the high places [altars] of Baal, to burn their sons in the fire for burnt-offerings unto Baal; which I commanded not, nor spoke it, neither came it into My mind. Therefore, behold, the days come, saith the LORD, that this place shall no more be called Topheth, nor The valley of the son of Hinnom, but The valley of slaughter
   Translation as in The Holy Scriptures JPS Tanakh 1917 retrieved from: https://biblehub.com/jps/jeremiah/19.htm
   I use "Gehenna" more in this, its most historical meaning, than as "hell" as in the later rabbinical literature. However, the ambiguity of the terms helps to highlight that Bodi's Gehenna has been built on evil and for evil's sake.
2. Spanish universities, for example, are overly politicised through a sound spoils system in the appointment of chairs, rectors and above all, vice-rectors (Sosa Wagner, 2004). The greatest exponent of this politicisation is the Universidad Rey Juan Carlos I (URJC), where some high profile politicians like Pablo Casado (former Spanish PM candidate) were delivered Masters without even presenting a thesis (Torres, 2018).
3. It was much more than the world's car production outside of the US only in NYC.
4. http://www.walden7.com/galeria-imatges/
5. Cf. Darryl Macer's references about Fromm and love in Bioethics is Love of Life (Macer, 1998, pp. 20, 22).
6. More information about meta-violence meaning in The Meta-violence of Trumpism in Saramo (2017).
7. And those yankee's political lap dogs as AMLO makes racism even more profitable.

Chapter 4
Transcendental Bioethics of Displacement:
Good v Evil III

ABSTRACT

To explore the question of what constitutes "good bioethics," the author begins by examining the criteria established by Hofmann and Magelssen for argumentative, empirical, and dialectic approaches. This leads to a meta-ethical definition of good bioethics that emphasises the importance of genuine communication and consideration of future generations. The author then introduces the transcendental bioethics of displacement, which focuses on energy and the psyche. By examining the relationship between energy and displacement, we can better understand the experiences of individuals with conditions such as autism spectrum disorder and advanced Alzheimer's dementia. Ultimately, we see how these concepts complement and expand upon one another, providing a more comprehensive understanding of what constitutes "good bioethics."

INTRODUCTION

Chapter 3 sets the stage for further exploration of bioethics in this chapter, where the focus shifts to defining "good" bioethics and introducing the concept of transcendental bioethics of displacement. This chapter seeks to provide a meta-ethical definition of good bioethics by emphasising the persistent pursuit of progress through genuine, open, and well-grounded personal communication that welcomes all concerned voices and considers the interests of future generations. It introduces the idea that understanding the energy and force underlying ethical decision-making is crucial and proposes the transcendental bioethics of displacement as a complementary perspective to further explore these dynamics. This chapter aims to deepen our understanding of bioethics and its connection to broader philosophical concepts while also highlighting the practical implications and ethical considerations that arise in the field.

DOI: 10.4018/978-1-6684-4808-3.ch004

WHAT IS GOOD BIOETHICS?

What is good bioethics? Hardly would we be able to go further in bioethics without this question. The bioethicists would turn into a melee of tabloid mud queens otherwise. Obviously it is not going to be so easy like writing a definition and ending this section. To start with, it's not proven that among bioethicists there is a greater ratio of supportive and upright individuals than there are among the Kolkata rickshaw riders or among the seafarers around the world. We are more self-inquisitive when talking about our own profession because we've more chances to do so, but it's a bias.

On this topic, Bjørn Hofmann and Morten Magelssen analysed the criteria for good bioethics and established three categories: argumentative, empirical, and dialectic (plus a healthy debate in order the categories don't get petrified like a dinosaur footprint) (Hofmann & Magelssen, 2018).

It leads us back again to the Table 1 where Darryl Macer and Joseph Fletcher's respective classifications are now expanded with the Hofmann and Magelssen ones and complemented, in turn, by the bioethics of displacement:

Table 1. Views of bioethics, ethical decision-making approaches and quality criteria categories

Main Views of Bioethics	Ethical Decision-Making Approaches	Category of Bioethics Quality Criteria
Prescriptive	Legalist	Argumentative
Descriptive	Antinomian	Dialectic
Interactive	Situationist	Empirical
Prospective	*Relativist*	*Communicational*

The relation between the prescriptive view of bioethics, legalist ethical decision-making, and argumentative quality criteria is that they are interdependent and complementary. The prescriptive view of bioethics sets the standard for the ethical treatment of the issue, and legalist ethical decision-making provides the framework for ethical decision-making. Argumentative quality criteria ensure that the debate is productive, expands our moral knowledge pool, and takes into account real-life experiences. Together, these components help to ensure that ethical decision-making is of high quality and respects the dignity of all stakeholders involved.

In turn, descriptive views, Antinomian decision-making and Dialectic quality criteria are interconnected because descriptive views of bioethics can inform antinomian ethical decision-making by providing insight into the actual ethical practices and beliefs of individuals and communities in the field. Dialectic quality criteria can be used to evaluate the quality of both traditional and antinomian ethical decision-making processes, helping to ensure that ethical decisions are well-reasoned and defensible. Furthermore, descriptive bioethics focuses on understanding how individuals and groups cope with ethical issues, beyond the minimum requirement of dignity. Antinomian ethical decision-making rejects the idea of absolute moral norms and instead considers the situation and context when making moral choices. Dialectic quality criteria require the bioethical debate to expand the moral knowledge pool by contrasting the choices with real cases and avoiding stalemate debates.

About the next three of a kind, Interactive bioethical views, Situationist ethical decision making and Empirical quality criteria.

The relation between Interactive views of bioethics, situationist ethical decision-making, and empirical quality criteria can be seen as forming a three of a kind in the field of bioethics because they all highlight the importance of taking into account the specific context and circumstances of a particular ethical dilemma or decision-making process. To begin with, Interactive views of bioethics recognize that ethical decision-making is a dynamic and interactive process that involves multiple stakeholders, perspectives, and values. This approach emphasises the importance of engaging in dialogue and deliberation with all relevant parties to arrive at ethical decisions that are both reasonable and acceptable. In turn, Situationist ethical decision-making acknowledges that ethical decision-making is influenced by contextual factors such as social norms, situational pressures, and individual differences. This approach stresses the importance of understanding the specific situation in which an ethical decision is being made and taking into account the psychological factors that may impact ethical reasoning and behaviour. Likewise, Empirical quality criteria pays attention to the empirical evidence to inform ethical decision-making. This approach points out the use of rigorous research methods to identify relevant empirical data and evaluate the quality and relevance of that data to the ethical decision at hand. Together, these approaches highlight the importance of understanding the context and circumstances of ethical decision-making, as well as the role of empirical evidence and deliberative processes in arriving at ethical decisions that are both reasonable and acceptable.

Finally, the prospective views of bioethics, relativist ethical decision-making, and communicational quality criteria can be seen as three of a kind in the sense that they all stress the importance of taking a future-oriented approach to ethical decision-making. Prospective views of bioethics focus on anticipating and preventing potential harms and ethical issues that may arise in the future, while relativist ethical decision-making highlights the need to consider the specific cultural and contextual factors that may impact ethical decision-making. In turn, communicational quality criteria emphasise the importance of clear and effective communication in ethical decision-making, which can help to ensure that all relevant parties have a shared understanding of the issues at hand and can work together to identify and address potential ethical concerns. Taken together, these three perspectives can help to provide a framework for approaching ethical decision-making in a way that is sensitive to both the potential risks and benefits of different courses of action, as well as the specific social, cultural, and communicative factors that may influence ethical decision-making in different contexts.

Table 2. Ethical coordinates and main focus of different bioethical paradigms

Three of a Kind	Main Focus
Prescriptive - Legalist - Argumentative	Making ethical decisions based on rules, laws, and arguments
Descriptive - Antinomian - Dialectic	Understanding ethical behaviour and decision-making rather than prescribing how to act
Interactive - Situationist - Empirical	Emphasise the situational and contextual nature of ethical decision-making
Prospective - Relativist - Communicational	Ethical decision-making from a forward-looking perspective, taking into account various perspectives and the context of the decision

Overall, these groups form a three of a kind because they offer different perspectives on ethical decision-making that complement each other and provide a comprehensive framework for understanding and addressing ethical issues. Of course every perspective has its own pros and cons.

Table 3. Pros and cons of different bioethical paradigms

	Pros	Cons
Prescriptive - Legalist - Argumentative	Unjams discussions easier thanks to objectivity More suitable during asynchronous communication	Unsuitable while exploring the ethical unknown
Descriptive - Antinomian - Dialectic	More suitable for complex discussions Prevents stalemate debates	Unsuitable for long-term moral choices
Interactive - Situationist - Empirical	Better for multiple groups' discussions Encourages mindfulness	Unsuitable for complex discussions
Prospective - Relativist - Communicational	More suitable for long-term moral choices and while exploring the ethical unknown Encourages open and transparent communication among all parties	Unsuitable for objective discussions due to its subjectivity

While all three combinations are interesting and important in their own right, the Prospective - Relativist - Communicational group is particularly groundbreaking because it addresses some of the most pressing ethical challenges we face today in a rapidly changing and globalised world. The Prospective view of bioethics acknowledges that new technologies and social changes constantly emerge and require us to anticipate and prepare for potential ethical challenges in advance. This approach is particularly relevant today as we face unprecedented technological advancements, such as AI and genetic engineering, and global challenges such as climate change, which will require us to anticipate and respond to ethical issues that are not yet fully understood. The Relativist ethical decision-making approach acknowledges the diverse and often conflicting cultural and moral values that exist across different societies and individuals. In today's multicultural and globalised world, it is increasingly important to understand and respect these differences and to develop ethical frameworks that can accommodate diverse perspectives. Finally, the Communicational quality criteria approach emphasises the importance of clear communication and dialogue in ethical decision-making. In a world where information is constantly being shared across borders and cultures, effective communication is essential to ensure that different perspectives are heard and understood, and that ethical decisions are made in a transparent and accountable manner. In summary, while all three groups are important, the Prospective - Relativist - Communicational group is particularly interesting and groundbreaking because it addresses some of the most pressing ethical challenges we face today in a rapidly changing and globalised world, emphasising the importance of anticipating and preparing for potential ethical challenges, respecting diverse cultural and moral values, and ensuring effective communication and dialogue.

This discussion leads to a meta-ethical definition of good bioethics because it touches upon the fundamental philosophical underpinnings of ethical decision-making in the field of bioethics. The different approaches and combinations of approaches discussed relate to broader meta-ethical questions about the nature of ethics, the role of cultural and contextual factors, and the importance of communication and dialogue in ethical decision-making.

By examining and evaluating these different approaches, we can arrive at a more comprehensive and nuanced understanding of what constitutes "good" bioethics. This understanding can inform the development of ethical frameworks and guidelines, as well as guide individual practitioners in their ethical decision-making. So we can try now to reply to the question at the beginning of this section with a meta-ethical definition of good bioethics:

Good bioethics entails a persistent pursuit of progress through genuine, open and well-grounded personal communication that welcomes all the concerned voices and considers the interests of future generations.

TRANSCENDENTAL BODI

The transcendental bioethics of displacement, with its focus on energy, offers a complementary perspective to the meta-ethical definition of good bioethics proposed in the previous chapter. While the meta-ethical definition emphasises the importance of genuine, open, and well-grounded personal communication that considers the interests of future generations, the transcendental bioethics of displacement highlights the importance of recognizing and addressing the energy and force that underlie ethical decision-making.

Transcendental Bioethics of Displacement is about energy. The displacement, together with condensation, are the behaviours that the Freudian psyche's free energy can adopt (Freud, 2008; Steiner, 1988). This Freudian dichotomy, this psychiatric view on the belonging and dis-belonging intimacy, has a postmodern echo. In 1984, Julia Kristeva used the displacement idea in her *Revolution in Poetics* while Jean-François Lyotard states that the messages displace the atoms of pragmatic relationships (Kristeva, 1984, Chapters 1, 2, 5; Lyotard, 1984, p. 16). The energy and force are, unsurprisingly, related with displacement even though they are new Bodi's entries. In a similar fashion to chemistry, the energy comes from something that tears, glues, explodes or implodes (a mind, a country, a home, a rock band…). A person's self-meaning goes from belonging to dis-belonging given that this energy is free, is not condensed. In some cases of Autism Spectrum Disorder (ASD) as well as advanced Alzheimer Dementia (AD) we can observe the 'wandering' behaviour: a person displacing through space without self-awareness (Solomon & Lawlor, 2018). In the case of a person with this condition the energy that came from the tearing of her psyche cannot condense and becomes locked in an embodied movement. They are displacing themselves because they are ultimately displaced but it stresses that "*the body is our general medium for having a world*" (Merleau-Ponty, 2013, p. 147). Thus, these patients are heartrendingly looking for their body and further interventions might consider it.

The illness, those moments in which we feel devoid of energy, are also a displacing experience. When driven to the limit, the fever shows the limit of embodiment: a quite underestimated philosophical source in academia (Carel, 2013). Someone can resort to therapeutic philosophy, but Merleau-Ponty yet warned us about the objectivist illusion that might be hidden in (Merleau-Ponty, 1973, pp. 64–65, 147). Even in the case of persons confined in psychiatric facilities, his phenomenological panoply like the lived-body and the embodied speech can be useful for a critical bioethics of body and place (Merleau-Ponty, 2013, pp. 92–99). Moreover, Merleau-Ponty's ideas can drive psychological qualitative research beyond the cognitive frame (Murray & Holmes, 2014). It means that Merleau-Ponty's ideas can expand psychological research beyond cognitive approaches to include the role of the body and lived experience in understanding mental health. Without going any further than the edge of cognitive experience and the French philosopher, the clinical condition of hemi-spatial neglect allows us to look for infuse-meaning body observations besides the usual interview-driven research (Klinke et al., 2014). So, thanks to Merleau-Ponty ideas, the study of a condition where a person ignores or is unaware of one side of their body or environment can help researchers observe the role of the body in mental health beyond traditional interview-based research.

Merleau-Ponty is the Ibn Batutta of embodiment, its main scholar and explorer in many different settings. But I feel Pink Floyd is closer:

Now I've got that feeling once again

I can't explain you would not understand

This is not how I am

I have become comfortably numb (Pink Floyd, 1980)

Besides energy, the force in the Bioethics of Displacement, unlike Newtonian physics, has no vector since this kind of displacement can't be expressed as a scalar value. In this hypothesis, the force is what determines if a displacement will start to happen, if the Wheel of Fate will *spin*. It determines if a person working in another country finally sees herself like a migrant or not; if a broken-hearted gets rid of a toxic relationship or not; in both cases the person has a re-built awaiting for her. It's not very different from the force that glues the nucleons in the atom. An example of a mathematical object is the arrow in your GPS and the coordinate transformations in space and time is what your GPS and its satellites do for you on the screen map. But, is it possible that there exists a tensor without the vectors that the scalars must compound? Well, it shouldn't be actually a tensor in the way I just put it but a non-scalar tensor, whose real name is *spinor*. Just like the tensor / GPS arrow, a spinor is a mathematical object but represents the rotation of an object in space. A spinor would represent the rotation of the Earth in three-dimensional.This whole 3D globe would rotate as you move.[1]

Let's recapitulate. Spinors are a mathematical object that represents the rotation of an object in space. They are related to the Bioethics of Displacement as a way to understand the force that determines if a displacement will happen or not. Given that the Bioethics of Displacement is a manifold from bioethics' moral space we can consider the dominion of force like a single spinor field of force. [2, 3] The idea of the Bioethics of Displacement is that the moral space of bioethics can be viewed as a manifold, or a mathematical space that can be bent and twisted in different ways. In this context, the dominion of force can be seen as a single spinor field of force, which represents the rotation of an object in space. This means that the force that drives displacement, such as the force that determines if a person working in another country finally sees themselves as a migrant or not, can be understood as a kind of rotation in moral space. In other words, just as a spinor represents a rotation in physical space, a spinor field can represent a rotation in moral space, which is what the dominion of force in the Bioethics of Displacement is thought to be. The idea of a broadest spinor field of force implies that all beings, regardless of species, have a shared connection to the force behind the Bioethics of Displacement, which allows all beings who are able to feel the love of life to connect with each other and belong. Thus, all beings, human and pre-human, able to feel the love of life share a very special élan vital. Even those beings who do not share it can get closer as far as we discern how to engage with them in the fittest spiritual sense (Barbiero, 2011). The love of belonging is transcendental because it goes beyond individual identities and connects all beings who are able to feel the love of life. The Bioethics of Displacement is the best way of thinking about it because it provides a framework for a deeper exploration about the nature of reality and human existence.

Human existence is a complex and multifaceted concept that has been explored in various ways throughout history. However, some philosophers have approached it differently than the Bioethics of Displacement, and their approaches may not resonate with everyone. For example, the aforementioned Jean-Paul Sartre, focused on the idea of freedom and the individual's ability to choose their own path in life but it doesn't necessarily address the idea of belonging and connection to others in a larger sense. Even Martin Heidegger explored the idea of "being-in-the-world," which emphasises the interconnectedness of humans and the environment. While this is a step closer to the idea of belonging than Sartre's, it lacks the emotional dimensions of the love of belonging -as could be expected from an overrated Nazi cocksucker (Nazi cocks, I mean). What sets the Bioethics of Displacement apart is its emphasis on the moral and spiritual dimensions of belonging, and its focus on the love of life as the basis for connection and belonging. The idea of a shared connection to the force behind the Bioethics of Displacement allows for a broader sense of belonging that extends beyond just humans and includes all beings that are able to feel the love of life. Furthermore, the Bioethics of Displacement offers a practical approach to fostering belonging through ethical decision-making and engagement with others. Even if transcendental Bioethics of Displacement is quite abstract, it enjoys a practical aspect that sets it apart from some other philosophical approaches that may be hopelessly abstract and theoretical. Philosophies that stink as mid-morning coffee and cigarettes break close to public institutions.

The anti-psychiatrist Ronald David Laing depicted humans like puzzled, numb, spaced out creatures catching a sanity that never condenses (Laing, 1990). *We all are coming through in waves*. Everybody rides her first wave when her mother breaks waters and starts the tide of the labour. Everybody stops being a *cellular* wave of probability and starts riding them. After we die, we become a probability again: the probability of being remembered. And later, we become another anonymous wave, a *soliton* perhaps, in a world in whose complexity nothing is lost -like in the Bergson's Élan Vital (Bergson, 1984, pp. 87–96, 209–217). Or perhaps there are turtles all the way down, I don't know. I just became comfortably numb.

CONCLUSION

Good bioethics is defined as a persistent pursuit of progress through genuine, open, and well-grounded personal communication that considers the interests of future generations. The transcendental bioethics of displacement underscores recognizing the energy and force that influence ethical decision-making and its impact on individuals' sense of belonging.

REFERENCES

Barbiero, G. (2011). Biophilia and Gaia. Two hypothesis for an affective ecology. *Journal of BioUrbanism*, *1*, 11–27.

Bergson, H. (1984). *Creative Evolution*. University Press of America.

Carel, H. H. (2013). Illness, phenomenology, and philosophical method. *Theoretical Medicine and Bioethics, 34*(4), 345–357. doi:10.100711017-013-9265-1 PMID:23836135

Freud, S. (2008). The Future of an Illusion. Academic Press.

Friar, J. L. (1993). Three-nucleon forces and the three-nucleon systems. *Czechoslovak Journal of Physics, 43*(3), 259–287. doi:10.1007/BF01589847

Hofmann, B., & Magelssen, M. (2018). In pursuit of goodness in bioethics: Analysis of an exemplary article. *BMC Medical Ethics, 19*(1), 60. doi:10.118612910-018-0299-9 PMID:29902999

Klinke, M. E., Thorsteinsson, B., & Jónsdóttir, H. (2014). Advancing Phenomenological Research: Applications of "Body Schema," "Body Image," and "Affordances" in Neglect. *Qualitative Health Research, 24*(6), 824–836. doi:10.1177/1049732314533425 PMID:24819692

Kristeva, J. (1984). *Revolution in Poetic Language*. Columbia University Press.

Laing, R. D. (1990). The Politics of Experience and The Bird of Paradise. Academic Press.

Lyotard, J.-F. (1984). *The Postmodern Condition: A Report on Knowledge*. University of Minnesota Press.

Merleau-Ponty, M. (1973). *The Prose of the World*. Northwestern University Press.

Merleau-Ponty, M. (2013). *Phenomenology of Perception*. Routledge. doi:10.4324/9780203720714

Murray, S. J., & Holmes, D. (2014). Interpretive Phenomenological Analysis (IPA) and the Ethics of Body and Place: Critical Methodological Reflections. *Human Studies, 37*(1), 15–30. doi:10.100710746-013-9282-0

Pink Floyd. (1980, June 23). *Comfortably Numb* [7"]. Harvest (UK), Columbia (US).

Schiffer, J. P. (2010). Implications of old physics simplify the understanding of nuclei. *Physics (College Park, Md.), 3*, 2. doi:10.1103/Physics.3.2

Solomon, O., & Lawlor, M. C. (2018). Beyond V40.31: Narrative Phenomenology of Wandering in Autism and Dementia. *Culture, Medicine and Psychiatry, 42*(2), 206–243. doi:10.100711013-017-9562-7 PMID:29368117

Steiner, R. (1988). "Paths to Xanadu …" Some Notes on the Development of Dream Displacement and Condensation in Sigmund Freud's "Interpretation of Dreams." *Int. R. Psycho-Anal., 15*, 415–454.

KEY TERMS AND DEFINITIONS

Belonging: The state or feeling of being connected, accepted, or included within a group, society, or environment.

Decision-Making: The process of selecting a course of action or choosing from available alternatives based on evaluation, reasoning, and judgment.

Energy: In Bioethics of Displacement is whatever causes changes in one's sense of belonging and self-meaning.

Ethical Frameworks: Guidance or direction for making ethical decisions that don't provide specific answers for each situation.

Ethics: The exploration of what is morally right or wrong and how we should behave towards others and nature.

Force: In Bioethics of Displacement, is what determines if a displacement will start to happen.

Good Bioethics: Persistent pursuit of progress through genuine, open, and well-grounded personal communication that considers the interests of future generations.

Meta-Ethics: The study of the nature of ethics and moral language, exploring what morality is rather than prescribing right and wrong actions.

Transcendental: Supernatural or beyond ordinary human experience, often associated with spirituality or metaphysics.

ENDNOTES

1. Spinors do not transform in the same way as tensors, so if GPS used spinors instead of tensors, the coordinate transformations would be much more complicated and difficult to compute.
2. The bioethics moral space was defined as an invariant group of the Euclid, Lobachevskii and Riemann geometries at Chapter 4, in the section titled Sensible Moral Space and Local Invariance.
3. The aforementioned glue for nucleons is called tensor force and is related with the nucleons' spin (Schiffer, 2010; Friar, 1993).

Chapter 5
On the Misery of Bioethics:
Good v Evil IV

ABSTRACT

The downstream burden of illness highlights the need for a bottom-up approach to solidarity and governance, emphasising discursive legitimacy, justice, and equality for vulnerable populations. The author delves into the interconnectedness of illness and health, advocating for the inclusion of structural theories in nursing education and the recognition of identity issues. Then, the author examines the dangers of crony capitalism and left-wing populism, promoting a bioethics of displacement that prioritises collective freedom and love of belonging. Additionally, the chapter critically evaluates the advantages and limitations of intersectionality as a research paradigm, highlighting the potential pitfalls of overgeneralization and identitarian labelling. Likewise, the author sheds light on the complexities and consequences of intersectional discourse, urging for a more nuanced understanding of vulnerability and social justice. Through these discussions, the chapter aims to stimulate critical thinking and contribute to a more inclusive and ethical approach to bioethics and social issues.

INTRODUCTION

In the last chapter, we explored the concept of "good" bioethics and discussed the importance of genuine, open, and well-grounded personal communication that considers the interests of future generations. This understanding forms the basis of a meta-ethical definition of good bioethics. Now, in this chapter, we delve deeper into the principles of Bioethics of Displacement and connect them with the ideas presented in Chapter 4.

OPPORTUNITIES ALWAYS FLOW UPSTREAM

From the perspective of Bioethics of Displacement, understood as love of belonging, the misery of bioethics lies in the cycles of inequality and the obscurantism coming from intersectionality talibans and the market-will-solve-it crusaders.

DOI: 10.4018/978-1-6684-4808-3.ch005

Let's start with inequality. Bioethics usually deals with informed consent but when the finances get involved both information and consent go to hell in a handbasket. I will elaborate it with an example. Plenty of deaths produced by car pollution and tobacco are insidiously imposed: even in well-developed countries the green and public transportation is so scarce many times that they must go to work driving an old used car while smoking, a premature vice promoted by the twenty years ago ubiquitous ads promising a fancy lifestyle. If they had had more info, many heavy smokers wouldn't have consented to their first puff. Likewise, if we would have more and better info on where our money goes, then greener mobility could be rightly assessed. This mess won't be repaired by Shell, Aramco and Philip Morris among others, quite the opposite, if you feel identified with the example above there are real chances that your black anointed lungs will be offered as sacrifice to these postmodern biz gods in an operating table that you will pay badly for. Or perhaps not. The support to euthanasia in places like the US is strong today (Brenan, 2018). And it's not only about dying with dignity, it's about bending down the cost curve of public healthcare! (Dobken, 2018) Many places have offered physician assisted death: California, Colorado, Oregon, Vermont, Hawaii, Montana, Washington State, the District of Columbia, the Netherlands, Switzerland, Germany and Canada (Dobken 2018). The Canadian academic supporters of this policy states that:

Providing medical assistance in dying in Canada should not result in any excess financial burden to the healthcare system, and could result in substantial savings. (Trachtenberg & Manns, 2017)

However, the money obtained from saving resources in persons hopelessly ill has not risen welfare simply because the end-of-life healthcare burden is not the reason behind the public administration straits when offering medical services (Silver & Hyman, 2018).

The core of financial bioethics is how finances are a pretext for *bending down bioethics* and stealing vulnerable people its share of healthcare opportunities. It was brightly exposed by Tom Koch when criticised the "*lifeboat ethics*" parable (Koch, 2012, pp. 77–79). Shortly, it means that public healthcare is like a lifeboat with fewer places than people in the water, so some of them will die. Let's play with this lifeboat with the parody mode on. The cox always starts up in the lifeboat (likely because her father was a cox, too) and must use his influence in order to obtain the desired crew. In the early versions of lifeboat games the cox hit with a row in the head of the unsuitable oarsmen trying to get or remain into the boat but in some rare cases the cox fell to the sea during the struggle. So, a cunning bioethicist, I mean a better tactician, was employed and the postmodern ruler, I mean the savvy cox, convinced the people floating around to choose those who must be left to drown. Now, the reason why the ship is sinking (does it? really?) and the reason why the lifeboats have no more places is a political capital. It can be spent to blame the people floating in the water for not smiling enough in their part-time job and, as a result, they have attracted a wave of universal-quantum-magneto-psycho-negativism that has put them in the place they deserve (Price, 2008, pp. 19–36; Damien, 2008; Kreidler, 2007).

A man said to the universe:

- *"Sir, I exist"*

- *"However," replied the universe,*

"the fact has not created in me

a sense of obligation." (Crane, 1972, p. 96)

The financial bioethics doesn't pay only attention to pretexts but also to solidarity. A sound governance of solidarity created from a bottom-up approach would increase the discursive legitimacy of justice and equality for vulnerable people and, therefore, it would offer a more sensible rationale behind the public health expenditure than the usual bovine attitude. Shortly, everybody must have a say to make sure everyone is treated fairly and it starts by being kind to others who need help.

BURDEN ALWAYS FLOWS DOWNSTREAM

The phenomenological analysis of illness has put it closer to health (Ostrowski & Sikorska, 2015, p. 35). As Havi Carel said *"illness is part and parcel of life, and on a continuum with health"* (Carel, 2016, p. 80). Delinating illness and health is not only a main topic in philosophy of medicine, bioethics and medical education but also in politics, namely health insurances (Ostrowski & Sikorska, 2015, p. 26). About this topic, the long-term care sector is generally characterised by low pay, difficult working conditions as well as deep embedded ethnic and gender pay differences that, in an ageing society, has led to a structural employment marginalisation to reflect the marginalisation of older people receiving care (Hussein & Manthorpe, 2014). Given that, the nurse theorists should incorporate structural theories and concepts into curricula to develop structural screening assessment tools (Drevdahl, 2018). On the other hand, the quashing fist of disability around children seems looser in the 21st century. The Life As A Disabled Child project (Watson et al., 2001), carried out at the universities of Edinburgh and Leeds, illustrates how ideas concerning ethnographic, reflexive and emancipatory research have been a successful critique of research with disabled children. Namely, it highlighted the previous common mistakes (J. Davis & Watson, 2017):

- Represent disabled children as a homogenous grouping (Shakespeare & Watson, 1998; Priestley, 1998)
- Over-emphasising the structural context of childhood disability and rarely discussing identity issues such as race and gender (Humphries & Gordon, 1992)
- A taken for granted perspective that disabled children would be unable to make choices and that they would require adults to make choices for them (J. M. Davis & Watson, 2001)

Besides the rising focus in the late stages of life, the burden of disease, impairment and disability is under the infotechnologies spotlight more than ever. On the Internet, anorexia has found a niche in websites like *Anagrrl* where a sustainable and proactive anorexic way of life is proposed.[1] A way paved by pills is it! Indeed, proactive anorexia grounds a drug's way of life, like the ones used by body-builders or the Viagra users that want to be the life of the orgy. Those TIC and pills anorexic rebels have subverted the biomedical and social model, but from an ethnographic point of view, pro-anorexia shows itself as a way of coping that rejects recovery "*as a simplistic solution to a symptom that leaves the underlying pain and hurt unresolved*" (Fox et al., 2005). In the same vein, ethical challenges in community-based participatory research (CBPR) are of increasing interest worldwide and extend to qualitative health research with communities (Wilson et al., 2018). Either in an online brokenhearted forum or within a researchers' community, the diseases are turning *3.0 speakable* (Lozano Rodríguez, 2018, pp. 73–75).

In contrast with the brand-new displacement to the digital plane, the places where disability ghettos have existed for a long time, like the ones inherited from the USSR to the postmodern Russia, will be mostly untouched due to the same constraining forces that shaped them once, but they have started to upload its geospatiality (Gaines, 2004). Indeed, the geospatial models can be useful to understand the dynamics of disability in the grey urban areas where the attention to persons with disabilities was centralised, that's the carpet below they were hidden, and it shows how the migration to the infosphere will reach even the *apparatchiks*' geographies of disability. Likewise, the inclusion of persons with disability in developing countries will keep being frustrating for a long time because the disability across cultures features discrimination: every day it is more accepted that disabled persons can contribute to society, but too often development plans bypass their interests (Turmusani, 2018). Technology is making it easier to understand and address disabled people's needs but also is helping to hide them away in certain areas. Bioethics of Displacement as love of belonging includes making sure that people with disabilities are included in society and not left behind in the digital age.

Junk food is going to be an American Empire the same that vodka was for the USSR (now Russia). One of the greatest displacements in the global burden of disease, impairment and disability before 2050 will happen in the US. In 2011, 54 million people in the USA were disabled and this figure will greatly rise in the next thirty years, as the baby-boom generation ages and many of today's children and young adults mature and experience complications related to overweight and obesity (Iezzoni, 2011). With this in mind, many states along the South Border are non-white majority (mostly Mexican people) and (besides Texas) they are below the US obesity rate (Desmond-Harris, 2015; The State of Obesity, 2018). Being a disempowered identity since the 1940s, the US Mexican, finally empowered, will have a motivation, a set of common values and a self-interest in contesting the monolithic US views on disease, impairment and disability in the very heart of private health insurances, which would have a worldwide echo. As more Americans will become disabled due to junk food-related illnesses, more motivated Mexican Americans will challenge US views on health insurance and advocate for themselves -and it is belonging, too.

Very likely the children in developed countries will be less burdened and the US baby-boomers and healthcare workers more, but the burden of disease, impairment and disability won't be lifted from their current wielders during the next generation. However, the global burden increasingly will be understood, lived and discussed in new digital ways rather than the deprecated vulnerability labels. Thus, the floor for bioethics discussion, especially in public health, will keep wide-open with old and new issues.

From a financial bioethics perspective, every social entrepreneur should ask herself those questions in order to prevent vulnerable persons (whether they are or not their current targets) from getting displaced from her social impact. On her side, the social entrepreneur faces some different ethical problems that may extend to other businesses (Lea, 1999). They can be sum up as follows:

- What ethical principle should I follow?
- How can I fulfil this ethical principle?
- How can fulfilling my duty become more meaningful?

Also it is important to avoid the *animalization* of target's thinking as happens with the contemporary neuromarketing trends (Nemorin, 2017). Let's now see two examples of Bioethics of Displacement related with finances. Firstly, I've found the case of "*agroecologies of displacement*" in which the farmers are shifting to a boundless agriculture more suitable for a climate of dispossession and crony capitalist land-grabbing (Glowa et al., 2019). The bright side comes from the employment fixation imposed by the global economy. Thanks to the postmodern reinsertion idea, many asylum patients (formerly tagged as disruptive workers or simply sick) can hold a grasp of agency since they are labelled now as potential workers (Parr, 2011). Let's see a different example. In the face of displacement caused by financial interests, a bioethics of love and belonging can offer solutions that prioritise solidarity and justice. One example of this is the community-led movement to protect the Standing Rock Sioux Tribe's sacred land and water from the Dakota Access Pipeline.[2] By standing together and asserting their right to their ancestral homeland, the tribe exemplified the power of belonging and the importance of protecting vulnerable communities from displacement. The protests led by Standing Rock Sioux Tribes about their sacred land and water from the Dakota Access Pipeline opened new economic opportunities for the tribe. In the wake of the protests, the tribe has been able to leverage its newfound political capital to pursue economic development projects that are consistent with its values and traditions. For example, the tribe is developing a sustainable energy plan that includes wind and solar power, as well as exploring the potential of hemp cultivation (Zimmerman, 2023).[3] By focusing on economic development that respects the land and water, the Standing Rock Sioux Tribes are demonstrating how the principles of Bioethics of Displacement can lead to positive outcomes for both people and the environment. The representative Biden and the inclusive US supreme court have pushed the burden of leakage risks on the Missouri river down from oil firms' bureaus to the Rock Sioux Tribes.

It is important here to shed light not only on the evils of crony capitalism, but also on the dangers of left-wing populism. Both ideologies promote the idea that the end justifies the means, leading to the displacement of vulnerable communities and the destruction of the environment. However, there is a way out of this vicious cycle. By adopting a Bioethics of Displacement and putting collective freedom and love of belonging first, we can create a better future for ourselves and our planet. As we have seen, the burden always flows downstream, from the powerful to the powerless, from the haves to the have-nots. But it doesn't have to be that way. By standing together and asserting our rights, we can create a world that works for everyone, not just for the privileged few. So let us embrace the principles of Bioethics of Displacement, and let us work together to build a better future for ourselves and for generations to come.

INTERSECTIONALITY MESSING EVERYWHERE

The theory of intersectionality holds that an individual or group is discriminated against by a given system of overlapping social categories, for example, a disabled black woman is not discriminated serially as disabled, black and woman but suffers from an emerging particular set of social hindrances instead. There are four advantages when using intersectionality as a research paradigm: simultaneity, complexity, irreducibility, and inclusivity (Carastathis, 2014). That's according the feminist theory as below:

However, this bagel is neither more true nor less indigestible than the 2030 Goals one. Let's follow the Black women example. Following the intersection properties, intersectionality only would happen if oppression is contingent for both Black and Women, that's if I can't conceive a Black person who is not oppressed and a woman who is not oppressed. Otherwise, those groups (Black and women) wouldn't inherit "oppression" and, therefore, a different degree of oppression never will arise. Actually, it's a blatant lie that Black women suffer any sort of aggravated oppression, at least in the US, the main exporter of all these bollocks. In a very different example that uses the same logic, I can't say that Whiteness in white rabbits makes them more vulnerable because not all rabbits live away from snowy forests and not all white fur animals are weaker samples of their respective species, for example, the polar bear is an all-white fearful apex predator among their own kind. Being wounded and undernourished for example, produces an intersectionality related to physical fitness since the wounds heal slower, if any, and the

Figure 1. An Intersectionality wheel diagram
Source: Talwar (2018)

chances to get food reduces as long as wounds are present. Thus, a wounded AND undernourished animal has far more chances to die than a wounded OR undernourished animal.

This intersectionality critique could be replied, resorting to fuzzy logic, that it's not needed till the last Black person and till the last woman gets oppressed. That there is enough with them having higher chances of being so. Indeed, a Black woman has more chances to meet the scorn of either a racist or a male chauvinist than she would be as a White man. But it leads to some issues: is it really so significant the number of racists who are not male chauvinists and the number of male chauvinists who are not racists? How to know if a White male is discriminating a non-White male because of race or gender instead of for being fat, old, ugly or dressing cheaply? Why does playing the group identity card only work so well for the staunchest White males? To skip the undecidability of prejudices, not to say of telling apart Whiteness and Blackness, the Woke has found two shortcuts: one is tagging issues as systemic and the other is tagging all they want to clutter as constructs. For example, racism is systemic, so Black people can't be racist. Of course, they can! Any person can have prejudices in their mind regarding skin colour along with many other things like age and body shape.

The toughest segregation, especially in the US, comes from the place you live, so much so that even if labelled Black people have raised their life expectancy, even in relation with labelled White people (partly thanks to suffering less homicides), Black neighbourhoods are more behind in life expectancy than Whites (Perry et al., 2021; Sanghi & Smaldone, 2022). That this residential segregation is followed from a history of miserable racial segregation and, therefore, is more relevant for non-WASP, it doesn't mean that the main problem today keeps being racism, but the lack of social mobility coming from real opportunities for the people born in a disfavored background. (And the same goes from colonialism and present-day corruption, by the way). It's true that being named Lakisha or Jamal makes you 50% less employable than Emily and Greg, but living in better served neighbourhoods makes you receive more callbacks and this effect does not differ by race (Bertrand & Mullainathan, 2003). On the other hand, being obese makes you face a similar handicap in employability (Bevan, 2019). Likewise, each inch above average height may be worth you $789 more per year, in turn (Judge & Cable, 2004). Even if the lack of Black employers and hirers and racial discrimination in hiring (and firing) is real, a shorty fatso White has less opportunities than a Black hunky when coming from the same neighbourhood. Danny DeVito Life Matters!

There is also the other version of the problem: White having more chances of getting only-White privileges, exclusive rights that Black can't obtain due to their blackness. Even in this case, these higher chances must be an inherent property of the collectives allegedly privileged. It leads to two obstacles. The first is how those exclusive privileges overlap and the second one is undecidability. If we take the cis-hetero-patriarchal ternary mantra, there will be an oppression pyramid with only eight echelons. So, according to intersectional feminism, trans-lesbian-women would be at the apex, suffering a living hell, and common jacks at the bottom surrounded by invisible privileges. (You can bet on that invisibility!). However, if we start to add most signalled privileges such as: White, Anglosaxon, not physically handicapped, not mentally handicapped, young, normative beauty bearer and living in a safe neighbourhood, then we will have a pyramid with over one million echelons. In comparison, the highest pyramid in Gizeh, Keops, has only 203 echelons. And this considering only mutually exclusive traits such as homosexual and heterosexual, but if we consider grey zones such a bisexual, asexual, etc. this one million echelons pyramid enlarges to a vanilla one with more than 635 millions, almost double the US population. Just by adding a few more privileges and oppressions, Twitter's bread and butter, such as: lacking gender neutral bathrooms or pronouns badges, having two parents and taking a knee, among many others, we'll have more categories than people in the world. This is how intersectionality implodes. If you have more categories than people there is no reason to have categories at all since, according to the Okkam's razor, it would be more true and easier considering individual cases only. Especially, when establishing VIP victims, as intersectionality does, is immoral. Implicit in this first obstacle is the second one. As you may have guessed, the list of privileges and oppressions can be very creative. Too creative. As far as there is no way to scientifically decide what is or not a privilege since it is the outcome of a complex and chaotic process involving psychology of masses and the success of fuzzy left-wing politics, categories are undecidable. Therefore, all this intersectionality jazz (critical race theory, critical gender theory, etc.) is plainly pseudoscience.[4]

Does it mean that a healthy idea of intersectionality cannot be conceived? No. For example, the movie *The Color Purple* (based on the Pulitzer Prize-winning 1982 novel of the same name by Alice Walker) describes a case in which the vulnerability of the protagonist is aggravated by the public disdain of Black women in early 20th century Georgia (Spielberg, 1986; Walker, 1982). A menacing Negroes vending machine for many White, a comfort woman for many males. Another movie, *Precious* (based also in a novel, *Push*) depicts the protagonist's realistic intersectionality when living in a poverty pocket in a better way even (Daniels, 2009; Sapphire, 1996). However, Afghan present-day men who were cross-dressed, prostituted and raped as a child following the *bacha bazi* Afghan tradition, surely were in a horrible situation, just like the *Color Purple* and *Precious* protagonist were and totally lacking of support as adults (Glinski, 2020). It's a crime against White males that the US government could have stopped but has been silent about till finally was banned by the Taliban (Borile, 2019; Noman, 2016). Likely, the White House didn't give a damn because some *distinguished members* were in the loop and, very likely, because those children just like the Black girls represented in the *Color Purple* and *Precious* represented no political gain. Yes, it could be the case that in this or that US county, a White minority of big cats are chauvinists and pay Black women less for no other reason that being Black and being women and, as I said, intersectionality can be useful in such fringe scenarios but never should become an overgeneralization depicting cis-hetero-White-males as the seed of every evil in the world or even as uniformly advantaged over the rest of people.

Booming since 2012, social media is brimming with wanna-be Celie Harris but it's not present-day reality, even *Precious'* fictitious crimes and disgraces happened in 1987. It's true that even controlling for education, employment, demographics and some other characteristics there is a 5-10% gender-gap in earnings in STEM in the US (Society of Women Engineers, 2023). However, being more prone to accept extreme jobs that unbalance social and family life, including last minute overtime, is a better explanation than sexism about what gives the edge to men, more even in stressed schedules sectors (Bolotnyy & Emanuel, 2022). If we compare US childless women, they are richer than men (Smith, 2022). Likewise, in jobs where overtime and long-distance travels are rare such as social workers, teachers and counsellors, women are better paid than men (US Department of Labor, 2021). More even, when comparing the earnings ratios by race and ethnicity of people who have attained a university degree, Whites are consistently the worst paid and Black and Hispanics the best ones (US Department of Labor, 2022). And turning back to STEM, there is an about $500 gap between Asian and Black women, more than the gap between men and women, and nobody complains about it. Indeed, Asians are wealthier than Whites in the US and nobody talks about anti-White discrimination or Asian oppression. Considering the lame starting conditions of many Asian families, it breaks the narrative of unsurmountable White privileges. Just like Jews do. Reality is that Black women are not doubly oppressed: conditional on their parents' income, black women actually outperform white women in terms of individual earnings (Chetty et al., 2020).

Figure 2. Individual income ranks are computed within a child's cohort pooling across race and gender

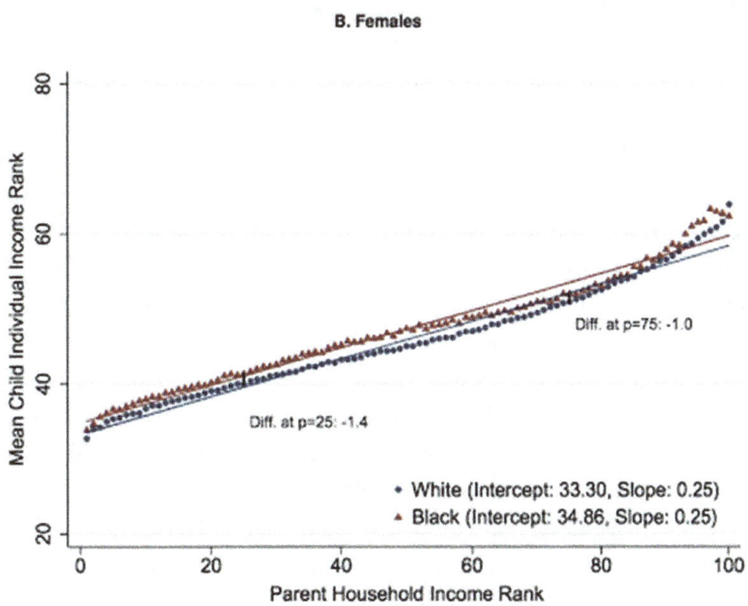

To cap it all, intersectionality really sheds no light on the indirect discrimination topic. Indirect discrimination happens when a regulatory practice rises inequity if applied equally amongst people. For example, a business moves its HQ to a country where Saturday is a working day. Despite that all the workers are affected equally, the Sabbath observant Jews are discriminated against. There is also a harder, structural indirect discrimination, for example, improving the labour conditions doesn't change the fact that the most part-time workers have a lesser status. The complexity of soft / hard indirect discrimination has led to confusions. In many studies, the results obtained are part of the diagnosis of discrimination instead of constituting the aim of indirect discrimination, therefore, they expose the obstacles rather than the outcome (Belavusau & Henrard, 2018, p. 93). On the other hand, if we tried to make a diagnosis of discrimination using intersectionality it hardly would be complete because of the number of collectives to consider, maybe half a dozen, maybe half a billion. And, in the previous examples, we have supposed only indirect discrimination when multiple factors are not involved. But it can be more difficult considering whether discrimination is intentional or unintentional, or whether it is the result of structural or individual biases. And why so much analysis? The result is always the same: the White man systemic evil doings.

Bioethics of Displacement as love of belonging opens better opportunities for change in the real world than intersectionality and indirect discrimination approaches. Instead of building a common enemy, it focuses on the importance of cultivating a sense of belonging and community. This approach acknowledges the complex social and historical factors that contribute to displacement and marginalisation, and emphasises the need for collective freedoms rather than political agendas.

Darryl Macer pointed out that no association between intersectionality and bioethical decision making has been found (Macer, 2019). Thanks to Bioethics of Displacement this bridge is not needed. Bodi approach uses meta-ethics to evaluate different approaches to bioethics and to clarify ethical arguments. By prioritising community-based solutions and recognizing the agency of those who have been forcibly displaced, this approach offers a more comprehensive understanding of the ethical challenges faced, for example, by refugee women. When using a mix of intersectionality and bioethics decision-making to address the experiences of forcibly displaced women, the focus is often on identifying and addressing various forms of oppression and discrimination. However, this approach may not fully capture the complexity of their experiences and may overlook the broader social and political context that led to displacement in the first place. The cultural dynamics that underlie these issues would get sunken down in cultural relativism.

In contrast, the Bioethics of Displacement approach recognizes the importance of addressing the root causes of displacement, such as political conflict, economic exploitation, and environmental degradation. It emphasises the need for community-based solutions that centre the voices and experiences of those who have been displaced. Moreover, intersectionalist approaches may sometimes pit different groups against each other in a competition for resources and recognition, which can ultimately undermine efforts to build a more just and equitable society.

Let's see an example of how intersectionality-fueled overgeneralization and identitarian labelling is making a fool of most people. For example, in January 2016, the US National LGBTQ Task Force announced that it had cancelled a post-Shabbat service reception at its annual Creating Change conference organised by the San Francisco-based nonprofit A Wider Bridge, which builds connections between LGBT communities in North America and Israel (Kirchick, 2016). Because of White cis-hetero bigots? No. Because of two (only two!) relevant public complaints: a trans prof and a non-White artistic duo. Non-White, non-cis bigots. Even if A Wider Bridge receives no funds from the State of Israel, collaboration was labelled as divisive because of intersectionality. Israelis are labelled as below in the scale of oppression than Arabs and A Wider Bridge guys were not in the mood this day of labelling themselves as Nazis and hitting their hearts while reciting *tajanunim*. It's how intersectionality compelled the US National LGBTQ Task Force to adopt such an insanely homophobic agenda as the Palestinian one against the only country in Middle East that not genocide homosexuals. Of course, Mizrahi Jews, who have overall darker skin than Arabs, don't exist in this story even if they are over the 40% of Israel population. No, if a Jew doesn't have a pale skin, German-like surname, goes to the synagogue in the Hamptons (NYC) and earns six-figures, wokism has decided that he doesn't qualify as a Jew. Namely, the US National LGBTQ Task Force didn't backpedal from its cancellation due to all this implicit racism, aporophobia and abuse of cultural relativism but due to political pressures coming from a different direction. And that's what gets on my nerves. I see intersectionality as commonly used as a lazy and pretentious try of getting a social Theory of Everything, but it only hiddens the most hideous banalization of true vulnerable people suffering, death and oblivion in favour of political carnage.

CONCLUSION

Many inequalities studied in Bodi are caused by crony capitalism and left-wing populism: by intersectionality talibans and holy-market crusaders. It's important not only solidarity but bottom-up governance for fair treatment of vulnerable populations. The downstream flow of burdens brings issues in long-term care and worse perspectives on disability. In turn, intersectionality promotes overgeneralization and divisive agendas. Instead of this, Bioethics of Displacement prioritises collective freedom and love of belonging to create a more inclusive and better future.

REFERENCES

Belavusau, U., & Henrard, K. (2018). *EU Anti-Discrimination Law beyond Gender*. Bloomsbury Publishing.

Bertrand, M., & Mullainathan, S. (2003). *Are Emily and Greg More Employable than Lakisha and Jamal? A Field Experiment on Labor Market Discrimination* (Working Paper No. 9873). National Bureau of Economic Research. doi:10.3386/w9873

Bevan, S. (2019, January 21). *Half of employers say they are less inclined to recruit obese candidates – it's not OK*. The Conversation. http://theconversation.com/half-of-employers-say-they-are-less-inclined-to-recruit-obese-candidates-its-not-ok-109821

Bolotnyy, V., & Emanuel, N. (2022). Why do women earn less than men? Evidence from bus and train operators. *Journal of Labor Economics*, *40*(2), 283–323. doi:10.1086/715835

Borile, S. (2019). Bacha Bazi: Cultural norms and violence against poor children in Afghanistan. *International Review of Sociology*, *29*(3), 498–507. doi:10.1080/03906701.2019.1672346

Brenan, M. (2018, May 31). *Americans' Strong Support for Euthanasia Persists*. Gallup.Com. https://news.gallup.com/poll/235145/americans-strong-support-euthanasia-persists.aspx

Carastathis, A. (2014). The Concept of Intersectionality in Feminist Theory. *Philosophy Compass*, *9*(5), 304–314. doi:10.1111/phc3.12129

Carel, H. (2016). *Illness: The Cry of the Flesh*. Routledge. doi:10.4324/9781315487410

Chetty, R., Hendren, N., Jones, M. R., & Porter, S. R. (2020). Race and economic opportunity in the United States: An intergenerational perspective. *The Quarterly Journal of Economics*, *135*(2), 711–783. doi:10.1093/qje/qjz042

Crane, S. (1972). *The Complete Poems of Stephen Crane*. Cornell University Press.

Damien, P. (2008). *Help! Debunking the Outrageous Claims of Self-help Gurus*. Synergy Books.

Daniels, L. (Director). (2009, November 20). *Precious* [Drama]. Lionsgate, Lee Daniels Entertainment, Smokewood Entertainment Group.

Davis, J., & Watson, N. (2017). Disabled children, ethnography and unspoken understandings: The collaborative construction of diverse identities. In Research with Children. doi:10.4324/9781315657349-13

Davis, J. M., & Watson, N. (2001). Where are the children's experiences? Analysing social and cultural exclusion in'special'and'mainstream'schools. *Disability & Society*, *16*(5), 671–687. doi:10.1080/09687590120070060

Desmond-Harris, J. (2015, February 26). *Here's when you can expect racial minorities to be the majority in each state*. Vox. https://www.vox.com/2015/2/26/8109947/minority-majority-states-race

Dobken, J. H. (2018). Physician-Assisted Suicide (PAS)/Physician-Assisted Death (PAD): The Rise of Lifeboat Ethics. *Journal of American Physicians and Surgeons*, *23*(4).

Drevdahl, D. J. (2018). Culture Shifts: From Cultural to Structural Theorizing in Nursing. *Nursing Research*, *67*(2), 146–160. doi:10.1097/NNR.0000000000000262 PMID:29489635

Fox, N., Ward, K., & O'Rourke, A. (2005). Pro-anorexia, weight-loss drugs and the internet: An 'anti-recovery' explanatory model of anorexia. *Sociology of Health & Illness*, *27*(7), 944–971. doi:10.1111/j.1467-9566.2005.00465.x PMID:16313524

Gaines, D. (2004). Geographical perspectives on disability: A socio-spatial analysis of the mentally disabled population in Russia. *Middle States Geographer*, *37*(1), 80–89.

Glinski, S. (2020, May 4). *As adults, Afghanistan's "dancing boys" lead double lives*. News.Trust.Org. https://news.trust.org/item/20200504031922-2nvol/

Glowa, K. M., Egerer, M., & Jones, V. (2019). Agroecologies of displacement: A study of land access, dislocation, and migration in relation to sustainable food production in the Beach Flats Community Garden. *Agroecology and Sustainable Food Systems*, *43*(1), 92–115. doi:10.1080/21683565.2018.1515143

Humphries, S., & Gordon, P. (1992). *Out of Sight: Experiences of disability 1900–50*. Northcote.

Hussein, S., & Manthorpe, J. (2014). Structural marginalisation among the long-term care workforce in England: Evidence from mixed-effect models of national pay data. *Ageing and Society*, *34*(1), 21–41. doi:10.1017/S0144686X12000785

Iezzoni, L. I. (2011). Eliminating Health And Health Care Disparities Among The Growing Population Of People With Disabilities. *Health Affairs*, *30*(10), 1947–1954. doi:10.1377/hlthaff.2011.0613 PMID:21976339

Judge, T. A., & Cable, D. M. (2004). The Effect of Physical Height on Workplace Success and Income: Preliminary Test of a Theoretical Model. *The Journal of Applied Psychology*, *89*(3), 428–441. doi:10.1037/0021-9010.89.3.428 PMID:15161403

Kirchick, J. (2016, January 21). How Intersectionality Makes You Stupid. *Tablet Magazine*. https://www.tabletmag.com/sections/news/articles/intersectionality-makes-you-stupid

Koch, T. (2012). *Thieves of Virtue: When Bioethics Stole Medicine*. MIT Press. doi:10.7551/mitpress/9079.001.0001

Kreidler, M. (2007, March 29). Secrets and Lies. *Skeptical Inquirer*. https://skepticalinquirer.org/exclusive/secrets_and_lies/

Lea, D. (1999). The Infelicities of Business Ethics in the Third World: The Melanesian Context. *Business Ethics Quarterly*, *9*(3), 421–438. doi:10.2307/3857510

Lozano Rodríguez, M. (2018). *Ethical Information Foresights: A journey across uncharted conflictology*. American University of Sovereign Nations.

Macer, D. R. J. (2019). Is There Cross-Cultural Evidence for an Association Between Intersectionality and Bioethical Decision Making? Not Yet, but Awaiting Advances in Mental Mapping. *The American Journal of Bioethics*, *19*(2), 34–36. doi:10.1080/15265161.2018.1558311 PMID:31543012

Nemorin, S. (2017). Neuromarketing and the "poor in world" consumer: How the animalization of thinking underpins contemporary market research discourses. *Consumption Markets & Culture*, *20*(1), 59–80. doi:10.1080/10253866.2016.1160897

Noman, A. (2016). The Tragedy of the Dancing Boys in Afghanistan: The US Silence on Bacha Bazi. *Seattle Journal for Social Justice*, *15*, 503.

Ostrowski, T. M., & Sikorska, I. (2015). *Health and Resilience*. Wydawnictwo UJ.

Parr, H. (2011). A brief spatial history of community and mental illness II: Community care. In *Mental Health and Social Space: Towards Inclusionary Geographies?* John Wiley & Sons.

Perry, A. M., Romer, C., & Barr, A. (2021, December 20). Why is life expectancy so low in Black neighborhoods? *Brookings*. https://www.brookings.edu/blog/the-avenue/2021/12/20/why-is-life-expectancy-so-low-in-black-neighborhoods/

Price, R. M. (2008). *Top Secret: The Truth Behind Today's Pop Mysticisms*. Prometheus Books.

Priestley, M. (1998). Childhood disability and disabled childhoods: Agendas for research. *Childhood*, *5*(2), 207–223. doi:10.1177/0907568298005002007

Sanghi, S., & Smaldone, A. (2022, January 27). *The Evolution of the Racial Gap in U.S. Life Expectancy*. https://www.stlouisfed.org/on-the-economy/2022/january/evolution-racial-gap-us-life-expectancy

Sapphire. (1996). *Push.* Vintage Contemporaries/Vintage Books.

Shakespeare, T., & Watson, N. (1998). Theoretical Perspectives on Research. *Growing Up With Disability, 34*, 13.

Silver, C., & Hyman, D. A. (2018). *Overcharged: Why Americans Pay Too Much for Health Care.* Cato Institute.

Smith, M. (2022, August 31). Women Who Stay Single and Don't Have Kids Are Getting Richer. *Bloomberg.Com.* https://www.bloomberg.com/news/articles/2022-08-31/women-not-having-kids-get-richer-than-men

Society of Women Engineers. (2023). *Earnings Gap.* https://swe.org/research/2023/earning-gap/

Spielberg, S. (Director). (1986, February 7). *The Color Purple* [Drama]. Warner Bros., Amblin Entertainment, The Guber-Peters Company.

Talwar, S. K. (2018). Identity Matters: Questioning Trauma and Violence through Art, Performance and Social Practice. In *Art Therapy for Social Justice: Radical Intersections.* Routledge. doi:10.4324/9781315694184-3

The State of Obesity. (2018, September). Adult Obesity in the United States. *The State of Obesity.* https://www.stateofobesity.org/adult-obesity/

Trachtenberg, A. J., & Manns, B. (2017). Cost analysis of medical assistance in dying in Canada. *Canadian Medical Association Journal, 189*(3), E101–E105. doi:10.1503/cmaj.160650 PMID:28246154

Turmusani, M. (2018). The Rise of Disability Politics. In *Disabled People and Economic Needs in the Developing World: A Political Perspective from Jordan.* Routledge., doi:10.4324/9781315257730

US Department of Labor. (2021). *Occupations with the smallest gender earnings gap.* https://www.dol.gov/agencies/wb/data/earnings/smallest-gender-earnings-gap

US Department of Labor. (2022). *Earnings ratios by race, ethnicity, and educational attainment (annual).* https://www.dol.gov/agencies/wb/data/earnings/earnings-ratio-race-ethnicity-education-annual

Walker, A. (1982). *The Color Purple.* Harcourt.

Watson, N., Shakespeare, T., Cunningham-Burley, S., Barnes, C., Corker, M., Davis, J., & Priestley, M. (2001). *Life as a disabled child: A qualitative study of young people's experiences and perspectives.* University of Edinburgh.

Wilson, E., Kenny, A., & Dickson-Swift, V. (2018). Ethical Challenges in Community-Based Participatory Research: A Scoping Review. *Qualitative Health Research, 28*(2), 189–199. doi:10.1177/1049732317690721 PMID:29235941

Zimmerman, Q. (2023, March 29). *Winds of Change: SWCA Assists Standing Rock Sioux Tribe with Environmental Services for Anpetu Wi Wind Farm.* SWCA. https://www.swca.com/news/2023/03/winds-of-change-swca-assists-standing-rock-sioux-tribe-with-environmental-services-for

KEY TERMS AND DEFINITIONS

Burden: The heavy load of responsibilities and negative impacts that weigh us down, including hardships and challenges in life.

Consent: The voluntary agreement to participate in an activity or treatment, including clear communication and respect for boundaries.

Equality: Human group property when everybody has the same rights, opportunities, and status, regardless of their unchosen differences or characteristics.

Inclusivity: The degree to which individuals feel a part of critical organisational processes, which can involve access to resources, decision-making, and so on.

Intersectionality: A critical race theory idea created by Kimberlé Crenshaw in 1989 aimed to explain how oppressed people are.

Justice: Treating everyone fairly, equally, and impartially, making sure they have what they need and deserve.

Marginalisation: The act of excluding or pushing someone or a group to the edges of society, denying them access to services and opportunities.

Oppression: Unfair treatment and abuse of power that restricts people's freedom, rights, and opportunities.

Structural Theories: Ways to understand society by looking at its systems, institutions, and how power influences people's behaviour.

ENDNOTES

1. The legacy of the original web is at: https://project-shapeshift.net/index.html
2. Link from the Smithsonian Institution: https://americanindian.si.edu/nk360/plains-treaties/dapl
3. Standing Rock Sioux Tribe's Industrial Hemp Production Plan https://www.ams.usda.gov/sites/default/files/media/StandingRockSiouxTribePlan.pdf
4. For a thorough and critical revision of feminism, divided in audiovisual little pills, I recommend YouTuber Xeno Shenlong work at: https://www.youtube.com/watch?v=FEftcMPvCi0&t=505s&ab_channel=Xeno (Spanish content)

Chapter 6
Bioethics of Displacement Across Time and Eras:
Genesis v Gehenna I

ABSTRACT

Displacement has a profound influence on both humans and other species across different periods of time. So, the author adopts a historical perspective to underscore the significance of considering the long-term consequences of our actions on future generations and the environment. The exploration extends beyond contemporary history and encompasses various epochs, commencing with the displacement triggered by the extinction of dinosaurs. It delves into the era of nomadic bowmen, the ancient Silk Road era, the era of transoceanic voyages, the era of new energies, and the era of data and computing. Each period brought about substantial transformations and alterations in power dynamics, emphasising the role of resource access, energy, and technology. Additionally, the author scrutinises the ascent and decline of empires, the impact of global debt, and the escalating significance of data in the modern world. By comprehending these historical patterns, we can gain valuable insights into the present global landscape and the forthcoming challenges we confront.

INTRODUCTION

Building on the principles of Bioethics of Displacement introduced in Chapter 5, this chapter takes a historical perspective to explore the concept of Bodi and its relevance to understanding displacement across time and eras. As discussed in the last chapter, the burden of displacement flows downstream, affecting both human and non-human species. In this chapter, we expand this perspective to include the potential long-term effects of our actions on future generations and the environment.

DOI: 10.4018/978-1-6684-4808-3.ch006

PRE-HUMAN BODI

A historical perspective in Bodi emphasises the need to consider not only the present consequences of our actions, but also the potential long-term effects they may have on future generations and the environment. As we have seen, resource extraction, pollution, and slavery by any name, have significant impacts on the natural world and can displace and disrupt the lives of both human and non-human species. But we should not limit ourselves to modern history and repeat again the same drag about transatlantic slavery and devilish factories, but go all-in with Bodi across time and eras.

We know that the beginning of Displacement is undetermined but very old, it is pre-human, just like bioethics is (Macer, 1994, pp. 2–4). One of the first species to migrate was a primitive shark that lived in rivers but swam down to the sea to breed and care for their young 300 million years ago (Braun, 2014; Sallan & Coates, 2014). Unlike in many cases today, in the beginning was love (at least lust) for life the main incentive for game changing. This happy shark mom with a pituitary on fire and surrounded by their babies after her otherworldly journey might be the first case of study for Bioethics of Displacement. However, for millions of years we'll only find animals somewhat locked on their scripted behaviours.

We could point to the dinosaurs' extinction process as, perhaps, the greatest displacement ever on our planet: a projectile above the quadrillion kilos impacting at 70.000 km/h, a blast that threw so much debris, sulphur and fire into the atmosphere that almost blocked the Sun light and splashed a 1.5 km wave high tsunami at hundreds km/h and, worse even, cooled the planet for decades (Gisler et al., 2004; Pope et al., 1998; Rodolfo, 2003). It was, definitively, what killed non-avian dinosaurs along with many other species. This 70.000 megatons impact changed Life on Earth forever and led to the rise of mammals' intelligence, but it's really difficult to put oneself in those scaly and feathered dino's shoes (I mean claws!) to tell if they felt displaced like us (Jay Gould, 1978, pp. 266–267). Likely there were some species with an intelligence similar to present-day ostriches or echidnas, even higher. However, there is no fossil registry of behavioural changes aiming to break the evolution game as later will happen with humans. No, among the Cretaceous fauna hardly we'll find talkative giant Angry Birds smarter than average Hollywood actors, as depicted in *sci-fi* (Ludwig & Faloyin, 2018).

NOMADIC BOWMEN ERA

The first creatures to invest an extraneous effort looking for a real life change in a non-instinctive way were some australopithecines and, foremost, the *Homo Ergaster* about 2 million years ago (Tattersall, 2007; Wood, 2011). In comparison, the first out of Africa *Homo Sapiens* adventures happened 180.000 years ago and the first mass migration dates 60.000 years back. The first migrations out of our African cradle were a blunder since the Neanderthal did a short work of us, or at least kept us at bay for more than 100.000 years, in present-day Israel (Groeneveld, 2017; Wolpoff & Lee, 2001). But we started to offer grave goods along with intentional burial just after meeting Neanderthals (Hublin, 2015; Pettitt, 2002; Phillips, 2011). That the oldest traces of our so bragged human dignity were, partly or totally, copied from non *homo sapiens* is a sobering lesson.

Human cognitive revolution happened 70.000 years ago and allowed us to solve more difficult problems and understand life, but also to become more inauthentic. Firstly, cognitive revolution's lies became cultural components for coexistence's sake, specially self-deception (Harari, 2015; Knight, 2010; Robotham, 2020). Homo sapiens was not the first species to lie, other primates lie like a rug on fire and lack our prosocial norms against fraud (Sun, 2023). But humans were the first beings to lie through stories. The first individuals to make a life of spreading crafted lies were humans.

Our mix of self-deception and afterlife beliefs to support life changes and the coldest-blood logic of power led to new meanings and findings that left Neanderthal obsolete. The ability to see our lives as stories that can be told and make-up allowed us to bear our personal divergences. As a result, our *homo sapiens* groups converged in a very primitive competitive and collaborative human web rather than stagnating in sparse tribes like the other humans (McNeill & McNeill, 2003, p. 5). The first arrowheads were found in Southern Africa around 60.000 years ago and only 7.000 years later, primitive human bowmen were so North as France and 6.000 years later as East as Sri Lanka (Langley et al., 2020; Lombard, 2005; Metz et al., 2023). It took 30.000 years for arrowheads to reach China; 40.000 years to reach Japan, America and North Europe from its origins (Altuntaş, 2022; Lorge, 2012, p. 16). This creative antagonism between the indifferent reality, plenty of microbes and nailed by natural hazards, and the human ability to tell themselves stories is the workbench of later History. We turned storytellers and bow wielders and reached the furthest parts of the planet -and decimated the Other humans.

After breaking into the old Neanderthal dominion, the humans kept leaving behind the unyielding severity of Sub-Saharan African geography and unquenchable fauna for more than 55.000 years. So much so that it wouldn't be crazy to suppose that, at some point between 15.000 and 10.000 years ago, Africa was for a few centuries the least populated continent, even less than Oceania. Geography and biology, at least, agreed that, after the last Ice Age, our species was more favoured between 25 and 45 parallel, where temperatures range between 1-35 ºC, humidity is average and water temperature is commonly above 18ºC in summer (Lavender, 2019; Raymond et al., 2020; Saycell et al., 2019).

Figure 1. Basic climate regions map based on the Köppen-Geiger classification system
Source: Retrieved from the University of Minnesota website[2]

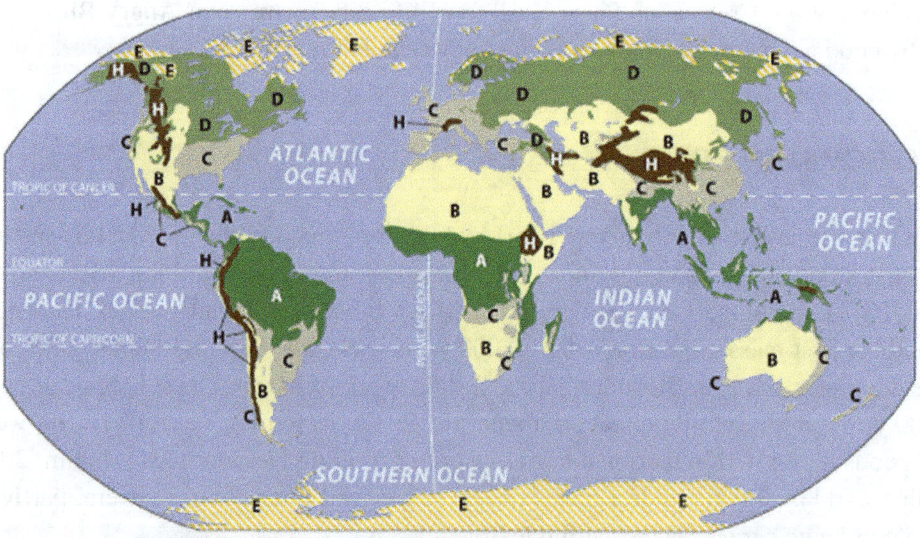

It's not by chance that the most relevant and perdurable civilizations had happened in C type weather. In every humble village was born an Alexandros and a Qin Shi Huang, but only a very few have thousands years of favourable climate backing them up.

Therefore, besides the Nile basin exception, humans spreaded from the Sinai strait to the fertile lands of Ukraine, Cyprus and the twice Fertile Crescent first and then, all across Eurasia and the world. The out of Africa groups who didn't incorporate the bow, just like the Neanderthal, Denisovan, Australian aboriginal or that incorporated the bow very late, e.g. the Inuit, never joined the former bow bearers' human web on their own terms -or didn't join at all.

Around 12.000 years ago, stories became more solid, Gobekli Tepe and Karahan Tepe sanctuaries were built in present-day Southern Turkey even before the pottery and husbandry. Karahan Tepe might be related with the bright star Deneb (α Cyg) in the Cygnus constellation and the Milky Way's Great Rift, while Gobekli Tepe, in turn, might be devore to praise the star Sirius, the brightest one across the European sky that, during this time, was first visible on Anatolia, or even a new solar cult (Collins, 2014; Magli, 2013). Anyhow, the strenuous effort of creating such structures with stone tools and the presence of modified human skulls without any nearby tomb witness the immense importance that those people gave to whatever they believed (Gresky et al., 2017). Along the Atlantic façade, 6500 years ago, megalithic tombs started to appear. The buried people belong to a kindred society and each megalith were built by non-sedentary hunter-gatherers and occupied by people from the same extended family across different generations, so they were visited periodically to bury a loved one and honour their ancestors (Sánchez-Quinto et al., 2019). 1500 years later the megalithic wonder of Stonehenge even if Gobekli Tepe stones, built 5000 years before, were more elaborate and reached a higher maximum weight pointing to a stronger demography. Anyhow, knowledge of the skies mixed with after-life beliefs created a sense of belonging beyond your own kind that opened the floor to the first cities both in Atlantic Europe, Egypt, the Crescent Fertile and wherever it happened. A feeling easy to exploit that soon got twisted.

Over time, as human groups flocked in larger numbers, they also tended to become less egalitarian and more hierarchical and human sacrifices became the way to go for stabilising societies as they became more stratified, by legitimating class distinctions and political authority (Watts et al., 2016). It is probably no coincidence that the victims of human sacrifices were often people who posed a threat to the elites, or who had fallen out of favour with them (Spinney, 2018). Yes, Stone Age USSR.

Likely, human sacrifices are older than Gobekli Tepe and the first city, Çatalhöyük, but in the most primitive societies, it was a sort of sorcery intended to deny the reality of death and incorporated sometimes ritual cannibalism, for example, ingesting the brain (Skinner, 1961). Even in early 2000s ritual cannibalism was practised in Liberia as a form of getting spiritual protection (Ellis, 1995; Thomas, 2017). However, in the earliest pyramids and Summerian tombs, 5000 years ago, we find a different human sacrifice, not motivated by survivalism or group leadership, but humans as burial goods, humans as things (Baadsgaard et al., 2011; Van Dijk, 2013). The self-domestication of Humankind reached a no-way back point. Judaism became the first monotheistic and transportable religion, the first to forbid human sacrifices, more than one thousand years before Republican Rome (Hasel, 1994; McNeill & McNeill, 2003, p. 61; Pliny the Elder, 1855, pp. 120–122). And, as History proves, Judaism is far better social glue than idolatry plus cutting a person's heart out. More even, Cyrus the Great, influenced by Zoroastrianism, engraved the first natural laws and human rights (including freedom of religion) that five centuries later will influence the Greco Roman world (Waters, 2022, p. vii). However, more than 3.000 years of bath-blooded bureaucracy marked the rest of human History. Since the first empires, the throne and altar got linked for millenia and the people out of the power circles became easily *sacrificable*.

After thousand years of roaming and bows, we splitted again. Now against our own species. This time the division would be between the sprouting urban world with organised and hierarchical religions and long-distance trade routes integrated into urban conundrums and domineering a hinterland versus the nomadic bowmen that spreaded humanity so far. One thousand years before Stonehenge, Sumeria in the Fertile Crescent pioneered civilization thanks to road and ship trade as never before (McNeill & McNeill, 2003, p. 48).

About 3600 years ago, about a century after the Summerian collapse, two clear poles of civilization existed: Ancient Egypt and Ancient China. It was not an Age of Bronze, first and foremost, it was the beginning of the Age of Taxes, which still endures. Both Tutankhamun tomb and the Yin dynasty leader Fu Shao's one were brimming with burial goods representing their power to be blindly obeyed and how they profited from it including, of course, human sacrifices (Ebrey, 1996, p. 26). Ancient rulers reserved for its own family the access to others' wealth tagging it as their own along with their very own lives. If classic states were going to be the new movement in the song of Humankind, durable (even intergenerational!) material profit became *the chorus director*. Leaders try to manage the emotions of others, not face them with harsh reality (Baños, 2020). Leaders solve, or at least soften, the nihilism and continuous conflict that preys a growing human group, even if some individual's welfare are sacrificed eventually. With the civilization and the raise of people bearing orders and the ability of doing evil by second hand, leaders become a new reality engine.

Ancient China and Egypt capitals, Thebes and Yin, were the home of a share of the human population similar to present-day London. But 1500-1350 BCE Egypt and China and even other regional hegemons before and then, such as Babylonia and the Harappan civilizations, were mostly ignorant of each other's importance even if having in common a disjointed set of spice trade routes. Out of China, Pharaonic Egypt fell before Alexandros just like Achaemenid Persian did. In turn, many Hellenic dominions and satrapies fell before the Romans. In China, many dynasties succeeded each other since the pyramids: Xia, Shang, Zhou and Qin. Back then, Rome was known by Chinese erudites as the mythical Daqin, a sort of upside down China located at the opposite end of their known world. In turn, Roman erudites speculated about the land of silk. Both were partly right.

ANCIENT SILK ROAD ERA

Even when cities were a demographic black hole periodically devastated by plagues, the human web that condensed in the Silk Road in 130 BCE became the way to spread human material progress, immune system updates and beliefs for sixteen centuries. In the opposing poles, Rome on the one hand and Xi'an, Hangzhou and Kaifeng on the other reached 1.000.000 habitants in the Silk Road dawn. Biggest cities' population before this first Eurasian axis of power, like Thebes and Babylon, had fitted into a crowded quarter in these megapolis of the classic age. The number of books and metal items, including swords and coins, rocketed. In the classic Western World, for the first time ever, Earth's circumference was known and the flow of time was measured by a geared clock. Even an insight on the atomic nature of matter and our position in the Solar System was stated. The first human glided through Cordoba's skyline and our present-day decimal system was used for the first time in India. In Imperial Song China, compass, gunpowder, papermaking, and printing appeared in this period to change Humanity forever. The Silk Road, just like it did with the Scythians, Huns and Turks, outlived Genghis Khan nomadic army, which proved itself so devastating on the Eurasian battlefields as lame culturally. Soon their leaders got assimilated into Chinese and Islamic cultures or packed to the Steppe again. Nomadic Peoples, regardless of how many bows and cannons used, were turned into the new Neanderthal.

It led to a rhythm in which the world will be immersed for around two millenia: re-spawning Chinese Empires at the Eastern pole and fleeting civilizations with Mediterranean dominions in the West. Western civs whose duration was measured by the Chinese beat for most of human History. For a few decades, during the Nerva-Antonine dynasty (2nd century CE), Rome was bigger than Han China. A similar thing happened during Justinian's Byzantine Empire and the brief Sui dynasty. Likewise, the Umayyad Caliphate (7-8th centuries) surpassed the influence of the early Tang Dynasty and started the Golden Age of Islam. But this caliphate was left behind, along with the Abbasids' one, when the library science and inventions during the Song Dynasty belittled the Grand Library of Baghdad, that was the greatest repository of books in the world (Algeriani & Mohadi, 2018; K. Chen, 1210; Neng-fu, 1991). The Southern Song Dynasty was so advanced that it could have started the industrial revolution in the 13th century (Edwards, 2013; Lin, 1995; Lozano Rodríguez, 2018, pp. 48–49). In my opinion, 13th century China has been one of the countries more ahead of its time, if not the most. Finally, the Ottoman Empire slowly withered the Silk Road in 1453 after the fall of Byzantine's capital Constantinople, the last Roman world living legacy.

The Roman Empire didn't last after a long climate adverse cycle, the end of Islamic Golden Age at the Mongol hordes' hands came from an increasing rain in the steppes that feeded more horses and riders than ever; even the most Chinese dynasties collapsed following volcanic activity (Gao et al., 2021; McCormick et al., 2012; Pederson et al., 2014). Gengis Khan genocide of Chinese Tangut belittles in comparison with the Black Death that, unwillingly, he might provoked while razing Tangut cities and later spreaded due to the rats and fleas parasiting his Hordes, a pandemic that started a reduced class wage gap in North Western Europe that will favour, more than four centuries later, the beginning of industrial revolution and further Anglo-Saxon global hegemony (Hymes, 2015; Pamuk, 2007). A similar thing goes for the Spaniards, Portuguese, Dutch and British that greatly benefited from a Silk Road hardened immune system during the early globalisation weakening natives populations wherever they went. If I tell you about dynasties and kings it is because of practical reasons, tagging time lapses, but the real drive of History never have been those gilded average Jacks, they only try to keep in power while the real leaders such as vulcans, dry spells, floods, zoonotic diseases, access to minerals and natural docks and, nowadays, solar storms, call the shots. Actually, the greatest degree of power that all historical leaders and dynasties referred to in this chapter have had is over the common people. For this reason they have used and abused it constantly, they need to exercise power in order the masquerade doesn't fall.

EARLY GLOBALISATION ERA

The early globalisation era started with transoceanic routes. It dragged people distant from the Silk Road world to a more and more extended human (and foremost geopolitical) web. The first of those travels were undertaken by 13th century Muslim merchants (acting as a sort of missionaries' too) at the end of the Islamic Golden Age. They joined the Red and Arabian Seas with the Pacific through the Indian Ocean and preached the Islam in lands far away from Eurasian coasts as Indonesia and Philippines. In the opposite direction, from Orient to Occident, Admiral Hen, at the early Ming Dynasty, led seven voyages (1405-1433) with a fleet unmatched by anything that had sailed before. It was the peak of a Chinese naval presence in the Indian Ocean that was yet formidable under the Song Dynasty in the 12th century (Ebai, 2021, p. 125). But it seems that size doesn't matter when it comes to discovering sea routes. The Portuguese Vasco de Gama circumnavigated Africa to reach the Indian subcontinent in 1498 with a fleet that, summed up, was lighter than Zheng He main ship. Only twenty-two years later, Portuguese reached the Moluccas Islands joining in their voyages three oceans. But this time there was no further voyage by Chinese admirals.

Rather than this, it was Castile, another kingdom in the Iberian Peninsula which got the jackpot with America's discovery -at least as America as a continent and idea to draw in a map. Only two years after the Portuguese reached the Moluccas Islands, the Magellan-Elcano expedition circumnavigated the world. Just like happened in the Ancient Silk Road Era, a Mediterranean power, the Spanish Empire, became more relevant in the global teather than a Chinese one, the Ming dynasty, even if its strength in numbers were superior. But the Spanish Empire was clearly overpowered in turn by the Qing Dynasty. Qing China had a share of world population similar to the demographic behemoths previous to the Silk Road such as the Achaemenid and Mauryan Empires, close to 40% and almost a third of the global economy (in comparison with only a 18% of European GDP by the Spanish Empire) (Y. Chen et al., 2021; Eckart, 2016). After the Qing golden age ended in 1799, the music of History diverged as never before during the last 2000 years.

During the Spanish and Portuguese early globalisation (16-18th centuries), empires in world regions traditionally away from the Silk Road, such as Africa and SouthEast Asia, became bigger than ever. It's the case of the Songhai and Toungoo Empires, two exotic giants out of the 25-45ºN range. The Mughal Empire, the most powerful empire below the parallel 25 and one of the wealthiest ever, also belongs to this era. Even the Aztec and Inca amerindian empires were at its peak, both in extension and dissent, just before the Spaniards. Those New World Empires were neither in the 25-45ºN and separated from the continuum formed by Eurasia and Africa for, at least, 5.000 years.

If the music of History would have been the same, the pendulum would have swayed to the Ottoman Empire as the new upside down of Qing's China. However, the best way to connect China and its Western upside down was not the Silk Road as usual. Furthermore, globalisation progressively turned the Atlantic Ocean into the new Mediterranean -and the Ottoman Empire had no access to it. With this in mind, the clear candidate was the Napoleonic Empire. But after half a century of industrialization, by 1815, over 2,000 miles of canals were in use in Britain, carrying thousands of tonnes of raw materials and manufactured goods; it was more than the Chinese Grand Canal network of water and taxes (Ji et al., 2018; White, 2009). France had nothing close to those infrastructures and by no means nobody close to Nathan Rothschild and his private intelligence service (Ferguson, 2008, pp. 73–86). Even if Rothschild family might play both sides, the most favoured by far was general Wellington, who could extend and replace their troops and supplies far more and more timely than any other army could do. While the British paid the war via a smart mix of taxation and loans, France still relied many times upon requisitions (Atkins, 1910; Bordo & White, 1991). Napoleon was less fit for the Age of Taxes than the British and a joke in big finances, so he was sent packing. It was the end of the Age of Taxes and started the Age of Global Debt. Why stop with grabbing a share of what people have today when you can grab a share of whatever anybody could have in the future even in places you never met?

NEW ENERGIES ERA

To make changes more radical even, the biggest present-day divergence debuted then: access to new energies. Till the 18th century there was a limit in the force and energy employed in a work, mostly human and animal ones. Of course, the force of the wind and the streams was used but rivers didn't flow and winds didn't arise at will.

All this changed with the carbon fueled steam engines. To begin with, it freed the population from doing a work machine that could be incorporated into the army and learn new skills and, more important even: new energies means new money and new outputs. Empires' strength were not mostly limited to Macchiavelian taxation over agrarian, trade and manpower outputs along with tributes but also to manage energy sources and profit from the boom of private companies sprouting from the incoming industrial revolutions. It led to a new set of empires not determined now by the favourable conditions of living between the 25-45ºN but by the easy access to high quality carbon: Netherlands, British, Belgian, United States, German and (to a lesser extent) French Empires. Only by geological chance, but all those empires' capitals but Washington are over 45ºN, which led to a wealthy North versus poor South dynamic that still endures. Shortly, just like nomads were turned into the new Neanderthals, the losers of industrialization were turned into the new heavily-indebted nomads. Vae Victis.

Among the industrialization early adopters, two empires had the edge: the British and the United States. The first one started industrial revolution decades ahead of the rest of the world and was a successful coloniser in the previous century; the United States, in tun, was able to join the two oceans across its natural geopolitical fortress -after becoming the top successful genocide ever and Hitler's musa. In the US, however, re-branding slavery was not so easy. Perhaps, many confederates felt that making Black and White the same would mean that all of them were going to become the New Whites' negroes. Just like happened with Napoleon, Confederates were less skilled raising taxes and unlucky with banking while the North had better infrastructures, specially thousands of railway kilometres. So Dixie was sent packing, too.

In less than 50 years after the victory of Napoleon, the British Empire trampled the once overpowering Mughal Empire and Qing Dynasty. It was a break with the traditional empires with Mediterranean dominions in the Silk Road time. However, besides the UK and the US, NorthWestern Europe as a whole was favoured in this new age of real pollution and imaginary money. Most European regions with a commercial tradition reached the industrialization train early and, to a lesser degree, Mediterranean Europe including the Ottoman Empire was also favoured.

Cunningly, furnace-powered colonialism switched slavery by coerced labour in order freedom couldn't be opposed to its interests, but with one hand offered abolition and with the other gave mass killing. Industrialised countries' needs multiplied when compared with non-industrialised ones, first iron and coal but then all types of things that the booming middle class could desire. After being the most protectionist countries in the world and benefiting from this, the US and UK were going to impose free-trade to countries that were not ready at gunpoint, even to the yet powerful Qing China (Guan, 1987; Shafaeddin, 1998). A scorch that still endures and explains, along with corruption, African poverty (Madeley, 2000). Before Africans, the global debt / poverty was going to be borne first by Iberoamerican peoples thanks to Bolivar, who ambitioned to be an entitled puppet with the help of the British and the US rulers (Perona Calvete, 2023). Bolivar was an entitled high-bureaucrat yet, so what was the point, then? More entitlement as well as racism and loath against their own kind. According to Bolivar, South America was going to have a hard time being a republic because: "... its population consist of but Indians and Negros who are more ignorant than the vile race of Spaniards we are just emancipated from. A country represented by such people must go to ruin" (Klooster, 2021, p. 49). He was useful to Anglosaxon interests and obtained a big profit, even if his genocidal plans were (thankfully) unfinished (Victoria Wilches, 2019). Anyhow, the poverty of South America is, mainly, because of him. Only in the first decade of the 21st century, Canadian mining corporations in Mexico grabbed more gold than the Spanish Empire in three centuries in all Iberoamerica without building roads, public buildings and granting security for both prosperous cities and traditional Indian communal lands -not even giving a fair price (Rodríguez Wallenius & Cruz Arenas, 2013, pp. 143–169). But nobody protests because the Iberoamerican leaders have followed Bolivar's path of servilism and homicidal scorn of their own most underserved subjects, more even than light beings like Justin Trudeau. And the same goes for some 20th century African countries where liberators turned into oppressors (Melber, 2016). They were the Others to bear the poverty that the UK and the US (and to a lesser extent the USSR) left behind. Quite the opposite, the countries who remained unsubordinated against UK and US economic exigences and found their own way to free trade (Japan, Korea, ...) prospered faster and sounder (Gullo, 2013).

Once the future of global poverty was externalised on the Other's shoulders the economy of the coal and carbon luckiest became all peaches and cream during the new energies era. But it was not going to last forever. As globalisation was closing all the blank spaces in the world map the bloodier industrialised land grabbing grew in order to ensure both strategic resources and a dynamic inner market. It was pioneered by the Dutch Company of East India as early as 1609-1621, when the Banda Islands natives were mostly exterminated in the first modern genocide to grant the species monopoly or during the 1740 Batavia massacre during which ethnic Chinese suffered a pogrom because they were prospering thanks to free market (Geerken, 2021, p. 13; Kemasang, 1985). This chauvinist fight against Others' prosperity bringing value to market became a constant during the industrial-driven colonisation. So frenzy the rush for having all that others grew that the 19th century was the century of corporate lobbies' genocide among coal and carbon luckiest. In 1770, the British East India Company were the architects of the Bengal famine that killed 10 million people (Bell, 2021; Rahman, 2020). Between 19 and 35 million Indians died of starvation under 200 years of British raj (1757-1947) while charitable activities and relief works were forbidden, for fear of interference with market laws (Davis, 2002; Monbiot, 2005; Paracka, 2018). In comparison, during three centuries of Mughal Empire's rule the famines' excess mortality was between 3 and 7 million even facing a little Ice Age (Damodaran et al., 2019). At the end of 19th century, Leopold II of Belgium killed over 10 million people in his Congolese private domains to hoard profits from red rubber (Friedman et al., 2017). The US genocide of Amerindians is difficult to calculate because figures are craftily mixed with the Spanish black legend and tags like "the European", "the colonizers" and counting the death toll always since 1492, but very likely the current global hegemon killed and sponsored the killing, neither by disease nor against other Amerindian empires but for the mere racial and cultural cleansing along with resource grabbing, more Amerindians and Mestizos than the British and Belgian together. Far more.

It would be untrue depicting industrialised European countries massacring non-White peoples as the monopoly of genocide. In 1804 the Afro-Haitian army genocided the Europeans living there. Along with the Queensland and Tasmanian aboriginals genocide, British also tried to wipe out pale White Boers in concentration camps to grab their lands in 1899-1902, 40 years ahead of Nazis. By 1862, Taranaki Maori completed one of the most successful genocides ever against another Islander ethnic group settled in Chatham Islands, the Moriori, including cannibalism and defecating in sacred Moriori lands (Gallant, 2003). Alexander II of Russia, the tsar that pioneered the Russian industrial revolution, killed around a million Circassian Muslim in 1864, a few years before Industry Minister Sergei Witte reforms (Laue, 1963). Lastly, till assassination in 1828, the Shaka Zulu led an ethnic cleansing that killed 1 to 2 millions Africans (Walter, 1969, pp. 111–112). He killed as much as the transatlantic Western slavery did in four centuries, but suspiciously not the same attention is paid. It's the same that happens when a Black kills another Black in the US media. Back to the topic, industrialization along with (fake) Enlightenment racial theories and Dutch Empire's corporate schemes maximised the effect of the rulers' lack of humanity, but it didn't put inside them anything that was not.

Billion people, even if not leaving their villages, became *displaced* as mostly WASP business and financial moguls played with their lives. For the first time in History, there was an ethnic background monopolising the wealthiest list. Furthermore, for the first time in History the wealthiest entities were not hegemonic crowned heads. At the time of his death in 1905 Nathaniel Rothschild's life savings were for sure higher than Queen Victoria's who died only four years later. This gap kept rising and, in 2023, Warren Buffet's Berkshire Hathaway assets are more than twenty times the British Crown's (Contreras & Murphy, 2023; Tognini & Porterfield, 2023). As a result, former lords of others' money tried to protect their old-privileges whatever their colour was and whatever it took switching the old alliance between the altar and the crown for a new one between the states and the banks being money the new heavenly mandate. Unsurprisingly, the global GDP output was displaced from East and South Asia to North Western Europe -quite the opposite than is happening nowadays.

Nevertheless, industrial countries' resources craving backfired at least two times, more violently every time, as the new industrial giants of the Soviet Empire and Imperial Japan appeared. It raised the land-grabbing competence along with Germany, lucky with coal and carbon but not so wealth in colonies. Shortly after the First World War, the British Empire, with 22 millions of km^2, became bigger than the Spanish and Qing empires together and the biggest empire ever, even if the United States was the most productive. The new energies' divergence keep widening thanks to the oil rigs and finally, the nuclear, which at the end of WW2 led to the ultimate empire: the United States of America. In a world where the other coal-lucky industrialists and Japan were razed, the US hoarded half of world production. Both the US' GDP and military projection of power multiplied the ones of the now shrinked British Empire and even the ones of the frozen and *fatso* Soviet Empire. Just like the British leaders turned the slavery into coerced labour, the Americans turned land-grabbing imperialism into a world of puppet states serving the US moguls' private interests. The Dutch and British East India Companies evolved into a myriad of less evident virulent companies spreading logos everywhere ruling over botched liberal democracies (Klein, 2001).

So successful was globalising externalities and privatise profits scheme that the global debt just before the 1914 Great War was around 1% of global GDP, rose to around 100% during the US peakest in VietNam and later degenerated further till the 349% of global GDP as in June 2022. In contrast, the first US Forbes list showed that the 30 wealthiest tycoons controlled 4.8% of the US economy, today it rises to 6.3%. Furthermore, considering that $50M in 1918 would be a billion today, in 1918 US the only billionaires would be the 30 ones in the first Forbes list, but in 2023 this list has grown to 800 members. Globally, the number rises to 2640 billionaires according to Forbes list. Staggering level of wealth transfer leading up to 2030, so inherited wealth concentrated among ultra-rich (Buchholz, 2022; X-Wealth, 2021). As Ed McCaffery explained, the public thinks the rich get away with paying no taxes because they can afford some exclusive financial scheme, but they aren't paying taxes because of perfectly legal reasons, which is not being prevented (McCaffery, 2016; Smagurauskaitė & Šarauskaitė, 2023). In spite of this constant x-wealthy demographic growth since Rothschild times, world famines only reduced after the

1970s, when vaccines and scientific agriculture became more affordable. There is no doubt, money is debt. Some get the money, others get indebted.

DATA AND COMPUTING ERA

But nothing lasts forever, sometimes things worsen. In 1956, the first commercial Hard Disk Drive ever was developed and shipped and a new movement in the song of History appeared and, progressively, the data turned into the new iron and coal. So much so that the Cloud's data explosion and China's rise happened in the 2010s. No other country has so many people connected to the Internet as China has, more than the US and India together. Unsurprisingly, China leads the data production, too (Radu, 2019). In addition to this, China, along with India and quite the opposite than the US, has one of the most competitive mobile prices plans in the world and a mobile speed and fixed broadband higher than the US one with Beijing among the 10 cities with the fastest internet connection.[3] In spite of this, apparently, the US is a clear leader since, in 2017, it had more broadband data consumption than China, all Europe, and Russia together. Furthermore, 39% of hyperscale data centres around the world are in the US, while China, Japan, UK, Germany and Australia only account for about 30% of the total (Vopson, 2021). Furthermore, the US investment in nanotechnology in the last 20 years dwarfs China's by about 30 to 1 (even if China in 2010 yet published more research papers than the US) and the number of US startups in nanotechnology are 836 against only 108 in China (Qiu, 2016; Subcommittee on Nanoscale Science, Engineering, and Technology, 2022, p. vii).[4,5] To put the US empire's power peak in perspective is important to consider it has been declining since the 1970s. Notwithstanding, remember that the sharpest phase of this decline, just like in Biden's brain, has happened since the 2010s: Chinese Tik Tok, MAGAs, Karens, intersectionality and BLM's wokism.

With the global hegemony of Washington and the Beijing boom, the 25-45ºN parallels are great again. Nobody knows the details of how Washington power is going to wither but the US government is closer to a civil war than preserving its global hegemony against China in 2030. The music keeps sounding and the metronome never stops. The US is a failed empire, built on a hypocritical narrative of fight against tyranny that, in an unsavoury poetic justice, is going to be smacked down by the biggest and most advanced form of authoritarianism ever. Just like the US did when it surpassed the British Empire industrial output and just like the British Empire when surpassed, in turn, the former Spanish Empire naval power, China's global hegemony will ground on what the US is paramount today: mass surveillance, AI and space race. And it doesn't rule out the partisan use of transhumanism and genetically tailored weapons… Orwell's *1984* is going to fall short.

FROM THE GENESIS OF BODI TO THE GEHENNOM OF POLITICS

Before following with the last section I want to rebuke historicism. Even if I have presented a very brief History of Bioethics of Displacement in a more or less rational way, it doesn't mean that we can predict History, even if we can make relatively accurate measures of future problems attending to readiness, sustainability and alike. History works more like if you are on a dancing floor, you may imagine how the song will continue because you have catched the rhythm but you don't know when the DJ is going to change the song. On the other hand, you know how patient the DJ is and how often songs change. Even if you are only guessing all the time you know that the DJ has grown impatient as eras go by. In this sense, Ray Dalio and the Spanish economist Niño Becerra warn about the incoming world cycle change, which includes Chinese hegemony and the shift away from liberal capitalism in favour of a different system yet to be known (Dalio, 2021; Niño-Becerra, 2022). My guess is that the 21st century world will see a similar paradigmatic crisis to 17th century Europe that led to modern Westphalian states (Lozano Rodríguez, 2023). In the period between 2030 and 2060 the music we have been dancing since the WW2 ending will fade away and the changes happening during this period may erase the way we are human or (quite improbable) the human species. Anyways, unless power distance shrinks a lot, civil society becomes paramount and collective freedoms become stronger than corruption everywhere, I wouldn't bet on our species as noteworthy on this planet beyond 2300.

History teaches us that there are two ways of being right. The first way is proving what you say, making it evident. When you can make predictions from your statement in a falsifiable way, then you are quite close to science. Primitive human bow shooting time after time led to technical progress that, in turn, led to new shooting capabilities (fire arrows, armour piercing, etc.). The second way of being right is to make others bear your wrongness, so you are *exclusively* right. Of course, this feature has also evolved through eras allowing to control in more and more sophisticated ways an increasing number of political NPCs. It started with the ancient kings, pharaohs and emperors who believed to be gods or god chosen (spoiler: they were not) and misled their subjects to believe that their stuff, efforts and lives were better employed in their royal tombs than serving to themselves. In exchange, many subjects got a slight chance of profiteering less organised peoples in turn, including lesser social echelons of their own kind. Today we have some kings and emperors but the ones calling the shots are the ones who own the countries' debt and data as well as other money masters such as bond rating agencies, central banks, IMF and the now classical nuclear-backed PMs. As in 2023, the biggest power conundrums are a loose set compounded by the US MIC along with data elites (from DARPA to Lockheed and Neuralink) and a more compact one centred in China's Party. As the Ukrainian War shows, as soon as Biden wanted more money, his European *viceroys* got their colonial subjects more indebted than ever. Just like in the Tutankhamun and Fu Shao times, the hard to earn stuff along with the precious blood, all will be devoted to a royal tomb. And regardless of how the US v China civilizational clash ends, none of their leaders will be wrong, not even the losers (if any), because really powerful people can't conceive a better world than this with them at the helm. Those authorities are quite the opposite than the bow artisan and the huntsman that constantly were hitted back by reality and were forced to remain psychologically flexible and resort to reason for improvement. In contrast, professional politicians must remain psychologically inflexible about themselves, about their deep righteousness, and over-rationalise the world. There is not enough with a life barking orders, because the life of those away from the social apex is less alive than a royal corpse. Even if randomly scattered resources and life-threatening natural phenomena (from dry spells to endogamy, rodents and pandemics) along with technical advances' bright and dark sides (from lead in fuels to vaccines and fertilisers), has shaped our history more than any ruler, the history of Humankind keeps stinking as elites. By elite I mean that minority who rarely feel how reality strikes back when choosing against the many or speaking on behalf of political or parapolitical subjects (nation, crown, god, oppressed, etc.). And the more insulated from the consequences of their actions the elites are, the more inauthentic they become. Tutankhamun's position might be compromised by an unheard flood, just like Joe Biden (no pun intended) tenure can reel before recession. However, Joe Biden's assets are far more safe than Tutankhamun's health, who died from malaria when he was 19. This insulation has always been key for psychological inflexibility at any pyramidal scheme of power.

History are the stories about the last stage of human self-domestication when atrocities arose alongside increasingly orchestrated cultures competing for land and resources while avoiding natural hazards. What makes the Human species a tragic one is the fact that we are lambs guided by lambs playing shepherds and butchers. The way of life of those in the shepherd and butcher roles always has been an upside down of the patient lambs' one. This upside down is the living politics of pain.

CONCLUSION

There is a connection between displacement and its long-term effects on the environment, future generations, and global power dynamics. We need to consider the consequences of our actions, not only in the present but also in the context of historical trends because power structures, inequality, and the exploitation of resources have shaped history and continue to do so. Reflecting on the lessons from the past and considering the ethical implications of Displacement can help to create a more sustainable future.

REFERENCES

Algeriani, A. M. A., & Mohadi, M. (2018). The Administrative and Technical Aspects in Baghdad and Cairo's House of Wisdom Libraries (a Comparative Analysis). *Arts and Social Sciences Journal*, *9*(5), 9. doi:10.4172/2151-6200.1000402

Altuntaş, L. (2022, December 24). Oldest known arrowheads uncovered in the Americas. *Arkeonews*. https://arkeonews.net/oldest-known-arrowheads-uncovered-in-the-americas/

Atkins, A. R. C. (1910). Requisitions in War. *Royal United Services Institution. Journal*, *54*(389), 837–860. doi:10.1080/03071841009423732

Baadsgaard, A., Monge, J., Cox, S., & Zettler, R. L. (2011). Human sacrifice and intentional corpse preservation in the Royal Cemetery of Ur. *Antiquity*, *85*(327), 27–42. doi:10.1017/S0003598X00067417

Baños, P. (2020). El dominio mental. Ariel.

Bell, K. (2021). Genocide or disaster: An investigation into the British Empire's role within the Great Bengal Famine of 1770. *Vulcan Historical Review*, 74.

Bordo, M. D., & White, E. N. (1991). A Tale of Two Currencies: British and French Finance During the Napoleonic Wars. *The Journal of Economic History*, *51*(2), 303–316. doi:10.1017/S002205070003895X

Braun, R. (2014, January 9). *Prehistoric sharks were earliest animals to migrate*. New Scientist. https://www.newscientist.com/article/dn24845-prehistoric-sharks-were-earliest-animals-to-migrate/

Buchholz, K. (2022, February 17). *Inherited Wealth Concentrates Among The Ultra-Rich* [Infographic]. Forbes. https://www.forbes.com/sites/katharinabuchholz/2022/02/17/inherited-wealth-concentrates-among-the-ultra-rich-infographic/

Chen, K. (1210, 1274). *Zhong xing guan ge lu.* [Compressed data]. Library of Congress. https://www.loc.gov/item/2021666333/

Chen, Y., Palma, N., & Ward, F. (2021). Reconstruction of the Spanish money supply, 1492–1810. *Explorations in Economic History*, *81*, 101401. doi:10.1016/j.eeh.2021.101401

Collins, A. (2014). *Karahan Tepe: Göbekli Tepe's Sister Site—Another Temple Of The Stars?* https://d1wqtxts1xzle7.cloudfront.net/34492368/Karahan_Tepe_double_column_with_pics_low-libre.pdf?1408528975=&response-content-disposition=inline%3B+filename%3DKarahan_Tepe_Gobekli_Tepes_Sister_Site_A.pdf&Expires=1684507578&Signature=dK2HL1N2Yin-N2QlDGX1KvhJI-cCbM9vCrppzHA5LjMBdYsdl5UcpM2zk9GWTQRgz7am03he-sQR8gtzAdcnsx5nr75MAjG7R~RW93xK68ZDvhAQo7QFjCK-Pf74IeIgCYM8AM9LThnhKlY1iheeMmGGEa4zE~UWAQ~qBFk-7i23OgYSnfYc2S2FnUXDAvE1wzIrTB4InGbWnBmwKYTN45hzBUZyVzX~4rPUBr2PVdHhfhodikxuMY-4hbHQJI-lC0MiVdvdu6JoYk383M83toX-LnHd~fhs4Hlpsep97ZkEXjCC6GOy~ertwu3DVQJktyoJGnZQhLLjZtLd0b5H-llQ__&Key-Pair-Id=APKAJLOHF5GGSLRBV4ZA

Contreras, I., & Murphy, A. (2023). *Forbes Global 2000 List 2022: The Top 200*. Forbes. https://www.forbes.com/sites/forbesstaff/2022/05/12/forbes-global-2000-list-2022-the-top-200/

Dalio, R. (2021). *Principles for Dealing with the Changing World Order: Why Nations Succeed and Fail* (1st ed.). Avid Reader Press / Simon & Schuster.

Damodaran, V., Hamilton, J., & Allan, R. (2019). Climate signals, environment and livelihoods in the long seventeenth century in India. In *A Cultural History of Famine* (pp. 52–70). Routledge. doi:10.4324/9781315316529-4

Davis, M. (2002). *Late Victorian Holocausts: El Niño Famines and the Making of the Third World*. Verso Books.

Ebai, P. N. (2021). *Nations Building*. Exceller Books.

Ebrey, P. B. (1996). *The Cambridge illustrated history of China*. Cambridge University Press., https://archive.org/details/cambridgeillustr00ebre_0

Eckart, J. (2016, June 23). Things you need to know about China's economy. *World Economic Forum*. http://gdelaurier.pbworks.com/w/file/fetch/130100685/8%20things%20you%20need%20to%20know%20about%20China%E2%80%99s%20economy%20_%20World%20Economic%20Forum.pdf

Edwards, R. A. (2013). Redefining Industrial Revolution: Song China and England. *2013 Meeting Papers*, Article 351. https://ideas.repec.org//p/red/sed013/351.html

Ellis, S. (1995). Liberia 1989-1994: A Study of Ethnic and Spiritual Violence. *African Affairs*, *94*(375), 165–197. doi:10.1093/oxfordjournals.afraf.a098806

Ferguson, N. (2008). *The Ascent of Money: A Financial History of the World: 10th Anniversary Edition*. Penguin.

Friedman, H. H., Friedman, L. W., & Adel, S. (2017). *Conscious Capitalism vs. Rapacious Capitalism: Lessons from King Leopold II* (*SSRN* Scholarly Paper No. 2988563). https://papers.ssrn.com/abstract=2988563

Gallant, P. (2003, April 11). A Moriori Lesson. *National Review*. https://www.nationalreview.com/2003/04/moriori-lesson-paul-gallant/

Gao, C., Ludlow, F., Matthews, J. A., Stine, A. R., Robock, A., Pan, Y., Breen, R., Nolan, B., & Sigl, M. (2021). Volcanic climate impacts can act as ultimate and proximate causes of Chinese dynastic collapse. *Communications Earth & Environment*, *2*(1), 1. Advance online publication. doi:10.103843247-021-00284-7

Geerken, H. H. (2021). *The Gold of the Bandas: The History of the Nutmeg: The forgotten islands that once made world history*. BoD – Books on Demand.

Gisler, G. R., Weaver, R. P., Mader, C. L., & Gittings, M. L. (2004). Two- and three-dimensional asteroid impact simulations. *Computing in Science & Engineering*, *6*(3), 46–55. doi:10.1109/MCISE.2004.1289308

Gresky, J., Haelm, J., & Clare, L. (2017). Modified human crania from Göbekli Tepe provide evidence for a new form of Neolithic skull cult. *Science Advances*, *3*(6), e1700564. doi:10.1126ciadv.1700564 PMID:28782013

Groeneveld, E. (2017, May 15). Early Human Migration. *World History Encyclopedia*. https://www.worldhistory.org/article/1070/early-human-migration/

Guan, S. (1987). Chartism and the first opium war. *History Workshop Journal*, *24*(1), 17–31. doi:10.1093/hwj/24.1.17

Gullo, M. (2013). *Insubordinacion fundante*. Editorial Biblos.

Harari, Y. N. (2015). *Sapiens: A Brief History of Humankind*. Harper Collins.

Hasel, M. G. (1994). Israel in the Merneptah Stela. *Bulletin of the American Schools of Oriental Research. American Schools of Oriental Research*, *296*, 45–61. doi:10.2307/1357179

Hublin, J.-J. (2015). The modern human colonization of western Eurasia: When and where? *Quaternary Science Reviews*, *118*, 194–210. doi:10.1016/j.quascirev.2014.08.011

Hymes, R. (2015). Epilogue: A Hypothesis on the East Asian Beginnings of the Yersinia pestis Polytomy. *The Medieval Globe*, *1*(1), 285–308. doi:10.17302/TMG.1-1.11

Jay Gould, S. (1978). Were Dinosaurs Dumb? In New Scientist. Reed Business Information.

Ji, S., Choi, Y., Lee, C.-K., & Mjelde, J. W. (2018). Comparing willingness-to-pay between residents and non-residents using a contingent valuation method: Case of the Grand Canal in China. *Asia Pacific Journal of Tourism Research*, *23*(1), 79–91. doi:10.1080/10941665.2017.1399919

Kemasang, A. R. T. (1985). How Dutch Colonialism Foreclosed a Domestic Bourgeoisie in Java: The 1740 Chinese Massacres Reappraised. *Review - Fernand Braudel Center*, *9*(1), 57–80.

Klein, N. (2001). *No Logo: No Space, No Choice, No Jobs*. Flamingo.

Klooster, W. (2021). *Spanish American Independence Movements: A History in Documents: (From the Broadview Sources Series)*. Broadview Press.

Knight, C. (2010). Language, Lies and Lipstick: A Speculative Reconstruction of the African Middle Stone Age "Human Revolution.". In P. M. Kappeler & J. Silk (Eds.), *Mind the Gap: Tracing the Origins of Human Universals* (pp. 299–313). Springer. doi:10.1007/978-3-642-02725-3_14

Langley, M. C., Amano, N., Wedage, O., Deraniyagala, S., Pathmalal, M. M., Perera, N., Boivin, N., Petraglia, M. D., & Roberts, P. (2020). Bows and arrows and complex symbolic displays 48,000 years ago in the South Asian tropics. *Science Advances, 6*(24), eaba3831. Advance online publication. doi:10.1126ciadv.aba3831 PMID:32582854

Laue, T. V. (1963). *Sergei Witte and the Industrialization of Russia.* Atheneum.

Lavender, N. (2019, January 12). *This is the point when cold weather becomes dangerous.* Ladders | Business News & Career Advice. https://www.theladders.com:443/career-advice/this-is-the-point-when-cold-weather-becomes-dangerous

Lin, J. Y. (1995). Why the Industrial Revolution Did Not Originate in China. *Economic Development and Cultural Change, 43*(2), 269–292. doi:10.1086/452150

Lombard, M. (2005). Evidence of hunting and hafting during the Middle Stone Age at Sibidu Cave, KwaZulu-Natal, South Africa: A multianalytical approach. *Journal of Human Evolution, 48*(3), 279–300. doi:10.1016/j.jhevol.2004.11.006 PMID:15737394

Lorge, P. A. (2012). *Chinese Martial Arts: From Antiquity to the Twenty-First Century.* Cambridge University Press.

Lozano Rodríguez, M. (2018). *Ethical Information Foresights: A journey across uncharted conflictology.* American University of Sovereign Nations.

Lozano Rodríguez, M. (2023, January 25). How COVID-19 is Changing Corruption Forever: The Crisis of the State-led Crisis Management and Why You Are Going to Pay for It. In *Conference on Disaster Management: Global Pandemic and Its Impact.* Binghamton University - State University of New York.

Ludwig, L., & Faloyin, D. (2018, June 20). We Asked a Dinosaur Expert How Realistic the New "Jurassic World" Really Is. *Vice.* https://www.vice.com/en_us/article/d3kkzy/we-asked-a-dinosaur-expert-how-realistic-the-new-jurassic-world-really-is

Macer, D. R. J. (1994). *Bioethics for the people by the people.* Eubios Ethics Institute.

Madeley, J. (2000). *Hungry for Trade: How the Poor Pay for Free Trade.* Zed Books.

Magli, G. (2013). *Sirius and the project of the megalithic enclosures at Gobekli Tepe.* https://doi.org//arXiv.1307.8397 doi:10.48550

McCaffery, E. J. (2016). Taking wealth seriously. *Tax L. Rev., 70*, 305.

McCormick, M., Büntgen, U., Cane, M. A., Cook, E. R., Harper, K., Huybers, P., Litt, T., Manning, S. W., Mayewski, P. A., More, A. F. M., Nicolussi, K., & Tegel, W. (2012). Climate Change during and after the Roman Empire: Reconstructing the Past from Scientific and Historical Evidence. *The Journal of Interdisciplinary History, 43*(2), 169–220. doi:10.1162/JINH_a_00379

McNeill, J. R., & McNeill, W. H. (2003). *The human web: A bird's-eye view of world history.* W.W. Norton & Company. https://archive.org/details/humanwebbirdseye00jrmc

Melber, H. (2016, April 11). *How liberators turn into oppressors: A study of southern African states*. The Conversation. http://theconversation.com/how-liberators-turn-into-oppressors-a-study-of-southern-african-states-57213

Metz, L., Lewis, J. E., & Slimak, L. (2023). Bow-and-arrow, technology of the first modern humans in Europe 54,000 years ago at Mandrin, France. *Science Advances*, *9*(8), eadd4675. Advance online publication. doi:10.1126ciadv.add4675 PMID:36812314

Monbiot, G. (2005, December 27). *How Britain Denies its Holocausts*. George Monbiot. https://www.monbiot.com/2005/12/27/how-britain-denies-its-holocausts/

Neng-fu, K. (1991). Chinese Library Science in the Twelfth Century. *Libraries & Culture*, *26*(2), 357–371.

Niño-Becerra, S. (2022). *Futuro, ¿qué futuro? Claves para sobrevivir más allá de la pandemia*. Academic Press.

Pamuk, Ş. (2007). The Black Death and the origins of the 'Great Divergence' across Europe, 1300–1600. *European Review of Economic History*, *11*(3), 289–317. doi:10.1017/S1361491607002031

Paracka, D. (2018). Year of India: Introduction to the Special Issue. *JGI*, *12*(1), 1–14.

Pederson, N., Hessl, A. E., Baatarbileg, N., Anchukaitis, K. J., & Di Cosmo, N. (2014). Pluvials, droughts, the Mongol Empire, and modern Mongolia. *Proceedings of the National Academy of Sciences of the United States of America*, *111*(12), 4375–4379. doi:10.1073/pnas.1318677111 PMID:24616521

Perona Calvete, C. (2023, January 7). *The Bolivar Legacy Part II: Debt and Regret*. https://european-conservative.com/articles/essay/the-bolivar-legacy-part-ii-debt-and-regret/

Pettitt, P. B. (2002). The Neanderthal dead: Exploring mortuary variability in Middle Palaeolithic Eurasia. *Before Farming*, *1*(4), 1–19. doi:10.3828/bfarm.2002.1.4

Phillips, J. C. (2011). *Symbolism, complexity and intentional burial among the Neanderthals*. Academic Press.

Pliny the Elder. (1855). *The Natural History of Pliny*. H. G. Bohn.

Pope, K. O., D'Hondt, S. L., & Marshall, C. R. (1998). Meteorite impact and the mass extinction of species at the Cretaceous/Tertiary boundary. *Proceedings of the National Academy of Sciences of the United States of America*, *95*(19), 11028–11029. doi:10.1073/pnas.95.19.11028 PMID:9736679

Qiu, J. (2016). Nanotechnology development in China: Challenges and opportunities. *National Science Review*, *3*(1), 148–152. doi:10.1093/nsr/nww007

Radu, S. (2019, February 14). *China Overtook the U.S. and Will Hold the Largest Share of World's Data at Least by 2025*. https://www.usnews.com/news/best-countries/articles/2019-02-14/china-overtook-the-us-and-will-hold-the-largest-share-of-worlds-data-at-least-by-2025

Rahman, A. U. (2020). *Reasons behind the Great Bengal Famine in 1770: British claim vs Reality*. Academic Press.

Raymond, C., Matthews, T., & Horton, R. M. (2020). The emergence of heat and humidity too severe for human tolerance. *Science Advances, 6*(19). doi:10.1126/sciadv.aaw1838

Robotham, M. (2020, July 30). Lying Is Fundamental to Who We Are as Humans. *CrimeReads*. https://crimereads.com/lying-is-fundamental-to-who-we-are-as-humans/

Rodolfo, K. S. (2003). *Tsunamites and seismites*. Access Science.

Rodríguez Wallenius, C. A., & Cruz Arenas, R. A. (2013). *El México bárbaro del siglo XXI*. Universidad Autónoma Metropolitana.

Sallan, L. C., & Coates, M. I. (2014). The long-rostrumed elasmobranch Bandringa Zangerl, 1969, and taphonomy within a Carboniferous shark nursery. *Journal of Vertebrate Paleontology, 34*(1), 22–33. doi:10.1080/02724634.2013.782875

Sánchez-Quinto, F., Malmström, H., Fraser, M., Girdland-Flink, L., Svensson, E. M., Simões, L. G., George, R., Hollfelder, N., Burenhult, G., Noble, G., Britton, K., Talamo, S., Curtis, N., Brzobohata, H., Sumberova, R., Götherström, A., Storå, J., & Jakobsson, M. (2019). Megalithic tombs in western and northern Neolithic Europe were linked to a kindred society. *Proceedings of the National Academy of Sciences of the United States of America, 116*(19), 9469–9474. doi:10.1073/pnas.1818037116 PMID:30988179

Saycell, J., Lomax, M., Massey, H., & Tipton, M. (2019). How cold is too cold? Establishing the minimum water temperature limits for marathon swim racing. *British Journal of Sports Medicine, 53*(17), 1078–1084. doi:10.1136/bjsports-2018-099978 PMID:30824457

Shafaeddin, M. (1998). How did developed countries industrialize. In *The History of Trade and Industrial Policy: The Cases of Great Britain and the USA*. Academic Press.

Skinner, J. (1961). Ritual Matricide: A Study of the Origins of Sacrifice. *The American Imago, 18*(1), 71–102.

Smagurauskaitė, A., & Šarauskaitė, G. (2023). Buy, borrow, die: How rich live off their paper wealth and avoid paying taxes and why we let them do it? *Vilnius University Open Series*, 6–14.

Spinney, L. (2018, February 27). *Did Human Sacrifice Help People Form Complex Societies?* The Atlantic. https://www.theatlantic.com/health/archive/2018/02/did-human-sacrifice-help-people-form-complex-societies/554327/

Subcommittee on Nanoscale Science, Engineering, and Technology. (2022). *The national nanotechnology initiative supplement to the president's 2022 budget*. National Science and Technology Council.

Sun, L. (2023, April 8). *What Animals Can Teach Us About Truth and Lies*. Time. https://time.com/6269499/animals-truth-lies-lessons/

Tattersall, I. (2007). Homo ergaster and its contemporaries. Handbook of Paleoanthropology, 3, 1633–1653.

Thomas, B. (2017, April 20). *Eating People Is Wrong—But It's Also Widespread and Sacred*. Sapiens. https://www.sapiens.org/biology/cannibalism-ritualized-sacred/

Tognini, G., & Porterfield, C. (2023, May 4). *How Rich Is King Charles III? Inside The New Monarch's Outrageous Fortune*. Forbes. https://www.forbes.com/sites/giacomotognini/2023/05/04/how-rich-is-king-charles-iii-inside-the-new-monarchs-outrageous-fortune/

Van Dijk, R. M. (2013). The use of bucrania in the architecture of First Dynasty Egypt. *Journal for Semitics*, *22*(2), 449–463. doi:10.10520/EJC147422

Victoria Wilches, P. (2019). *El terror bolivariano: Guerra y genocidio contra España durante la independencia de Colombia y Venezuela en el siglo XIX*. Academic Press.

Vopson, M. M. (2021, May 4). *The world's data explained: How much we're producing and where it's all stored*. The Conversation. http://theconversation.com/the-worlds-data-explained-how-much-were-producing-and-where-its-all-stored-159964

Walter, E. V. (1969). *Terror and resistance; a study of political violence, with case studies of some primitive African communities*. Oxford University Press. https://archive.org/details/terrorresistance0000walt

Waters, M. W. (2022). *King of the World: The Life of Cyrus the Great*. Oxford University Press. doi:10.1093/oso/9780190927172.001.0001

Watts, J., Sheehan, O., Atkinson, Q. D., Bulbulia, J., & Gray, R. D. (2016). Ritual human sacrifice promoted and sustained the evolution of stratified societies. *Nature*, *532*(7598), 7598. Advance online publication. doi:10.1038/nature17159 PMID:27042932

White, M. (2009, October 14). *The Industrial Revolution*. https://www.bl.uk/georgian-britain/articles/the-industrial-revolution

Wolpoff, M. H., & Lee, S.-H. (2001). The Late Pleistocene Human species of Israel. *Bulletins et Mémoires de La Société d'Anthropologie de Paris. Bulletins et Memoires de la Société d'Anthropologie de Paris*, *13*(3–4), 3–4. doi:10.4000/bmsap.6220

Wood, B. (2011). Did early Homo migrate "out of" or "in to" Africa? *Proceedings of the National Academy of Sciences of the United States of America*, *108*(26), 10375–10376. doi:10.1073/pnas.1107724108 PMID:21677194

X-Wealth. (2021). *Preservation And Succession: Family Wealth Transfer 2021*. https://go.wealthx.com/download-preservation-and-succession-family-wealth-transfer-2021

KEY TERMS AND DEFINITIONS

Computing Era: Historical period started in the mid-20th century featuring computing advancements that transformed industries and ushered in the Information Age.

Data: Information in a structured or unstructured format consisting of individual facts or statistics that can be moved, processed, and analysed.

Early Globalisation: The period of oceanic navigation, expanded trade and new cultural and biological exchanges, between distant civilizations, occurring roughly between 1300 and 1800.

History: The study of past events, people, societies, and civilizations through a systematic record of significant occurrences.

Natural World: The Earth's ecosystems, biodiversity, and environments, comprising all living organisms and their interactions as untouched by humans.

Silk Road: Ancient trade routes connecting China to the Mediterranean, enabling the exchange of goods, ideas, and microbes.

Slavery: Economic system where individuals are treated as property at the master's bidding without consent or fair treatment.

ENDNOTES

1. The greatest extinction on Earth ever is (by far) the Precambrian one, although the most complex organisms in this time, the Ediacaran biota, were a sort of jelly giant protozoa.
2. https://open.lib.umn.edu/worldgeography/chapter/1-2-the-environment-and-human-activity/
3. https://www.speedtest.net/global-index
4. https://tracxn.com/explore/Nanotechnology-Startups-in-China
5. https://tracxn.com/explore/Nanotechnology-Startups-in-United-States

Chapter 7
Examining Biopolitics and Bioethics Through Abusive Pain Management:
Genesis v Gehenna II

ABSTRACT

The author examines the politics of pain through various examples and stories that illustrate the overlapping of biopolitics and bioethics. Animal experimentation presents a contradiction between framing suffering as necessary while limiting unnecessary pain. The Tuskegee experiment in the US and the Guatemala study exemplify how decision-makers have used vulnerable populations for scientific purposes without regard for their well-being. The psychological inflexibility of individuals in high-income societies has contributed to a reliance on prescribed drugs and inadequate medical practices for pain management. These issues highlight how material progress and the politics of pain intersect, perpetuating a society of winners and losers.

INTRODUCTION

This chapter delves into the politics of pain and explores the unethical practices carried out in laboratories where research on "victims of science" is conducted. By examining historical displacements and disruptions caused by resource extraction, pollution, and slavery, Chapter 6 sets the stage for a deeper exploration of the lab as a site where biopolitics and bioethics intersect. It sheds light on notorious experiments such as the Tuskegee experiment in the Jim Crow US and the parallel and equally heinous experiments conducted in Guatemala. This chapter connects these themes with the historical perspective presented in Chapter 6, illustrating the ongoing impact of displacement and human actions on individuals and society.

DOI: 10.4018/978-1-6684-4808-3.ch007

THE LAB

The best lab to study the politics of pain is the lab itself, namely those where research with 'victims of science' is being performed (Ryder, 1983, pp. 66–68). Even the real practices of exemplary researchers as Pavlov were ethically fishy sometimes (Todes, 2000, pp. 39–40, 77). In the experiments with animals, biopolitics and bioethics mutually overlap. Framing animal suffering as a necessary evil while ensuring the experimentation doesn't involve unnecessary pain makes the living lab a contradictory facility.

A society focused on material progress (and that relies on it for almost anything else) works like a giant lab where the decision-makers are scientists trying to know how to put to work a lot of guinea pigs. Many experiments like the labyrinthine malls or the colour discrimination during a soccer match on TV doesn't seem like a mess. Others are obvious politics of pain, like the drones' wicked game in the Cover Predator War in Pakistan (Williams, 2010) or the public health expenditure cuts in Spain with a polemic alleged death toll of more than a half a million people from 2011-2015 (Gerrard, 2018).

Modern politics of pain came from the multiplication of forces that industrialization put in a coalition of politicians and industrial captains (aka baron robbers). This superhuman power for shaping the Earth's crust and dictating the schedule of billions brought reckless ambitions around the globe that ended in genocidal late colonialism and two World Wars. The philosophical background of the politics of pain is the over-rationalization of the top decision-maker righteousness. It happens most times under the pretext of greater good or lesser evil. For example, Hiroshima bombing fits into a scheme in which the Enola Gay pilots were the lab technicians, Washington the research director and the big experiment was the Pax Americana. The Hiroshima civilians nuked were the rats (sic). It was sold as a necessary evil for the greatest good. Of course, preventing the Japanese from seeing Hirohito's true face and his role in Japanese army war crimes was part of this new and bigger imperial experiment with a new American research director (Pye, 2000; Bergamini, 1971; Harada, 1946; Bix, 2008).

Far recent and not so cinematic, is the way how factories moved away from liberal democratic countries to others with no labour rights. It has been bragged in neoliberal circles as a way to spread prosperity across the world and an opportunity for the fired workers to freely devote their vocation. A quest for the greater good, a new Forbes Pax, rather than obvious greed and short-term profit rush. In this case, factory displacement is an experiment led for the few wealthy enough to be bored of funding political campaigns and tweaking public opinion. They are choosing between feeding this sort of enlightened despotism that Western social democracy is and the Chinese brand of national socialism, that's plain despotism but tax-friendly and light in labour rights. Considering that, in neoliberalism's eyes, taxes are the greatest evil and labour rights bring the gulag home, we'll see how the Western economic titans destroy the Anglo-Saxon system that gave room for them. As soon as the Pax Americana experiment results started being disappointing for the GDP-risers, aka research directors, a new global experiment centred in Beijing started. It's the humankind experiment of our time and, just like with the Hiroshima people, the lab rats / guinea pigs are going to be the have-nots while the many Hirohitos of Western politics, the contemporary manipulators who will play the dunce tomorrow, will cope quite well and will go scot-free for crippling human rights.

But, back to the present 2023, present-day monetization and politicisation of health is adulterating the love for life. It is more relevant even for the human body as a set of systems in a changing environment. Economic analysis is valuable and beneficial, but it leads to ethical concerns when a value is placed on human life. In contemporary economic systems, the value of life is too often reduced to a calculation of cost-benefit analysis. This approach reduces human life to a commodity that can be measured in terms of monetary value, and this can have serious consequences for how we treat ourselves and others. For example, in the realm of healthcare, cost-benefit analysis can lead to the devaluation of certain types of lives, particularly those of individuals who are deemed to be less economically productive or who have pre-existing conditions. Even if human preferences for the provision of health care or other life-saving interventions are probably too complex to be adequately represented by means of a single monetary value expressing the benefits of life-saving, the value will be fixed anyways, by the usual pen pusher on steroids. This can lead to decisions to withhold or ration care based on economic, or even ideological, factors rather than medical necessity. Similarly, in the realm of environmental policy, cost-benefit analysis can lead to decisions that prioritise short-term economic gains over long-term environmental sustainability and the health of future generations. This can have serious consequences for the human body as a system in a changing environment, as environmental degradation can lead to a range of health problems, from respiratory illness to cancer. The biggest mass killing ever was such a banal evil as adding lead to gasoline, which may have killed over 100 million people and poisoned more than 800 million children only in 2022 (World Health Organization, 2022; Rees & Fuller, 2020). Of course none of the suits covering lead poisoning up, those leading the lead poisoning big profiteers (Exxon, Chevron, Standard Oil, etc) had suffered a well-deserved Nuremberg trial of their own (Kitman, 2000; Kovarik, 2021). Quite the opposite, their greed has been rewarded with unprecedented political power. As a result, we are under an extremist philosophy of domineering, that has nothing to envy the Al-Qaeda chauvinism. It brings us to put a big buck in protecting perimeters and borders but not giving a damn of treasuring the human microbiome that we bring within, in our gut, mouth, skin, genitalia against food additives, tobacco cancerous invasion and avoidable environmental pollution at large.

The dominant globalisation ideologies negates the human body as a composite of many systems cooperating for creating their own bounds within an ever-changing world (Moser, 2018). Doing so, a genuine green behaviour is avoided and the main obstacle for an unleashed and risky material progress is removed. To address these issues, it is important to recognize that the value of life cannot be reduced to a simple economic calculation. Instead, we must acknowledge the complex and multifaceted nature of human life and the ways in which our economic decisions can impact our health, well-being, and the sustainability of the natural world. Too many rewards for those making money from resources, data and workers, instead of protecting nature has led us to forget that the microbiome is more entitled and ready to rule over the human body than any ministry, religious leader, tycoon or big pharma but receives no honours. Industrialised societies don't pay enough attention to the human body as it is made up of many different systems working together. Doing so, the necessity of caring for the environment to keep ourselves healthy is not sufficiently stressed. But we really need to remember that our health and well-being are connected to the health of the environment around us. Even though those tiny organisms in our body do the dirtiest work, they pave the way our feelings and emotions are connected to the world we live in. Microbiome is a world-brain that we need to think about how our actions affect others and the planet along with our mind-brain. Both our world-brain and mind-brain sustain reality in a way that their weakness leads to an alienating loss of meaning, that's displacement as Bodi understands it.

Instead of observing the multi-layered nature of human bodies, psychological inflexibility has fetishised them and given some parts and byproducts a special status. First and foremost, blood has been treated as both old sorcery and new scientism commonplace, but also semen, vaginal flux, vulva and lips, both in the face and across the vulva. As we'll see psychological inflexibility is politically linked with hidden at plain sight extremism.

Let's start with underground racism in the Jim Crow US. The Tuskegee experiment consisted of fooling about six hundreds of Afro-American men to obtain data on the development of untreated syphilis. The infected men remained untreated and uninformed for forty years (1932-1972) despite the fact that penicillin could have cured them (Fairchild & Bayer, 1999). Most of them were illiterate sharecroppers turned into guinea-pigs. The wickedness of this experiment accommodated by the highest institutional hierarchies was so inhuman that the Black rhetoric still echoes the infamy when facing clinical trials (Laws, 2018; Lisio, 2012, p. 198). Sadly, a Tuskegee alone was not enough to satisfy the curiosity of many US authorities about the syphilis. While the most notorious Nazi doctors were being hanged to death in 1948 (partly thanks to other US authorities), a parallel and unholier *Tuskegee* was running in Guatemala. Ironically, Adolf Pokorny, the Nazi physician who was in charge of researching syphilis and other venereal diseases, was acquitted during the Nuremberg Doctors' trial.[1] The new Tuskegee's victims were the Guatemalan indigenous persons and they suffered from a more sadistic do-harm principle than the Afro-American community even (Walter, 2012). *Most of them were illiterate sharecroppers turned into guinea-pigs!*

BAD BLOOD

Syphilis was one of the sicknesses tagged as *bad blood* ailments. It's very significant for the psychological inflexibility idea since it's linked with the over-rationalization required by the politics of pain. Sexual behaviour in humans is a complex one where many systems are involved: endocrine, canons of sex appeal and religious systems of belief. Equating bad blood with a complex behaviour like this, follows the need of reducing to a component (the blood) a complex and evolving reality.

The most blatant politics of pain are related with the vulva, vagina and womb. There is where inhuman over-rationalization becomes more evident. The epidemic sexual violence-related pregnancies (SVRPs) during wars, like the ones that have been studied in Bukavu, Democratic Republic of the Congo, shows off the multi-faceted influences on decision-making for raped pregnant women (J. Scott et al., 2018). For the women suffering from SVRPs the ethnic and social identity of the foetus, stigmatisation and religious views are worrisome, their children might inherit *bad blood.* During the sack of cities (aka liberation) the women suffer politics of pain twice. The incentive in this war crime is truly perverted: the cost opportunity of utterly destroying the identity and demographic growth of the enemy (even when risking international reputation) *pays dirt* for the over-rational military leaders who drive those mass rapes. In addition to this, war rape is also in line with other politics of pain. As we can see in the Guatemalan genocide from 1981-83, when mass repression began so did mass rape (Higonnet, 2011, p. 53). Of course, men also suffer politics of pain twice (first when drafted and then when they come back) and they are the target of inhuman sexual and genital abuse under the same inflexible logic as women (Féron, 2018, pp. 23–45).

It's important to highlight that we are not discussing war crimes in countries where human rights are neither present nor expected anytime soon but also on *good countries'* war crimes. The WW2 poster in the figure above is very interesting because it represents the militarised-gendered views on women: a menacing flying giant head against which you-the-man must be aware of. In this case, the woman shows a Southern-European ethnicity, which is not fortuitous given that the GI-Joe in the bottom of the poster was going to be sent to France. The message, otherwise, is clear: "*the women who accept presents or money for sex are corrupt, better if you rape*". Mass-rapes at gunpoint by thousands followed the libera-

Figure 1. Poster titled "She May Look Clean - But Pick-Ups 'Good Time' Girls Prostitutes Spread Syphilis and Gonorrhea" (ca. 1940)[2]

tors' track across France while the US generals were more worried about how close the brothels were to the city and how to hold the *color line* (Roberts, 2013; Stahl, 2013). The reason why war related sexual violence figures differs between men and women can be explained by bellic effort conveniences better than sexism, which doesn't rule out the military's misogyny.

But let's sail away from the sacked cities and go to the apothecary. A good chunk of the present-day Western culture of therapeutics comes from the uneasiness of high-income earners who rely on prescribed drugs to face their contradictions in life (Taylor, 2007, p. 622).

SHARP PSYCHOLOGICAL INFLEXIBILITY AND MODERN POLITICS OF PAIN

The psychological inflexibility of the people who don't accept that the money has not brought to them the happiness they expected is a powerful engine within the machinery of capitalism rather than a flaw. Beyond the psychological suffering, the plainest physical ache neither can run away from politics. The daily medical practice on pain is inconsistent and inadequate, it's littered with inappropriate diagnosis of addiction to opioids, even in patients who suffer from acutely painful conditions; likewise, it's plenty of bias and disbelief of pain reporting according to racial and ethnic stereotyping (Heidelbaugh & Yen, 2008; Roehrs & Roth, 2011; Rupp & Delaney, 2004). We've jumped from oligocracy to oligoanalgesia.

It can be seen in and out of clinic facilities, for example, in pharmacies. In a study on the difference in analgesics availability between Michigan *white* and *minority* pharmacies (according zip codes and racial composition) was found that (Green et al., 2005):

- There it was 52 times more likely to carry sufficient opioid analgesics in white pharmacies than minority ones,
- Regardless of income, corporate pharmacies were less likely to carry sufficient opioid analgesics in minority areas

Having your neighbourhood tagged as Black means bad news for everybody, regardless of colour because it means you live in a cheap place with cheap services. It reaches the delivery room. The social disparity has a weight in the decision of using or not intrapartum epidural analgesia. This was the conclusion of a 2010 study about the Canadian publicly funded health care system (Liu et al., 2010). Likewise, ethnicity was still a significant predictor of epidural analgesia after the researchers controlled for age, rural-urban residence, and availability of anesthesiologists for Medicaid eligible women in Georgia (Rust et al., 2004). This stinginess in analgesia comes from the same source as the war rapes: it's all about decision-makers allowing others' suffering to the far extent of their possibilities.

But the politics of pain can be even more insidious. We're witnessing a surge of slum tourism and *poverty porn* in the news (Weiner, 2008; Wilson, 2010). However, this voyeurism is nothing new. At the end of the 19th century, the wealthy people of London and New York, the winners of commercialism, went to do *slumming* in Jack the Ripper's Whitechapel and the Bowery Boys' home (Saint-Upéry, 2010). The slum tourism replicates within the downtown the *bad blood* criteria, it means more inequality (Hall, 2007, pp. 40, 150). From the Victorian era the social side of slumming, mostly charities at the beginning, has been counterweighted by booming prostitution (O'Brien, 2011). But the prostitutes today likely suffer more from their stigma than the menace of Jack the Ripper. In & out the fog, male prostitutes, in turn, turned into a public health problem as the 20th century went on as their mental health was on decline (Minichiello & Scott, 2014; W. Scott et al., 2014). No surprise there if we consider that HIV research found that the link between stigma and pain relies on anxiety (Wong et al., 2018). And we live in an age scorched by the effects of anxiety, debt and the anxiety coming from debts (Drentea, 2000; Twenge, 2000). This vicious circle harms especially those who are obviously not under a respectable business umbrella, like street prostitutes (by the way, traditionally *debt-bound* women) (Bouche & Crotty, 2018; Ondrášek & Kajanová, 2018). The rush for whoring coming from slumming is the *civilized* way to mass rape and debt-owners and pimps are the surrogated daily rapists.

Most war rapes and related pregnancies, needless painful deliveries as well as the bulk of sex tourism and slumming are politics of pain around the woman genitalia. Even if men also share some of their experiences, the women who can't reach safety traditionally become the *displaced sex* as soon as violent inflexibility rips a country. And it happens through the politics of pain.

The second largest opening in the human body is the mouth, some kind of horizontal vulva delivering ideas, indeed. Thus, the communication about the many bodies should have a higher status than markets. It's not the case. Regardless of how abject this practice is, the organ trafficking is not the darkest market shadow over the bodies but the insurance companies along with many public health communication campaigns.

Let's delve into it:

- Around 5-10% of global transplant activity comes from Transplant Tourism (TT) (Shimazono, 2007). Even a country with outstanding results in organ donation like Spain barely covers the half of its waiting list for kidneys while other countries where donation is more difficult, like Japan, only get the 10% of the waiting list covered (Akoh, 2017; Fujita et al., 2010).
- 1 in 10 organs transplanted comes from the black market (Campbell & Davison, 2012).
- The worth of all illegal organ traffic in the world in 2017 was between $840 millions and USD 1.7 billion (May & Clough, 2017).
- The global transplantation market was estimated at USD 23.5 billion in 2016 which is coherent with an illegal organ traffic of 5-10% of the total volume.[3]

Therefore, some claim that the main (if not the only) source of organs received by transplant tourists in some countries is the black marketeers. But, where do those organs come from? In Qatar, for example, 80-100% of procured organs are from working class expatriates (Broumand & Saidi, 2017). In China a decade ago, several studies reported up to 90% came from executed prisoners (Kennedy et al., 2005). In some international studies, 81% of vendors do it to pay off financial debts and they usually spend the money within 5 months without getting any improvement in their quality of life (Budiani, 2006). Being a *displaced* person can turn you into a box of spare parts.

Here we face a blunt logic, an over-rationalist one. Despite that a person cannot offer an appealing added-value work to the market, she has a value *in the flesh*. Quite macabre, but the private insurance firms foster medical tourism in spite of the fact that 1 in 3 USD comes from the TT (Pafford, 2009). The US healthcare system could save more than USD 1 billion sending patients to have some medical services abroad (Smith et al., 2011).

If nothing changes, a *laissez-faire* is likely to happen with the organ commerce and the laws will be on the money side. For example, there is no *bad blood* for the firms dealing with blood thanks to the laws shielding them in an alleged non-profit industry (Goodwin, 2006, p. 152). The public health communication campaigns will do their bit to help. Their messages are poisoned with culpability and labelling while promoting health as an individual value rather than as a universal human right (Guttman & Salmon, 2004). When it comes to the lack of communication, not hearing the pain narratives of a patient is a way of isolating her, of displacing her away from the humanitarian focus of clinical practice (Malpus, 2018). The non-impoverished people, as we've seen before, have more access to chemical friends, aka pills, so they are more resistant against the isolation's displacement -at least in the short-run.

In addition to this, we need to consider two factors about psychological inflexibility:

- Is related with the need of analgesia and attitude before the pain (Galán et al., 2018; Guildford et al., 2018).
- Is linked with conservatism and lack of empathy (Tackett, 2018)

In short, only in 2016, 40.3 million people were raped, butchered, sold and marooned for the profit of an almost paranoid minority that paid dirty for it to happen in a *conscious and calculated* way as we have never seen before.[4] In a world-wide spreading society of fluid winners and losers what the winning class needs is an inflexible self-asserting, not to progress. The material progress is what the guinea pigs in the social lab *do*!

CONCLUSION

Politics of pain materialises in the unethical practices carried out in laboratories where research on "victims of science" takes place. Psychological inflexibility is connected with that modern politics of pain and the lab is where biopolitics and bioethics intersect. Social failure to grant pain relief (or giving too much) perpetuates the suffering and Displacement of individuals, reinforcing the power dynamics of an unjust system.

REFERENCES

Akoh, J. A. (2017). State of the Trade: World Transplant Tourism. In J.-D. Rainhorn & S. El Boudamoussi (Eds.), *New Cannibal Markets: Globalization and Commodification of the Human Body* (pp. 199–214). Éditions de la Maison des sciences de l'homme. https://books.openedition.org/editionsmsh/10770

Bergamini, D. (1971). *Japan's Imperial Conspiracy*. Morrow.

Bix, H. P. (2008, May 3). War Responsibility and Historical Memory: Hirohito's Apparition. *The Asia-Pacific Journal: Japan Focus*. https://apjjf.org/-Herbert-P.-Bix/2741/article.html

Bouche, V., & Crotty, S. M. (2018). Estimating demand for illicit massage businesses in Houston, Texas. *Journal of Human Trafficking*, *4*(4), 279–297. doi:10.1080/23322705.2017.1374080

Broumand, B., & Saidi, R. F. (2017). New Definition of Transplant Tourism. *International Journal of Organ Transplantation Medicine*, *8*(1), 49–51. PMID:28299028

Budiani, D. (2006). *Consequences of living kidney donors in Egypt. Presentation At The Middle East Society On Organ Transplants (Mesot) Meetings. The Middle East Society On Organ Transplants (Mesot)* Meetings.

Campbell, D., & Davison, N. (2012, May 27). Illegal kidney trade booms as new organ is "sold every hour." *The Guardian*. https://www.theguardian.com/world/2012/may/27/kidney-trade-illegal-operations-who

Drentea, P. (2000). Age, Debt and Anxiety. *Journal of Health and Social Behavior, 41*(4), 437–450. doi:10.2307/2676296

Fairchild, A. L., & Bayer, R. (1999). Uses and Abuses of Tuskegee. *Science, 284*(5416), 919–921. doi:10.1126cience.284.5416.919 PMID:10357678

Féron, É. (2018). *Wartime Sexual Violence against Men: Masculinities and Power in Conflict Zones*. Rowman & Littlefield.

Fujita, M., Slingsby, B. T., & Akabayashi, A. (2010). Transplant tourism from Japan. *The American Journal of Bioethics*, *10*(2), 24–26. doi:10.1080/15265160903506384 PMID:20131168

Galán, S., Roy, R., Solé, E., Racine, M., de la Vega, R., Jensen, M. P., & Miró, J. (2018). Committed Action, Disability and Perceived Health in Individuals with Fibromyalgia. *Behavioral Medicine (Washington, D.C.)*, *0*(0), 1–8. doi:10.1080/08964289.2018.1467370 PMID:29671691

Gerrard, J. (2018, July 30). Study claims austerity has killed more in Spain than those who died in the Civil War. *Euro Weekly News Spain*. https://www.euroweeklynews.com/2018/07/30/study-claims-austerity-has-killed-more-in-spain-than-those-who-died-in-the-civil-war/

Goodwin, M. (2006). *Black Markets: The Supply and Demand of Body Parts*. Cambridge University Press. doi:10.1017/9781107051041

Green, C. R., Ndao-Brumblay, S. K., West, B., & Washington, T. (2005). Differences in Prescription Opioid Analgesic Availability: Comparing Minority and White Pharmacies Across Michigan. *The Journal of Pain*, *6*(10), 689–699. doi:10.1016/j.jpain.2005.06.002 PMID:16202962

Guildford, B. J., Daly-Eichenhardt, A., Hill, B., Sanderson, K., & McCracken, L. M. (2018). Analgesic reduction during an interdisciplinary pain management programme: Treatment effects and processes of change. *British Journal of Pain*, *12*(2), 72–86. doi:10.1177/2049463717734016 PMID:29796259

Guttman, N., & Salmon, C. T. (2004). Guilt, Fear, Stigma and Knowledge Gaps: Ethical Issues in Public Health Communication Interventions. *Bioethics*, *18*(6), 531–552. doi:10.1111/j.1467-8519.2004.00415.x PMID:15580723

Hall, C. M. (2007). *Pro-poor Tourism: Who Benefits?: Perspectives on Tourism and Poverty Reduction*. Channel View Publications. doi:10.21832/9781845410766

Harada, K. (1946). *The Saionji-Harada Memoirs*. Academic Press.

Heidelbaugh, J. J., & Yen, G. (2008). Lifestyle Risks. In J. J. Heidelbaugh (Ed.), *Clinical Men's Health* (pp. 321–337). W.B. Saunders. doi:10.1016/B978-141603000-3.10017-6

Higonnet, E. (2011). Quiet Genocide: Guatemala 1981-1983. Transaction Publishers.

Hilberg, R. (2003). *The Destruction of the European Jews*. Yale University Press.

Kennedy, S. E., Shen, Y., Charlesworth, J. A., Mackie, J. D., Mahony, J. D., Kelly, J. J. P., & Pussell, B. A. (2005). Outcome of overseas commercial kidney transplantation: An Australian perspective. *The Medical Journal of Australia*, *182*(5), 224–227. doi:10.5694/j.1326-5377.2005.tb06671.x PMID:15748132

Kitman, J. L. (2000). The secret history of lead. *Nation*, *270*(11), 11–11.

Kovarik, B. (2021, December 8). *A century of tragedy: How the car and gas industry knew about the health risks of leaded fuel but sold it for 100 years anyway*. The Conversation. http://theconversation.com/a-century-of-tragedy-how-the-car-and-gas-industry-knew-about-the-health-risks-of-leaded-fuel-but-sold-it-for-100-years-anyway-173395

Laws, T. (2018). Tuskegee as Sacred Rhetoric: Focal Point for the Emergent Field of African American Religion and Health. *Journal of Religion and Health*, *57*(1), 408–419. doi:10.100710943-017-0505-y PMID:29064071

Lisio, D. J. (2012). *Hoover, Blacks, and Lily-Whites: A Study of Southern Strategies*. UNC Press Books.

Liu, N., Wen, S. W., Manual, D. G., Katherine, W., Bottomley, J., & Walker, M. C. (2010). Social disparity and the use of intrapartum epidural analgesia in a publicly funded health care system. *American Journal of Obstetrics and Gynecology*, *202*(3), 273.e1–273.e8. doi:10.1016/j.ajog.2009.10.871 PMID:20045506

Malpus, Z. (2018). Guest editorial for May issue. *British Journal of Pain*, *12*(2), 69–71. doi:10.1177/2049463718769938 PMID:29796258

May, C., & Clough, C. (2017). *Transnational crime and the developing world*. Global Financial Integrity. https://www.gfintegrity.org/wp-content/uploads/2017/03/Transnational_Crime-final.pdf

Minichiello, V., & Scott, J. (2014). *Male Sex Work and Society*. Columbia University Press. doi:10.17312/harringtonparkpress/2014.09.msws

Moser, K. (2018). The Sensorial, Biocentric Philosophy of Michel Serres and Michel Onfray: Rehabilitating the Human Body. *Pacific Coast Philology*, *53*(1), 92–110. doi:10.5325/pacicoasphil.53.1.0092

O'Brien, E. (2011). Fuelling traffic: Abolitionist claims of a causal nexus between legalised prostitution and trafficking. *Crime, Law, and Social Change*, *56*(5), 547–565. doi:10.100710611-011-9334-1

Ondrášek, S., & Kajanová, A. (2018). *Financial aspect of working in private sex business*. https://www.narodacek.cz/wp-content/uploads/2019/10/Ondrasek_Kajanova.pdf

Pafford, B. (2009). The third wave—Medical tourism in the 21st century. *Southern Medical Journal*, *102*(8), 810–813. doi:10.1097/SMJ.0b013e3181aa8ce4 PMID:19593301

Pye, L. W., & Bix, H. P. (2000). Hirohito and the Making of Modern Japan. *Foreign Affairs*, *79*(5), 128. doi:10.2307/20049894

Rees, N., & Fuller, R. (2020). *The toxic truth: Children's exposure to lead pollution undermines a generation of future potential*. UNICEF.

Roberts, M. L. (2013). *What Soldiers Do: Sex and the American GI in World War II France*. University of Chicago Press. doi:10.7208/chicago/9780226923123.001.0001

Roehrs, T., & Roth, T. (2011). Medication and Substance Abuse. In M. H. Kryger, T. Roth, & W. C. Dement (Eds.), *Principles and Practice of Sleep Medicine* (5th ed., pp. 1512–1523). W.B. Saunders., doi:10.1016/B978-1-4160-6645-3.00132-8

Rupp, T., & Delaney, K. A. (2004). Inadequate analgesia in emergency medicine. *Annals of Emergency Medicine*, *43*(4), 494–503. doi:10.1016/j.annemergmed.2003.11.019 PMID:15039693

Rust, G., Nembhard, W. N., Nichols, M., Omole, F., Minor, P., Barosso, G., & Mayberry, R. (2004). Racial and ethnic disparities in the provision of epidural analgesia to Georgia Medicaid beneficiaries during labor and delivery. *American Journal of Obstetrics and Gynecology*, *191*(2), 456–462. doi:10.1016/j.ajog.2004.03.005 PMID:15343221

Ryder, R. D. (1983). *Victims of Science: The Use of Animals in Research*. National Anti-Vivisection Society Limited.

Saint-Upéry, M. (2010, October 21). *Left at the Crossroads: Ogling the poor*. https://sputniknews.com/analysis/20101021161035393/

Scott, J., Onyango, M. A., Burkhardt, G., Mullen, C., Rouhani, S., Haider, S., Albutt, K., Greiner, A., VanRooyen, M., & Bartels, S. (2018). A qualitative analysis of decision-making among women with sexual violence-related pregnancies in conflict-affected eastern Democratic Republic of the Congo. *BMC Pregnancy and Childbirth*, *18*(1), 322. doi:10.118612884-018-1942-7 PMID:30089448

Scott, W., McCracken, L. M., & Trost, Z. (2014). A psychological flexibility conceptualisation of the experience of injustice among individuals with chronic pain. *British Journal of Pain*, *8*(2), 62–71. doi:10.1177/2049463713514736 PMID:26516537

Shimazono, Y. (2007). The state of the international organ trade: A provisional picture based on integration of available information. *Bulletin of the World Health Organization*, *85*(12), 955–962. doi:10.2471/BLT.06.039370 PMID:18278256

Smith, R., Martínez Álvarez, M., & Chanda, R. (2011). Medical tourism: A review of the literature and analysis of a role for bi-lateral trade. *Health Policy (Amsterdam)*, *103*(2–3), 276–282. doi:10.1016/j.healthpol.2011.06.009 PMID:21767892

Stahl, R. Y. (2013, March 28). *Stop Rape: A WWII Chaplain's Advice*. Nursing Clio. https://nursingclio.org/2013/03/28/stop-rape-a-wwii-chaplains-advice/

Tackett, K. (2018). The Composition of Worldviews: The Relationships between Conservatism, Religiosity, Empathy, Dogmatism, and Psychological Flexibility. *Morehead State Theses and Dissertations*. https://scholarworks.moreheadstate.edu/msu_theses_dissertations/151

Taylor, C. (2007). *A Secular Age*. Harvard University Press.

Todes, D. (2000). *Ivan Pavlov: Exploring the Animal Machine*. Oxford University Press.

Twenge, J. M. (2000). The age of anxiety? The birth cohort change in anxiety and neuroticism, 1952–1993. *Journal of Personality and Social Psychology*, *79*(6), 1007–1021. doi:10.1037/0022-3514.79.6.1007 PMID:11138751

Walter, M. (2012). Human experiments: First, do harm. *NATNews*, *482*(7384), 148–152. doi:10.1038/482148a PMID:22318585

Weiner, E. (2008, March 9). Heads Up: Poverty Tours - Slum Visits: Tourism or Voyeurism? Travel. *The New York Times*. https://www.nytimes.com/2008/03/09/travel/09heads.html

Williams, B. G. (2010). The CIA's Covert Predator Drone War in Pakistan, 2004–2010: The History of an Assassination Campaign. *Studies in Conflict and Terrorism*, *33*(10), 871–892. doi:10.1080/1057610X.2010.508483

Wilson, B. (2010, February 4). What is 'Poverty Porn' and are we guilty of indulging in it? *Trespass Magazine*. https://www.trespassmag.com/what-is-%e2%80%98poverty-porn%e2%80%99-and-are-we-guilty-of-indulging-in-it/

Wong, C. C. Y., Paulus, D. J., Lemaire, C., Leonard, A., Sharp, C., Neighbors, C., Brandt, C. P., Lu, Q., & Zvolensky, M. J. (2018). Examining HIV-Related stigma in relation to pain interference and psychological inflexibility among persons living with HIV/AIDS: The role of anxiety sensitivity. *Journal of HIV/AIDS & Social Services*, *17*(1), 1–15. doi:10.1080/15381501.2017.1370680 PMID:30034300

World Health Organization. (2022, October 23). *Almost 1 million people die every year due to lead poisoning, with more children suffering long-term health effects*. https://www.who.int/news/item/23-10-2022-almost-1-million-people-die-every-year-due-to-lead-poisoning--with-more-children-suffering-long-term-health-effects

KEY TERMS AND DEFINITIONS

Biopolitics: The regulation of human populations through politics focused on fostering and maintaining control upon life.

Capitalism: An economic system where individuals own wealth and can grow it through satisfying real or fabricated needs along with favourable government interventions.

Clinical Practice: The application of medical knowledge and treatment to provide care for patients in a healthcare setting.

Inflexibility: Refusing to recognize own mistakes or alter one's ways or mindset.

Pain: Unpleasant physical or emotional sensation caused by injury, illness, or distress.

Politics: Actions and strategies aimed to obtain and exercise power over a country, society, or organization.

War Crime: Causing unnecessary harm or mistreatment of civilians or prisoners.

ENDNOTES

1. In spite of that he was a Nazi to the bone who suggested to Himmler the forced sterilisation of Russian prisoners of war (Hilberg, 2003, p. 1006).
2. Retrieved from the US National Library of Medicine: https://www.nlm.nih.gov/exhibition/visualculture/infectious21.html
3. According Grand View Research report on markets in: https://www.grandviewresearch.com/industry-analysis/transplantation-market
4. https://www.globalslaveryindex.org/2018/findings/global-findings/

Chapter 8
Critiquing the Psychological Inflexibility and the Politics of Pain in Woke Politics and Neoliberalism's Moral Governance:
Genesis v Gehenna III

ABSTRACT

In this thought-provoking chapter, the author explores the concept of psychological inflexibility, highlighting its presence not only in far-right and far-left ideologies but also in woke politics and neoliberalism. So the author delves into the dangerous consequences of inflexible commitment to ideas that prioritise political or economic gain over ethical considerations, perpetuating inequalities and injustices. It also means a critique of the abuse of intersectionality theory and the creation of a human pyramid that trans-values Jim Crow WASP America. The author concludes by cautioning against the narrow definition of morality and the dangers of technological and financial moguls leading the way in moral governance. Instead, lesser social and economic stakeholders should take the lead in shaping a win-win deal before the present political complexity and geopolitical stage.

INTRODUCTION

This chapter continues the exploration of psychological inflexibility and its relationship to the politics of pain, building upon the themes discussed in the last chapter. While Chapter 7 delved into the historical and contemporary manifestations of psychological inflexibility and its impact on various aspects of society, this chapter takes a broader perspective by examining its connection to postmodern politics of pain.

DOI: 10.4018/978-1-6684-4808-3.ch008

FUZZY PSYCHOLOGICAL INFLEXIBILITY AND POSTMODERN POLITICS OF PAIN

Psychological inflexibility is by no means a monopoly of old far-right and far-left, Juches, Unabombers, capital cronies and chauvinist people at large but also belongs to woke politics and neoliberals. Their inflexible commitment to ideas that allegedly seek to drastically improve society is exactly the same. If the markets-over-the-people ideology reduces the value of life to a simple economic calculation first and political power then, wokism reduces the value of life to a simple political calculation first and monetary gains then. Neoliberals enjoy the status quo granted by the state that publicly they despise. In turn, Wokes obtain money from the taxes paid by those middle-class men who publicly scorn, especially if White. It's a vicious circle in which the people who have neither influence nor money, regardless of any condition, will have less and less everyday and, in exchange, only will get broken promises of freedom and equity.

Abusing intersectionality theory and perverting Popperian open society ideals, wokism has created a human pyramid with the more politically victimised minorities and historically disadvantaged identities on top and the historically advantaged and privileged minorities in the bottom. Wokism is nothing but the transvaluation of Jim Crow WASP America just like Christianism was the transvaluation of Roman values and it's extremely likely to have analogous results for the American Empire. Just like Hirohito or the big lead poisoning firms owners that neglected human life value without being punished, wokism doesn't lack impunity either.

The true mark of political power is committing crimes with impunity.

Of course, impunity needs a tight mass control -don't miss with controlling incivility and vandalism (Lozano Rodríguez & Choclán Roca, 2019). In the markets-over-people ideology it is obtained through the ownership of media. On the woke side, in turn, it's obtained via social media censorship and cyber bullying. But also infiltrating activists (nuns and inquisitors at the same time) in formerly WASP media. It can be seen in Disney, the Marvel Cinematic Universe and Hollywood at large. In this case, wokism is overtly and nauseatingly racist and sexist. For example, the woke publication The Cut shared a video in which a bunch of non-white people ranted about Whites as superior in pretending bravery, ignorance, stealing, genocide and being "a dick" in general.[1] Well, you can find an only-Boer city in South Africa called Orania where statues of apartheid architects enjoy a reserved place, I mean historical figures who represent the worst of Western European ethnicity after WW2 along with the US Jim Crows and their *dictadores* exposed in a good light.[2] Notwithstanding, there is a Berkeley University off-Campus in the US core where White people are banned and scorned, insulted and called racists and homophobes only due to their skin colour (Nerozzi, 2022). Even if the White supremacists, who actually are a fringe fraction of all Caucasian admixture, enjoy a bigger real world community in Orania, Black-self-labelled supremacists and anti-White racists have a bigger community in main social media, where White supremacists are quickly banished to shitty forums. For example, Ayo Kimathi, a US Department of Home Security officer was not fired at the spot after his anti-White and antisemitic stand-up racist rant calling for genocide became viral (Terry, 2013). Rather than this, he was being paid over $115.000 a year from the taxpayers' money during four months before being fired.[3] It happened in 2014, today DHS wouldn't have fired him anyways. On the other extreme of the rope, in 2020, Emmanuel Cafferty was expeditiously fired from his job mapping underground utility lines for making the OK gesture with the hand, which was then trendy as a white supremacist symbol (Castañares, 2021). The firm where

Emmanuel worked, SDG&E, never gave his job back even after the moron who started the cancellation in Twitter recognized that it hasn't racist intentions. But his name was linked to this pic and Emmanuel couldn't find a job because every time a potential employer looked for info about him on the Internet, he appeared as a notorious racist.

Figure 1. If you are looking for a racist, Emmanuel Cafferty is not a Racist. Try luck with your electoral roll and amidst SDG&E suits.
Source: Castañares (2021)

Just like White supremacists, those social media Black racists as Ayo Kimathi are a fringe fraction of African admixture ethnic heritage, living most of them in the US. In 2016, US Blacks were more likely than Whites to see (and post) race-related content on social media (Anderson, 2016). Later, in the 2020 election year, BLM was on fire with its violent social uprising and Facebook overhauled the algorithm that detects hate speech and deprioritized hateful comments against Whites, men and Americans in relation with minorities (Dwoskin et al., 2020). The BLM and similar businesses are selling Black people (fake) safety because they "are in mortal danger each and every moment of every day, whether in a public space or in their own home" (Orlu, 2021). And the only reason is *systemic racism* ™. Of course, they are selling White people an Ivory Tower to look down on the rest of so labelled Whites who are not so moral to join the BLM. Actually, there is a systemic ZIP classism in the way many US public schools are partly funded by property taxes. No other measured aspect of schools is nearly as important in determining student achievement as professors' quality (Hanushek, 2011). Of course, better funded schools in safe places are more attractive and better professors, at least on paper, have more chances to get employment there. Having better professors means an average increase of between $110,000 and $230,000 in lifetime earnings (Hanushek, 2011). Many people in the US, especially politicians, know that they should make schools in impoverished districts more attractive for talented professors. Meh! It's better to bicker about the racial composition of the school. The US has a racism and resentment legacy and wants to impose its new racism and resentment upon the rest of the West and whoever is paying attention; don't let BLM fool you.

And, just like most hardline neoliberal doctrines is a haven for tax evaders, staunch wokism is a haven for sexual offenders. As a transvaluated chauvinism worthy of this name, wokism is deeply sexist, in this case, brimming with misandry. For example, the Spanish secretary of state for Equity Angela Rodríguez, said that Spanish men were all rapists and quite rapists (Huarte, 2023). From what she says, the men around her must all be rapists, abusers, paederasts and perverts but she should look beyond her own *safespace* to do meaningful work. Actually, Spain, in spite of being the worst country in Europe for robbery and cocaine consumption, is (in other crimes and even by European standards) a modellic country. Rape is one of these crimes. So much so that Spain has been the country with least rapists per capita in Western Europe in the 2010-2020 period.[4] However, when Spanish political parties are involved it's another bucket of fish. In Vilassar de Mar, a nice touristic town governed by a Spain ruling party partner, ERC, the city hall organised a sex gymkana aiming at 12-30 years old (sic) participants. It included such educational activities as buttering a banana with cream and licking it, putting condoms on a stick with the mouth and replicating kamasutra positions with a balloon between the 12-30 years old participants' genitalia (López Tovar, 2022). To cap it all, the sex gymkana images were uploaded to the Internet without parents' consent as the whole gymkana itself was being perpetrated. The public denunciation was immediate and the country got shocked, but after 8 months no public officer and less even a politician has entered prison yet. But even this belittles when compared with the legislative reform undertaken by the aforementioned Angela Rodríguez's party, Podemos. The *Ley Montero* named after the (In)equality Monster (I mean Minister) Irene Montero that has benefited more than 1000 sexual aggressors and has freed more than 100 rapists due to the reduction of sentence in the first few months (González Gómez & Álvarez, 2022). Thousands of sexual offenders will be on the street prematurely thanks to the politicians who tag all the men as rapists. A similar free speech double standard than applied for racism can be seen here. For example, the Spanish capital mayor 2015-2019 Manuela Carmena stated that: "...violence is embedded in what is the *DNA of masculinity*" (Ollero, 2019). As a result of the lack of resistance before derogatory misandry, in 2023 Spain, Podemos' fascist agenda has advanced and men have the burden of proof when accused of gender violence (an extremely loose penal type, by the way) (Álvarez, 2022). Indeed, just like in the 1920-1945 fascist regimes, there is a guilty presumption based only on innate characters that are, allegedly, dangerous and antisocial. To make things closer to the Nazis, the Woke movement enjoyed burning books in Canada (Warmington, 2021). It's the hegemonic feminism business, to endanger and belittle women to sell them (fake) safety later. It's not morally different from pushing drugs, but it's legal, pusher friendly, applauded and not less lucrative.

Both the old, chauvinist and concrete politics of pain and the new, hate-grounded and fuzzy ones, share the same psychological inflexibility. The old politics of pain were all about not contaminating the social, political and economic apex. In the US case, preserving it wealthy and WASP. Kennedy was the first Catholic POTUS and the only one who took distance from the MIC (Military Industrial Complex). But he didn't take distance from a bullet in the head. Ironically, since Kennedy the US never reached the 40% of global share of GDP again (Bhutada, 2021). JFK was the first symptom of the US WASP Empire weakness that started an overall slow decline. Obama and Biden are the red flag for an incoming unprecedented crisis since the Civil War. Staunch wealthy families WASP were only divided between ol' school Democrats and Republicans. That's about imposing the US way of life globally or only worrying about trade routes, dollarization, beef, Bible and bullets.

The new fuzzy politics of pain, neoliberalism and wokism, don't ground in the idea of avoiding race and gender pollution but in promoting the new man. I mean the new no-man, she who will cleanse the absolute power's house from rats. In this case, the division between the new and fuzzy politics of pain is who must be the pseudo-messianic cleanser and who must be the rat. For hardline neoliberalism the cleanser must be an Elon Musk template object. However, as in the Argentinian Milei case, they will get enough with a sycophant of the staunchest crony capitalism as long as he insults and humiliates the people receiving subsidies. As you may have guessed, the rats in the rawest neoliberalism are the people who don't contribute to the market, those who don't raise the GDP and don't *freely* choose to die by starvation. If you are not a person with private income (e.g. a landlord) or you don't have a successful career in private enterprise then you are a kept, a parasite, and a minorised rodent. In turn, in the Woke case the anti-rodents are the oppressed subjects from intersectionality and critical race and gender theories. Shortly, whoever besides cis-hetero White men. The rats, in this case, are, well, cis-hetero White men regardless of assets origin and walk of life.

The psychology grounding both neoliberalism and wokism is an inflexible axis that goes from magic voluntarism to Goebblelian blaming. In the neoliberal case, you are poor because you want or you deserve it. If you want to be on the good side of God Market you must change your mindset according to any pseudo-coaching drag (e.g. Rich Dad Poor Dad), tune in to wealth and embrace the survivor fallacy as it were the Thermodynamic laws. And don't forget about smiling till death while exploiting yourself because attitude is everything. In this case, the Goebblelian blame, the enemy everybody must rally around to tear it down are taxes since they represent the state swimming upwards the Market profiteers' river. On the other hand, wokism tells that you must tune your sexual identity and pronoms at will and according to the heat of the moment. But, first and foremost, wokism exhorts you to get awake of oppression. Indeed, just like happens with neoliberals and taxes, you don't live in fucking lollipop-land because of oppression. In this case, oppression is what cis-hetero white men emanate, maybe via radiation, spores or plain overgeneralization fallacy and historical revisionism. So, is there any country in the world that never suffered White man colonial scorch, is almost entirely composed of non-Whites and where there is virtually neither taxes nor subsidies? Yeah, Liberia. It includes ethnic groups with a matriarchal tradition such as the Golla people and was ruled by a female PM till 2018: Ellen Johnson Sirleaf, who ruled longer than Margaret Thatcher in the UK. In spite of this, Liberia is one of the countries with a smaller share of female representatives in the parliament -and not the best place to be born as a woman by far.

Both woke and neoliberal secular religions have questionable dogmas. All-in neoliberalism supposes that there will be a better market without state. Actually, the seed of the present-day global market, the Silk Road, required such strong ancient states as Republican Roma and Han China. Roma and China were able to offer so many hundreds of kilometres of safe roads and tons of coined gold that trade was still appealing even during the Mediterranean Punic Wars, Middle Eastern Seleucid Empire decline and Indian Mauryan Empire collapse. In addition to this, the power abhors the void and nothing would prevent the wealthiest people from reaping the political power in addition to the productive and speculative factors. It would somewhat reset history to a point of little and medium family-owned kingdoms but concentrated around strategic zones rather than sparse as in ancient times. In the bottom of neoliberal mathematical models and fishy indexes identifying prosperous countries with neoliberalism and neoliberalism with prosperity, there is the belief that the markets are not only an autonomous sphere of human activity (classic liberalism) but an autonomous sphere itself. A sort of present-day singularity. However, if economics is a science, it's a social science. It's not a science whose research subjects are away from human existence such as gems or distant suns. When dealing with poverty, starker neoliberalism repeats the same mistake that those who think that poor people remain poor because they have plenty of children. It's quite the opposite, as families lift out of poverty, they start to reduce their household size (Axinn & Barber, 2001; Bloom et al., 2001; Snopkowski et al., 2016; Wang & Sun, 2016) For this reason balancing aid, especially women's education, with family planning is so important to get the best results. By the same token as families get misery behind when corruption, inflation and unemployment reduces, they become more productive and not the other way around. No country is poor because of its lazy and spineless citizens overburdening the neoliberal Atlas.

The wokism dogmas are even worse. To begin with, 24/7 social media woke followers believe in race as a sort of solid filing cabinet opposed to a flowing sexuality. Put it easy, the BIPOC filing cabinet has three drawers: Black, Indigenous and People of Colour but nothing prevents you from defining yourself as heterosexual, bisexual, pansexual, tortillasexual or whatever in the heat of the moment (very practical when looking for casual sex, for sure). Facts, however, point in quite the opposite direction. To begin with, there are no human races, we are not garden plants of exclusive lap dogs, sorry (Livingstone & Dobzhansky, 1962). In addition to this, woke historical revisionism is overwhelming and infiltrated across most cultural products, including those aiming to minors, e.g. Disney's *Proud Family* (Iacono, 2019; Logan, 2022). Woke revisionism equates, for example, Blackness with a familiar history of slavery and oppression at White hands. However, Africans didn't come to the US via transatlantic slavery exclusively and slavery owning was not an only-White thing (Pressly, 2006). Furthermore, more Africans have migrated to the US since 1990 to 2005 than in the preceding two centuries (Roberts, 2005). But the Oscars in Woke historical revisionism goes to *Queen Cleopatra*, a fake documentary aired on Netflix in which the Ptolemaic leader is depicted as a sort of Black gladiatrix in an uchronic stage, served cold by the hand of Jada Pinkett Smith.[5] The degree of manipulation is so simple and the fake documentary is so deranged that even the former Egyptian Antiquities Minister Zahi Hawass (the kind of guy who will be unnoticed in a White Power meeting, as you may have guessed) resumed it as: "This is completely fake." (Egypt Independent, 2023) Some Egyptians are so angry with this mockery (Jada Pinkett Smith's *Queen Cleopatra* was presented as a documentary, not as a biopic or adventures movie) that an Egyptian lawyer has invoked his Constitution to compel authorities to stop this mess (Egypt Independent, 2023). However, the most dangerous part of this mockumentary by negligence, in my opinion, appears in the trailer itself and is an old Black woman who says that her granny told her when a child that no matter what she learned at the school, Cleopatra was a Black woman.[6] I doubt that even her granny was old enough to check it and, likely, she was saying something like Cleopatra knew how to be fiercely on her own rather than speculating on Ptolemaic Egypt's haplogroups and pharaonic cuckoldry. But in the way it is presented, it is another instance of feelings over falsifiable facts, close to magic voluntarism.

We are all a stockpile of genetic admixtures from populations that came far before us, even pre-human in many cases, along with a more or less hodgepodge of other present-day human populations. In practice, it leads us to hilarious cases such as racial activist Angela Davis exposed as a Mayflower descendent or the Paul Craig Cobb case, a white nationalist and white supremacist who learned he is 14% sub-Saharan admixture.[7,8]

To cap it all, the boldest Wokes are obsessed with sexual tastes and genitalia cuts and body shaping and rare sex-related conditions. Indeed, sexual orientation is not well determined since childhood and it flows more along young adulthood but it doesn't mean that sexual orientation can be changed at will or that it's the normal (most likely) outcome (Mock & Eibach, 2012; Stewart et al., 2019). Sexual orientation across life rarely changes and never because somebody dug her heels in. In spite of this, political lesbianism, the belief in feminists' superpower of switching her sexual orientation and the right to tag heterosexual women as "collaborators with the enemy", keeps being a discussed topic after more than forty years (Leeds Revolutionary Feminist Group, 1982). Hold on, don't let plain facts stain your ideology.

Out of radfems, hegemonic feminism becomes ambivalent since it's more inclusive but also more politically exploitative of non-standard human bodies and medical conditions. In the woke language, oppression and binary are synonymous because binary points to the old man and wokeness is a new non-man system of beliefs. So looking for somebody who might have difficulties to create a traditional family (besides common infertility) and capture their fight is an obvious goal. The first group are intersexual people, those whose bodies differ from standard male or female. Actually, intersexuality is an umbrella gathering Klinefelter and Androgen insensitivity syndromes, congenital adrenal hyperplasia and some others. Even if the prevalence of most atypical intersexuality-related conditions is as rare as 1 in 150.000 births, the many intersexualities sum up a 1.7% of newborns (UN Human Rights Office, 2015).[9] The other group are transexual people. Feeling from a different sex than the one you are born in was, essentially, the reason to start a sex transition. According to the DSM-IV, the prevalence of gender dysphoria is 0.005-0.014% for adult natal males and 0.002-0.003% for adult natal females. It means that 2-3 people in every 100.000 will face this life challenge. In despite of being about 30-50 times less prevalent among general population than intersexual individuals, transexual collective has gained a disproportionate relevance in woke ranks. The reason for this is activism's cookie *cutter* machine. Intersexuality-related chromosomal syndromes in their mildest forms could remain undetected and, at most, produce infertility. In their most severe forms, they can reduce life expectancy significantly. More even, these syndromes cannot be induced, they are even prenatal. Quite the opposite, a sex transitioner can be either a gender dysphoric person or a teen disoriented about her sexual identity that doesn't require reassignment. Furthermore, transgenderism lacks the *diversity* of treatments and special needs of intersexuality. Rather than this, either trans men or women, the problem can be tackled with well-known hormone pills and surgery. Therefore, they are far more fit to follow an agenda, becoming activists and being politically herded. So much so that the woke movement went so far. In 2015, Swedish health authorities labelled puberty blockers and cross-sex hormones as safe when they were experimental at its best, inaugurated an open bar with them and invited minors (Grossman, 2023; Swedish Agency for Health Technology Assessment and Assessment of Social Services, 2019). But some of the serious side effects of such puberty blockers were exposed in the 2019 documentary "The Trans Train" (Sweden's National Public Television Broadcaster, 2019). As a result, in February 2022, Swedish govt put brakes on treatments for trans minors such as puberty blockers and mastectomies (Socialstyrelsen, 2022; Swedish Agency for Health Technology Assessment and Assessment of Social Services, 2022). In addition to this, from 2008 to 2018, sex reassignment surgery of 13-17 years old female-born girls, unlike male-born boys, rocketed 1.500% (Socialstyrelsen, 2020). What has changed so much in Sweden? Social media, 4th wave feminism and woke politics. In spite of the Swedish fiasco, Spaniards might become the dumbest of the 5 Big Fooled (Argentina, Canada, Spain, Sweden and the US) and allow its glorious leaders to repeat the Swedish experiment again (Jones, 2023). Don't let the health, childbearing, parenthood and regrets for life of minors stop your political greed. Do you know a thing? Spanish government members who live in a large family are about three times the Spanish general population and the female government members' fertility rate is above 50% higher than the Spanish women's. Even the Spanish baroness of feminism and [In]Equality minister, Irene Montero, is mother of three. To cap it all, the Spanish minister of Consumption, Alberto Garzón, who also has a large family, published in April 2023 a video stating that Spanish people should think twice about having children due to climate change.

Few lines above, I've expressed my deep concern and mistrust for allowing teens as young as 13 years old to take puberty blockers and transition. Some treatments almost entirely rely upon physician criteria such as a transplantation but others do the same with patients' will, for example, going to palliative care or keeping with chemotherapy. All those choices ground on maturity and success chances that can be reasonably assessed. But maturity is what a 13 years old teen lacks. You cannot condition a no wayback major surgery for a really rare condition to how an underage patient feels about herself and less even to guessing what her self-image will be 5 years ahead. And by no means in present-day circumstances. At this juncture, another intervention than surgery is needed; psychological, for example. Notwithstanding, we shouldn't skip over the fact that there are more mature (and more immature) 13 year old teens. The harms of whether intervening or not, in turn, must be assessed considering non-surgical factors such as family and social context among others. I've delved into this now and not before in order to highlight that I'm going to be considered homophobic because of not being aligned with the woke's magic voluntarism.

Honestly, I would understand that an activist who earns her life from activism makes a mountain from a molehill after reading my opinion. If a person like Dylan Mulvaney, for example, privately thinks I am a Nazi monster eagering to send to the gas chamber any non-cis-hetero living being it would be pathetic but nothing to write home about. Not to mention that the Budweiser's *sex symbol* has the perfect right to think anything of anybody. But wokeness is fuzzy. In order to work, snowflaking must fall upon every (self-appointed) social justice warrior. It means stigmatising anybody out of the woke's echo chamber as whatever-phobic and inviting everybody to express how bad they feel about dissent, how wicked and offensive these words are and how badly she should regret. Out-and-out wokism is the rape and abuse of natural human compassion.

Hardline neoliberalism, in turn, uses an inverse-victimisation to delegitimize any critique. To begin with, any remote idea about "to be" or "to do" head and shoulders with "to have" swiftly becomes a red flag. Only *if you can pay, you can stay* kind of rhetoric is acceptable. The Neoliberalism slogan and mantra is "private property, human life and freedom". However, the two last ones in this three of a kind list are actually different shades of private property. You could say "private property, private property and private property" and it would be closer to the truth. In Hebrew, truth is אֶמֶת (emet) and it does not extinguish in telling A while thinking B. Emet implies a sense of loyalty and trustworthiness. Perhaps the truth of neoliberalism is "private property, human life and freedom" but its *emet*, its deep truth, its loyalty, is on private property. This worldview would be better named as *privatepropertism* rather than neoliberalism, which is a misleading tag. Here, life is the private property of one's own body and freedom is choosing how to spend one's private wealth, time and skills. Shortly, a human becomes a freelancer for herself. Thus, any form of collective freedom or right is suspicious for neoliberals and plainly evil if it touches their pocket. Just like wokism is all about draining political power for collective needs, neoliberalism does the same with the collective build of those who want to be the only collective: the collective of individualists. Most times collective needs arise from the lack of material means in order to face a burden or adversity. So, there is a need for professionals to put the problem under the spotlight and help the affected collective before it becomes a humanitarian crisis. In the best case scenario, it's what taxes are for along with preventing any other moral horror. Actually, it is not even close in most countries because of corruption, but neoliberalism's main target is not corruption but taxes. If it weren't by taxes, most vulnerable needy collectives would rely on charities. And, quite the opposite than taxes and collective rights, charities can have a name and surname and belong to individuals. Neoliberalism would invest the wealthiest people with a political power that they don't have yet. Therefore, neoliberalism is not necessarily opposed to modern slavery or authoritarian regimes but to welfarism. Indeed, just like wokism is the transvaluation of US Jim Crow laws, present-day mediatic neoliberalism is a transvaluation of the European welfare state.

When it comes to harassment, neoliberal mobs are *straighter* than woke, which is not difficult, by the way. The strategy used by most Gadsden flag bearing youtubers, tiktokers, instagramers, etc. before dissent is labelling you as a woke and depicting you in the same way. If you don't wake up with Murray Rothbard and go to bed with Ayn Rand, then you are a loser, green with envy every time Elon Musk appears on the news. If you think that taxes, as far as the ruler is a People-fearing one, are always better than fancy philanthropy, then you are a state-kept crybaby. If you don't dance around the City of London with "Rich Dad, Poor Dad" and "Think and Grow Rich" books over your head and you don't firmly believe that the wealthier the more self-made, then you are a wuss grabbing the nanny-state's leg. Bluntly, just follow Margaret Thatcher, the milk snatcher:

She destroyed too many good things in society, and created too many bad ones, then left a social and moral vacuum in which the selfishly rich and unimaginatively fortunate could too easily destroy still more of what they don't need and can't see that everyone else does need. (Pasha-Robinson, 2016)

Essentially, both woke and neoliberal postmodern politics of pain are not so different from their older versions from the Modern Age. All of them want your belly hollow, your bank account hollow, your home hollow, your family life desvirtuated and even your community, collective and social circles meaningless or even hostile. Then, your only option will be to buckle under the trendy great leader till a new one enters to line her pockets or inflate her ego. For wokism, it means to become an activist or regret alone. In the neoliberalism case, the dilemma is to join the precariat and live an existence without predictability and security or get ashamed. It's related to magical voluntarism and the way fuzzy psychological inflexibility manages dissent. That's the true difference between modern and postmodern politics of pain. From a sharp psychological inflexibility and modern politics of pain, dissent is an unsurprising reaction among bad blood beings to be managed by dedicated institutions: reeducation camps, policial tortures, educational curriculum propaganda, etc. The desirable outcome is reducing dissent just like any other human form of resistance. However, fuzzy psychological inflexibility needs dissent because it's the main ingredient of victimisation and fear mongering identitarian glue. Everytime dissent appears an attention-whore crow's feast starts. Postmodern politics of pain will try by all means fool you in order you bear all the world's problems. If the modern politics of pain milked your agony, the postmodern one success relies upon your regret and shame of being born.

Facing this problem requires a broader perspective that takes into account ethical, historical, and cultural considerations alongside economic ones. By doing so, we can work towards a more just and equitable society that values human life and promotes the health and well-being of all individuals. Besides Bioethics, as we have seen, Displacement has a lore of its own that we should expand now. Even the logic itself has been displaced when losing its use (this is its *meaning*) in favour of new ones. It was the case of the Hegelian one before the analytic philosophy of Frege and Russell. And every culture, even the littlest one, has its own inner logic. It's heart-rending to realise how whole populations are treated like deprecated theories. Like unsatisfying replies to the question of how to coordinate the human effort in more efficient ways pursuing a common goal. It happened countless times regardless of others' willingness to cooperate or how whimsical is this goal for the *civilised* ones that impose it. But even the Hegelian logic has a new chance today (Redding, 2007). Therefore, if a bore like this has a second opportunity, the different communities deserve far more than this.

The psychological inflexibility and politics of pain aims to trap a complex creature like the human one that is both singular in features and plural by nature in a flat screen framed by pain and gain. Partisan politics maximise liars and quickly shapes ordinarily honest people into dishonest leaders who lead the society in a devaluating truth cycle that, in turn, produces even worse rulers (Curtis & Hart, 2022; Hart & Curtis, 2023). If this attitude prevails, bioethics won't. A society that prioritises gain over ethical considerations and perpetuates inequalities and injustices will get ridden by inauthenticity and, as a result, will fall into the Gehenna upside down of bioethics.

Gehenna sets prevarication as a way of life and corruption as the main governance factor.

MORAL ENHANCEMENT, NEUROPOLITICS… AND ZOMBIES

A moral enhanced person is one whose reasons to act are morally better than they would have been otherwise. Despite that it's not a bad theoretical meaning, it's useless in practice since 'morally better' and 'goodness' have different interpretations depending on the circumstances and most important virtues in the observer's eyes. Even a virtue widely engraved in our genes like friendliness (Walker, 2003) could be a hindrance in many environments when not packed with other social skills. So, a moral enhanced person comes to be one whose feelings (greed, jealousy, aversion…) don't prevent her to be positively motivated to share, trust, care, etc. For example, a man who has a great aversion to women won't offer a nurturing relationship and vice versa. Sadly, moral enhancement tends to narrow morals to a lack of motivation. To cap it all, the "*neuroliberalism*" implicit in this crunch of neurosciences with moral governance is closer to the sociological models than ever. It may lead to riskier outcomes on biocitizenship as sociology loses its critical perspective (Wade, 2019). The evolutionary advantage whether having morality or not is an improved connection between actions and their likely outcomes, that's a keener sense of risk (Bugliarello, 2008). This is important because tech and financial moguls have shown a proverbially self-delusional perception of the crisis they can provoke and, therefore, putting Forbes list gurus in front is the second worst possible state of things when it comes to the morality of future technologies, barely better than political parties' *intelligentsia*.

Anyhow, the appeal of moral enhancement is so intense that hardly the risks or knowledge gaps are going to play a major role in this tech development, so what's the next? Well, many ways of moral enhancement were feasible in the beginning of the 21st century, mostly drugs like oxytocin (Kosfeld et al., 2005) and psychological therapies. Therefore, the most likely moral enhancers in the near future apothecaries and detention centres will aim to alleviate illegal behaviour proneness but no morality pill is at sight (Shook, 2012). Thankfully.

Despite this, the most radical paths to moral enhancement are the incoming genetic modification and BCIs for moral enhancement. From a bioethics perspective it exposes some evident issues, for example, who is it to say if we are unfit for the future as a society and in need of a compulsory enhancement? (Deane-Drummond, 2017, pp. 177–180) From a bioethics of displacement position: what would be the fate of the un-enhanced (rebels and unfitted for treatment)? The border between enhancement and therapeutic intervention is blurred when it comes to morals. And it will be since the moral enhancement "*do not fully account for the complexity of moral agency*" (Jotterand & Levin, 2019). To make things more complex even, since the dawn of the 21st century, the transhumanism bound with soteriology and neurotheology has gained momentum (Hughes, 2007).

The pharmacological fading of traumatic experience (e.g. traumatic experiences) could also erode our moral sensitivity and twist our identity towards rogue persons devoid of moral responsibility senses (Glannon, 2013, pp. 138–139). If a drug turns a vet suffering PTSD into a well-trained psycho, who is to blame? The vet, the hospital, the big pharma...? If taking a drug to erase a rape's trauma makes the victim to lose interest in bringing the rapist to the court of law and he gets released and again in the streets, should the victim have the right to take this drug before her testimony? (Sandler, 2016, p. 179)

Neuroreality is reality. Neuropolitics is politics. (Dunagan, 2004)

The development of disruptive neural technologies can change the current grounds of self and society but makes the human species vulnerable against a brain to money masters' taste as happens with politics. Besides the corporate desires, there are hidden memory masters, a brain branch of NSA-like agencies behind the doctrine of shock and it may reach an unheard penetration with neuropolitics: either behind *prevention & preemption* catchphrases or with an endless "*war on terror*" (Dunagan 2004). The politics professionals have paid attention yet to how neuroimaging joins to the political psychology's methodology regardless that it's a fertile ground for hype and overstatement (Theodoridis & Nelson, 2012). Therefore, the neuro-enhancement and mainstream neuroethics ground talks on a financialist "Second Enlightenment". During this new zeitgeist the mind concerns will shift into brain ones and the welfare could be pinned amidst the synapses leaving to the individual means the task of finding it (Wade, 2016).

From a different perspective, cognitive enhancement has been pointed out as a public health topic in the near-future. Considering, for example, that 1 out of 4 US students in 2008 were using nootropic cognitive enhancers such as Ritalin, "*neuroenhancing health*" may result in a public health improvement (Shaw, 2014). The use of neuroenhancers could be narrowed to the segment of population where the inequity hits harder but it's actually the segment where the informed consent is least free. If you give a slight chance of getting out of the slum to a person living in Karachi or Lagos taking drugs, is it very different from the freedom of selling them? Wouldn't it be rather the chance to experiment in order to obtain a second generation of commercial neuroenhancers, more expensive and with lesser side effects?

Anyhow, all that we know of human moral excellence comes from our experience with non-enhanced humans, so this rush for a new man undermines the logic behind the enhancement proposal. In other words, "*the project of moral enhancement is self-defeating: its fulfilment would spell the abolition of the very conditions that allow us to describe it as an 'enhancement'*" (Skidelsky, 2018). Thus, the laws on cognitive liberty allowing the citizen's sovereignty over her mind, which would include the legality of future enhancements should be redacted as soon as possible (Bublitz, 2013). But what will happen with the individuals with a non-citizen status as illegal migrants? Could be they forcefully *enhanced* in order to satisfy hate policies or avoid deportation?

The alleged therapeutic forgetting is the real moral enhancement we can aspire to in this book's time. However, if thought on a biopolitical scale, this kind enhancement would mean to turn populations into zombies. Nowadays, in 2019, there are no zombies as crowds of roaming and muttering stupid cannibals, but there is zombie meat (engineered lifeless living tissues) in the market (Stephens, 2010). Besides the funny movies on this topic (along with a tide of copy-paste drags), there is in the US and abroad a rising fear of the living dead that makes the moral serious issues worthy. How do the international laws of war apply here? Those slobbering enemies would constitute an invaluable and tireless military asset behind enemy lines, indeed. According to Jessica Engler, the weaponized zombies would break the Hague Law principles of necessity, proportionality, and distinction and depict them as "*an abhorrent weapon that should not be used in armed conflict*" (Engler, 2012). She also highlighted the need to prohibit this weapon since, likely, it would turn into an indispensable weapon when used the first time. I fully agree and I want to stand out, from both a logic of power and a bioethics of displacement perspective, the great advantages of using the non-allied persons like dissidents, objectors and prisoners of war (POWs) confined in camps as zombies during scorched earth tactics. Furthermore, even if the Z-virus is banned from war and curable it could be used to cheapen the refugees' allowance through reversible zombification.

On the other hand, a person in a permanent vegetative state could include in her advances health care directives the wish of becoming a wimpy zombie (Cattorini, 2009). It would allow her to save her moral agency awaiting for the cure of irreversible coma cases, cheapen the economic burden on her family and even realise menial tasks to contribute with. In this case, the legal status of the zombie should determine whether she is a person or not and a clear statement on her dignity, capacity, rights and responsibility (Arnold, 2016). Anyhow, a person zombified in this way would have her brain closer to brain death than deep-sleep and the ascription of self-regarded interest should only be done about well-known facts (Gillett, 1992). For example, if the Zombie Anna tirelessly expressed her/its disgust with flurry animals it would be unethical putting a cat in her/its room or storage. But what happens if Anna, when *alive*, was trying to become a mother, expressed this wish as a last will and her family supported her and still does? If Bennie was a well-known KKK wizard, should Carmen Sandiego share his/its storage room? Likely, the persons in wheelchairs could get an offence of sharing accessible toilets with undead individuals, would there be even different toilets for those civic zombies?!

Of course the idea of "*shambling deads*" allows us, as walking amateur meds, to project and diagnose them as consumerists, anti-capitalists, refugees, colonised, poor… (Schmeink, 2017) The zombies are, in pop culture, a kind of monsters like Freddy Kruger (Craven, 1984), Dracula (Coppola, 1992) or even far bigger like Godzilla (Dougherty, 2019). They can attack alone, in hordes or both, just like the H. R. Giger's Aliens (Cameron, 1986; Giger, 1994) or the insectoids in Starship Troopers (Verhoeven, 1997). The monsters can return after defeated like many *slashers*: It (Muschietti, 2017, 2019), Jason (Cunningham, 1980), Freddy again (Bayer, 2010), Jason again (Nispel, 2009), Freddy vs Jason (Yu, 2003)… oh, I'm sure you've got the idea. To cap it all, they can strike back through time holes like Thanos (Russo & Russo, 2019). Sometimes its appearance is masked behind the sculptural bodies of Arnold Schwarzenegger in *Terminator* (Cameron, 1984) and Natasha Henstridge in *Species* (Donaldson, 1995) but they only fully unleash their crazy rage when unveiled as *true* monsters. But, in any case, the monsters always are displaced due to its appearance before its vicious behaviour even (Smith, 2008, p. 11). The *Starship Troopers* main characters or the manga *Goblin Slayer* (Ozaki, 2018) gang are a bunch of haughty xenocides with other species but the evilness and bitter specism falls only in the *monsters*' side. However, unlike what happens in the Frankenstein novel (Shelley, 1869), the scientists and corporations are now labelled as monsters like in *Maximum Ride* (Patterson, 2007) and *The Island* (Bay, 2005) where those institutions' moral ugliness becomes impossible to blend with human society (Smith, 2008, p. 12). Crichton's Jurassic Park novel and further movie (Crichton, 1991; Spielberg, 1993) shows a halfway in this path, presenting a non-evil corporation led by an endearing grandpa that get screwed by a monstrously obese data analyst bribed by an evil-corp with no further relevance. The playing god narratives are a common thread among the aforementioned movies turning the popular concern with this tech into a great deal.

Ironically, the zombies themselves may be a way of moral enhancement… for a few lucky ones. E. L. Quarantelli and Russel R. Dynes identifies several factors related with the lack of community conflict in a natural disaster situation (Dynes & Quarantelli, 1975):

- Natural disasters involves an external threaten
- The disaster agent can be pointed out for the most people

- There's high consensus in the priorities during natural disasters
- The disaster brings people back to present
- Social distinctions get levelled during disasters
- Disaster aftermath leads to a greater community identification

So, the surviving communities, likely would be a selection of the morally fittest and the survivors would be more like selfless hoplites rather than Mad Max punks -at least for a while. In the post-Z world, the infected under treatment in order not to turn zombies likely will be displaced as many other persons who felt in a new foreignness. Indeed, the future of humankind will be determined by the amount of surviving children able to create a new sense of communality of its own (Thompson & Thompson, 2015, p. 203) as well as the existence of an anchor to a supranational past to learn about and the resilience of communication systems. It can be deduced from the principles stated by John D. DeLamater and Daniel J. Myers about overcoming the hard insider-outsider dichotomy (DeLamater & Myers, 2011, pp. 424–429).

In conclusion, the moral enhancement away from the one coming from sustained reflection and active awareness is unlikely to help the human species. To begin with, it's useless against the extinction threats of climate change nor weapons of mass destruction (Agar, 2015). Put in bluntly, turning the moral judgement in LEGO pieces and oversizing one political shape over others won't make the game better. To cap it all, the most powerful leaders and institutions of human society are unfit to face the present complexity: they boast on a sensible governance of emerging technologies but are unable to reach an agreement to stop the deviant development that causes climate change. It's up to the lesser social and economic stakeholders (including the zombie meat producers!) to figure out a win-win deal while adapting to the 2020s geopolitical complexity (Battaglia et al., 2018).

CONCLUSION

Psychological inflexibility is not limited to extreme ideologies but also exists within woke politics and neoliberalism, perpetuating inequalities and devaluing ethical considerations. In turn, moral enhancement that lacks sustained reflection and active awareness will fail when facing major threats like weapons of mass destruction. Even if the surviving communities in post-apocalyptic scenarios may exhibit temporary unity and moral fortitude, it's better to collaborate with diverse stakeholders to improve our present.

REFERENCES

Agar, N. (2015). Moral Bioenhancement and the Utilitarian Catastrophe. *Cambridge Quarterly of Healthcare Ethics*, *24*(1), 37–47. doi:10.1017/S0963180114000280 PMID:25473856

Álvarez, G. M. (2022, July 6). *El peligro de sustituir la presunción de inocencia por la de culpabilidad*. Cinco Días. https://cincodias.elpais.com/cincodias/2022/07/05/opinion/1657038118_685321.html

Anderson, M. (2016, August 15). *Blacks more likely than whites to see – and post – race-related content on social media.* Pew Research Center: Internet, Science & Tech. https://www.pewresearch.org/internet/2016/08/15/blacks-more-likely-than-whites-to-see-and-post-race-related-content-on-social-media/

Arnold, B. B. (2016). Is the Zombie My Neighbour: The Zombie Apocalypse as a Lens for Understanding Legal Personhood. *Canberra Law Review*, *14*, 25.

Axinn, W. G., & Barber, J. S. (2001). Mass Education and Fertility Transition. *American Sociological Review*, *66*(4), 481–505. doi:10.2307/3088919

BattagliaE. M.MeiJ.DumasG. (2018). *Systems of Global Governance in the Era of Human-Machine Convergence*. https://arxiv.org/abs/1802.04255

Bay, M. (Director). (2005). *The Island*. https://www.imdb.com/title/tt0399201/

Bayer, S. (Director). (2010). *A Nightmare on Elm Street*. https://www.imdb.com/title/tt1179056/

Bhutada, G. (2021, January 14). *The U.S. Share of the Global Economy Over Time*. Visual Capitalist. https://www.visualcapitalist.com/u-s-share-of-global-economy-over-time/

Bloom, D. E., Canning, D., & Sevilla, J. P. (2001). *Economic growth and the demographic transition*. Academic Press.

Bublitz, J.-C. (2013). My Mind Is Mine!? Cognitive Liberty as a Legal Concept. In E. Hildt & A. G. Franke (Eds.), *Cognitive Enhancement: An Interdisciplinary Perspective* (pp. 233–264). Springer Netherlands. doi:10.1007/978-94-007-6253-4_19

Bugliarello, G. (2008). Reflections on Morality, Ethics, and Bioethics Decisions. *Journal of Long-Term Effects of Medical Implants*, *18*(1), 85–90. Advance online publication. doi:10.1615/JLongTermEffMedImplants.v18.i1.590 PMID:19348615

Cameron, J. (Director). (1984). *The Terminator*. https://www.imdb.com/title/tt0088247/

Cameron, J. (Director). (1986). *Aliens*. https://www.imdb.com/title/tt0090605/

Castañares, A. (2021, April 12). *SDG&E Employee Fired Over Alleged White Supremacist Hand Sign is a Latino*. La Prensa. https://laprensa.org/sdge-employee-fired-over-alleged-white-supremacist-hand-sign-latino

Cattorini, P. (2009). Shall I Become a Zombie? Stories of Illness, Ethical Dilemmas and Visions of Society. *Medicina Nei Secoli*, *21*(3), 1017–1035. PMID:21560776

Coppola, F. F. (Director). (1992). *Bram Stoker's Dracula*. https://www.imdb.com/title/tt0103874/

Craven, W. (Director). (1984). *A Nightmare on Elm Street*. https://www.imdb.com/title/tt0087800/

Crichton, M. (1991). *Jurassic Park*. Ballantine Books.

Cunningham, S. S. (Director). (1980). *Friday the 13th*. https://www.imdb.com/title/tt0080761/

Curtis, D., & Hart, C. L. (2022). *Pathological Lying: Theory, Research, and Practice*. American Psychological Association.

Deane-Drummond, C. (2017). The Myth of Moral Bio-enhancement: An Evolutionary Anthropology and Theological Critique. In T. J. Trothen & C. Mercer (Eds.), *Religion and Human Enhancement: Death, Values, and Morality*. Springer. doi:10.1007/978-3-319-62488-4_11

DeLamater, J. D., & Myers, D. J. (2011). *Social Psychology*. Wadsworth.

Donaldson, R. (Director). (1995). *Species*. https://www.imdb.com/title/tt0114508/

Dougherty, M. (Director). (2019). *Godzilla: King of the Monsters*. https://www.imdb.com/title/tt3741700/

Dunagan, J. F. (2004). Neuro-futures: The brain, politics and power. *Journal of Futures Studies*, *9*(2), 1–18.

Dwoskin, E., Tiku, N., & Kelly, H. (2020, December 3). Facebook to start policing anti-Black hate speech more aggressively than anti-White comments, documents show. *Washington Post*. https://www.washingtonpost.com/technology/2020/12/03/facebook-hate-speech/

Dynes, R. R., & Quarantelli, E. L. (1975). *Community Conflict: Its Absence and Its Presence In Natural Disasters*. http://udspace.udel.edu/handle/19716/382

Egypt Independent. (2023, May 8). Lawyer demands Egypt suspend Netflix's 'Cleopatra' documentary. *Egypt Independent*. https://www.egyptindependent.com/lawyer-demands-egypt-take-action-against-cleopatra-documentary/

Engler, J. (2012). *War of the Living Dead: An Analysis of the Permissibility of the Zombie Weapon in Modern International Armed Conflict* (*SSRN* Scholarly Paper ID 2171477). Social Science Research Network. doi:10.2139/ssrn.2171477

Giger, H. R. (1994). Giger's Alien (2nd ed.). Morpheus International.

Gillett, G. (1992). Coma, Death and Moral Dues: A Response to Serafini. *Bioethics*, *6*(4), 375–377. doi:10.1111/j.1467-8519.1992.tb00214.x PMID:11652098

Glannon, W. (2013). *Brain, Body, and Mind: Neuroethics with a Human Face*. Oxford University Press.

González Gómez, J. A., & Álvarez, D. G. (2022, November 17). *El balance escalofriante de la ley del "solo sí es sí": Ya son más de 1.000 los delincuentes sexuales beneficiados*. Periodista Digital. https://www.periodistadigital.com/politica/justicia/20221117/mapa-delincuentes-sexuales-beneficiados-ley-irene-montero-noticia-689404809015/

Grossman, H. (2023, April 26). *Top Swedish doctors blow whistle on trans puberty-suppressing drugs affecting children's bones: "Experimental."* Fox News. https://www.foxnews.com/media/top-swedish-doctors-blow-whistle-on-trans-puberty-suppressing-drugs-affecting-childrens-bones-experimental

Hanushek, E. A. (2011). Valuing teachers: How much is a good teacher worth. *Education Next*, *11*(3), 40–45.

Hart, C. L., & Curtis, D. A. (2023). *Big Liars: What Psychological Science Tells Us about Lying, and How You Can Avoid Being Duped*. American Psychological Association.

Huarte, P. (2023, March 11). *Ángela "Pam" dice que en España los hombres "son bastante violadores"*. The Objective. https://theobjective.com/espana/2023-03-11/angela-pam-hombres-violadores/

Hughes, J. (2007). The compatibility of religious and transhumanist views of metaphysics, suffering, virtue and transcendence in an enhanced future. *Institute for Ethics and Emerging Technologies*. https://www.researchgate.net/profile/James_Hughes2/publication/237467243_The_Compatibility_of_Religious_and_Transhumanist_Views_of_Metaphysics_Suffering_Virtue_and_Transcendence_in_an_Enhanced_Future/links/02e7e5318b7ab3a5a5000000/The-Compatibility-of-Religious-and-Transhumanist-Views-of-Metaphysics-Suffering-Virtue-and-Transcendence-in-an-Enhanced-Future.pdf

Iacono, C. (2019, September 28). *No, Slavery Did Not Make America Rich | Corey Iacono.* https://fee.org/articles/no-slavery-did-not-make-america-rich/

Jones, S. (2023, February 16). Spain passes law allowing anyone over 16 to change registered gender. *The Guardian.* https://www.theguardian.com/world/2023/feb/16/spain-passes-law-allowing-anyone-over-16-to-change-registered-gender

Jotterand, F., & Levin, S. B. (2019). Moral Deficits, Moral Motivation and the Feasibility of Moral Bioenhancement. *Topoi*, *38*(1), 63–71. doi:10.100711245-017-9472-x

Kosfeld, M., Heinrichs, M., Zak, P. J., Fischbacher, U., & Fehr, E. (2005). Oxytocin increases trust in humans. *Nature*, *435*(7042), 673–676. doi:10.1038/nature03701 PMID:15931222

Leeds Revolutionary Feminist Group. (1982). Political Lesbianism: The Case Against Heterosexuality'. In The Woman Question. Fontana London.

Livingstone, F. B., & Dobzhansky, T. (1962). On the Non-Existence of Human Races. *Current Anthropology*, *3*(3), 279–281. doi:10.1086/200290

Logan, T. (2022, February 24). Slavery Was Never an American Economic Engine. *Bloomberg.Com.* https://www.bloomberg.com/opinion/articles/2022-02-24/slavery-was-never-an-american-economic-engine

López Tovar, A. (2022, August 5). *Padres indignados por una gincana 'porno' con niños en Vilassar de Mar.* La Vanguardia. https://www.lavanguardia.com/vida/20220805/8450732/padres-indignados-gincana-porno-ninos-vilassar-mar.html

Lozano Rodríguez, M., & Choclán Roca, C. (2019). Plataforma digital para atajar el incivismo: el valor de lo intangible en la governanza metropolitana. *V Congreso de Ciudades Inteligentes*, 34-40.

Lozano Rodríguez, M. (2021, January 9). On Privileges—A reflection on the worshipping and demonization of privileges part 3. Thirteenth International Public Health Ambassador Conference.

Mock, S. E., & Eibach, R. P. (2012). Stability and change in sexual orientation identity over a 10-year period in adulthood. *Archives of Sexual Behavior*, *41*(3), 641–648. doi:10.100710508-011-9761-1 PMID:21584828

Muschietti, A. (Director). (2017). *It.* https://www.imdb.com/title/tt1396484/

Muschietti, A. (Director). (2019). *It Chapter Two.* https://www.imdb.com/title/tt7349950/

Nerozzi, T. (2022, August 20). *White people banned from off-campus UC Berkeley student housing common areas.* Fox News. https://www.foxnews.com/us/white-people-banned-off-campus-uc-berkeley-student-housing-common-areas

Nispel, M. (Director). (2009). *Friday the 13th.* https://www.imdb.com/title/tt0758746/

Ollero, D. J. (2019, January 12). *Cuando Carmena criticaba la Ley de Violencia de Género con los mismos argumentos que Vox.* ELMUNDO. https://www.elmundo.es/espana/2019/01/12/5c392b3b21efa00b688b4710.html

Orlu, S. (2021, March 14). *Why I do not support the Black Lives Matter movement—Troy Media*. https://troymedia.com/lifestyle/why-i-do-not-support-the-black-lives-matter-movement/

Ozaki, T. (Director). (2018). *Goblin Slayer*. https://www.imdb.com/title/tt8690728/

Pasha-Robinson, L. (2016, October 18). *Margaret Thatcher is the worst prime minister in 100 years*. The Independent. https://www.independent.co.uk/news/uk/politics/margaret-thatcher-worst-prime-minister-pm-david-cameron-century-100-years-a7367481.html

Patterson, J. (2007). *The Angel Experiment: A Maximum Ride Novel*. Academic Press.

Pressly, T. J. (2006). "The Known World" of Free Black Slaveholders: A Research Note on the Scholarship of Carter G. Woodson. *Journal of African American History*, *91*(1), 81–87. doi:10.1086/JAAHv91n1p81

Redding, P. (2007). *Analytic Philosophy and the Return of Hegelian Thought*. Cambridge University Press. doi:10.1017/CBO9780511487620

Roberts, S. (2005, February 21). More Africans Enter U.S. Than in Days of Slavery. *The New York Times*. https://www.nytimes.com/2005/02/21/nyregion/more-africans-enter-us-than-in-days-of-slavery.html

Russo, A., & Russo, J. (Directors). (2019). *Avengers: Endgame*. https://www.imdb.com/title/tt4154796/

Sandler, R. (2016). *Ethics and Emerging Technologies*. Springer.

Sax, L. (2002). How common is intersex? A response to Anne Fausto-Sterling. *Journal of Sex Research*, *39*(3), 174–178. doi:10.1080/00224490209552139 PMID:12476264

Schmeink, L. (2017). New Ways to Think about Zombies. *Science Fiction Studies*, *44*(3), 623–625. doi:10.5621ciefictstud.44.3.0623

Shaw, D. (2014). Neuroenhancing public health. *Journal of Medical Ethics*, *40*(6), 389–391. doi:10.1136/medethics-2012-101300 PMID:23793059

Shelley, M. W. (1869). Frankenstein, or, The Modern Prometheus. *Dent, Contemporary Dentistry*.

Shook, J. R. (2012). Neuroethics and the possible types of moral enhancement. *AJOB Neuroscience*, *3*(4), 3–14. doi:10.1080/21507740.2012.712602

Skidelsky, E. (2018). Moral Enhancement and the Human Condition. *Royal Institute of Philosophy*, *83*(Supplements, 83), 109–120. doi:10.1017/S1358246118000310

Smith, T. (2008). *Bioethics for the masses: The negotiation of bioethics in film and fiction*. https://search.proquest.com/openview/4f3f42eca10d14c80f751711fbbb8250/1?pq-origsite=gscholar&cbl=18750&diss=y

Snopkowski, K., Towner, M. C., Shenk, M. K., & Colleran, H. (2016). Pathways from education to fertility decline: A multi-site comparative study. *Philosophical Transactions of the Royal Society B: Biological Sciences, 371*(1692). doi:10.1098/rstb.2015.0156

Socialstyrelsen. (2020, February 12). *Vanligt med flera psykiatriska diagnoser hos personer med könsdysfori*. Socialstyrelsen. https://www.socialstyrelsen.se/om-socialstyrelsen/pressrum/press/vanligt-med-flera-psykiatriska-diagnoser-hos-personer-med-konsdysfori/

Socialstyrelsen. (2022, February 22). *Uppdaterade rekommendationer för hormonbehandling vid könsdysfori hos unga*. Socialstyrelsen. https://www.socialstyrelsen.se/om-socialstyrelsen/pressrum/press/uppdaterade-rekommendationer-for-hormonbehandling-vid-konsdysfori-hos-unga/

Spielberg, S. (Director). (1993). *Jurassic Park*. https://www.imdb.com/title/tt0107290/

Stephens, N. (2010). In Vitro Meat: Zombies on the Menu. *SCRIPTed: A Journal of Law, Technology in Society*, *7*, 394.

Stewart, J. L., Spivey, L. A., Widman, L., Choukas-Bradley, S., & Prinstein, M. J. (2019). Developmental patterns of sexual identity, romantic attraction, and sexual behavior among adolescents over three years. *Journal of Adolescence*, *77*(1), 90–97. doi:10.1016/j.adolescence.2019.10.006 PMID:31693971

Sweden's National Public Television Broadcaster (Director). (2019, April 2). The trans train. In Uppdrag granskning. Author.

Swedish Agency for Health Technology Assessment and Assessment of Social Services. (2019, December 20). *Gender dysphoria in children and adolescents: An inventory of the literature*. https://www.sbu.se/en/publications/sbu-bereder/gender-dysphoria-in-children-and-adolescents-an-inventory-of-the-literature/

Swedish Agency for Health Technology Assessment and Assessment of Social Services. (2022, February 22). *Hormonbehandling vid könsdysfori – barn och unga* [Systematisk litteraturöversikt; HTA; Systematic review]. https://www.sbu.se/sv/publikationer/SBU-utvarderar/hormonbehandling-vid-konsdysfori/

Terry, D. (2013, August 21). *DHS Employee Promotes Race War in Spare Time, Advocates Mass Murder of Whites*. Southern Poverty Law Center. https://www.splcenter.org/hatewatch/2013/08/21/dhs-employee-promotes-race-war-spare-time-advocates-mass-murder-whites

Theodoridis, A. G., & Nelson, A. J. (2012). Of BOLD Claims and Excessive Fears: A Call for Caution and Patience Regarding Political Neuroscience. *Political Psychology*, *33*(1), 27–43. doi:10.1111/j.1467-9221.2011.00860.x

Thompson, A. L., & Thompson, A. S. (2015). *But If a Zombie Apocalypse Did Occur: Essays on Medical, Military, Governmental, Ethical, Economic and Other Implications*. McFarland.

UN Human Rights Office. (2015, September 25). *A step forward for intersex visibility and human rights*. OHCHR. https://www.ohchr.org/en/stories/2015/09/step-forward-intersex-visibility-and-human-rights

Verhoeven, P. (Director). (1997). *Starship Troopers*. https://www.imdb.com/title/tt0120201/

Wade, M. (2016). *Intensive Labours, Expansive Visions: Emerging ideals of the ethical subject amidst the rise of cognitive neuroscience*. https://openresearch-repository.anu.edu.au/handle/1885/105355

Wade, M. (2019). Risky disciplining: On interdisciplinarity between sociology and cognitive neuroscience in the governing of morality. *European Journal of Social Theory*. Advance online publication. doi:10.1177/1368431018810330

Walker, M. (2003). *Genetic virtue*. Institute for Ethics and Emerging Technologies.

Wang, Q., & Sun, X. (2016). The Role of Socio-political and Economic Factors in Fertility Decline: A Cross-country Analysis. *World Development*, *87*, 360–370. doi:10.1016/j.worlddev.2016.07.004

Warmington, J. (2021, September 10). *WARMINGTON: School board says it got burned in Indigenous book burning project*. Torontosun. https://torontosun.com/news/national/warmington-school-board-says-it-got-burned-in-indigenous-book-burning-project

Yu, R. (Director). (2003). *Freddy contra Jason*. https://www.imdb.com/title/tt0329101/

KEY TERMS AND DEFINITIONS

Biocitizenship: The rights, responsibilities, and social belonging of individuals in relation to their biology and healthcare choices.

Climate Change: Long-term alterations in Earth's weather patterns, including temperature rise and shifts in rainfall, primarily caused by high levels of carbon dioxide and greenhouse gases.

Community: A group of people with shared interests, living in the same area, supporting and interacting with one another.

Corruption: The relation whereby the worshipping of privileges is learned" (Lozano Rodríguez, 2021).

Governance: The process of governing, including decision-making, control, and leadership in a country, organisation, or institution.

Institution: An established organisation or system that serves a specific purpose and can influence the public opinion.

Neoliberalism: A secular religion grounded on the belief that taxes are robbery that venerates the logic of the market (first and foremost, the financial market) as an ultimate truth / reality.

Resilience: The strength to bounce back from tough situations and grow through adversity.

Society: People with shared institutions and living in the same area.

ENDNOTES

1. https://www.youtube.com/watch?v=03FFPCg6Bh8&ab_channel=MegynKelly
2. https://www.karoo-southafrica.com/eastern-upper-karoo/orania/
3. https://www.splcenter.org/fighting-hate/intelligence-report/2014/irritated-genie-race-war-enthusiast-finally-leaves-dhs
4. https://ec.europa.eu/eurostat/web/products-eurostat-news/-/EDN-20171123-1
5. At least it is not the hand of her husband Will Smith, PAM!!
6. https://www.youtube.com/watch?v=IktHcPyNlv4&t=9s&ab_channel=Netflix
7. https://twitter.com/HenryLouisGates/status/1628210213236142080
8. https://www.youtube.com/watch?v=ptSZnTtGCQA&t=78s&ab_channel=HuffPostLive
   Bonus: his neo-Nazi former friends burned his property down, lol.
9. This figure has been discussed to be exaggerated, if intersex would be restricted to those conditions in which chromosomal sex is inconsistent with phenotypic sex, or in which the phenotype is not classifiable as either male or female the prevalence would be 0.018%, almost 100 times lower (Sax, 2002).

Chapter 9
Individual Mindset Change and Civil Disobedience in Shaping Societal Re-Evolution:
Praxis I

ABSTRACT

The author plays with bioethics of displacement as a means of addressing modern society's caste systems through civil disobedience. Using the Nepali case as an initial example, the author exposes the atrocities that arise from social structure and the dual global pyramids perpetuating inequality. On the other hand, the author emphasises the need for patient rebels to create a sense of belonging and draw attention to injustices. In a global context, both injustices and threats require a global response, and individuals have the moral agency to rebel against injustice. For this to happen, rebellion must ground on moral legitimacy rather than opportunism. Otherwise, whatever is gotten won't bring a more just society.

INTRODUCTION

Present chapter mirrors the previous one on fuzzy psychological inflexibility and postmodern politics of pain opposing to them the need for individual mindset change as a catalyst for meaningful societal transformation. When facing those politics of pain is important mastering nonviolent resistance and cultivating a sense of belonging to challenge unjust systems[1]. Technology, particularly in educational settings, must be used for offering alternative perspectives and empowering individuals to peacefully resist those unjust systems or the same tech will be used against critical thinking.

DOI: 10.4018/978-1-6684-4808-3.ch009

ON THE PRAXIS OF BIOETHICS OF DISPLACEMENT

Let's think about a society of castes. As we can see in the Nepali case, the atrocity is related to social structure (Chalise, 2018, pp. 1–2). Put it bluntly, the castes can drive you crazy about what you can do to the people in the bottom. For example, the rape is more frequent among *untouchable women* (Chalise, 2018, pp. 41–42). To a certain degree, we live in a dual global society of castes. On the one hand we have a pyramid in which the wealthiest are the top echelon, the potential consumers are in the middle and those who are so poor that result unfit for further personal indebtedness are one kind of *glocal dalits*. On the other hand, there is another pyramid with managers of blame and regret above all, their faithful followers in the middle and the cancelled, unproductive heretics or unfaithful sleepers to be stigmatised, forming another layer of glocal dalits. Of course, most governments lack incentives to alleviate their dalits' born-debt beyond public aid dependence or to cease public harassment fearing of getting exposed, in turn, as a sort of McCarthyists profiteers. Through nonviolent resistance, individuals can draw attention to the injustices they face and cultivate a sense of belonging that transcends caste divisions and encourages resistance to unjust systems.

One of the main challenges of Bioethics of Displacement is using two different ways of thinking (spiritual and scientific) at the same time. Joining Bodi's metaphysical and positivist duality in a fashion that doesn't get itself out of science, for example when it comes to talk about spirit and matter. The differences in methods and methodologies applied by exact scientists dealing with the human being and the ethicists' ones are often a source of problems, misunderstandings and work opportunities. As the Chaos specialist Prof. Andrew Bosworth told me in a personal communication: "we all reconcile duality, or else we become displaced". The postmodern loss of order we live into also can be a chance for a boundless tolerance. The bioethics of displacement can be a way out of the heavy influence of money in academia given that the self-awareness of our own displacement in the world will highlight the empowerment we can reach through *a practical ethical theory*.

The Bodi rebels will hardly live adventures like the ones of the Katniss' bunch in *The Hunger Games* (Collins, 2008). Thankfully, it's not because of any oppressive brain-washing government but due to true reasons for optimism. Let's put it in perspective. Recent real examples of gene splicing hybridization include the Enviropig that incorporates mouse DNA to improve digestion of phosphorus and experimental mice being given a "humanised" version of a gene linked to speech (Cook et al., 2014). Even if the last experiment is for me comic and uneasy at the same time, these animals are far less frightening than the jabberjays and tracker jackers from the dystopian novel. On the other hand, it's true that neoliberalism is a sort of masquerade that wrongly presents the individuals as autonomous *units* (Samuels, 2016). Actually an individual can be disunited (e.g. fractured due to schizophrenia) and yet remaining as an individual. The point is avoiding a new political thinking that questions the neoliberalism vision of individual responsibility regarding her group, society, culture and the whole Earth.

Thus, rather than guerrilla skills, a bioethics' rebel must have a knack for education and mediation and the character of individuals like Albert Camus (Blenkinsop & Morse, 2017; Corbett, 2012). A person who, rather than resign herself, exalts life in an expanding meaning. Sadly, our current postmodern academic *zeitgeist* wraps a class of intellectuals who don't speak up, march and fight along with marginalised groups (Kahn & Lee, 2010, p. 34). Scholars of terracotta who don't throw the truth in the big cats' faces, who are afraid to disturb and don't care about where humanity is going. Cultural referees who must be counterbalanced for the common good (Said, 2014).

However, a rebellious person, despite being usually loyal and engaging, is often full of contradictions which create self-doubt and indecisiveness, too (Taylor, 2006, p. 9). Therefore, in addition to character, the other main quality of a bioethics' rebel is mindfulness. Mindfulness teaches that being a bioethics rebel implies being first a revolutionary in your mind. It's a knowledge obtained from self-discipline and that's all that requires mindfulness to lead us back to the present. In turn, self-discipline is gained when you value your inner voice as a sharp, clear and keen intelligence worthy of the time it needs to mature. Mindfulness, by the way, is a powerful resource vastly used by therapists (Birnbaum & Birnbaum, 2004).

The third quality of a rebel is non-con*form*ism. Those who rule when we develop our personality since childhood want to give a meaning of life to the new generation, increasing their chances of remaining in power and not necessarily a meaning streamlined with our personal interests. A rebel first questions it and then constructs a meaning of life through her individual moral agency. A rebel doesn't accept the shape in which the Gehenna side wants to melt her into, rather than this she chooses one of her own (Camus, 1984).

The more bioethics rebellion spreads, the more annoying and extremists the rebels will be found. The ethical use of technology will play a major role here. Let's start by its use in educational settings. The digital tools, including surrogate realities, must change the way modern political philosophy is learned. The rebel educators must to use them to offer (Affolter, 2015):

- Alternative visions of the partitocratic state (including single-party)
- Explaining why this kind of state has been taken for granted so often
- A big picture of the most important dissidents' work, e.g. Étienne de La Boétie

Of course, when psychological inflexibility rules, asking is betraying. Attacking the presumption of innocence is a mark of necrophilia and a trait of the Gehenna side. But it is better to preserve life than hoping not to die. Therefore, to keep the Gehenna side at bay, Bodi needs a vanguard made up by patient rebels.

WHY IS BODI'S PRAXIS MEANT TO BE GLOBAL?

Because the greatest threats to humankind are global and provoked by humans or grounds on their acquaintance: global warming, WMD escalation… We are also the species that drinks sugar-sweetened drinks, makes puffs of arsenic and nicotine and swallows XXL meals. On the other hand, we have improved many times in History, usually after hitting rock bottom, like after WW2. If we have this capacity of changing for better is because most of us have within the moral agency that takes "*to rebel against injustice and malignancy*" (Hastings, 2017). On this topic, Stuart Mill and Edmund Burke warned us not to allow evil to be committed in our name and how easy it is to happen through inactivity (Hall, 2009, p. 123). Mill and Burke views are close to Camus writings, from which we can conclude that our errors of omission can be as blameworthy as our misdeeds (Ashby & Rich, 2014). Patrick E. Haggerty, co-founder of the ground-breaking technology company Texas Instruments, said in an interview that "*a small group of individuals can change the world if they really want to*" (Altwegg, 1963). The goal of the rebellion is making those little groups to ground on moral legitimacy rather than opportunism and inherited social advantages.

The overt discrimination is more likely to fall when facing the rule of law; it's not the case of the insidious one. Against the subtle discrimination is needed that the people in charge encourage a proactive engagement against (Sturm, 2001). Sadly, the leaders are chipped with technocracy when they are in the school and, henceforth the shared responsibility on social justice turns into calculus (Hargreaves & Shirley, 2009, p. 29). Is in this point when the exclusion of values, beliefs, plural identities happens in the leader's curricula (Crow & Møller, 2017). And, in the long run, it causes overt discrimination to come again.

Unlike the racial and gender marginalisation, the disability (a more global phenomenon) is too often forgotten by the sociology disclosing ethos (Inckle, 2018). Impairment and disability had their major change after WW2. The greatest theoretical apology of medical killing is *Die Freigabe der Vernichtung lebensunwerten Lebens* (The Permission to Destroy Life Unworthy of Life) originally published in 1920 (Binding & Hoche, 2006). The two distinguished German university professors of law and psychiatry, Karl Binding and Alfred Hoche argued in the book that a policy of killing was compassionate and consistent with medical ethics. He pointed to situations where doctors were obliged to destroy life -interrupting a pregnancy to save the mother, for example. However, in the light of inhuman abuses committed by the most developed countries (which were put on the blast only after the Nazi fall) these cravings for unwilling merciful killing vanished -at least in Europe and the Anglosphere. Yet under an ableist attitude, this way of looking over the knowledge created in the margins in favour of those created in the mainstream is a global way of wasting knowledge. Despite the scientific literature on daily life medicalization, not enough is said about the sociology of distress around this phenomena. We can see how this lack of harmony between sociology and bioethics replicates in too many places when it comes to the home accessibility in the refugees' camps (Ivan, 2019, p. 72). As was evidenced during the 63rd session of the United Nations departments, the MDGs are not streamlined enough to face this issue (General Assembly United Nations, 2008, pp. 3–4, 10. 22; Hirsch et al., 2019). The accommodation also can be turned into displacement (Kandylis, 2019).

Furthermore, the current scientific trend points to synthetic genomics as a platform technology, which is suitable for the information age but present two main vulnerabilities:

- The moral obligations might be too liquid when it comes to the platform exploitation
- There is no grant that the processes undermining livelihoods will be faced, however, the processes "*more beneficial than the status quo may not*" (Attfield, 2012)

Thus, a bioethics ready for the incoming changes in human identity needs to shift to biocentrism. Thankfully, the disruptive techs, namely those that can alter the human species identity can turn the genuine discussions back to bioethics. For example, the human improvement in its intergenerational side enjoys today (in 2019) both a well-known prenatal screening and the latest gene editing techniques, such as CRISPR. But who is to say what sorts of people will populate our future?

Another global issue is the widespread cleavage between the redistributive justice of goods and the one of respect and consideration. In this arbitrary separation grounds many politics of pain. Regardless of the clarity that this double standard could bring to the ethical analysis, it shatters the idea of justice. In other words, creating castes and logics around a limited resource unyielding leads to politics of pain since this inequity is mirrored in the space for personal growth and interaction. Therefore, discrimination can happen when the discriminator is *not* present, even globally.

Anyhow, merely identifying vulnerable groups is a poor tool for protecting people from possible harms of research (Hurst, 2008; Levine et al., 2004). In addition to this, as Henk ten Have (2015) warned, the bioethical attention to vulnerability has led to insufficient attention to the social structures that shape human life. The origins of vulnerable groups in bioethics are in *The Belmont Report* (US Department of Health and Human Services, 1976) whose role is enduring when it comes to human subjects research regulations, although it was heavily criticised from its publishing in 1976 (Beauchamp, 2008). The main issue in *The Belmont Report* is its fixation with labels, which conceptually was a blunder. Thankfully, there has been in bioethics a global displacement from labels to layers, which conceptually aims to find solutions and to discern how and by whom those solutions should be implemented (Walker & Fox, 2018).

Identifying the human experience with the rational subjects bioethics imposes a norm on what qualifies as a human and declaring the equality of all human beings "*in dignity and rights*" (United Nations, 1948, p. 2) is insufficient to recognize the lack of social justice and political voice (Rawlinson, 2016, p. xviii). Indeed, *The Universal Declaration of Human Rights*' Preamble also talks about the rightful resort to rebellion (United Nations, 1948, p. 1), but it only really applies when you replace the former regime, so what's the point anyways? Given that "*all's fair*" approach to human rights any country where the women were publicly stoned could merrily sign the *Convention on the Elimination of All Forms of Discrimination Against Women* (CEDAW) (UN General Assembly, 1979) and keep doing it forty years after the declaration.

Besides the globalisation of structural unfairness there are other transcendental reasons for a global bioethics quest. The first of them is that, regardless of an ubiquitous mistrust, the identity of our species is no longer as solid as it was some decades ago, before the feasibility of chimaeras, cloning and tailored children. The second reason is that the burden of disease, impairment and disability is now displaced globally thanks to worldwide vaccination campaigns, a greater life expectancy and both more accessibility and awareness of the lack of it in quick developing economies. The third reason, the future role not of the women, but political-collectivist feminism in bioethics, is such a great game-changer that deserves its own section below this one. The global fertility industry, for example, can only be understood from the structural forces on women's reproductive choices (Rawlinson, 2017).

All this *l'État c'est moi* kind of stuff is stunning bioethics on the basis of social contracts, conflicting interests and glass-hearted structures of power; therefore, bioethics it's getting close and closer to debates on liability and property rather than… ethics (Rawlinson, 2016, p. xix). But humanity is shifting and the rebellion must start.

YES, REBELLION MUST START BUT INDIVIDUAL MINDSET CHANGE COMES FIRST

The most crucial and transformative aspect of a re-evolution lies in the profound change that occurs within each individual's mindset. The Bioethics of Displacement ideology recognizes this fundamental truth and places significant emphasis on the internal transformation of individuals as a catalyst for meaningful societal change.

By challenging the narratives and imposed meanings that have been placed upon them, individuals can claim identities, values, and beliefs of their own, away from vested political interests. Quite the opposite, this transformative process of individual change in mindset enables individuals to critically engage with civic issues. An internal re-evolution of thought to break free from the constraints of societal expectations and genuinely participate in shaping our communities as catalysts for positive change by sceptical questioning of established systems. In summary, the Bioethics of Displacement recognizes that the most vital aspect of a revolution lies in the profound transformation of individuals' mindsets.

The Bioethics of Displacement ideology combines individual mindset change with collective freedoms in an original way that distinguishes it from traditional philosophical approaches. While there are philosophers who have explored related concepts, the Bioethics of Displacement offers a distinct perspective. Let's know more.

Emmanuel Levinas is known for his ethical philosophy centred on the face-to-face encounter with the Other and the responsibility it entails. His work stresses the significance of ethics and the recognition of the other's vulnerability (Levinas, 1979, pp. 256–259). While Levinas provides insights into the ethics of interpersonal relationships, the Bioethics of Displacement expands on these ideas by incorporating the broader context of displacement and its implications for collective freedoms and social structures.

Hannah Arendt explores the relationship between power, authority, and freedom, and emphasises the importance of collective action and public space in the realm of politics. Her insights on the loss of life meaning and the potential for tyranny resonate with the concerns raised in the Bioethics of Displacement (Arendt, 2019, pp. 151–193). However, the Bioethics of Displacement goes beyond Arendt's perspective by incorporating a bioethical and transcendental framework.

Bioethics of Displacement follows the line of Levinas and Arendt and departs from the postmodern authors who came later and their fuzzy leftism that degenerated in the woke ideology. The most relevant of those philosophers is the one who sits on the postmodernity fence. Michel Foucault addressed the combination of individual mindset change and collective freedoms. Foucault's work on biopolitics and the relationship between power, knowledge, and subjectivity emphasises how power operates through disciplinary techniques and societal norms, shaping individuals' beliefs, behaviours, and identities (Foucault, 2019). Foucault's ideas highlight the interconnectedness of individual experiences and collective power structures. However, the Bioethics of Displacement goes further by explicitly addressing the concept of love of belonging, focusing on the meaning of life, and highlighting the unspeakable nature of displacement.

Nussbaum's capabilities approach emphasises the importance of human flourishing and the development of individuals within a just society. Her work delves into the moral and political aspects of social justice (Nussbaum, 2013, pp. 113–124). While Nussbaum's approach touches on collective freedoms and individual capabilities, the Bioethics of Displacement broadens the scope by addressing new specific challenges and ethical considerations related to displacement as loss of life meaning and its impact on both individual mindsets and collective freedoms.

A bit more recently, Seyla Benhabib's work on cosmopolitanism and the ethics of global democracy explores the tension between universal human rights and the preservation of cultural diversity. Her ideas on the relationship between individual agency and the collective are relevant to the discussion of individual mindset change and collective freedoms but Bioethics of Displacement, in addition, delves more into the phenomenological approach of change (Benhabib, 2006, pp. 18–64).

Bodi brings something new to this philosophical table thanks to its unique principles and approaches that combine individual mindset change with collective freedoms. It presents the concept of Preemptive bioethics, which emphasises considering the ethical implications of current trends and future possibilities, going beyond existing frameworks and providing a proactive approach to ethical decision-making. Additionally, the Bioethics of Displacement introduces the Syncretist approach to ethical decision-making, blending different ethical approaches to create a context-dependent and comprehensive framework. This approach recognizes the strengths and weaknesses of various ethical perspectives, striving for a more refined and balanced assessment of moral dilemmas through a novel perspective on ethical decision-making in the face of Displacement and Belonging.

CONCLUSION

Meaningful ethical decision-making, proactive thinking, and the blending of diverse perspectives can help to overcome Displacement issues effectively. But it also applies to social macrocosmos. The success in challenging imposed narratives and societal expectations to reclaim personal identities and values relies upon the transformative power of individual mindset change, it's the real societal progress' engine.

REFERENCES

Affolter, J. (2015). Challenging the State: Teaching Alternative Historiographies in Early Modern Politics. *Metaphilosophy*, *46*(3), 398–413. doi:10.1111/meta.12140

Altwegg, A. (1963, May 12). Why Busy Men Go Extracurricular. *The Dallas Morning News*.

Arendt, H. (2019). *The Human Condition*. University of Chicago Press.

Ashby, M. A., & Rich, L. E. (2014). Two Deaths and a Birth: Reminiscing and Rehashing Principles in Biomedical Ethics. *Journal of Bioethical Inquiry*, *11*(1), 1–4. doi:10.100711673-014-9509-4 PMID:24647806

Attfield, R. (2012). Synthetic Biology, Deontology and Synthetic Bioethics. *Ethics, Policy & Environment*, *15*(1), 29–32. doi:10.1080/21550085.2012.672681

Beauchamp, T. L. (2008). The Belmont report. *The Oxford Textbook of Clinical Research Ethics*, 149–155.

Benhabib, S. (2006). *Another Cosmopolitanism*. Oxford University Press. doi:10.1093/acprof:oso/9780195183221.001.0001

Binding, K., & Hoche, A. (2006). *Die Freigabe der Vernichtung lebensunwerten Lebens: Ihr Maß und ihre Form*. Bwv - Berliner Wissenschafts-Verlag.

Birnbaum, L., & Birnbaum, A. (2004). In Search of Inner Wisdom: Guided Mindfulness Meditation in the Context of Suicide. *TheScientificWorldJournal*, *4*, 216–227. doi:10.1100/tsw.2004.17 PMID:15105961

Blenkinsop, S., & Morse, M. (2017). Saying Yes to Life: The Search for the Rebel Teacher. In B. Jickling & S. Sterling (Eds.), *Post-Sustainability and Environmental Education: Remaking Education for the Future* (pp. 49–61). Springer International Publishing., doi:10.1007/978-3-319-51322-5_4

Camus, A. (1984). *The Fall*. Vintage Books.

Chalise, A. (2018). *Nepalese social structure, atrocities and mental health* [PhD Thesis]. Tribhuvan University.

Collins, S. (2008). *The Hunger Games*. Scholastic Inc.

Cook, K., Keller, D., & Myers, A. (2014). Bioethics in the Hunger Games. *Science Teacher (Normal, Ill.), 81*(1), 31. doi:10.2505/4/tst14_081_01_31

Corbett, J. (2012). *Rebel with a Cause: The Making of a Community Mediator* (*SSRN* Scholarly Paper ID 2030438). Social Science Research Network. https://papers.ssrn.com/abstract=2030438

Crow, G. M., & Møller, J. (2017). Professional Identities of School Leaders across International Contexts: An Introduction and Rationale. *Educational Management Administration & Leadership, 45*(5), 749–758. doi:10.1177/1741143217714485

Foucault, M. (2019). *Discipline and Punish: The Birth of the Prison.* Penguin UK.

General Assembly United Nations. (2008, July 28). Fifth quinquennial review and appraisal of the World Programme of Action concerning Disabled Persons. In *Social Development, Including Questions, Relating to the World Social Situation and to Youth, Ageing, Disabled Persons and the Family*. 63rd Session General Assembly United Nations. https://documents-dds-ny.un.org/doc/UNDOC/GEN/N08/442/48/PDF/N0844248.pdf?OpenElement

Hall, I. (2009). *British International Thinkers from Hobbes to Namier*. Springer. doi:10.1057/9780230101739

Hargreaves, A. P., & Shirley, D. L. (2009). *The Fourth Way: The Inspiring Future for Educational Change*. Corwin Press. doi:10.4135/9781452219523

Hastings, G. (2017). Rebels with a cause: The spiritual dimension of social marketing. *Journal of Social Marketing, 7*(2), 223–232. Advance online publication. doi:10.1108/JSOCM-02-2017-0010

Hirsch, A., Maylea, C., David, C., & Nipperess, S. (2019). The changing face of disability and refugee services and policy in Australia: Implications for social work. *Social Work and Policy Studies: Social Justice, Practice and Theory, 2*(1). https://openjournals.library.sydney.edu.au/index.php/SWPS/article/view/12775

Hurst, S. A. (2008). Vulnerability in research and health care; describing the elephant in the room? *Bioethics, 22*(4), 191–202. doi:10.1111/j.1467-8519.2008.00631.x PMID:18405317

Inckle, K. (2018). Irrational Perspectives and Untenable Positions: Sociology, Madness and Disability. In Subcultures, Bodies and Spaces: Essays on Alternativity and Marginalization (Vol. 1–0, pp. 169–188). Emerald Publishing Limited. doi:10.1108/978-1-78756-511-120181011

Ivan, M. K. (2019). *Protective Obligations Owed by States to Persons with Disabilities under International Human Rights Law, Before, During and After Situations of Armed conflicts* [PhD Thesis]. Cardiff University.

Kahn, S., & Lee, J. (2010). *Activism and Rhetoric: Theories and Contexts for Political Engagement.* Routledge.

Kandylis, G. (2019). Accommodation as Displacement: Notes from Refugee Camps in Greece in 2016. *Journal of Refugee Studies, 32*(1), i12–i21. Advance online publication. doi:10.1093/jrs/fey065

Levinas, E. (1979). *Totality and Infinity: An Essay on Exteriority*. Springer Science & Business Media. doi:10.1007/978-94-009-9342-6

Levine, C., Faden, R., Grady, C., Hammerschmidt, D., Eckenwiler, L., & Sugarman, J. (2004). The limitations of "vulnerability" as a protection for human research participants. *The American Journal of Bioethics, 4*(3), 44–49. doi:10.1080/15265160490497083 PMID:16192138

Nussbaum, M. C. (2013). *Creating Capabilities: The Human Development Approach*. Harvard University Press. doi:10.2307/j.ctt2jbt31

Rawlinson, M. C. (2016). *Just Life: Bioethics and the Future of Sexual Difference*. Columbia University Press. doi:10.7312/rawl17174

Rawlinson, M. C. (2017, March 21). *The Concept of a Feminist Bioethics: IJFAB at Ten (world)* [Letter]. doi:10.3138/ijfab.10.1.1

Said, E. W. (2014). *Culture And Imperialism*. Random House.

Samuels, A. (2016). *I Rebel, Therefore We Are*. New Political Thinking on Individual Responsibility for Group, Society, Culture, and Planet.

Sturm, S. (2001). Second Generation Employment Discrimination: A Structural Approach. *Columbia Law Review*, *101*(3), 458. doi:10.2307/1123737

Taylor, L. E. (2006). The Enneagram: A tool for self-reflection, critical appraisal and introducing spirituality in the classroom. Presented at the *First North American Conference on Spirituality and Social Work*.

UN General Assembly. (1979). *Convention on the elimination of all forms of discrimination against women*. Author.

United Nations. (1948, December 10). *Universal Declaration of Human Rights*. https://www.un.org/en/universal-declaration-human-rights/

US Department of Health and Human Services. (1976). *The Belmont Report*. Office of the Secretary.

Walker, A. K., & Fox, E. L. (2018). Why Marginalization, Not Vulnerability, Can Best Identify People in Need of Special Medical and Nutrition Care. *AMA Journal of Ethics*, *20*(10), 941–947. doi:10.1001/amajethics.2018.941 PMID:30346922

KEY TERMS AND DEFINITIONS

Agency: The power of an individual to act and make choices raising her intentional binding.

Freedom: The ability to choose between meaningful options, without hindrance or external control, allowing personal growth and self-realisation.

Identity: 1. Unique characteristics that make a person or group distinct. 2. The state of being oneself.

Individual: Referring to a single person or thing, separate from a larger group.

Mindset: A person's established set of attitudes, beliefs, and assumptions.

Resistance: The act of opposing or refusing to accept a plan, idea, or authority.

Values: Shared principles or beliefs about what is right, wrong, and important in society and individual lives.

ENDNOTE

1. To know more about pacific resistance and civil disobedience, I recommend Rubén Gisbert YouTube Channel: https://www.youtube.com/@RubenGisbert

Chapter 10
Dignity and Engagement for Making Bioethics of Displacement Real:
Praxis II

ABSTRACT

The author delves into the concept of dignity and how it is under attack in different forms, from the forced donation of a kidney to the underserved population of the Global South. The author explores how the majority population displaces responsibility for a certain outcome and places it on the disadvantaged themselves. It also looks at how political dignity is attacked when individuals are required to reveal their disadvantage to qualify for compensation. However, this chapter also highlights how the concept of dignity can be successfully challenged, as seen in the recognition of three extensive geographical regions in New Zealand as legal persons to Mātauranga Māori, a happy marriage of personhood plus nonhuman dignity. Along with dignity, the author studies engagement because both concepts are essential in Bodi. They bring to light the ways in which displacement can harm individuals and communities and highlight the importance of recognizing and respecting their humanity, agency, and knowledge.

INTRODUCTION

The connection between this chapter and the last one lies in their shared goal of promoting a comprehensive understanding of bioethics and addressing the ethical implications of Displacement and Belonging. Building upon the rebellion and mindset change introduced in Chapter 9, this chapter provides further insight into the complex nature of human dignity and its significance in ethical decision-making.

DOI: 10.4018/978-1-6684-4808-3.ch010

ON DIGNITIES

Individual Dignity

There are countless attacks that can be inflicted against individual dignity. Nevertheless, measuring a person's freedom in her body parts is a wound of a kind in the victims' dignity. Mississippi governor Haley Barbour promised to release the inmate Gladys Scott, who was serving two life sentences for armed robbery after sixteen years of prison with the condition that she donate one of her kidneys to her sister Jamie, sentenced for life along with Gladys (Jefferson-Jones, 2013). The sisters (19 and 21 years old then) had no previous criminal record, nobody was hurt and the robbery netted no more than eleven dollars (Herbert, 2010). Perhaps, the life sentence was caused by the aggravation of walking by a Black skin in Mississippi (Donaldson, 2010). Perhaps, as the sisters' mother Evelyn Rasco declared, it was a payback for her family's testimony against a corrupt sheriff (Pitts Jr., 2010). It was a true bioethics debate on moral repugnance.

Figure 1. Gov. Haley Barbour overcoming repugnance thanks to the Trumpist mindset: "These sisters brought drugs. These sisters brought crime. These sisters are rapists."
Source: Skidmore (2011)

Either from a position of power or powerless, human beings daily bitterly belittle each other in order to give oneself an extra value and when this value can be showed off without being shrunk-shun by others, it's (in a common, non-philosophical sense) our dignity (Cabrera, 2018). For example, a garbage collector who sings and shows himself joyful in the work and doesn't show any concern about if the others think that he is surrounded in crap and flies, has indeed a great deal of dignity. The tricky side is that he has the same dignity even if he is playing the happy shiny people while, privately, sees himself as a dung beetle. When the quest for dignity at work starts with a low salary in a non-glamorous position you are against all odds, besides the odds of being humbled and even pornified (Lozano Rodriguez, 2021). For most people, dignity is in the street.

Group Dignity

A very common way to twist the dignity of a group of displaced persons is a populationist approach. For example, in 2016, during the Zika epidemic, it was found that pregnant women with the virus faced the risk of having babies with microcephaly (Mlakar et al., 2016). The rescue stakeholders, governments and international organisations, encouraged women to delay pregnancy and sent personnel to survey their domestic spaces and water management practices. However, neither the quality reproductive health barriers nor the women's particular vulnerabilities to climate change were criticised (Rivera-Amarillo & Camargo, 2019). In this surreptitious way, a population of underserved women from the Global South *were put in charge* of keeping at bay an international epidemic by no salary while maintaining the house clean and dry. And the sum of the cynicism is that the Zika virus was soon associated with global warming! (Paz & Semenza, 2016)

As a general rule, there will be a technological gap between the displaced group of population and the logic of power will try to cheapen the expenditure of the biopolitics applied over. One of the options will be employing autonomous weapon systems (AWS) or killer-robots. It has been argued that their use in war is against human dignity, but the AWS (besides their singularity) are not against human dignity in an unseen form when compared with existing weapons (Sharkey, 2019). So the AWS use for controlling displaced people should be criticised on a case-by-case basis using not only dignity-related arguments.

The displaced persons and the transhumanists (whether they are displaced individuals or not) are the most prone subjects to undergo the most daring DNA and DNA germline modification. The disempowered individuals offer a higher chance of mistakes being ignored or dealt with a favourable economic agreement, so the displaced persons may result in more appealing for the people in charge of these experiments. Yes, I mean illiterate sharecroppers turned into guinea-pigs. The transhumanists are more likely to accept being a subject of research on ideological grounds or as a previous step before adopting a new enhancement. The main exploits of human dignity and the most obvious case in which the modification must not to be performed are:

- Whenever it is not crystal clear who is entitled to accept and perform the modification and why
- There is any suspicion on designers or physicians -*any*
- One of the main drives to participate is the advice or example of an influential figure
- It's justified in a significant degree by the benevolence, goodness, prestige or trustworthiness of researchers and funders
- Grounds on "*abstaining from intervention is already a form of intervention*" and similar reasoning aimed to lightening the consent

Very likely, the cyborgs will be at any point a displaced group with the special vulnerability that their corpses can be worth a wealth of money in technology but also in data from the corpse analysis in order to create better enhancements. Given that the death of a cyborg is an issue in which the clinical research and bioethics won't agree easily, the professionals must wisely balance her dignity after death and the educational value of the corpse to the cybernetic researchers, cyborg community and general public.

Another way of sapping the dignity of a person and badly displacing the lowest economic segment is to allow bureaucracy to multiply. For example, if a person has been exposed to a carcinogen in the work and suffers a cancer should be presumed that this substance mainly contributed to it, in addition to this, it's unethical and plain overconsumption to put her in line with the other thousands people awaiting for an identical case to be solved (Shrader-Frechette, 1991, p. 202).

The most harming way of belittling the dignity of a population group, however, is publicly judging the value they give to life. Let's put an example of this. The General William Westmoreland, the US commander in VietNam in the peak of the conflict said:

The Oriental doesn't put the same high price on life as does a Westerner... We value life and dignity. They don't care about life and human dignity. (Jackson, 2005).[1]

In Phi Phu, Trieu Ai, My Luoc and so many other hamlets, decades after the conflict ended, villagers still mourn loved ones who were slain in "*the American War*" (Turse, 2013).

Political Dignity

Too many times, the majority population displaces responsibility for a certain outcome, as well as responsibility for wider social structures, and places it on the disadvantaged themselves. Given that the respect is dictated by the opinions of the dominant group, the options are not very open, either. More even considering that the dominant group can't see the inhuman conditions in which the poor live... in a clinical sense. The individuals affected with "*poverty blindness*" disorder can see the poor with the eyes while bouncing poverty from consciousness (Ventres & Gusoff, 2014). Considering that the prevalence of poverty blindness rises with income the persons who suffer this blindness mostly don't suffer it at all! In addition to this, the in-ward postmodernity turns the shame (and many, if not all, costs borne by disabled persons) to the politically disadvantaged ones. To cap it all, there is the indelible harm of the state sponsored disrespect of requiring individuals to reveal and bear her disadvantage in order to qualify for compensation. The mistrust and intrusiveness are not openly discussed layers of welfare that act against disempowered people's dignity (mostly disabled and migrants) (Popescu, 2018, p. 111). This kind of attack against political dignity is like that suffered by the impoverished women who faced the Zika virus but in a different context.

The *Western-more-land* hawkish mindset doesn't exhaust itself in group dignity violations. It's soaked in political hypocrisy. The kind of political hypocrisy, for example, that allows a hunger strike in Guantanamo to the same people who would be extrajudicially killed if not in the US (Kelly, 2015). A harsher example. An all-curing privileged position of medicine only can be obtained eliminating those who cannot be cured, mainly mentally handicapped persons, that's eugenics policies (McKenny, 1997, p. 147). The Nazi doctors (unlike the Soviet or Chilean) differ in their biological visions that link eugenics and national healing in which they were a sort of self-appointed key (Lifton, 1988, p. 42). If the bioethics turn its back to the medically disenfranchised it would contribute to its fragility before authorities blinded to their own partiality (International Association of Bioethics, 1999, pp. 8–9). The victims of medical atrocities were not generics Jews, Gypsies, Afro-American, etc. but they were political targets who previously were targeted as vile.[2]

I don't want to end the section of dignity without a cheery content. The dignity, as so many things, is colonised and it's a hindrance for inclusive democracy and sometimes is successfully challenged. For example, the Māori philosophical concepts of *mauri*, *tapu* and *mana* are linked with an unique event happened in New Zealand in which three extensive geographical regions (Te Awa Tupua, Te Urewera and Mount Taranaki) were recognized as legal persons to Mātauranga Māori (Winter, 2019). It's, from a Western perspective, an unheard happy marriage of personhood plus nonhuman dignity.

ENGAGING BIOETHICS OF DISPLACEMENT

Solid Engagement

The tormented Wittgenstein asks once how we know that the indigenous habitants of an island are understanding us when not speaking a common language. Then, he said: "*The common behavior of mankind is the system of reference by means of which we interpret an unknown language*" (Wittgenstein, 2009, para. 206). Many contents covered for the first time in this book are a newfound island with plenty of natives with an *uncommon behavior*. Thankfully, the references for understanding this aboriginal language are not only their actions, but also the action inside ours (our bioethical baggage), which we may share or not (Condé, 2014, p. 54). Therefore, we can learn more about this uncommon behaviour by studying the natives' diet, demography, institutions or endemic maladies. Also we can go to the square, buy some fruits or dance with them at their parties. The first, is a solid engagement whose results are expected to be lasting and falsifiable. Recollecting scattered experiences like in the second way always has an abstract and subjective outcome. Since we are going to live on the island with the natives for a while we should learn their language, that's the logic behind the events. To do this, we will need both engagements.

A solid engagement requires data analysis readiness. Public health increasingly relies on Big Data and in the near-future won't be longer feasible to rely continually and solely on individual consent, so a governance of data collection and use across labs, clinics, and health care settings is required (Knoppers & Thorogood, 2017). Likewise, the research in this branch of Bioethics requires expertise handling and incorporating the best value tools for the systematic, unbiased analysis of large amounts of data, very specially the underestimated qualitative ones (Martínez-García et al., 2019).

Given that a Bioethics of Displacement solid engagement can contribute in new ways to many fields, it will make losing focus far easier. Losing focus eventually will lead to bottlenecks in debate and a drought of new insights. In those cases, my proposal goes around three points:

- To advocate for open data and to share pre-competitive ones about refugees, chems, cybernetics, surrogate realities…
- To mine social media in order to make debates proactive and more appealing
- To keep the ball rolling: foster public-private partnerships in areas of interest through publications, podcasts, etc.

Even in lean years, a solid engagement with Bioethics of Displacement also needs researchers *go social*. All branches of bioethics benefit from the development of different forms of interdisciplinary learning that is obtained to a notable degree through conferences (Chrysanthou & Nottingham, 2018).

In order to prevent annoying ethical dilemmas like the ones in which resolving a precept leads to violating another, the bioethicists of displacement should be open to inquire about the hierarchy of field research. Neither conceding the need of control nor being disrespectful to the persons in charge could be a good attitude to prevent principles getting biassed due to managerial interests. Let's see it with an example. Figure two researches on abortion, one funded by the conservative The Heritage Foundation and the other by the National Abortion Rights Action League (NARAL).[3,4] The comparison between abortion and killing or removal very likely will be on the table. Indeed, applied bioethics engages with this kind of concrete moral issues and one of its most used philosophical tools is the comparison. But who makes those comparisons in research? Likely, the persons who manage this research, in our case, the report editors from The Heritage Foundation and NARAL. So, when it comes to discerning between killing and letting die, like in this example, the decision-maker's comparisons will be close to their affine circle and, therefore, the ethics of the research may result, at least, skewed. To sum up, the comparisons attempting to produce cases that are exactly alike (except for the moral issue at hand!) are "*doomed to failure*" and hinders the debate progress (Asscher, 2007).

Besides research hierarchies, the Bodi settings of research themselves will prove to be quite challenging. Those researchers who enter confinement settings might be shocked by their very own institutional and relational dynamics (Gomes & Duarte, 2018). Given that unusual settings will be the rule when it comes to Bodi, a solid engagement also will require a great deal of flexibility. It means switching between formal and informal procedures, knowing how to distance from the social actors involved (it includes proxemics) and dealing with strangeness.

Liquid Engagement

The liquid engagement is streamlined with Edmund Husserl's criticism about how science has been losing touch with the intuitive evidence that would legitimise it due to its lame dependence on a formal-mathematical conceptual apparatus (Husserl, 1970, pp. 49, 125–135). Besides that, there are questions like what is human life and also when does it begin and when does it end for which rational thinking is out of place and some other type of research is needed if we want to grasp new meanings (Iftime, 2006). Thankfully, there is an alternative but well-documented synergy between technological progress in medicine and philosophy of care with artistic vanguards (Cattorini, 2019). Thus, new insights in clinical bioethics can be drawn from aesthetic contributions.

The cinema and series are an invaluable way of liquid engagement with Bodi. A philosophical filmmaker influenced by Husserl was Orson Welles (Singer, 2004). When he walked into a movie set, he had the gift of pinpointing the best place for the camera without ever showing a moment of doubt and getting the greatest impact on narration (Kapell & McVeigh, 2011, p. 44). The intellectually amphibious Wittgenstein is also present in the liquid approach as he was in the solid to highlight the power of a "*picture*" to regulate action and thought (Wittgenstein, 2007, pp. 25–43, 63–66). Great is the moral importance of non-moral descriptions (Nash, 2011).

The liquid engagement with Bioethics of Displacement issues should be this *orsonwellian* gifted way to lock on the most sensitive quandaries in a way any new insight provokes the greatest social impact. In the same manner that a solid approach to Bodi requires a visual participatory research, the liquid one requires a smart use of visual bioethics. Bioethics would get denatured without visual arguments (Chubak, 2008; Lauritzen, 2008). For example, a heart transplantation can be viewed from arts, biosciences and humanities but we can also merely behold the three perspectives at once, allowing a full visual layer of expression. In this case, a "*dynamic entangling of multiple data produces a constitutive assemblage of elements in which no one can claim priority*" and "*a fuller comprehension of the significance and experience of heart transplantation in both theory and practice*" happens (Shildrick et al., 2018).

A striking feature of any inquirer is imagination. Relaxing and daydreaming while looking for counterexamples for a justifiable true belief analysis of knowledge is an old friend for the armchair philosophy (Williamson, 2005). Imagination is a must for a successful liquid engagement. Many biotechnological creations are proper of a sci-fi work, till point that the last is a good and clear analogy of the former (Brem & Anijar, 2003). If you were a wealthy half-baked transhumanist that has paid for every guru talk, you'd tear your hair out after becoming aware of how much of those contents have come from 1950s inexpensive classics works of science fiction and how much from publications by distinguished researchers (Clayton, 2013).

And vice versa: Star Wars films have been used to teach about psychiatric disorders ranging from "*obsessive-compulsive personality disorder in C3PO to psychopathy in Jabba the Hutt to malingering by Yoda*" (Friedman et al., 2018).[5] Even NASA has embraced Star Trek! (Anijar, 2000) If cinema is the main medium in a liquid engagement, then sci-fi with shades of space opera and cyber-punk is its genre.

Besides films, another medium of engagement with the Bioethics of Displacement is the novel. The role that literature might play in bridging gaps between bioethics, neuroscience, and the humanities has been underrated. That's unfair. While many researchers on neurological memory augmentation pays only lip service to human enhancement, fiction writers such as Larissa Lai and William Gibson have explored issues with women without bodily autonomy, men severed from corporeal reality, governments in shambles, mass amnesia, slavery, and "*mind death.*" (Giaimo, 2016) And more importantly, has generated the need to learn more about bioethics.

But also twisted things can be learned. The Russian conservative revolutionaries (a mix of neo-Nazis and Bolsheviks) have successfully entertained its geopolitical vengefulness with their own imperial sci-fi sub-genre, which paradoxically contains a sort of postcolonial sensibility (Suslov, 2016). For example, the Kremlin insider Konstantin Rykov's *Etnogenez* series of science fiction and fantasy novels are an imperial nostalgia blockbuster with a strong Russian exceptionalist bias (Bassin & Kotkina, 2016). It's an example of how technology can be the new opiate of the masses (Villaverde & Carter, 2013).

Despite that those *nazbol* novels are a deviant influence where a culture is above the rest, but thankfully, social protest sci-fi subgenre, can be written and drawn in funny, thrilling and inclusive ways. To revert this evil the youth should be the main target of a Bioethics of Displacement liquid engagement creations. We must expose how the teenagers' body and mind are a site of resistance and control at the heart of strongly hawkish representations and policies. It can be done through parody, denunciation and open scorn of the tension between the banality and the threats implicit in the hawkish contents. A good starting point to learn more about banality and threat is the 1950s adolescent short educational films on how to become successful members of American society (Ervin-Ward, 2009).

However, a cure of reality is integral with the liquid engagement. At the end of the day, it's a book on Bioethics not about making your own blockbuster! A dedicated kind of journalism is the last key in Bioethics of Displacement. Modern bioethics flourished such lust as discipline thanks to investigative science reporters and whistle-blowers in the Tuskegee case (Rosenfeld, 1999). In the turn of the millennium, bioethical issues kept gaining newsworthiness as the media influenced public policy and informed personal decision making, but there was room for improvement (Goodman, 1999). Back to the present 2019, the Bioethics of Displacement needs a watchdog political journalism focused in finding the responsible of humanitarian disasters and research blunders gratuitous vulnerability.

Gel Engagement

The place between the solid and liquid engagements is not void. In the states of the matter this place is occupied by the gel state, so I've named this engagement over it. Besides ethical dilemmas and strangeness, another joint point between the solid and liquid engagements might be the participatory visual research. Being participatory and visual is important for a research since it pushes it beyond the "*do no harm*" principle (democratising research and avoiding the risk of retraumatization, for example) and has been used to help displaced persons before (Weber, 2018).

As I see it, the transhumanists and non-political feminists (if there is any left) are the closest groups to enrich the Bodi pool of knowledge with an alternative metaphysics of research suitable for this proposal. The transhumanist approach to a non-essentialist ontology in which the object and subject of study are one, does *"not displace established research practices of inquiry*" but leads to a non-teleological framework (Mauthner 2018, 63). It would be especially disruptive for critical bioethics since this framework turns qualitative research into metaphysical practices (Mauthner, 2018, p. 65). In addition to this, the feminist-materialist view recognizes all sorts of bodies (even non-human ones) as having agency, which "*displaces the human as the principle ground for knowledge*" (Ivinson & Taylor, 2013, p. 666).

Turning back to the Wittgenstein example, in the future we might not need to understand the natives' languages, but the meta-language that the island itself talks. It seems far complex and looks like the place where we are moving to. Will the human displacement lead to a *many everythings* world? Is this new perspective with no up and down, no gravity, a natural change in a specie about to travel the outer space? What a gas or plasma engagements would mean? What would they bring?

CONCLUSION

The different dimensions of dignity: individual, group, and political can be undermined through continuous instances of power imbalances and injustices. For this reason, it's important to recognize and upholding the dignity of all individuals and communities. In this task, Bodi can be engaged through solid engagement, liquid engagement, and gel engagement. Solid engagement involves actively learning from underserved communities, while liquid engagement incorporates artistic perspectives. Gel engagement encourages participatory and visual research.

REFERENCES

Anijar, K. (2000). *Teachings towards the twenty-forth century: Star Trek as social curriculum*. Routledge.

Asscher, J. (2007). Killing and Letting Die: The Similarity Criterion. *Journal of Applied Philosophy*, *24*(3), 271–282. doi:10.1111/j.1468-5930.2007.00381.x

Bassin, M., & Kotkina, I. (2016). The Etnogenez Project: Ideology and Science Fiction in Putin's Russia. *Utopian Studies*, *27*(1), 53–76. doi:10.5325/utopianstudies.27.1.0053

Brem, S. K., & Anijar, K. Z. (2003). The Bioethics of Fiction: The Chimera in Film and Print. *The American Journal of Bioethics*, *3*(3), 22–24. doi:10.1162/15265160360706787 PMID:14594476

Cabrera, J. (2018). *Discomfort and Moral Impediment: The Human Situation, Radical Bioethics and Procreation*. Cambridge Scholars Publishing.

Cattorini, P. M. (2019). Narrative bioethics and film. Some historical remarks. *Medical History*, *3*(2), 110–123.

Chrysanthou, A., & Nottingham, E. (2018). Conferences as a Tool for Interdisciplinary (Postgraduate) Learning in Bioethics. *JAHR*, *9*(2), 234–246. doi:10.21860/j.9.2.6

Chubak, B. (2008). Visual Bioethics: Seeing is Believing? *The American Journal of Bioethics*, *8*(12), 58–60. doi:10.1080/15265160802521047 PMID:19085481

Clayton, J. (2013). The Ridicule of Time: Science Fiction, Bioethics, and the Posthuman. *American Literary History*, *25*(2), 317–343. doi:10.1093/alh/ajt005

Condé, M. (2014). Wittgenstein's grammar and the contemporary scientific rationality. *Analytische Und Kontinentale Philosophie: Perspektiven Und Methoden Analytical and Continental Philosophy: Methods And*, 52. https://www.alws.at/alws/wp-content/uploads/2018/06/papers-2014.pdf#page=53

Donaldson, S. (2010, December 30). Sisters Given Life for $11 Crime Released. *ABC News*. https://abcnews.go.com/US/scott-sisters-released-16-years-prison-kidney-deal/story?id=12508754

Ervin-Ward, A. (2009). Doing It for the Kids: Rebels and Prom Queens in the Cold War Classroom Film. *Colloquy*, *18*, 160–161.

Friedman, S. H., Hall, R. C. W., & Appel, J. M. (2018). The Last Jedi Takes His Own Life: Rational Suicide and Homicide-Suicide in Star Wars. *Academic Psychiatry*, *42*(4), 503–509. doi:10.100740596-018-0938-y PMID:29797244

Giaimo, G. N. (2016). Memory, Brains, and Narratives? The Humanities as a Testing-Ground for Bioethical Scenario-Building. *Literature and Medicine*, *34*(1), 53–78. doi:10.1353/lm.2016.0003 PMID:27545476

Gomes, S., & Duarte, V. (2018). What about ethics? Developing qualitative research in confinement settings. *European Journal of Criminology*. Advance online publication. doi:10.1177/1477370818801305

Goodman, K. K. W. (1999). Philosophy as News: Bioethics, Journalism and Public Policy. *The Journal of Medicine and Philosophy*, *24*(2), 181–200. doi:10.1076/jmep.24.2.181.2533 PMID:10344425

Herbert, B. (2010, October 12). So Utterly Inhumane. *The New York Times*. https://www.nytimes.com/2010/10/12/opinion/12herbert.html

Hughey, M. W. (2017). Bad hombres? The implicit and explicit racialization of immigration. *Humanity & Society*, *41*(1), 127–129. doi:10.1177/0160597616680016

Husserl, E. (1970). *The Crisis of European Sciences and Transcendental Phenomenology: An Introduction to Phenomenological Philosophy*. Northwestern University Press.

Iftime, O. (2006). Life sciences, apophatism and Bioethics. *European Journal of Science and Theology*, *2*(1), 21–46.

International Association of Bioethics. (1999). *Embodying Bioethics: Recent Feminist Advances*. Rowman & Littlefield.

Ivinson, G. M., & Taylor, C. A. (2013). Material Feminisms: New Directions for Education. *Gender and Education, 25*(6), 665–670. doi:10.1080/09540253.2013.834617

Jackson, D. Z. (2005, July 22). The Westmoreland mind set. *The New York Times*. https://www.nytimes.com/2005/07/22/opinion/derrick-z-jackson-the-westmoreland-mindset.html

Jefferson-Jones, J. (2013). The Exchange of Inmate Organs for Liberty: Diminishing the Yuck Factor in the Bioethics Repugnance Debate. *SSRN, 16*, 105. doi:10.2139srn.2251248

Kapell, M. W., & McVeigh, S. (2011). *The Films of James Cameron: Critical Essays*. McFarland.

Kelly, M. G. E. (2015). *Biopolitical Imperialism*. John Hunt Publishing.

Knoppers, B. M., & Thorogood, A. M. (2017). Ethics and Big Data in health. *Current Opinion in Systems Biology, 4*, 53–57. doi:10.1016/j.coisb.2017.07.001

Lauritzen, P. (2008). Visual bioethics. *The American Journal of Bioethics, 8*(12), 50–56. doi:10.1080/15265160802559146 PMID:19085479

Lifton, R. J. (1988). *The Nazi Doctors: Medical Killing And The Psychology Of Genocide*. Basic Books.

Lozano Rodríguez, M. (2021). Don't Let Kellys Go to Hell: From Welfarism to Misery in the Cleaning Ladies' Sector in Spain. Handbook of Research on the Impact of COVID-19 on Marginalized Populations and Support for the Future.

Martínez-García, M., Vallejo, M., Hernández-Lemus, E., & Álvarez-Díaz, J. A. (2019). Novel methods of qualitative analysis for health policy research. *Health Research Policy and Systems, 17*(1), 6. doi:10.118612961-018-0404-z PMID:30642358

Mauthner, N. (2018). A Posthumanist Ethics of Mattering: New Materialisms and the Ethical Practice of Inquiry. In R. Iphofen & M. Tolich (Eds.), *The SAGE Handbook of Qualitative Research Ethics*. SAGE. doi:10.4135/9781526435446.n4

McKenny, G. P. (1997). *To Relieve the Human Condition: Bioethics, Technology, and the Body*. SUNY Press.

Mlakar, J., Korva, M., Tul, N., Popović, M., Poljšak-Prijatelj, M., Mraz, J., Kolenc, M., Resman Rus, K., Vesnaver Vipotnik, T., Fabjan Vodušek, V., Vizjak, A., Pižem, J., Petrovec, M., & Avšič Županc, T. (2016). Zika Virus Associated with Microcephaly. *The New England Journal of Medicine, 374*(10), 951–958. doi:10.1056/NEJMoa1600651 PMID:26862926

Nash, M. (2011). Wittgenstein and Aesthetic Reasoning with Stories in the Bioethics Classroom. *Doctoral Dissertations*. https://trace.tennessee.edu/utk_graddiss/1105

Paz, S., & Semenza, J. C. (2016). El Niño and climate change—Contributing factors in the dispersal of Zika virus in the Americas? *Lancet, 387*(10020), 745. doi:10.1016/S0140-6736(16)00256-7 PMID:26850984

Pitts, L., Jr. (2010, November 22). A tragedy of 'justice'. *The Spokesman-Review*. https://www.spokesman.com/stories/2010/nov/22/a-tragedy-of-justice/

Popescu, D.-E. (2018). *Dynamic injustice: Interlocking recognition and distribution* [PhD Thesis]. The London School of Economics and Political Science (LSE).

Rivera-Amarillo, C., & Camargo, A. (2019). Zika assemblages: Women, populationism, and the geographies of epidemiological surveillance. *Gender, Place and Culture*, *0*(0), 1–17. doi:10.1080/0966369X.2018.1555518

Rosenfeld, A. (1999). The Journalist's Role in Bioethics. *The Journal of Medicine and Philosophy: A Forum for Bioethics and Philosophy of Medicine, 24*(2), 108–129. 10.1076/jmep.24.2.108.2532

Sharkey, A. (2019). Autonomous weapons systems, killer robots and human dignity. *Ethics and Information Technology*, *21*(2), 75–87. doi:10.100710676-018-9494-0

Shildrick, M., Carnie, A., Wright, A., McKeever, P., Jan, E. H.-C., Luca, E. D., Bachmann, I., Abbey, S., Bo, D. D., Poole, J., El-Sheikh, T., & Ross, H. (2018). Messy entanglements: Research assemblages in heart transplantation discourses and practices. *Medical Humanities*, *44*(1), 46–54. doi:10.1136/medhum-2017-011212 PMID:28972037

Shrader-Frechette, K. S. (1991). *Risk and Rationality: Philosophical Foundations for Populist Reforms*. University of California Press. doi:10.1525/9780520320789

Singer, I. (2004). *Three Philosophical Filmmakers: Hitchcock, Welles*. Renoir.

Skidmore, G. (2011). *Haley Barbour speaking at CPAC in Washington D.C. on February 12, 2011*. https://commons.wikimedia.org/wiki/File:Haley_Barbour_by_Gage_Skidmore.jpg

Suslov, M. (2016). Of Planets and Trenches: Imperial Science Fiction in Contemporary Russia. *The Russian Review*, *75*(4), 562–578. doi:10.1111/russ.12102

Turse, N. (2013, October 9). For America, Life Was Cheap in Vietnam. *The New York Times*. https://www.nytimes.com/2013/10/10/opinion/for-america-life-was-cheap-in-vietnam.html

Ventres, W., & Gusoff, G. (2014). Poverty Blindness: Exploring the Diagnosis and Treatment of an Epidemic Condition. *Journal of Health Care for the Poor and Underserved*, *25*(1), 52–62. doi:10.1353/hpu.2014.0025 PMID:24509012

Villaverde, L. E., & Carter, R. A. (2013). Singularity, Cyborgs, Drones, Replicants and Avatars: Coming to Terms with the Digital Self. *Science Fiction and Speculative Fiction*, 119–131. https://doi.org/doi:10.1007/9789462093805_008

Weber, S. (2018). Participatory Visual Research with Displaced Persons: 'Listening' to Post-conflict Experiences through the Visual. *Journal of Refugee Studies*. Advance online publication. doi:10.1093/jrs/fey038

Williamson, T. (2005). I *—Armchair Philosophy, Metaphysical Modality and Counterfactual Thinking. *Proceedings of the Aristotelian Society*, *105*(1), 1–23. doi:10.1111/j.0066-7373.2004.00100.x

Winter, C. J. (2019, February). *Decolonising Dignity for Inclusive Democracy*. https://doi.org/info:doi/10.3197/096327119X15445433913550

Wittgenstein, L. (2007). *Wittgenstein: Lectures and Conversations on Aesthetics, Psychology and Religious Belief*. University of California Press.

Wittgenstein, L. (2009). *Philosophical Investigations*. John Wiley & Sons.

KEY TERMS AND DEFINITIONS

Engagement in Bioethics of Displacement: Active participation in research, storytelling, and imaginative exploration of ethical dilemmas.

Group Dignity: The unconditional recognition of collective contributions and suffered harms and persecutions.

Human Rights: The fundamental rights and freedoms that every person should have. In practice, there is one core human right far away from being satisfied: the power to prevent and avoid ill-treatment by any government.

Individual Dignity: The inherent and unconditional worth of every person, beyond usefulness or abilities.

Political Dignity: The sense of worth and respect influenced by not being blamed for disadvantages and being entitled to challenge societal values.

Poverty: A state of extreme lack, where people don't have enough money or services for meeting basic needs, developing intellectually and enjoying a decent life.

Power: A situation in which an individual can influence or control others, often through authority, dominance, or control of resources.

Respect: Treating others with consideration, and acknowledging their rights and dignity as fellow individuals.

Social Justice: the moral duty of fighting the arbitrary deprivation of opportunities.

ENDNOTES

1. Cf. with "*They're bringing drugs. They're bringing crime. They're rapists.*" Donald Trump, on June 16, 2015. Quoted from Matthew W. Hughey (Hughey, 2017).
2. Even if some eugenics proponents were Jews, their programmed extermination was solely on political not eugenics grounds.
3. https://www.heritage.org/
4. https://www.prochoiceamerica.org/
5. Furthermore, the Disney's Star War trilogy is itself a form of psychiatric disorder.

Chapter 11
The Paradigmatic Displacement of Migration:
Belonging v Set Apart I

ABSTRACT

The author traverses the complex issues surrounding migration in the context of the current refugee crisis. The author argues that the main reason for migration is the lack of economic diversity, which leads to displacement of local human networks and infrastructure. To cap it all, the study of migration is often limited by assumptions and neglect from liberal democracies, resulting in a lack of understanding about the true impact of migration on global social change. Thus, the author discusses the philosophical grounds of migration, the biopolitics of migration, and the phenomenology of migration, highlighting the challenges faced by immigrants and the structural barriers they encounter. The author also examines the public health implications of migration and argues that policies aimed at reducing discrimination are needed. It takes a call for a deeper understanding of the human experience of migration and a subsequent more humane response to the refugee crisis.

INTRODUCTION

This chapter continues the discussion on dignity and displacement from Chapter 10, exploring the context of migration and its philosophical, public health, biopolitical, and phenomenological aspects. Building upon the understanding of individual and group dignity, the chapter delves into the challenges faced by displaced individuals and communities in the wake of forced migration. It highlights the complex dynamics between economic diversity, global markets, and the stigmatisation of displaced persons as unemployment bearers. By linking this chapter with Chapter 10, we enjoy a comprehensive examination of the multifaceted nature of displacement, dignity, and the challenges faced by displaced individuals and communities.

DOI: 10.4018/978-1-6684-4808-3.ch011

THE PHILOSOPHY AND PUBLIC HEALTH IMPLICATIONS OF MIGRATION

Context

The refugee movement and forced migration happening since the 2008 global financial crisis is surging. In 2017 the United Nations High Commissioner for Refugees (UNHCR) recorded the highest ever figure of displaced people:

Figure 1. Forcibly displaced persons trend 2003-2017
Source: Retrieved from the UNHCR website[1]

The common views both in secular and some religious traditions is that the borders have *a moral weight* although it doesn't excuse people from their moral responsibilities towards the displaced people who come in (Hollenbach, 2016). However, the people displaced within their own borders are almost three fold those who arrive in other countries (11.8 to 4.4 million according to UNHCR data referenced above).

A new layer of complexity is added when considering the possible consequences of a country changing its ethnicity. For example, according to the 2015 US Census Bureau projections, in 2020 non-Hispanic white children were going to be a minority and in 2044 persons of non-Hispanic white ethnicity will no longer be the majority in the USA.[2] Of course these changing demographics may alter the use of the term 'minority'. In 2015, about 1 out of 8 residents were born out of the USA.

While the materially developed countries try their best to keep the displaced population away, their dominant admixture is also displaced by a majority of racial minorities:

In 2010, more than 15 million of the 25 million displaced persons worldwide came from African countries (Deng, 2010, p. 109). The major shift currently in Europe is Brexit, which is the result of the conservatives' insane animus against the migrants (Berry, 2018; Laurent, 2018). Let's see one of the most significant nonsensical remarks of Theresa May about migrants:

Figure 2. Projecting majority-minority in the US by age and ethnicity group
Source: Retrieved from a US Census Bureau poster[3]

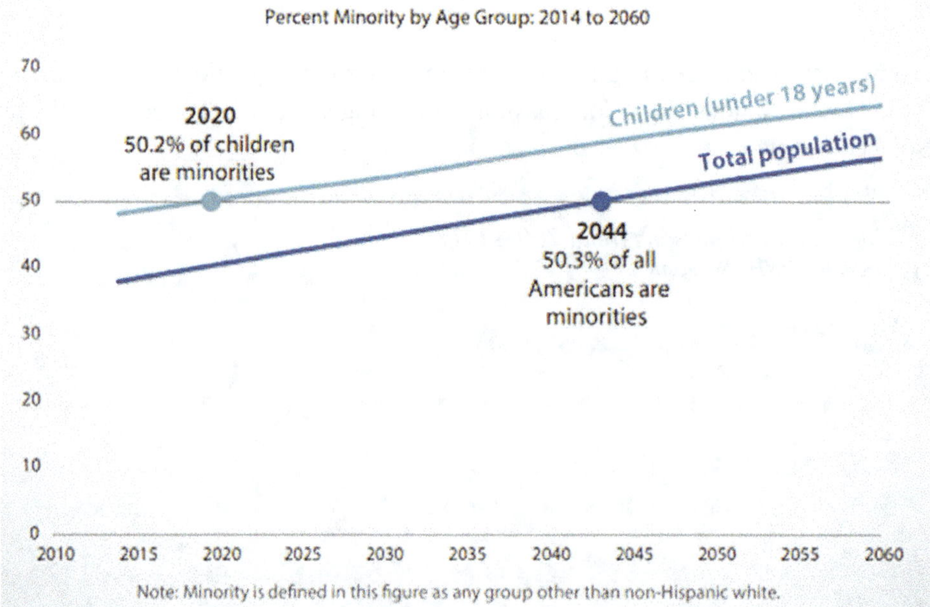

Figure 3. The French soccer team during France's World Cup 1998
Source: Shergold (2016)

Figure 4. The French soccer team poses for photographers before a match against Cameroon on May 30, 2016 in Nantes, France
Source: Tharoor (2016)

When immigration is too high, when the pace of change is too fast, it's impossible to build a cohesive society. (Dominiczak et al., 2015)

Indeed, the *Brexiteers* promised their voters the better of the *two worlds*: a Hobbiton in one hand and a City of London in the other. Obviously, it's not the first time that a bunch of hardhats promise the moon in a stick and the outcome in this case was envisaged: Brexit soon became a "*small, boring and stupid*" topic (Heath, 2018). Brexit-like views thrive thanks to misinformation. What happened in the UK was not so different from what happened in the US in the middle of the 19th century when nativism preached a fear-inducing mythology like the one preached today by the brexiteers. If we look at the figure of a theoretically antagonistic figure of May like Millard Fillmore, 13th US president, a whig (the last one) and, unlike May, a foreign policy skilled individual who shared, nevertheless, her anti-immigration views, then we'll find a common pattern: the clear stroke that goes from the mediocrity to the xenophobia (Daftary, 2018; McKinstry, 2018; Prior, 2010).

The overstated dubious morality stressed that the main issue in the immigration is not the displaced people and refugees. In spite of the fact that humanitarian crises are in our face, it's an error to think that there is *a* political solution for this, not even that it's *a* political or geopolitical issue. To be a geographically displaced person is not only a social condition but also a theological and ethical one (which stresses the importance of interfaith dialogues, by the way) (Orobatar, 2010). As far as I know, the same goes for any person in need of hospitality, which includes (usually to a lesser extent) the most non-geographically displaced persons. So, I don't mean only scriptural hospitality but also an attentive recreation of communal membership and, in some cases, of transnational human rights (Heyer, 2012). Besides that, Katherine Philipps advocated for a Theology of Migration that she contextualised in the US Catholic Church and the geographical migrants (Philipps, 2015). Likewise, some insightful links between theology and ageing can be found in Ryo Takahashi's book *The Science, Philosophy and Bioethics of Gerontology in the Context of Our Future* (Takahashi, 2019, pp. 47–53). So, there is plenty of room for other traditions and disciplines (like bioethics!) to join.

The truth is that in the 2020s dawn most people are migrants, newcomers in a digital world, in a society with an increasingly fuzzy ethnicity, in an old-age brimming with new needs to deal with. They may live in both an emotionally cold but an economically rich country, in an ocean of plastic shelters, in a world of adults that behave like children… We must face immigration not only in *space and hurries*, but also considering the many *zeitgeister*, the different lifespans and biological clocks and how we have driven crazy the geological time (what means incoming waves of climatic refugees, for example). In short, we must deal with the full spectrum of human displacement using the whole human pool of knowledge.

The main immigration pump today is the lack of economic diversity. The sovereign democratic agency is under the macroeconomic actors' boot. The oil, war and agriculture business as well as the IMF and alike, in unaccountable ways, displace investments and whiter local human networks and infrastructures leading to many silent forced migrations (Rawlinson, 2016, p. 193). To cap it all, these barely forcible displaced persons are stigmatised as unemployment bearers. Wrong. Most people work producing goods and services that will be used elsewhere, so their jobs rely upon global markets and are, therefore, uncorrelated with a growth of population due to immigration (Chomsky, 2018, p. 9).

Philosophical Grounds

The most philosophical research on migration has been historically burdened by the many things we've given for granted (White, 1980). Either naive or ill-disposed, the liberal democracies didn't foresee a citizenship able to crawl beyond the solid walls of nationality and territory when needed (Bauböck, 1994). Perhaps a great share of the Western rulers never wanted to be true to their speeches on freedom, anyways. However, it must be recognized that it's very easy to get lost in a global and omnipresent social change narrative that blinds us of a clear insight of the migration-driven changes (Portes, 2010). Broadly. the study of migration has often been limited by assumptions and neglect from liberal democracies, resulting in a lack of understanding about the true impact of migration on global social change.

In addition to this, the wide-open social sciences are not suitable for the usual experiments of many other natural science fields (Iosifides, 2011). Likely, we're not yet fully aware of it. When it comes to philosophy of migration it has turned us into more fragile targets to be swayed by epistemological tides: realism, cognitivism, relativism… Therefore, the moral foundations of territoriality as we can see in Locke and Simmons have been underscored. To sum up, there is some kind of confusion among ethics and political philosophy on the one hand and global and social justice on the other that we cannot release (Lemmons, 2017).

All this mess has allowed the last gen Western populist parties to claim a right to culture against the tidal immigrants while they only have been thriving for casting persons aside upon *their* cultural background (Blake & Risse, 2007). Paradoxically, there are regional differences of migration patterns within the sending country regarding its foreign "*spheres of educational influence*" from the countries where migrants prefer to go (Alexandrov et al., 2018).

As the philosophy beachhead in the migration shores was abandoned, the quack theories surged. If we picture the philosophy of immigration as a device, then we jump to… the new (o man, I mean new!) Post-New-Age Biofeedback Quantum Prosperity brand-new Device: this on-sale EPEX/SCIO artefact consists of a harness that fastens around the ankles, wrists and head and allegedly guides the user to heal herself and get rich. Indeed, social media are a political *self-help* machine (Burwell, 2018, pp. 196–197). Once plugged in, you can hear the ones like Bolsonaro and Trump echoing and sometimes mooing the *perfect* solution against the immigration: don't let the *aliens* enter (García-Peña, 2019).

Greasy and small-time academicians and leaders chase the easy gain of relevance around the pandemic retro-macho opportunistic movement while contributing to the full displacement of any philosophical intention about migration by more or less slick hate speech. Perhaps it's inaccurate to talk about a philosophy of immigration and we should limit ourselves to talk about a methodology of migration. A plain and simple nationalist methodology based upon sedentariness, state sovereignty, territorial borders, and membership; a place of justice (the well-knitted nation-state) and a dis-place for in-justice where "*people enter by birth and exit by death*" (Sager, 2018).

To cap it all, the migration is strongly influenced by a colonised view of international relations. Its lexicon (e.g. 'sovereignty', 'multipolarity', 'hard, soft and balance of power') is soaked in positions of privilege the most times (Gofas et al., 2018). From both a philosophical and common perspective, the real world has become, in turn, an alien to international relations' one. A streamlined and deeper view is shared by the Jewish Sigmund Freud and the Palestinian Edward Said, who agreed in the Hagar's displacement, disbelonging, and foreignness (Freud, 1939; Said, 1990). The Egyptian slave is too exotic, too alien, for a paternalistic and identitarian Abraham-Europe to make a *true* family-citizenship. The conclusion is clear:

Our philosophy of migration is in the Iron Age.

Public Health And Immigration

The image of public health and immigration from the most recipient countries is a blur. In the consulting room the rising complexity of a more diverse society can be highlighted. Despite that patients with a Turkish and Albanian migration background and their German physicians described similar issues and their perspective mainly diverging when it came to (Würth et al., 2018):

- Patients' behaviour in relation to doctors' advice
- Relationship issues

In the EU there is a gap in the availability of high quality information and research in the field of migration health because the data has not been gathered properly and the exposures suffered during their travel remain unaccountable (Padilla & Miguel, 2009).

When going from the examining room to the governors' boards the things grow hairy. The multicultural health research suffers from being short of funds, consequently, the mental health care of immigrants, for example, is limited to psychiatry (Carta et al., 2005). Due to this straitjacketed toolbox we will be cornered to deal with a big pile of academic inconsistencies (Vega et al. 1987). The migrants' mental health is not a secondary topic and neither a rare condition: the discrimination effects are one of the main health differences between them and the non-migrants (Finch et al., 2001).

The acculturative stress and discrimination perceived by many migrants could lead to depression, as has been proved in a study of Mexican migrants in the US (Finch et al., 2000). A 2007 study on Asian populations concluded that individuals who reported discrimination were at a twofold greater risk of having a DSM-IV mental disorder within the past 12 months, and a threefold greater risk of having two or more disorders (Gee et al., 2007).

That's a very rooted evil in the case of refugees. As we can see from the IHARP longitudinal survey and the 1991 Rubén Rumbaut's study on Indochinese people in San Diego, the refugees' mental health was less worthy to be considered by policy-makers than dependency welfare and economic self-sufficiency (Rumbaut, 1991). Twenty-seven years later, a rights-based approach for mental health and psychosocial wellbeing policies are also claimed for the Syrian refugees in Turkey (Cantekin, 2018).

Away from mental conditions and focused on infectious-contagious diseases, a blatant case study can be gathered from the US-Mexico border. These immigrants, as usual in the populations that face social disparities in the access to medical care, are expectedly to become more vulnerable against tuberculosis and HIV (Moya et al., 2012). The demagogue Donald Trump uses this vulnerability (and a lot of money from the Federal Fund) to break migrant families (Straube, 2018). Mexicans don't bring HIV (their country's prevalence is one of the lowest in the world), rather than this, the most of them get infected when in the US (Strathdee et al., 2012). In Europe, there are other politicians streamlined with Trump, like the ministers Michalis Chrisochoidis and Andreas Loverdos, who jointly defined the immigrants like a "*health bomb ready to explode*" (TO BHMA, 2012).

Also the effects of displacement in health must be considered. For example, the Sub-Saharan migrants that come to Europe, mainly to the Southern buffer, have stopped previously in the Maghrebi migration hubs where there is a 90% chance of being victimised (Keygnaert et al., 2014). The immediate consequence is the ill-health of the many newcomers, namely gang-raped women. A public health system is globally considered a first-rank tool for preventing sexual violence, it makes more striking if possible the invisibility of those responsible for this burden on health, again (Gnes, 2013; van den Ameele et al., 2013).

When poor people migrate to richer countries they endanger their health twice: the first time to come and then when they stay. The largest expenditure in the migrants' health care is unlikely to be produced by them, but by the dehumanising attitudes on the immigration together with a willful ignorance of the social determinants of health. If we want to be humanitarian, it's meant to be through policies aimed to reduce discrimination rather than stigmatising the discriminated. About this, as Prof. Darryl Macer commented, it is different experiencing a great deal of suffering coming from an impersonal evil than facing the daily life obstacles that perhaps came from many different sources -including ourselves. Hopefully, even under the harsh displacement's boot there is a chance for getting rid of the usual hindrances that spoil the fun to common people and attaining a whole maturity.

BIOPOLITICS OF MIGRATION

Many of the most sadistic behaviours against immigrants are acts of structural injustice. A structural injustice is an unintended outcome from a widely accepted moral norm. For example, the prolonged encampment of refugees is not an explicit policy anywhere but the states have the sovereign right of choosing who enters and who uses its resources and services. Given this, when a state commits an immoral act like depriving the refugees of some personal freedoms, it can be *just-ified* as protecting its own citizens from terrorism or epidemic diseases. These structural injustices lead to structural barriers: fences, gates, sentry boxes, inaccessibility… elements that create a place for citizens and a *dis-place* for refugees.

A remarkable example of the structural injustice usual background can be found in *Territories of Difference: Place, Movements, Life, Redes*:

By the late 1990s, however, the regime of imperial globality had forcefully reasserted itself, and the region [of the Colombian Pacific] became submerged in a quagmire of violence, merciless capitalist expansion, and massive displacement that has affected black and indigenous communities and the environment with particular violence -a reassertion of the coloniality of knowledge, power and nature. (Escobar, 2008, p. 4)

In the same vein, Rosemary Rojas Martínez, member of the Yaqui and Apache peoples and representative of La Via Campesina's (LVC) Women's Committee in North America, briefed the biopolitics in the current 2019 (Vargas, 2019):

- Exploitation by agribusiness
- Exposure to agrochemicals
- Women's body as spoil of capitalism
- Proactive restrictive migration policies even in Mexican soil through the Goodfellow and Fort Bliss US military bases

Some collectives, like the Roma migrants in Europe, face a devalued citizenship patched with boundary practices (Baar, 2017). In those cases, hubs of human networks like *Alle bleiben!* struggle with the more and more stifling extra and intra-territorial securitization over the non-wealthy minorities.[4] It doesn't rule out that some pro-migrant NGOs are ethically washed human traffic networks. To put the securitization in perspective: when Turkey requested via Interpol about 700 German firms backing terrorist groups it was put to rest as a "*communications problem*" (Foundation Thomson Reuters, 2017). To cap it all, while populism rises in Europe a blabbermouth like Tobias Plate, the overseer of the German security apparatus, says: "*discrimination does not necessarily mean that people are also persecuted or mistreated*" (Brenner, 2016). Considering the meagre transcendence of this declaration among the European representatives, it seems that the psychological mistreatment that happens during discrimination doesn't fit as a *true* mistreatment when the targets are… *aliens*. Sadly, this attitude has been silently rooting in education (Bobis, 2013). 9,431 km away, in El Salvador, the borders grow thicker and this "*carceral quality is a dimension of national territory*" today (Coutin, 2010).

Neoliberalism and associated policies and practices of constraining and confining are adversely impacting minorities. An example is the Irish Travelers, Roma and asylum-seekers under Irish neoliberalism, who have been dragged through evolving modes of "*impoverishment, containment and quasi-incarceration*" (Garrett, 2015). Terms like "*development-induced displacement*" are misleading (Penz et al., 2011, pp. 41–42). Many people migrate *to* develop themselves, some are displaced *by* the construction of development related infrastructures like a dam and others displace themselves *away from* hindering development calamities (war, floods, droughts or simply draconian economic policies and processes). "*Development-induced displacement*" is ethically fishy since it skips the most important questions: when will this person's displacement end? Who is entitled to end it? How can I set the development-induced one apart from the forcible displacement? Should I?

A structural violence worthy of special consideration by the Bioethics of Displacement is the insufficient medicalized recognition of the racialized discrimination burden on health, as we can see in the Latino population in the US since the 1980 onwards (Quesada et al., 2011). The health sciences in the Global North are not sufficiently aware of how the scarcity of natural resources shapes the human experience (Robertson & Hoffman, 2014). It prevents the health scientist turning into a humanitarian biopolitical stakeholder empowered to deal with the psychosocial health consequences of forced migration and the consequent disparities.

On the other hand, the foreign aid acts as a gravity force when it comes to the flow of migrants (Lanati & Thiele, 2018). Poverty and war are global problems that cannot be solved with the carrot and the stick. The international Human Rights laws allegedly limit the states' sovereignty when dealing with persons of foreign nationality or residence, however, they rely upon the state power to realise, so the right to health becomes compromised when the people are displaced (Flegar, 2018).

The solutions offered to refugees share a sedentary bias (repatriation, local integration and, re-settlement) that fails to solve the forced migration problem because it does not consider the refugees' own transformative solutions (Long, 2014). The biopolitics of migration creates a category of persons grounding on dislocation rather than displacement. To cap it all, the local integration, that allows refugees to become full members of the host country, is forbidden first and forgotten then because it's too close to the autonomy and belonging red lines (Hovil, 2014).

The dislocation pervades the way we understand basic human rights. The causes of displacement (why are you out of your box?) are oversized detrimental to the existential threats that a person can suffer. It's feasible to set a threshold for threatened human rights able to remove the arbitrary casuistry of asylum (Betts, 2013, p. 188). For example, war refugees from Iran, Iraq, Libya, Somalia, Sudan, Syria, and Yemen coming to the US must to face the recent Executive Order 13769, titled "*Protecting the Nation from Foreign Terrorist Entry into the United States*" in despite that these nationals have killed zero people in terrorist attacks on U.S. soil between 1975 and 2015 (Friedman, 2017; Lemmons, 2017).

The sedentary bias extends to the time-frame of displacement. It's seen as something temporary, but human displacement is a process that endures on average 17 years (Parekh, 2016).[5] Considering that it's a long process, the accountability and moral scrutiny of procedures, governments and institutions should be more solid. As Angela Naimou, echoing Hannah Arendt, states:

In a strange spatiotemporal mode, these migrant refugees approach Europe as the not-quite-afterimages of unending wars elsewhere; their incorporation into the European Union weaves the effects of these wars into the future "social texture" of Europe in ways that the financial cost of war, death tallies, or even returning veterans and service workers do not. (Naimou, 2016)

On the other side of the fence, most "*existing narratives on and about development in regions of the global South, the majority of which are enshrined in social-political disenfranchisement and cultural erasure*" (Espinal, 2018, p. 119). Sometimes, the acculturation process starts in the origin country. Besides forcibly displaced persons, the ones that decide to emigrate, like the participants in a 2011 study of British migrants in New Zealand, highlights the importance of the pre-departure process (Tabor & Milfont, 2011). From my own experience, I can witness that, in a very different setting, it's also true. On the other end of the rope, anthropology shows us that those who have been subjects of the evil sovereign hygiene of deportation, start to glue their identities, names, histories as soon as they reach the post-deportation, a kind of solid displacement (De Genova, 2018).

A similar phenomena that happens in acculturation can be also translated into an imposed dis-ability. For example, the illegal refugees that come to Australia are displaced to Nauru in order to be *processed* where their impairment is *produced* (Guillaume & Finlay, 2018). Sadly, beyond particular cases, we don't know how the impairment is produced when the Global North and South meet. This theoretical gap can be partly explained by attending to the role of the experts when explaining how this geographical ableism makes prey upon the refugees' bodies.

Even in close well-developed countries, the immigration experts' knowledge is integrated differently in policy-making, for example Sweden allows the experts to be influential in rethinking former policies while Denmark politicians are more prone to pick and choose from the current research to legitimate their political decisions (Jørgensen, 2011). However, it must be put in perspective. The social construction of the Afro-Swedish *being-towards-the-nature* as an urban anomalous slow-paced stalking in the margins of biopolitics warns us about the hereditary burden of scientific bias even in a relatively fair system (Skinner, 2019).

For good or evil, in a world with plenty of areas badly plagued by Gehenna biopolitics new opportunities for an interdisciplinary venture appears. For example, the 2017 US legal crackdown on the parents bringing their children during an illegal entry became a potential data faucet for studying the relation between queer migration and human trafficking (Luibhéid, 2019). By the same token, new areas of interest in which gender is the key, such as gendered transnational communities, the geographical and spatial contours of immigrant occupational sex segregation, and the inclusion of youth and children in gendered analysis of immigrant communities has come under the spotlight with the new millennium (Hondagneu-Sotelo, 1999).

On the other hand, the experts want to be published and become somewhat influential *even when* it means facing deeply-held beliefs with an honest data gathering. It's more than an egoist or childish attitude. Regardless of the researcher's honesty, goodwill and wish to help this or that vulnerable people, nobody can grant that your analysis will be used in a humanitarian fashion when it goes to the political arena. Further than this, the donors, sponsors, governments, etc., are the ones that in-forms the research when selecting the topics. Social research in Africa, for example, has a lot of *scientific myths* (Landau, 2006; 220). The same goes for the many Africas.

Till now, I have considered the migrants as the only displaced but there is also a double standard here. Wokes has celebrated as a success the blackwashing of Disney characters because representation. However, if a French child would want to see more white soccer players in order to feel represented then she would be a racist and cancelled. It extends to the management of immigration as a whole. Opening borders sound cool, humanitarian and inclusive but not regularising migration favours the mafias and radical groups because it is easier to get a country when you have a network devoted to helping you rather than the occasional help of a relative that migrated before, if any. To cap it all, the bulk of migration usually concentrates in very determined cheap towns relatively close to big cities, with the subsequent suffering of persons who are limited to deal with changes (e.g. some elders) and feel displaced, passively migrating, in turn. Of course, this ghetto-like scheme favours the very minority of radicals who can hide at plain sight. Not regulating migration is not respecting the honest majority of migrants painting them with the same brush that those who come to a new country to be criminals and freeriders.

Namely, the migration from MENA countries to Europe has two added difficulties. Sharia as a source of right, not necessarily as a creed, is incompatible with the European sources of right since it does not even include freedom of religion, not to say of women's rights as understood in Europe (Brzozowski, 2018). It is partly assimilable in the private sphere but not as a political option unless erasing European identity becomes acceptable. But Europe doesn't realise what Sharia, politically, means. Intolerance, if not fought, in the long run always wins for mere probabilities (Taleb, 2016). Worse even, the MENA countries are well aware of how its demographic excess is an invaluable asset in hybrid warfare (Shotter et al., 2021). The best example of this is Moroccan crown expansionist politics. For example, in May 2021, Morocco flooded the Spanish city of Ceuta (which Morocco's ruler wants, by the way) with irregular migrants because Spain sheltered the Polisario Front leader, who defends the independence of another region that the Moroccan ruler also wants (Martín, 2021). About two years later, in 2023, the Moroccan President of the Senate cheered the Moroccan migrants in Spain to enter into politics and institutions "to defend the homeland" -and he didn't mean Spain (Cembrero, 2023). Spain is an outlier because the whole migratory policy is a botch and the people who could change this mess would sell their granny for two bucks. So much so that the situation is better understood by looking at the flow of the money between the Moroccan king, EU parties and biz at both sides of the border than using a map (Brigada Antifraude, 2023; Zornoza, 2021). All Europe faces in a lesser degree a similar dilemma than Spain and must find an intolerance useful to preserve the European rule of law or be washed away by a source of law that more than one billion people take seriously and are not eagering to change to look more inclusive and representative.

The tragedy of human forcible and quasi-forcible displacement is that the few ones who could solve the bulk of the problem think that everything can be solved with a bit more hypocrisy.

PHENOMENOLOGY OF MIGRATION AND MIGRATION SIGHTING

Approaching the phenomenology of migration considering space as an architecture for displacement eventually leads us to face a displacement-as-out-migration entelechy which can be avoided focusing on the lived experience of constructive space (Davidson, 2009). However, a constructive space doesn't necessarily lead to displacement, so, rather than this, I'd consider the human corridors as a vehicle of displacement. One from which you can't get out, like those mediaeval cages on a chariot where the villagers throw rotten vegetables to the prisoner.

In my mind, the most rotten of these issues is the institutional "*treason*" that some policies actually are, e.g. the late acculturative US pressures that migrants experience (Vesely et al., 2017). For the sake of asylum status, many agencies impose over the migrants a fake dichotomy between voluntary and forced migration when actually it shows as a continuum as soon as we sail off from a despotic attitude; the most compelling evidence of the stated is the wide academic gap between the migration and the conflicted-migration studies (Erdal & Oeppen, 2018). For example, the migration of a Salvadoran working-class family to the US, and of a Chinese middle-class to Australia walk through very different "*institutional and discursive landscapes*" up to the point that these differences challenge the common scholarly literature on families and immigration (Landolt & Wei Da, 2005).

A common place among the so-called voluntary migration is the dough cravings. It's a myth! In China, from 1980 to 1995, ten million people migrated from rural areas to the cities, even when incurring an irregular situation, mainly running away from abusive taxes and officers rather than pursuing a better salary (Keidel, 2006; Zhao, 1999). The same non-monetary pursuit happened in the interregional migration to US metropolitan areas where it is highlighted that the goals of migrants are an improvement in the quality of social life and environments (Hsieh & Liu, 1983). It's not quite different from a crystal clear forced migration. In an archive located at Houghton Library, Harvard University, entitled *My Life in Germany before and after January 30, 1933*, collected in 1939-40 hold the three main reasons the Jews gave for deciding to emigrate (Liebersohn & Schneider, 2001):

- Loss of employment opportunities
- A feeling of moral repugnance (for the Nazi regime, in this case)
- An experience of physical threat

Regardless of the Nazi high-profile bureaucratized machine of statutory mass killing, the German migrant's feelings in 1933 are not quite different from the other cases exposed in this section. Either if they can be tagged as forcibly displaced or not, a similar human pattern appears.

In addition to this, the Harvard's archive holds that even the very idea of the migrant's mindset shift is a slow-paced process rather than a springing turning-point (Anstey, 2017). The phenomenology of migration doesn't allow us to pinpoint the insight of leaving your home but allows the underpinning of it. The replies to 'what is home?' and 'where do I belong?' are built as the days go by away from the origin and eventually are shattered by the deportation quake (Gomez & Vannini, 2017).

Mixing the migrants' personal paradigmatic changes with their life satisfaction leads us to a complex interpretation. As exposed in a 2008 study published in *Social Indicators Research* on a Turkish-Dutch population, being an ethnic minority lowers the joy of life regardless of education and income (Verkuyten, 2008). On the other hand, continues the paper, a strong group identification is associated with a higher life satisfaction but also with a greater perception of structural discrimination, which, in turn, sinks the migrants' happiness. To cap it all, different minorities have different views on discrimination, which are likely shaped, in turn, by the social stratification of minorities themselves. A clear example of this can be seen in the US, where the Afro-Caribbean sense of identity is associated with a higher perception of discrimination than the African American or any other ethnic minorities (Gong et al., 2017).

Seeing through the evident complexity is a must when delving into the migrants' consciousness and direct experience. The nowadays asylum seekers in Australia, for example, forms an entangled human web that transcends the usual bipartisan semi-structured surveys model in favour of an aspiration (an *élan*) kind of drive (Rowe, 2018). The aspiration, desire and drivers of migration are relevant enough to occupy a special issue in the *Journal of Ethnic and Migration Studies* focused in the new semantic field that is replacing the old causes and consequences view (Carling & Collins, 2018). The rancid old-school solutions we can see daily makes more evident even the indifference of immigration agencies for the state-of-the-art theories.

Figure 5. A migrant woman jumps over a puddle after bathing. A daily life stamp in a Tijuana waterlogged sports complex turned into a camp
Source: Smith and Ashford (2018)

In *Displacement, Asylum, Migration: The Oxford Amnesty Lectures*, Kate Tunstall explains the late idea of displacement in Freud with the example of waking from a dream about a burning house to avoid seeing the little child abandoned by his mother inside (Tunstall, 2006, p. 269). Thanks to populism, the peoples from the hottest borders, chiefly South Europe and US, have been awake too much, for such a long time that they don't remember how the conflagration started. This bias is not as trivial as it seems, not for most people. There is abundant academic literature that assumes that we compare our standards against another set by a paper or an index, however, social psychology teaches us that we compare with the closest people to us (Falk & Knell, 2004). Sadly, most people are not so lucky to have friends abroad, as a result, it's very difficult to realise how hard-you have become since all your folks and even your town major are going to shift with you. Put in bluntly, our gaze on the migrants is toxic.

CONCLUSION

Philosophy of migration needs an urgent update as it remains rooted in outdated notions and fails to capture the complexities of the current global situation. It's confined in simplistic and often hateful rhetoric that ignores the true impact of migration on social change, either desirable or undesirable. To begin with, shedding light on the structural injustices and barriers faced by refugees and the complexities arising from conflicting legal systems may reduce discrimination as a social determinant of health.

REFERENCES

Alexandrov, D., Karepin, V., Musabirov, I., & Chuprina, D. (2018). Educational Migration from Russia to the Nordic Countries, China and the Middle East. Social Media Data. *Companion Proceedings of the Web Conference 2018*, 49–50. 10.1145/3184558.3186923

Anstey, J. (2017). *Decision processes of emigrants from Nazi Germany*. University of Rochester.

Baar, H. van. (2017, July 6). *Boundary practices of citizenship: Europe's Roma at the nexus of securitization and citizenship*. Within and Beyond Citizenship. doi:10.4324/9781315268910-10

Bauböck, R. (1994). *Transnational Citizenship: Membership and Rights in International Migration*. Edward Elgar Publishing.

Berry, C. (2018, November 16). Theresa May's deal is almost exactly the Brexit the UK voted for. *The Conversation*. http://theconversation.com/theresa-mays-deal-is-almost-exactly-the-brexit-the-uk-voted-for-107040

Betts, A. (2013). *Survival Migration: Failed Governance and the Crisis of Displacement*. Cornell University Press. doi:10.7591/cornell/9780801451065.001.0001

Blake, M., & Risse, M. (2007). *Migration, Territoriality, and Culture* (*SSRN* Scholarly Paper ID 963130). Social Science Research Network. https://papers.ssrn.com/abstract=963130

Bobis, Z. (2013, October 29). *Stand up for equality in German schools*. Open Society Foundations. https://www.opensocietyfoundations.org/voices/german-schools-quiet-deep-discrimination-problem

Brenner, Y. (2016, June 5). *Germany: Roma march against asylum-seeker crackdown*. https://www.aljazeera.com/news/2016/06/germany-roma-march-asylum-seeker-crackdown-160605092954463.html

Brigada Antifraude (Director). (2023, April 9). *Empresas y políticos españoles apoyan a Marruecos y su ocupación*. https://www.youtube.com/watch?v=dh7VcLkg4Uk

Brzozowski, W. (2018). *Is Islam Incompatible with European Identity?* (*SSRN* Scholarly Paper No. 3301195). doi:10.2139/ssrn.3301195

Burwell, J. (2018). *Quantum Language and the Migration of Scientific Concepts*. MIT Press. doi:10.7551/mitpress/11353.001.0001

Cantekin, D. (2018). Syrian Refugees Living on the Edge: Policy and Practice Implications for Mental Health and Psychosocial Wellbeing. *International Migration (Geneva, Switzerland)*. Advance online publication. doi:10.1111/imig.12508

Carling, J., & Collins, F. (2018). Aspiration, desire and drivers of migration. *Journal of Ethnic and Migration Studies*, *44*(6), 909–926. doi:10.1080/1369183X.2017.1384134

Carta, M. G., Bernal, M., Hardoy, M. C., & Haro-Abad, J. M. (2005). Migration and mental health in Europe (the state of the mental health in Europe working group: Appendix 1). *Clinical Practice and Epidemiology in Mental Health*, *1*(1), 13. doi:10.1186/1745-0179-1-13 PMID:16135246

Cembrero, I. (2023, April 9). *El presidente del Senado marroquí insta a sus compatriotas a entrar en la política española para favorecer a Rabat*. elconfidencial.com. https://www.elconfidencial.com/espana/2023-04-09/presidente-senado-marruecos-ceuta-melilla_3607689/

Chomsky, A. (2018). *"They Take Our Jobs!": And 20 Other Myths about Immigration*. Beacon Press.

Coutin, S. B. (2010). Confined within: National territories as zones of confinement. *Political Geography*, *29*(4), 200–208. doi:10.1016/j.polgeo.2010.03.005

Daftary, A.-M. H. (2018). Confronting Immigration Myths with the Reality: A Necessary Perspective for Culturally Grounded Social Work Practice. *Smith College Studies in Social Work*, *88*(2), 96–114. doi:10.1080/00377317.2018.1437995

Davidson, M. (2009). Displacement, Space and Dwelling: Placing Gentrification Debate. *Ethics Place and Environment*, *12*(2), 219–234. doi:10.1080/13668790902863465

De Genova, N. (2018). Afterword. Deportation: The Last Word? In S. Khosravi (Ed.), *After Deportation: Ethnographic Perspectives* (pp. 253–266). Springer International Publishing. doi:10.1007/978-3-319-57267-3_13

Deng, F. M. (2010). *Protecting the Dispossessed: A Challenge for the International Community*. Brookings Institution Press.

Dominiczak, P. (Ed.). (2015, October 5). *Theresa May: Mass immigration making "cohesive society" impossible*. https://www.telegraph.co.uk/news/uknews/immigration/11913392/Theresa-May-Mass-immigration-making-cohesive-society-impossible.html

Erdal, M. B., & Oeppen, C. (2018). Forced to leave? The discursive and analytical significance of describing migration as forced and voluntary. *Journal of Ethnic and Migration Studies*, *44*(6), 981–998. doi:10.1080/1369183X.2017.1384149

Escobar, A. (2008). *Territories of Difference: Place, Movements, Life, Redes*. Duke University Press.

Espinal, S. A. (2018). *Enabling Geographies: Neurodivergence, Self-Authorship, and the Politics of Social Space* [PhD Thesis]. California Institute of Integral Studies.

Falk, A., & Knell, M. (2004). Choosing the Joneses: Endogenous Goals and Reference Standards. *The Scandinavian Journal of Economics*, *106*(3), 417–435. doi:10.1111/j.0347-0520.2004.00370.x

Finch, B. K., Hummer, R. A., Kol, B., & Vega, W. A. (2001). The Role of Discrimination and Acculturative Stress in the Physical Health of Mexican-Origin Adults. *Hispanic Journal of Behavioral Sciences*, *23*(4), 399–429. doi:10.1177/0739986301234004

Finch, B. K., Kolody, B., & Vega, W. A. (2000). Perceived Discrimination and Depression among Mexican-Origin Adults in California. *Journal of Health and Social Behavior, 41*(3), 295–313. JSTOR. doi:10.2307/2676322

Flegar, V. (2018). Vulnerability and the Right to Preventive Health Care for Asylum-Seeking and Undocumented Migrant Children. *Children's Rights in Health Care*, 334–356. doi:10.1163/9789004327573_015

Foundation Thomson Reuters. (2017, July 24). *Turkey withdraws list of German firms suspected of backing terrorism-Germany*. News.Trust.Org. http://news.trust.org/item/20170724100356-1gk0g/

Freud, S. (1939). *Moses and monotheism*. Knopf.

Friedman, U. (2017, January 30). *Where America's Terrorists Actually Come From*. The Atlantic. https://www.theatlantic.com/international/archive/2017/01/trump-immigration-ban-terrorism/514361/

García-Peña, L. (2019). Immigrant Rights are Human Rights. *NACLA Report on the Americas*, *51*(1), 101–104. doi:10.1080/10714839.2019.1593700

Garrett, P. M. (2015). Constraining and confining ethnic minorities: Impoverishment and the logics of control in neoliberal Ireland. *Patterns of Prejudice*, *49*(4), 414–434. doi:10.1080/0031322X.2015.1074785

Gee, G. C., Spencer, M., Chen, J., Yip, T., & Takeuchi, D. T. (2007). The association between self-reported racial discrimination and 12-month DSM-IV mental disorders among Asian Americans nationwide. *Social Science & Medicine*, *64*(10), 1984–1996. doi:10.1016/j.socscimed.2007.02.013 PMID:17374553

Gnes, D. (2013). *Maghnia: Crossing the Uncrossable Border*. https://euromedrights.org/wp-content/uploads/2013/01/Maghnia_Crossing-the-uncrossable-border_EN.pdf

Gofas, A., Hamati-Ataya, I., & Onuf, N. (2018). Positionality and the Pluriverse. In *The SAGE Handbook of the History, Philosophy and Sociology of International Relations*. SAGE. doi:10.4135/9781526402066

Gomez, R., & Vannini, S. (2017). Notions of Home and Sense of Belonging in the Context of Migration in a Journey through Participatory Photography. *The Electronic Journal on Information Systems in Developing Countries*, *78*(1), 1–46. doi:10.1002/j.1681-4835.2017.tb00569.x

Gong, F., Xu, J., & Takeuchi, D. T. (2017). Racial and Ethnic Differences in Perceptions of Everyday Discrimination. *Sociology of Race and Ethnicity (Thousand Oaks, Calif.)*, *3*(4), 506–521. doi:10.1177/2332649216681587

Guillaume, L. S., & Finlay, E. (2018). Disabled mobility and the production of impairment: The case of Australia's migration policy framework. *Asia Pacific Viewpoint*, *59*(1), 119–131. doi:10.1111/apv.12182

Heath, R. (2018, December 13). *Brexit Britain: Small, boring and stupid*. POLITICO. https://www.politico.eu/article/brexit-britain-small-boring-and-stupid-theresa-may-eu-withdrawal-deal/

Heyer, K. E. (2012). Reframing Displacement and Membership: Ethics of Migration. *Theological Studies*, *73*(1), 188–206. doi:10.1177/004056391207300109

Hollenbach, D. (2016). Borders and Duties to the Displaced: Ethical Perspectives on the Refugee Protection System. *Journal on Migration and Human Security*, *4*(3), 148–165. doi:10.1177/233150241600400306

Homer. (2011). *The Odyssey*. Collector's Library.

Hondagneu-Sotelo, P. (1999). Introduction: Gender and Contemporary U.S. Immigration. *The American Behavioral Scientist*, *42*(4), 565–576. doi:10.1177/00027649921954363

Hovil, L. (2014). *Local Integration*. The Oxford Handbook of Refugee and Forced Migration Studies., doi:10.1093/oxfordhb/9780199652433.013.0042

Hsieh, C., & Liu, B. (1983). The Pursuance of Better Quality of life: In the Long Run, Better Quality of Social Life Is the Most Important Factor in Migration. *American Journal of Economics and Sociology*, *42*(4), 431–440. doi:10.1111/j.1536-7150.1983.tb01730.x

Iosifides, T. (2011). Theoretical and Epistemological Issues in Qualitative Research. In *Qualitative Methods in Migration Studies*. https://www.taylorfrancis.com/

Jørgensen, M. B. (2011). Understanding the Research–Policy Nexus in Denmark and Sweden: The Field of Migration and Integration. *British Journal of Politics and International Relations*, *13*(1), 93–109. doi:10.1111/j.1467-856X.2010.00441.x

Keidel, A. (2006). China's social unrest: The story behind the stories. *Carnegie Endowment for International Peace, 48*.

Keygnaert, I., Dialmy, A., Manço, A., Keygnaert, J., Vettenburg, N., Roelens, K., & Temmerman, M. (2014). Sexual violence and sub-Saharan migrants in Morocco: A community-based participatory assessment using respondent driven sampling. *Globalization and Health*, *10*(1), 32. doi:10.1186/1744-8603-10-32 PMID:24885537

Lanati, M., & Thiele, R. (2018). The impact of foreign aid on migration revisited. *World Development*, *111*, 59–74. doi:10.1016/j.worlddev.2018.06.021

Landolt, P., & Wei Da, W. (2005). The Spatially Ruptured Practices of Migrant Families: A Comparison of Immigrants from El Salvador and the People's Republic of China. *Current Sociology*, *53*(4), 625–653. doi:10.1177/0011392105052719

Laurent, L. (2018, December 17). *The Colossal Price of Theresa May's Immigration Obsession*. https://www.bloomberg.com/opinion/articles/2018-12-17/brexit-the-colossal-price-of-theresa-may-s-immigration-obsession

Lemmons, T. (2017). Justice in Migration: A Case Study for War Refugees. *CMC Senior Theses*. https://scholarship.claremont.edu/cmc_theses/1562

Liebersohn, H., & Schneider, D. (2001). "My Life in Germany before and after January 30, 1933": A Guide to a Manuscript Collection at Houghton Library, Harvard University. *Transactions of the American Philosophical Society, 91*(3), 1. doi:10.2307/3655110

Long, K. (2014). *Rethinking 'Durable' Solutions*. The Oxford Handbook of Refugee and Forced Migration Studies., doi:10.1093/oxfordhb/9780199652433.013.0013

Luibhéid, E. (2019). Heteronormativity: A Bridge Between Queer Migration and Critical Trafficking Studies. *Women's Studies in Communication, 0*(0), 1–5. doi:10.1080/07491409.2018.1544001

Martín, L. J. V., & Francisco Peregil, M. (2021, May 17). *5.000 personas llegan a nado a Ceuta en plena escalada de tensión diplomática con Marruecos*. El País. https://elpais.com/espana/2021-05-17/un-millar-de-personas-llegan-a-nado-a-ceuta-en-plena-escalada-de-tension-diplomatica-con-marruecos.html

McKinstry, L. (2018, November 16). How can mediocre Mrs May compare herself to cricket icon Geoffrey Boycott? *The Telegraph*. https://www.telegraph.co.uk/politics/2018/11/16/can-mediocrity-like-mrs-may-compare-cricket-icon-geoffrey-boycott/

Moya, E. M., Loza, O., & Lusk, M. (2012). Border Health: Inequities, Social Determinants, and the Cases of Tuberculosis and HIV. In M. Lusk, K. Staudt, & E. Moya (Eds.), *Social Justice in the U.S.-Mexico Border Region* (pp. 161–178). Springer Netherlands. doi:10.1007/978-94-007-4150-8_9

Naimou, A. (2016). Double Vision: Refugee Crises and the Afterimages of Endless War. *College Literature, 43*(1), 226–233. doi:10.1353/lit.2016.0004

Orobatar, A. E. (2010). Justice for the Displace: The Challenge of a Christian understanding. In H. David (Ed.), *Driven from Home: Protecting the Rights of Forced Migrants*. Georgetown University Press.

Padilla, B., & Miguel, J. P. (2009). Health and migration in the European Union: Building a shared vision for action. *Health and Migration in the European Union: Better Health for All in an Inclusive Society, 1a*, 15–22.

Parekh, S. (2016). *Refugees and the Ethics of Forced Displacement*. Taylor & Francis. doi:10.4324/9781315883854

Penz, P., Drydyk, J., & Bose, P. S. (2011). *Displacement by Development: Ethics, Rights and Responsibilities*. Cambridge University Press. doi:10.1017/CBO9780511973499

Philipps, K. (2015). *Hospitality and Emerging Populations: Toward a Theology of Migration in the Context of the Catholic Church in the United States* [PhD Thesis].

Portes, A. (2010). Migration and Social Change: Some Conceptual Reflections. *Journal of Ethnic and Migration Studies, 36*(10), 1537–1563. doi:10.1080/1369183X.2010.489370

Prior, A. (2010, February 18). No Joke: Buffalo and Moravia Duke It Out Over Millard Fillmore. *Wall Street Journal*. https://www.wsj.com/articles/SB10001424052748703444804575071322401554694

Quesada, J., Hart, L. K., & Bourgois, P. (2011). Structural Vulnerability and Health: Latino Migrant Laborers in the United States. *Medical Anthropology*, *30*(4), 339–362. doi:10.1080/01459740.2011.576725 PMID:21777121

Rawlinson, M. C. (2016). *Just Life: Bioethics and the Future of Sexual Difference*. Columbia University Press. doi:10.7312/rawl17174

Robertson, C. L., & Hoffman, S. J. (2014). Conflict and Forced Displacement: Human Migration, Human Rights, and the Science of Health. *Nursing Research*, *63*(5), 307–308. doi:10.1097/NNR.0000000000000058 PMID:25171556

Rowe, E. A. (2018). *Life-saving and life-changing: The decision-making processes of people seeking asylum* [PhD]. Queensland University of Technology. https://eprints.qut.edu.au/116152/

Rumbaut, R. G. (1991). *Migration, Adaptation, and Mental Health: The Experience of Southeast Asian Refugees in the United States* (*SSRN* Scholarly Paper ID 1886652). Social Science Research Network. https://papers.ssrn.com/abstract=1886652

Sager, A. (2018). Political Philosophy, Migration, and Methodological Nationalism. In A. Sager (Ed.), *Toward a Cosmopolitan Ethics of Mobility: The Migrant's-Eye View of the World* (pp. 17–36). Springer International Publishing. doi:10.1007/978-3-319-65759-2_2

Said, E. (1990). In the shadow of the West: Edward Said. *Discourses: Conversations in Postmodern Art and Culture*, 93–104.

Shergold, A. (2016, June 9). France's World Cup win in 98 was not appreciated until 12 years later. *Mail Online*. https://www.dailymail.co.uk/sport/football/article-3632753/France-s-World-Cup-win-1998-not-properly-appreciated-12-years-later-squad-went-strike-claims-Frank-Leboeuf.html

Shotter, J., Fleming, S., & Hall, B. (2021, December 5). How migration became a weapon in a 'hybrid war.' *Financial Times*. https://www.ft.com/content/83ece7e4-cc71-45b5-8db7-766066215612

Skinner, R. T. (2019). Walking, talking, remembering: An Afro-Swedish critique of being-in-the-world. *African and Black Diaspora: An International Journal*, *12*(1), 1–19. doi:10.1080/17528631.2018.1467747

Smith, J., & Ashford, B. (2018, November 29). *Rainbow lightens up migrants' view of US at the Mexican border*. Mail Online. https://www.dailymail.co.uk/news/article-6444247/Rainbow-lightens-sky-migrant-camp-border.html

Strathdee, S. A., Magis-Rodriguez, C., Mays, V. M., Jimenez, R., & Patterson, T. L. (2012). The Emerging HIV Epidemic on the Mexico-US Border: An International Case Study Characterizing the Role of Epidemiology in Surveillance and Response. *Annals of Epidemiology*, *22*(6), 426–438. doi:10.1016/j.annepidem.2012.04.002 PMID:22626001

Straube, T. (2018, July 11). *Trump Agency Is Using Federal HIV Funds to Separate Immigrant Families*. POZ. https://www.poz.com/article/trump-agency-using-federal-hiv-funds-separate-immigrant-families

Tabor, A. S., & Milfont, T. L. (2011). Migration change model: Exploring the process of migration on a psychological level. *International Journal of Intercultural Relations*, *35*(6), 818–832. doi:10.1016/j.ijintrel.2010.11.013

Takahashi, R. (2019). *The Science, Philosophy and Bioethics of Gerontology: An Individual and Community Journey from Japan*. LAP LAMBERT Academic Publishing.

Taleb, N. N. (2016). *The most intolerant wins: The dictatorship of the small minority*. Incerto.

Tharoor, I. (2016, June 8). Europe's top soccer teams reflect the power of multiculturalism. *Washington Post*. https://www.washingtonpost.com/news/worldviews/wp/2016/06/08/europes-top-soccer-teams-reflect-the-power-of-multiculturalism/

TO BHMA. (2012, April 1). *Λοβέρδος - Χρυσοχοΐδης: Υποχρεωτικό πιστοποιητικό υγείας για μετανάστες*. TO BHMA. https://www.tovima.gr/2012/04/01/society/loberdos-xrysoxoidis-ypoxrewtiko-pistopoiitiko-ygeias-gia-metanastes/

Tunstall, K. E. (2006). *Displacement, Asylum, Migration: The Oxford Amnesty Lectures 2004*. OUP Oxford.

van den Ameele, S., Keygnaert, I., Rachidi, A., Roelens, K., & Temmerman, M. (2013). The role of the healthcare sector in the prevention of sexual violence against sub-Saharan transmigrants in Morocco: A study of knowledge, attitudes and practices of healthcare workers. *BMC Health Services Research*, *13*(1), 77. doi:10.1186/1472-6963-13-77 PMID:23442386

Vargas, M. (2019, March 13). *The globalisation of border control and peoples' resistance* (K. Lang, Trans.). Globalisation of Border Control and the Peoples' Resistance. https://www.tni.org/en/article/the-globalisation-of-border-control-and-peoples-resistance

Verkuyten, M. (2008). Life Satisfaction Among Ethnic Minorities: The Role of Discrimination and Group Identification. *Social Indicators Research*, *89*(3), 391–404. doi:10.100711205-008-9239-2

Vesely, C. K., Letiecq, B. L., & Goodman, R. D. (2017). Immigrant Family Resilience in Context: Using a Community-Based Approach to Build a New Conceptual Model. *Journal of Family Theory & Review*, *9*(1), 93–110. doi:10.1111/jftr.12177

White, S. E. (1980). A Philosophical Dichotomy in Migration Research. *The Professional Geographer*, *32*(1), 6–13. doi:10.1111/j.0033-0124.1980.00006.x

Würth, K., Langewitz, W., Reiter-Theil, S., & Schuster, S. (2018). Their view: Difficulties and challenges of patients and physicians in cross-cultural encounters and a medical ethics perspective. *BMC Medical Ethics*, *19*(1), 70. doi:10.118612910-018-0311-4 PMID:29973220

Zhao, Y. (1999). Labor Migration and Earnings Differences: The Case of Rural China. *Economic Development and Cultural Change*, *47*(4), 767–782. doi:10.1086/452431

Zornoza, M. G. (2021, November 14). *Cuáles son las empresas europeas "cómplices" de los "crímenes de guerra" de Marruecos en el Sáhara Occidental*. https://www.publico.es/internacional/son-empresas-europeas-complices-crimenes-guerra-marruecos-sahara-occidental.html

KEY TERMS AND DEFINITIONS

Asylum: The protection given by a country to people fleeing persecution or danger in their home country.

Citizenship: The legal status of being a member of a country, with rights and responsibilities.

Diversity: The coexistence of the wide range of human differences.

Forced Migration: The coerced removal of a human group from their quotidianities due to political causes, including external interventions, crime and corruption, and lack of natural disasters preparedness.

Migration: The movement of people from one place to another, either within a country or across borders. In Bodi, migration extends to shifting from their each day belonging either due to body changes or different causes.

Rebellion: Defiance or opposition to authority or established norms, often involving organised actions by a large group.

Structural Injustice: Unfairness caused by systems and institutions that harm certain groups and benefit others due to their design and operation.

Xenophobia: Strong fear, dislike, or hostility towards people from different countries or cultures, based on their perceived foreignness.

ENDNOTES

1. https://www.unhcr.org/globaltrends2017/
2. https://www.census.gov/newsroom/press-releases/2015/cb15-tps16.html
3. https://www.census.gov/content/dam/Census/newsroom/releases/2015/cb15-tps16_graphic.pdf
4. http://www.alle-bleiben.info/
5. An average result for a Homeric Odyssey, so it's by no means a temporary situation (Homer, 2011, p. 295).

Chapter 12
Displaced, Aging, and Cyborgs' Migration Through Borders and Beyond Borders:
Belonging v Set Apart II

ABSTRACT

The author delves into migration in a different way, not only in moving to another country but to another stage of life and even within your own posthuman body. The author starts exploring migration and bioethics in the context of climate change and increasing displacement. We need moral imagination and the ability to plan justice to ensure solidarity with displaced populations, especially the elderly and forcibly displaced. To begin with, increasing awareness of social, political, and cultural dimensions of disability can lead to a more culturally humble rework of the international classification of functioning, disability, and health. Furthermore, we'll know more about migration, ageing and the perpetuation of poverty through ethnic background inheritance. We'll also examine the potential of near-future healthcare devices in detecting health problems of refugees and even the use of chemical analysis to detect ethnic backgrounds at the border. Finally, the author speculates about the concept of cyborgs and transhumanism and the potential for abuse by those in power.

INTRODUCTION

This chapter builds upon the themes explored in the last chapter delving deeper into the intersections of migration and bioethics. While Chapter 11 examined various aspects of migration, including its philosophical grounds, public health implications, and biopolitical dimensions, this chapter broadens the bioethical dimensions of migration and Displacement. It delves into the complexities of displacement, examining the role of bioethics in promoting justice, understanding the experiences of migrants, and addressing the challenges posed by the new migration within cyborg embodiment. By intertwining migration and

DOI: 10.4018/978-1-6684-4808-3.ch012

bioethics, this chapter offers valuable insights into the ethical considerations that arise in the context of Displacement and the potential for a more compassionate approach to healthcare and human rights.

BODI AND MIGRATION AS USUAL

The expected degree of solidarity relies on our place-making towards the other, so it requires moral imagination. Also it requires the ability to plan justice. So much so that in bioethics, training in this mental gym can lead us to become more proficient when dealing with the forcibly displaced and elderly people's problems, among many others (Eckenwiler, 2018). Both the high regulatory markets and the generous welfare for *legal migrants* like we see in Sweden, Norway and Denmark are the self-provoked migratory concerns that burden the disincentivized migrants (Brochmann & Hagelund, 2011). It's not a byproduct of the lack of ethics, but the lack of bioethics.

The neurons' ability to mirror the behaviour of others has given a say to neuroscience in the philosophical discussion about the "*other minds*" tackled before by Locke and Wittgenstein, among others (Hamauzu, 2018, p. 287). Empathy is human, desirable and unavoidable in most cases. However, separating mobility from migration leads to an uncritical discrimination between us and them and an inhuman line of thought (Sager, 2018). From our previous section in phenomenology we can see that an ethics of migration is impossible, useless or blatantly unneeded. Quite the opposite than a bioethics of displacement. For example, even if we consider that the gap in health care due to different languages will never be fully closed, the bioethics' cornerstone of cultural humility will enhance our communication in a way that weighting the morals behind crossing borders never had given us.

The bioethics of displacement in migration goes beyond the human casuistry, reaches whatever we reach. The Beach Flats Garden Community in the greater Santa Cruz, close to the San Lorenzo River, is the scenario for "*agroecologies of displacement*" where the multidimensional dispossession have shaped creole agricultural practices bringing migration to veggies (Glowa et al., 2019). The most diminutive life forms can't slip away from the aura of human displacement … even when they try their best. The Koch's bacilli, responsible of tuberculosis, is responsible to political adjustment, but it's needed a quite multidisciplinary approach to envisage its response including (Mason et al., 2017):

- Medical anthropologists
- Health sociologists
- Social cognitive psychologists
- Social geographers
- Policy analysts
- Health economists

It means that the multidisciplinary work the bioethics is more used to deal with, happens to be its soundest moral ground when compared with ethics. I'm not apologising about sending ethics packing. Key moral values for the bioethics of displacement like inclusiveness and respect have a lifelong work in ethics that must be accommodated within. In the case of the Asian diaspora in the UK, for example, bioethics has performed greatly as a vehicle for embedding faith-based values and needs of migrants at the end of life (Samanta et al., 2018).

Climate change is linked with the global displacement of people flowing as climatic refugees, mainly coming from (Heslin et al., 2019):

- The Pacific Island States suffering the sea level rise
- Bangladesh populations affected by great and greater cyclonic storms
- West African victims of desertification
- South America's Southern Cone deforested rural settings

Namely, most of the small island developing states are failing to integrate displacement in disaster risk reduction and climate change adaptation policies nor with national sustainable development plans (Thomas & Benjamin, 2017).

A challenge in migration and health research, that will affect the aforementioned populations first, is relying upon categorizations that lead us to false generalisations and us / them kind of dichotomies rather than producing relevant knowledge (Rask, 2018). Quite otherwise, the regional integration process can play a major role in a smooth progressive extension of both social protection and social security rights to migrant workers (Fornalé, 2017).

Let's see two insights about the healthcare opportunities on the verge of climate change massive displacement. The near-future healthcare devices will "*see*" almost intangible particles like the ones producing odour in order to detect body abnormal reactions before tomography, magnetic resonance or ultrasound and even before they start to materialise (Dixit & Prasad, 2017, p. 189). Ironically, the decreased ability to identify odours is a mortality harbinger (Pinto et al., 2014). What about using the near-future chemical analysis power to detect ethnic backgrounds in the border? Today it only exists as studies on cosmetics about axillary odour and volatile compounds in cerumen able to discriminate between African American, Caucasian and East Asian (Prokop-Prigge et al., 2014, 2016). However, its potential for biopolitics amidst a conservative revolution facing a new broader race identity (Mercer & Cunningham, 2003) is evident. On the other hand, the chances for affordable and personalised healthcare to climatic refugees will rise when including the data gathered in the borders. And if it weren't enough, nursing education, practice and research might benefit from an increasing awareness of the social, political and cultural dimensions of disability as the displaced population grows (Kearney & Pryor, 2004). It may lead to a more culturally humble bottom-up rework of the International Classification of Functioning, Disability and Health (ICF).

THE MANY MIGRATIONS

The Migration to the Old Age

In Homer's *Odyssey*, a masterwork of deep and ancient Mediterranean lore, we read about the return to home of the crafty Ulysses after twenty years of sailing, killing, cheating, loving and surviving. Only his dog recognized him just before he was to die in the saddest scene of the book. Regardless of Athena tricks, Ulysses was difficult to recognize because he grew old. In the Bioethics of Displacement, the Migration is reflected in Aging in a philosophical sense.

In the same sense that Ulysses or Phileas Fogg were almost professional migrants, there is also a gilded ageing readiness. From the Sandra Torres works on ethnogerontology and ethnogeriatrics follows that there is a kind of successful ageing for every culture (Torres, 1999, 2003). For example, in many Western countries a successful ageing is an independent one but in rural Jamaica about not having to do the things by yourself, anyways there is a clear cultural idea of first-class old age (Bourne & McGrowder, 2009).

The reflection of a person who migrates with a sound economic backup is a healthy ageing, free at least of the serious and preventable issues. We can see in this case a perpetuation of poverty through the ethnic background inheritance that defines a person like belonging to a minority. In the US, for example, the government is now unable to deny the weight that segregation in every shape has had in the health across the lifespan in spite of the fact that segregating has been a long-time tagged as criminal practice, namely since the Brown v. Board of Education decision (Chief Justice Warren, 1954). In the structural increased risks of having growth up in the ghettos we can find the intersection of the Bodi migration and ageing.

Cyborgs, the Migration Within the Body

A cyborg is a cybernetic organism that blends electronic devices with living-tissues. In this section the definition will apply to a tech-hybridised human with non-human capacities. Regardless of where their body parts came, the cyborg advocates expected they would become full *political bodies.* Therefore, a Bill of Rights and a Declaration has been proposed (Furness, 2018). The cyborgs' advocates are not very popular among other excluded collectives and those collectives, in turn, are seen by transhumanists as opportunists and shortsighted (Baard, 2003). Also according Erik Baard, Rabbi Norman Lamm, chancellor of Yeshiva University, replied to the risen doubts about cyborgs' personhood:

I would say if a creature is both sentient and intelligent, and has a moral sense, then that creature should be considered a human being irrespective of the genesis of that person. (Baard 2003)

Transhumanism is a social, political and intellectual movement that advocates for the freedom of use of technologies that eventually could transform the human body into a posthuman cyborg one. The term transhuman is shorthand for transitional human. In turn, a posthuman, according to the World Transhumanist Association, is:

A being whose basic capacities so radically exceed those of present-day humans as to no longer be unambiguously human by our current standards.[2]

Transhumanism is important for bioethics not only because it has its own brand, but also because the techs that transhumanism focuses on are imminent (Porter 2017). And they focus on a lot of them! Besides the foreseeable BCI, the transhumanists embrace: cryonics, nanotechnology, cloning, psychopharmacology, genetic enhancement, artificial intelligence, brain chips, robotics, space colonisation… (Elliott, 2003) You'd get the point: whatever promising tech-enhancement for the human condition. Thus, with the exception of genetic medicine, the enhancements available to date (cosmetic surgery, synthetic hormones, psychoactive medications and lifestyle drugs) are not included in a transhumanist agenda. Tacitly, the persons who only wear prostheses giving back their former capabilities shouldn't be considered posthumans but human cyborgs. It leads to further bioethical discussion on identity depending if the body parts replaced are the face or the genitalia. In the same vein we should ask ourselves which is the minimum tissue body or set of organs that a human cyborg must have before becoming posthuman. Should a human body incompatible with life not be considered posthuman? Is not the capacity for staying alive after being squished a superhuman one? Would those persons be displaced from both posthuman and antagonist circles?

The antagonist to transhumanism, by the way, is bioconservatism and it's a very populated movement. Most people see cyborg enhancements as a change for the worse and have raised their mistrust about them, which can discourage potential transhumanists (Trothen & Mercer, 2017, p. 166). Paradoxically, the desire as drive of a genuine human transcendence should support the cyborg's way, but very few people actually wish to follow it. On the other hand, artificial intelligence studies have gone so far that some 2019 people can influence what the human (or human-like intelligence) will be, so the human species has been dragged from the digital technologies culture and placed in a cyber-tech present, regardless of conservative desires (Uğur & Kurubacak, 2019).

The clear red line between public value and descriptive scientific discourses has been disappearing since the 2010s (Cheshko et al., 2017). To make things more difficult, heterogeneity is usually ignored when a major human network grows, for example, the use of aesthetics and critical theory applied to interfaces are both marginal and uneven in quality (Bardzell, 2009). This careless attitude could shake our future foundations while leading to issues like splitting selves and multiple displacement as in the case of cyborgs (Star, 1990).

The use of technology to shape the human body functionalities turn the body into information (Gray et al., 1995, p. 366) and leads to a historic convergence of the past of both humans and the machines (Waldby, 2000, pp. 46–47). The subject path to computer data goes through the digital *dismemberment* of the body (Scholl, 2018, pp. 162–171) and a further moral, spiritual and existential crisis of corporeality due to massive biotech control required to reach the obsolescence of the only embodiment our specie has known (Williams, 1997). At which point will we stop considering our life as not of our own but from another one? And what about cyborg-children enhanced to overcome a disability? (Prusak, 2005) Will an unrequested body enhancement result in a metaphysical killing? And if a pregnant mother dead is turned into a cyborg-womb? Would it be different from a zombie mother case?

On the other hand, the cyborg prostheses are bringing, will bring and will bring back more capabilities and happiness to formerly disabled people. From a transhumanist view, information is a prosthesis as Stelarc stands out (Stelarc, 1997). These feats, allegedly, challenge the normative status of the human body of irreplaceability but it's more likely that persons will have similar relations to prostheses that they had with their former body parts (Walker, 2018). Might we will share the Ronald Bailey's joy in *Liberation Biology* when he says:

People who benefit from the fruits of biotechnological progress in the future will be neither Frankenstein's monsters nor genetic robots. Rather, they will be our grateful descendants to whom we have eased the burdens of disease, disability, and early death—if only we chose not to slow or kill the development of this new technology. They will look back in wonder, and perhaps in horror, at those who would have denied them the blessing of biomedical progress. (Bailey, 2013, p. 181)

Dignity will be the "*rhetorical feel-good concept that everyone can applaud*" (Sadler, 2010) but it's all the way too fuzzy when it comes to cyborg bioethics. By the same token, I dismiss the critiques of bioconservatives stating that enhancements undermine our human dignity and "*erodes something that is*

Figure 1. Cover of Time magazine of the 21st of February, 2011
Source: Toledano and Myers (2011)

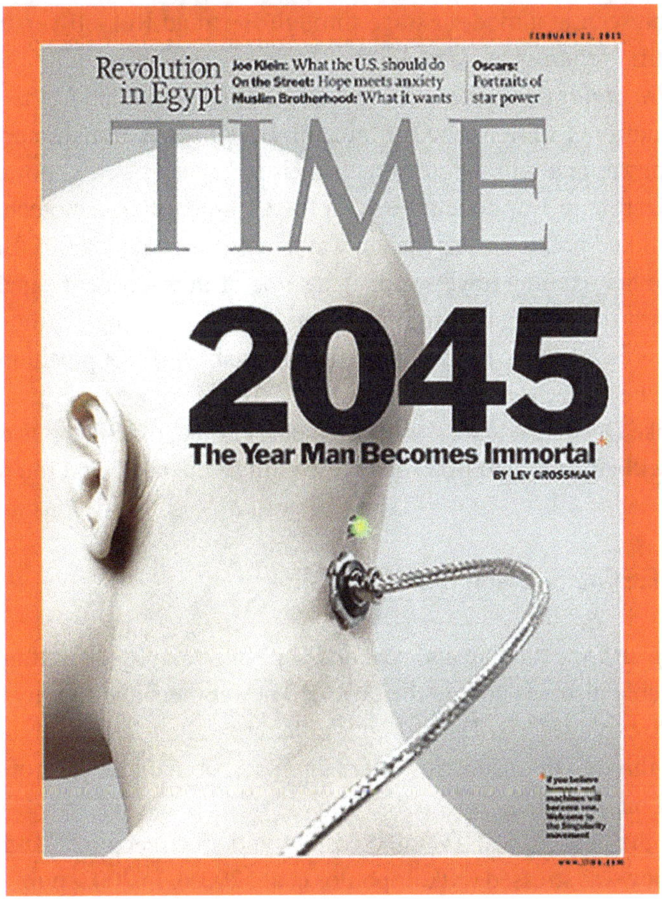

very valuable about being human but that is difficult to put into words" (Bostrom, 2015). At least until they find the words. In addition to this, transhumanism has a shade of religious dignity thanks to its radically different soteriology, a DIY salvation closer to a machine-magic though where death has been put aside (Trothen & Mercer, 2017, p. 293). We must be aware that smart technologies not implies smarter behaviour but rather new kinds of ignorance and violence, that's brutality (Bertolotti et al., 2017). As suggested in the previous Chapter, the dignity of human effort itself could grant the minimum quality expected for an output. Moreover, the hybridization with an AI rather than a mere device, could blend any human motivation with the effortless information available. What would come to be a cyborg if the human results happened to be the weakest link in the chain human-AI-machine? What dignity would await us then?

Let's see the bright side now. Whatever happens, due to transhumanism we're close to the borders of our own biological and cultural identity. And, likely, many people of the first generation of cyborgs would be marooned in this border and beyond. But perhaps, instead of cherishing human dignity just like the most Western Bioethics Committees and other bioconservatives do, the posthumanist are really giving to "human" and "dignity" a broader unheard meaning (Smith, 2008, pp. 8–10). And might it be that some of them turn into supermen coming back from the agrammatical abroad (Deleuze, 1988,

p. 132). A group of undertakers of Anthropocene treasuring the force of being human plus new ones (Deleuze, 2006, p. xix; Weinstein & Colebrook, 2017). So be it.

This dilemma can be seen in a modern stage through Bertrand Russell versus John B. S. Haldane. Haldane saw the scientific progress as some kind of butler of humanity, quite the opposite, Russell replied to the Haldane's *Daedalus* (Haldane, 1926) with one of its own titled *Icarus* (Russell, 1924) where foresaw an unsavoury future of subjugation for humanity. As I see it, transhumanism will undoubtedly serve those who hold power and will magnify their abuses for some years. For example, Chris Gray's analysis of the cyborgization of war concludes with the argument that nanotechnologies and infowars will soon blend business and politics competition, war and crime (Gray, 2000). But when the elites find themselves sated or another exciting novelty appears some of their coveted capabilities will start to be less exceptional. The rest of people, according to their closeness to the posthuman elite, will start to acquire with less and less effort as the time goes by the invaluable gifts of posthumanity. We may enjoy a dream-like healthcare where death and disability will be a true rarity. Or perhaps those first posthumans' life results will be not like Daedalus, Icarus or Prometheus, but like the Sisyphus one (Umbrello & Lombard, 2018). Then we will get smashed like ants in the path of the myth and the rock countless times...

ZOOM IN THE GREECE'S SANITARY BOMB

Let's go into something less speculative and real and sad. In 2012, Greek authorities got surpassed by the surge of communicable diseases among the crowded newcomers and rapidly tagged the migrants as a sanitary bomb about to explode (Nikolas, 2012). We'll now see how keen the bioethics of displacement is when it comes to see through this scenario. Get ready for a zoom on the lack of belonging through the loath of life. Let's get a bit of lore on this.

From 1989 to 2003 the number of HIV infected migrants in Greece rose and the need for a group-based public health policy was stressed (Nikolopoulos et al., 2005). In 2002, public health regulations on prostitution were enacted to cluster prostitution in determined neighbourhoods to attain a better control, however the Greek law-makers copy-pasted the practices of other European countries away from the buffer zone in despite of their legal tradition in this topic (Korasidou, 2002; Themis, 2017). Despite this careless attitude no major HIV crisis happened until 2010. Also in 2002, Greece joined the European Monetary Union and its capitalism resulted to be far more inefficient than the one from most European countries, which led to a rising debt-to-GDP ratio (Sakellaropoulos, 2016). The same year, the Greek government debt manager Christoforus Sardelis started to buy from his Goldman Sachs' old friend Antigone Loudiadis complex financial products that he didn't care to understand to hide the country's debt away from the EU sight (Richards, 2012).

Nick Dunbar disclosed all this legal murkiness in an awesome investigative article published in 2003, meanwhile Standard & Poor's (S&P) raised the Greece long-term debt credit rate from A to A+ (Dunbar, 2003). The S&P CEO at this time and during the Greek crisis, Harold McGraw III, became the leader of the CEOs' lobby known as Business Roundtable that controls "*U.S. companies with $4.5 trillion in annual revenues, more than 10 million employees, and nearly a third of the total value of the U.S. stock markets*".[3] In 2015, Harold McGraw III became Chairman Emeritus of McGraw Hill Financial (S&P owner) and was gushed over by the Board Presiding Director in these terms: "*a visionary and dedicated leader of this Company since he became President and CEO in 1998 and he has earned the respect of the global business community for his many achievements*".[4]

Figure 2. Antigone Loudiadis, the architect of the Greek crisis. She got filthy rich after getting her country on a Euro-mess.
Source: Armitage (2015)

In 2019, the three major rating companies, Standard & Poor's, Moody's Investors Service and Fitch Ratings responsible for greatly embittering the financial crisis, kept being astonishingly profitable and hoard 95% of the global rating market (Mutize, 2019). As the Pulitzer Prize-winning journalist and author Thomas Friedman warned in 1996, twelve years before the 2008 global financial scam:

There are two superpowers in the world today in my opinion. There's the United States and there's Moody's bond rating service [in the 2008 context it would go for S&P and Fitch likewise]. The United States can destroy you by dropping bombs, and Moody's can destroy you by downgrading your bonds. And believe me, it's not clear sometimes who's more powerful. (Friedman, 1995)

Unsurprisingly, the Harold McGraw III family and the president George Bush have been *transgenerational friends* and made money also during the 2008 crisis (Davis, 2016; Trelease, 2006). What gives a few trillions wiped out when you're among your buddies? The European Commission, led by Romano Prodi, in turn, didn't lift a finger to avoid the Greek disaster. Thus, between greedy spoiled brats, loafers and arse-lickers, it was how thousand lives went to hell.

In 2008 the financial crisis struck a badly enfeebled Greece economy, then the ordinary people's access to healthcare and social welfare got severely crippled (Kentikelenis et al., 2011). Soon the bribes given to medical staff to jump queues in overstretched hospitals started (Telloglou & Kakaounaki, 2011). The suicides rose 17% from 2007 to 2009 and likely more even in 2010 when both the heroin and the extortionists were rampant (Avgenakis, 2011).

In 2010, a surge of 620 unlicensed brothels sprayed Athens, most of them in Metaxourgio, a gentrified town where any middle-class racist has an immigrant to finger (Alexandri, 2015; Nikolas, 2012).

Figure 3. Harold McGraw III, S&P CEO during the 2008 heist, fostering a volunteers' day in 2013[5]

Meanwhile, the budget cuts meant a major reduction of street programs aimed at drug users: in the fall of 2010, 85% of Athenian drug users weren't in a program (EKTEPN, 2010).

In 2011, the loath of life in the once joyful Greece got to such an extent that some people self-inflicted HIV infections to get access to a 930 USD social aid and a hastened access to rehab (EKTEPN, 2011). The reports of HIV-1 cases among injecting drug users rose to 1600% in Athens, mostly among the most vulnerable people, like the homeless and immigrants (Paraskevis et al., 2013).

After supporting for ten years the Greek ruling class laid-back attitude about the access to health care of immigrants, HIV and prostitution, the then ministers Michalis Chrisochoidis and Andreas Loverdos profited the new wave of migrants from the Syria War started the year before to say in 2012 that they were a biological menace to Greece (Nikolas, 2012). Not the troika, not Goldman Sachs, not the phoney credit rating companies, not the irresponsible financial policies everywhere, not the lack of transparency in the access to health care, not the cuts in the public health budget, not the abandonment of the injecting drug users in the streets, not the unrealistic measures on prostitution…

The deaths attributable to austerity in 2012 were about 2000 people, the AIDS-related deaths according UNAIDS have remained stable and below 100 since 1992 onwards, even the number of new-infected from 2008-2012 is not close even to the half of the victims of *pure* economic choices (Vlachadis et al., 2014).[6] From 2009 to the 2012 elections in Greece, the foreign banks got rid of the about 300 USD billion debt detrimental to the European Union taxpayers, mainly the Greek citizens themselves; meanwhile the most profitable firms folded their profits and the great fortunes doubled their share of the country's GDP (Sakellaropoulos, 2016). The plundering of Greece wouldn't be possible without misreporting economics statistics to the EU (White, 2010). Neither it hadn't been possible without Goldman Sachs and Wall Street backing up and magnifying the Greek budget managers' misinformation (Balzli, 2010; Story et al., 2010).

In 2016, after seven years of austerity, all that cock and bull story about migrants and sanitary bombs sounds like a mix between chasing a scapegoat and breaking a country to better control it. The "*criminal consequences*" of the austerity in Greece has aged and diminished the population, from 2010 to 2016: "*Greece was faced with a five times greater rate of annual all-cause mortality increase... compared with pre-austerity*" -that's three times the Western Europe expected figure (Tyrovolas et al., 2018). About a fifth of the women born in 1970 will remain childless as Greece dies out over imposed austerity (Flood, 2017).

Figure 4. Christoforos Sardelis, the developer of the Greek crisis, became the Chairman of BoD, Ethniki Asfalistiki (National Insurance). In 2018, he received the award for the Most Admired Enterprise of Greece.
Source: Sardelis (2018)

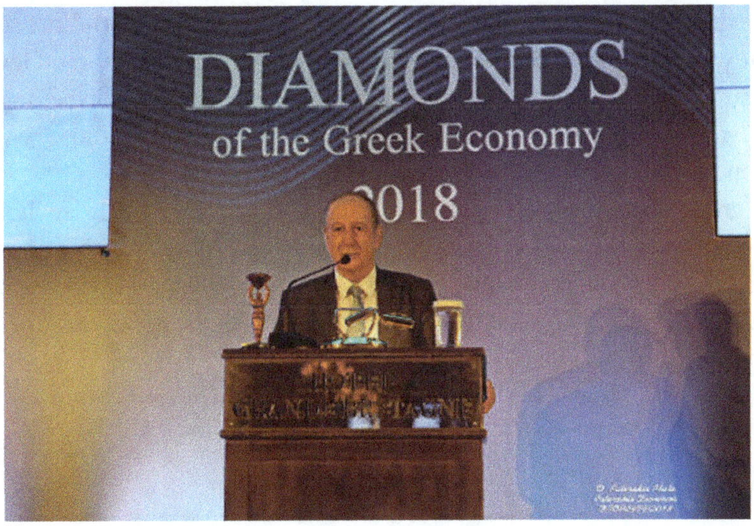

Let's go closer to the truth and the present 2022. In the Filis street in Athens 23 USD caché street prostitutes accept sex without condom to take revenge upon the kind of people who infected them (Magra, 2018). Ten minutes away by car, in the Syntagma square, Syrian penniless boys, many of them suffering psychological conditions from their traumatic displacement, give themselves to prostitution for as little as 2 USD while they come and go from the Turkish refugee camps contributing to the STD epidemics through politics of pain (Avilov, 2018; Cantekin, 2018; Samuels, 2016).

Figure 5. Still from Russia Today's documentary titled Sex, Drugs & Refugees. Syrian teenagers in Athens resort to prostitution to survive.
Source: Avilov (2018, pt. 25:44)

In the turn of the new millennium the derivatives allowed the countries to lower their debt without raising taxes or reducing expenses. It means to kick the problems ahead in time but turning them into a killing and growing snowball. It means displacing the last bit of responsibility from the current decision-makers to the further ones and the most spineless ruling classes from Southern Europe like Greece, Italy and Spain fell for it while the hegemonic corruption centred in the Benelux and Germany finally reap the profit of their EU scam. As expected, this displacement also brought a disbelonging, the incoming high-profile bureaucrats will inherit the debt but the common citizens are those who will face it. The way in which it happens is the politics of pain (e.g. austerity).

The democidal greed of the ones like Sardelis, Loudiadis, McGraw, Chrisochoidis and Loverdos endangers the rest of us since the epidemics of diseases like the HIV adapts to a rising human mobility fueled by crisis -it's not only about receiving a population with a higher STDs prevalence. The bioethics of displacement makes clear as water that the harshest sanitary bomb we face is the marabunta of pen pushers ready to make things explode, not immigrants.

CONCLUSION

There is an urgent need for understanding the experiences of migrants in the approach to healthcare and human rights. Promoting justice will raise our readiness to face the challenges posed by old forms of migration such as slowly moving towards old age and unexpected forms of migration such as moving to a cyborg embodiment that also needs to be understood. Lack of justice, in turn, would lead to perpetuation of poverty and health disparities not only through ethnic background inheritance but during the many migrations begging for a bioethical framework that recognizes dignity and rights as well as equality in healthcare.

REFERENCES

Alexandri, G. (2015). Unravelling the yarn of gentrification trends in the contested inner city of Athens. In L. Loretta & S. Bang Hyun (Eds.), *Global gentrifications: Uneven development and displacement.* Policy Press. doi:10.2307/j.ctt1t894bt.8

Armitage, J. (2015, July 10). *Meet the Goldman Sachs banker who got rich getting Greece into the euro.* The Independent. https://www.independent.co.uk/news/uk/home-news/greek-debt-crisis-meet-the-goldman-sachs-banker-who-got-rich-getting-greece-into-the-euro-10381951.html

Avgenakis, E. (2011). *Question: Dramatic increase in the number of suicides due to the economic crisis and rumors on the operation of networks of usurers and blackmailers.* Academic Press.

Avilov, A. (Director). (2018, June 19). *Sex, Drugs & Refugees.* Russia Today. https://www.youtube.com/watch?v=m4HpLitVRK8

Baard, E. (2003). Cyborg Liberation Front. *Village Voice, 29.*

Bailey, R. (2013). Liberation Biology: The Scientific And Moral Case For The Biotech Revolution. *Prometheus.*

Balzli, B. (2010, February 8). Greek Debt Crisis: How Goldman Sachs Helped Greece to Mask its True Debt. *Spiegel Online*. https://www.spiegel.de/international/europe/greek-debt-crisis-how-goldman-sachs-helped-greece-to-mask-its-true-debt-a-676634.html

Bardzell, J. (2009). Interaction Criticism and Aesthetics. *Proceedings of the SIGCHI Conference on Human Factors in Computing Systems*, 2357–2366. 10.1145/1518701.1519063

Bertolotti, T., Arfini, S., & Magnani, L. (2017). Of Cyborgs and Brutes: Technology-Inherited Violence and Ignorance. *Philosophies*, *2*(1), 1–14.

Bostrom, N. (2015). In Defense of Posthuman Dignity. In H. Kuhse, U. Schüklenk, & P. Singer (Eds.), *Bioethics: An Anthology*. John Wiley & Sons.

Bourne, P. A., & McGrowder, D. A. (2009). Rural health in Jamaica: Examining and refining the predictive factors of good health status of rural residents. *Rural and Remote Health*, *9*(2). Advance online publication. doi:10.22605/RRH1116 PMID:19338392

Brochmann, G., & Hagelund, A. (2011). Migrants in the Scandinavian Welfare State. *Nordic Journal of Migration Research*, *1*(1), 13–24. doi:10.2478/v10202-011-0003-3

Cantekin, D. (2018). Syrian Refugees Living on the Edge: Policy and Practice Implications for Mental Health and Psychosocial Wellbeing. *International Migration (Geneva, Switzerland)*. Advance online publication. doi:10.1111/imig.12508

Cheshko, V. T., Glazko, V. I., & Kosova, Y. V. (2017). Bioethics: Reincarnation of Natural Philosophy in Modern Science. *Biogeosystem Technique*, *4*(2), 111–121.

Chief Justice Warren, E. (1954). Brown v. Board of education. *United States Reports, 347*(1954), 483.

Davis, O. (2016, December 19). *No Test Left Behind:How Pearson Made a Killing on the US Testing Craze*. http://talkingpointsmemo.com/features/privatization/four/

Deleuze, G. (1988). *Spinoza: Practical Philosophy*. City Lights Books.

Deleuze, G. (2006). *Nietzsche and Philosophy*. Columbia University Press.

Dixit, S., & Prasad, R. (2017). *Human Bond Communication: The Holy Grail of Holistic Communication and Immersive Experience*. John Wiley & Sons. doi:10.1002/9781119341451

Dunbar, N. (2003, July 1). *Revealed: Goldman Sachs' mega-deal for Greece*. Risk.Net. https://www.risk.net/node/1498135

Eckenwiler, L. (2018). Displacement and solidarity: An ethic of place-making. *Bioethics*, *32*(9), 562–568. doi:10.1111/bioe.12538 PMID:30450599

EKTEPN. (2010). *Annual report on the state of the drugs and alcohol problem*. Greek Documentation and Monitoring Centre for Drugs.

EKTEPN. (2011). Report of the ad hoc expert group of the Greek focal point on the outbreak of HIV. Greek Documentation and Monitoring Centre for Drugs.

Elliott, C. (2003). Humanity 2.0. *The Wilson Quarterly*, *27*(4), 13–20.

Flood, R. (2017, April 29). *Shock birth rates shows Greeks may be dying out over imposed austerity*. Express.Co.Uk. https://www.express.co.uk/news/world/798318/Greece-birth-rate-dying-economy-austerity-EU-debt-population-crisis

Fornalé, E. (2017). Global–regional interaction to extend access to social protection for migrant workers: Insights from ASEAN and MERCOSUR. *International Social Security Review*, *70*(3), 31–52. doi:10.1111/issr.12140

Friedman, T. L. (1995, February 22). Opinion | Foreign Affairs; Don't Mess With Moody's. *The New York Times*. https://www.nytimes.com/1995/02/22/opinion/foreign-affairs-don-t-mess-with-moody-s.html

Furness, D. (2018, July 4). *Cyborg Law: What Laws Should Govern a Society of Cyborgs*. Digital Trends. https://www.digitaltrends.com/cool-tech/cyborg-law-and-rights-of-augmented-humans/

Glowa, K. M., Egerer, M., & Jones, V. (2019). Agroecologies of displacement: A study of land access, dislocation, and migration in relation to sustainable food production in the Beach Flats Community Garden. *Agroecology and Sustainable Food Systems*, *43*(1), 92–115. doi:10.1080/21683565.2018.1515143

Gray, C. H. (2000). Cyborg Warriors. In *Cyborg Citizen: Politics in the Posthuman Age*. Routledge. doi:10.4324/9780203949351

Gray, C. H., Mentor, S., & Figueroa-Sarriera, H. (1995). *The cyborg handbook*. Routledge.

Haldane, J. B. S. (1926). *Daedalus, Or, Science and the Future*. Kegan Paul.

Hamauzu, S. (2018). On Development from Husserl's phenomenology: Between Phenomenology of Intersubjectivity and Clinical Philosophy of Caring. *大阪大学大学院文学研究科紀要*, *58*(2), 1–311.

Heslin, A., Deckard, N. D., Oakes, R., & Montero-Colbert, A. (2019). Displacement and Resettlement: Understanding the Role of Climate Change in Contemporary Migration. In R. Mechler, L. M. Bouwer, T. Schinko, S. Surminski, & J. Linnerooth-Bayer (Eds.), *Loss and Damage from Climate Change: Concepts, Methods and Policy Options* (pp. 237–258). Springer International Publishing. doi:10.1007/978-3-319-72026-5_10

Kearney, P. M., & Pryor, J. (2004). The International Classification of Functioning, Disability and Health (ICF) and nursing. *Journal of Advanced Nursing*, *46*(2), 162–170. doi:10.1111/j.1365-2648.2003.02976.x PMID:15056329

Kentikelenis, A., Karanikolos, M., Papanicolas, I., Basu, S., McKee, M., & Stuckler, D. (2011). Health effects of financial crisis: Omens of a Greek tragedy. *Lancet*, *378*(9801), 1457–1458. doi:10.1016/S0140-6736(11)61556-0 PMID:21988763

Korasidou, M. (2002). *Hotan hē arrōstia apeilei: Epitērēsē kai elenchos tēs hygeias tou plēthysmou stēn Hellada tou 19ou aiōna*. Topothētō Giōrgos Dardanos.

Magra, I. (2018, October 28). 'They Don't Have Money': Greece's Prostitutes Hit Hard by Financial Crisis. *The New York Times*. https://www.nytimes.com/2018/10/27/world/europe/prostitutes-greece-crisis.html

Mason, P. H., Roy, A., & Singh, P. (2017). Reciprocity-building and the importance of interdisciplinary collaboration in tuberculosis research. *Journal of Biosocial Science*, *49*(4), 559–562. doi:10.1017/S0021932016000626 PMID:27869060

Mercer, S. H., & Cunningham, M. (2003). Racial Identity in White American College Students: Issues of Conceptualization and Measurement. *Journal of College Student Development*, *44*(2), 217–230. doi:10.1353/csd.2003.0021

Mutize, M. (2019, June 23). *Why credit rating agencies are still getting away with bad behaviour*. The Conversation. http://theconversation.com/why-credit-rating-agencies-are-still-getting-away-with-bad-behaviour-117549

Nikolas, K. (2012, April 2). *Greece to tackle problem of migrant communicable diseases*. Digital Journal. https://www.digitaljournal.com/article/322251

Nikolopoulos, G., Arvanitis, M., Masgala, A., & Paraskeva, D. (2005). Migration and HIV epidemic in Greece. *European Journal of Public Health*, *15*(3), 296–299. doi:10.1093/eurpub/cki064 PMID:15923211

Paraskevis, D., Nikolopoulos, G., Fotiou, A., Tsiara, C., Paraskeva, D., Sypsa, V., Lazanas, M., Gargalianos, P., Psichogiou, M., Skoutelis, A., Wiessing, L., Friedman, S. R., Jarlais, D. C., Terzidou, M., Kremastinou, J., Malliori, M., & Hatzakis, A. (2013). Economic Recession and Emergence of an HIV-1 Outbreak among Drug Injectors in Athens Metropolitan Area: A Longitudinal Study. *PLoS One*, *8*(11), e78941. doi:10.1371/journal.pone.0078941 PMID:24265730

Pinto, J. M., Wroblewski, K. E., Kern, D. W., Schumm, L. P., & McClintock, M. K. (2014). Olfactory Dysfunction Predicts 5-Year Mortality in Older Adults. *PLoS One*, *9*(10), e107541. doi:10.1371/journal.pone.0107541 PMID:25271633

Prokop-Prigge, K. A., Greene, K., Varallo, L., Wysocki, C. J., & Preti, G. (2016). The effect of ethnicity on human axillary odorant production. *Journal of Chemical Ecology*, *42*(1), 33–39. doi:10.100710886 015-0657-8 PMID:26634572

Prokop-Prigge, K. A., Thaler, E., Wysocki, C. J., & Preti, G. (2014). Identification of volatile organic compounds in human cerumen. *Journal of Chromatography. B, Analytical Technologies in the Biomedical and Life Sciences*, *953–954*, 48–52. doi:10.1016/j.jchromb.2014.01.043 PMID:24572763

Prusak, B. G. (2005). Rethinking "Liberal Eugenics": Reflections and Questions on Habermas on Bioethics. *The Hastings Center Report*, *35*(6), 31–42. doi:10.2307/3528563 PMID:16396203

Rask, S. (2018). *Diversity and Health in the Population: Findings on Russian, Somali and Kurdish origin populations in Finland*. https://helda.helsinki.fi/handle/10138/233970

Richards, H. (2012, February 21). *How Goldman Sachs helped mask Greece's debt*. The Bureau of Investigative Journalism. https://www.thebureauinvestigates.com/stories/2012-02-21/how-goldman-sachs-helped-mask-greeces-debt

Russell, B. (1924). *Icarus: Or, The Future of Science*. Dutton.

Sadler, J. Z. (2010). Dignity, Arête, and Hubris in the Transhumanist Debate. *The American Journal of Bioethics*, *10*(7), 67–68. doi:10.1080/15265161003728845 PMID:20582838

Sager, A. (2018). Toward a Political Philosophy of Mobility. In A. Sager (Ed.), *Toward a Cosmopolitan Ethics of Mobility: The Migrant's-Eye View of the World* (pp. 91–95). Springer International Publishing. doi:10.1007/978-3-319-65759-2_6

Sakellaropoulos, S. (2016). Winners and Losers in the Greek Financial Crisis. *Global Dialogue*, *6*(4). https://globaldialogue.isa-sociology.org/winners-and-losers-in-the-greek-financial-crisis/

Samanta, J., Samanta, A., & Madhloom, O. (2018). A rights-based proposal for managing faith-based values and expectations of migrants at end-of-life illustrated by an empirical study involving South Asians in the UK. *Bioethics*, *32*(6), 368–377. doi:10.1111/bioe.12447 PMID:29882999

Samuels, G. (2016, June 7). *Refugee men are selling sex for less than price of a sandwich to survive.* The Independent. https://www.independent.co.uk/news/world/europe/dozens-of-refugee-men-sell-sex-for-as-little-as-2-to-survive-in-athens-a7068741.html

Sardelis, C. (2018, June). Christoforos Sardelis: The role of the insurance sector in the post-program era. *Economy Watch.* http://www.economywatch.gr/christoforos-sardelis-the-role-of-the-insurance-sector-in-the-post-program-era/

Scholl, L. (2018). *Medicine, Health and Being Human*. Routledge. doi:10.4324/9780203732700

Smith, T. (2008). *Bioethics for the masses: The negotiation of bioethics in film and fiction.* https://search.proquest.com/openview/4f3f42eca10d14c80f751711fbbb8250/1?pq-origsite=gscholar&cbl=18750&diss=y

Star, S. L. (1990). Power, technology and the phenomenology of conventions: On being allergic to onions. *The Sociological Review*, *38*(S1), 26–56. doi:10.1111/j.1467-954X.1990.tb03347.x

Stelarc. (1997). From psycho to cyber strategies: Prosthetics, robotics and remote existence. *Cultural Values, 1*(2), 241–249. doi:10.1080/14797589709367146

Story, L., Thomas, L., Jr., & Schwartz, N. D. (2010, February 13). *Wall St. Helped Greece to Mask Debt Fueling Europe's Crisis—NYTimes.com.* https://www.nytimes.com/2010/02/14/business/global/14debt.html

Telloglou, T., & Kakaounaki, M. (Directors). (2011). Three weeks and one night on duty. In *Folders*. Greece: Skai TV. https://scholar.google.com/scholar_lookup?hl=en&author=Telloglou+T&author=Kakaounaki+M&title=Three+weeks+and+one+night+on+duty

Themis, A. (2017, May). Brothels. *Athens Social Atlas*. https://www.athenssocialatlas.gr/en/article/brothels/

Thomas, A., & Benjamin, L. (2017). Policies and mechanisms to address climate-induced migration and displacement in Pacific and Caribbean small island developing states. *International Journal of Climate Change Strategies and Management*, *10*(1), 86–104. doi:10.1108/IJCCSM-03-2017-0055

Toledano, P., & Myers, D. (2011, February 21). *TIME Magazine Cover: 2045: The Year Man Becomes Immortal - Feb. 21, 2011*. https://content.time.com/time/covers/0,16641,20110221,00.html

Torres, S. (1999). A culturally-relevant theoretical framework for the study of successful ageing. *Ageing and Society*, *19*(1), 33–51. doi:10.1017/S0144686X99007242

Torres, S. (2003). A Preliminary Empirical Test of a Culturally-Relevant Theoretical Framework for the Study of Successful Aging. *Journal of Cross-Cultural Gerontology*, *18*(1), 79–100. doi:10.1023/A:1024830629385 PMID:14617959

Trelease, J. (2006). *All in the Family: The Bushes and the McGraws*. http://www.trelease-on-reading.com/whatsnu_bush-mcgraw.html

Trothen, T. J., & Mercer, C. (2017). *Religion and Human Enhancement: Death, Values, and Morality*. Springer. doi:10.1007/978-3-319-62488-4

Tyrovolas, S., Kassebaum, N. J., Stergachis, A., Abraha, H. N., Alla, F., Androudi, S., Car, M., Chrepa, V., Fullman, N., Fürst, T., Haro, J. M., Hay, S. I., Jakovljevic, M. B., Jonas, J. B., Khalil, I. A., Kopec, J. A., Manguerra, H., Martopullo, I., Mokdad, A., ... Kotsakis, G. A. (2018). The burden of disease in Greece, health loss, risk factors, and health financing, 2000–16: An analysis of the Global Burden of Disease Study 2016. *The Lancet. Public Health*, *3*(8), e395–e406. doi:10.1016/S2468-2667(18)30130-0 PMID:30055996

Uğur, S., & Kurubacak, G. (2019). Artificial Intelligence to Super Artificial Intelligence, Cyber Culture to Transhumanist Culture: Change of the Age and Human. Handbook of Research on Learning in the Age of Transhumanism, 1–16. doi:10.4018/978-1-5225-8431-5.ch001

Umbrello, S., & Lombard, J. (2018). Silence of the Idols. *Appropriating the Myth of Sisyphus for Posthumanist Discourses.*, *9*(4), 98–121. doi:10.18662/po/47

Vlachadis, N., Vrachnis, N., Ktenas, E., Vlachadi, M., & Kornarou, E. (2014). Mortality and the economic crisis in Greece. *Lancet*, *383*(9918), 691. doi:10.1016/S0140-6736(14)60250-6 PMID:24560049

Waldby, C. (2000). *The Visible Human Project: Informatic Bodies and Posthuman Medicine*. Psychology Press. doi:10.4324/9780203360637_chapter_1

Walker, M. J. (2018). On Replacement Body Parts. *Journal of Bioethical Inquiry*. Advance online publication. doi:10.100711673-018-9889-y PMID:30565032

Weinstein, J., & Colebrook, C. (2017). Preface: Postscript on the Posthuman. In *Posthumous Life: Theorizing Beyond the Posthuman*. Columbia University Press. doi:10.7312/wein17214

White, A. (2010, October 7). *EU Stats Office: Greek Economy Figures Unreliable*. ABC News. https://web.archive.org/web/20101007143658/http://abcnews.go.com/Business/wireStory?id=9541636

Williams, S. J. (1997). Modern medicine and the "uncertain body": From corporeality to hyperreality? *Social Science & Medicine*, *45*(7), 1041–1049. doi:10.1016/S0277-9536(97)00031-2 PMID:9257396

KEY TERMS AND DEFINITIONS

Cultural Humility: Recognizing and respecting cultural differences while maintaining an open and humble attitude.

Injustice: Unfair treatment, violation of rights, or lack of fairness, often targeting marginalised groups and individuals.

Moral Imagination: The ability to empathise, envision ethical solutions, and consider various perspectives in decision-making processes.

Public Health: The science of keeping communities healthy, preventing disease, and promoting well-being.

Refugee: An individual forced to leave their country due to danger, seeking safety and protection in another nation.

Solidarity: Unity, mutual support, and agreement among individuals or groups with shared interests, values, or goals.

ENDNOTES

1. https://beachflatsgarden.org/
2. https://humanityplus.org/philosophy/transhumanist-faq/
3. https://www.wharton.upenn.edu/wp-content/uploads/125anniversaryissue/mcgraw.html
4. https://www.prnewswire.com/news-releases/harold-mcgraw-iii-to-become-chairman-emeritus-of-mcgraw-hill-financial-300041543.html
5. https://www.youtube.com/watch?v=In05W6HzBRU
6. https://www.unaids.org/en/regionscountries/countries/greece

Chapter 13

Exploring Human Identity From Social, Cultural, Philosophical, and Biopolitical Perspectives:
Free v Confined I

ABSTRACT

The author travels across the human identity regions, including the social, cultural, philosophical, and biopolitical. First, the author meets the concept of social identity and how it influences an individual's self-concept and body-image. Then a few strokes of the historical and cultural context of human identity, highlighting how technological progress and social structures shape our understanding of identity. By the same token, the author also touches on the ethical dilemmas that may arise from merging body and information technologies, including the possibility of widening income inequality and genetic editing. The philosophical grounds of human identity are then examined, with a focus on the narrative nature of selfhood and the significance of personal identity in bioethics. Finally, the author assesses the impact of public health and biopolitics on identity, including the creation of fictive ethnicities and the challenges of addressing displacement.

INTRODUCTION

This chapter delves deeper into the concept of identity in the context of migration, yet explored in the last chapter. But, while Chapter 12 examined the philosophical and social aspects of migration and its impact on societies, This chapter focuses specifically on human identity and how it is shaped by social structures, cultural traditions, and technological progress.

DOI: 10.4018/978-1-6684-4808-3.ch013

DIGGING FOR HUMAN IDENTITY IN POLITICAL AND PHILOSOPHICAL GROUNDS

Social Identity

The social identity is an individual self-concept as a member of a social group that influences how this person thinks, acts, feels as well as her self-view and her body-image (Falvo & Holland, 2017, p. 21). When identity mainly grounds on politics a dichotomous mindset rules and the experience of oppression fades away (Talwar, 2018). *The political identities are tricky.*

The identity expression is versatile. For example, the soccer chants shift from group-reinforcement to group-threatening when one team or another scores (Scheepers et al., 2003). The discrimination intergroups responds to a desire for cognitive coherence and a need for positive self-esteem (Abrams & Hogg, 1988). We dislike the self-blurriness coming from a hasty game-changing, we abhor the *displacement* and intuitively identify it with the melancholy of disbelonging.

In my opinion, the goal of the social sciences is empowering us to deal with our place within a very big picture and spans of time that are subsumed in our daily experience. But for making the social sciences humanly understandable is needed to suppose the existence of "*managerial identities*" among our many identities that gives cohesion to the set, that look both inwards and outwards (Watson, 2008).

We have overlooked the transformative practices able to change the oppression, masochism and guilt social schemes around the politics of identity (Grossberg, 1996). The architecture and natural surroundings partly determines the identity of a person influencing her self-view, shaping its manifestation (Hauge, 2007). Thus, identity can be shattered due to a displacement. Indeed, the forced migration due to the lack of support to local networks resulting from the global financial stakeholders' unwillingness to change their role as masters of humankind is paramount over any other identity crash (Rawlinson, 2016, pp. 193–194).

In other words, the modern cravings for control, chiefly over the nature, has led to a social fragmentation that makes the contemporary construction of identity both situational and incongruent since our identity has been reduced to profiles and *profilicity* (D'Ambrosio, 2018). Therefore, Bodi would also advocate for *the right to be forgotten* like a fundamental one (Abbasi et al., 2017).

The widening income inequality and the impact on indigenous populations also raises questions about the social and cultural dimensions of human identity. Additionally, the displacement caused by policies and decisions made about our cities and sexuality highlight the role of social structures and power dynamics in shaping our identity. Therefore, we'll reflect now on the ways in which technological progress, social structures, and cultural traditions shape our understanding of human identity.

Progress: Religion and Identity Throughout History

While History and stories moved at the pace of chariots, camels and sailboats, meaning of life was mostly dictated by local traditions and bureaucratic religions. On the other hand, the meddling in inner life was not so prevalent as today. Most people believe that those who lived 1500 years ago were driven by fanaticism and superstition. It's true that the means for fighting aberrant proselytism were meagre and weak and the public education, as far as it came with other freedoms as well, has greatly reduced the illiteracy intrinsic vulnerability. However, if you look at the violence arising from what a few people think about what the markets' needs are or what the providence and national destiny brings, then you'll realise how pathetic our progress towards serenity has been.

Mainly, we've assimilated the technical progress of those many who came before us. It's true that, 1500 years ago, when they lacked the Internet, many peoples, especially in Europe, attended religious services and local ceremonies far more often than in 2023. Notwithstanding, many mediaeval people were not as religiously devout as some might assume, and that religious observance and beliefs varied widely depending on time, place, and social context (Southern, 1970; Swanson, 1995). For most people, religion was a part of daily life, but not necessarily the most important or central aspect of their lives. So, yes, many thought that god, demigods and saints were looking and ready to punish them, but… *what the hell*, divinity was going to forgive them the little guilty pleasure that they so badly need. Nevertheless, in the 21st century, fulfilling the obey-consume-repeat commandments wastes more time of our lives than any religion before given that you can consume contents, too. It's easy, as long as misery is not widespread, to fall into the delusion that personal and collective freedoms have been and will be on the rise. But this mesmerism blinds us to see the present.

I also mean the consequences that are going to face a Humankind that is tricked by some moguls and govts while migrating to the digital plane, dragged away due to elites choices on climate changes and lured to migrate to other planets by messianic tycoons (Lozano Rodríguez, 2019). All this in a time in which body borders are getting blurrier thanks to the former decision-makers. To do so, we must challenge established beliefs and encourage critical thinking on hierarchies.

Bodi takes a look at the big picture. It's not all about predatory politics as usual. Since the 1820s the human networks have been condensing thanks to the ICTs, but they barely have not expanded (McNeill & McNeill, 2003). However, from the 2010s onwards two major new borders, Space and the human body, gained momentum. Our views on Space will change in the next few years, but the way information integrates with us seems increasingly boundless today. Such a new academic curriculum on Bioethics at the highest level must address the new roles of body and culture in the virtual and augmented reality. Not to mention when body and information technologies merge. Almost always the new material progress has produced a rising inequality before spreading any benefit for the most people just like happened with the telegraph. But consider how the gap will widen if inheritable improvements through genetic editing remain prohibitive for 20 years. Or prohibited in one country and not another. And what about the income earning gap after biohacking plays a major role in productivity? In Bodi we'll delve into the present-day predatory mindset that is writing the bioethical dilemmas to be faced tomorrow because we need this preemptive feature in the discipline.

Ethical Dilemmas in the Future of Body and Information Technologies

The potential consequences of merging body and information technologies and the possible ethical dilemmas that may arise are discussed in Bodi, including the potential for widening income inequality and inheritable improvements through genetic editing, which could be prohibited in some countries and not others, highlighting the ongoing impact of colonialism on indigenous populations as a possible example of the complexity of recognizing and addressing displacement. As we have seen, extended meaning for displacement reaches not only the Reason, Logic and the nature of Reality, but also the Ethics. *For good or evil*, here is where the learning curve peaks. Like Derrida's warnings on undecidability of the future (or postponed) ethical choices, we should be aware not to mess with recent historical displacement events. For example, Glen Coulthard holds that the recognition policies from the Canadian government to the First Nations are a way of neutralising displacement and violence while legitimising colonialism (Coulthard, 2014). He lay emphasis on colonialism as an *ongoing* process. Likely, the recognition policies in Canada are a step ahead in a democratic country (Clegg et al., 2007). But it cannot help building a paved road for the most recalcitrant post-truth meanwhile.

Indeed, even the definition of rigour can be displaced by a colonist-friendly one. For example in India, the traditional ways of disseminating knowledge have been squeezed by the British methodology-centrism first and the neo imperialism then (Bell et al., 2017). Even the most kindhearted approach to the colonised peoples from the colonists will always be a type of *monologue*.

Even in a decolonized scenario, the displacement provoked by policies is close to us. Most likely, you live in a city where generations of people have been so close to deciding its fate like the trees in the park. Surely a good chunk of your city is a place where the decision-makers have decided not to spend money, not to make architecture. It means displacing both old residents and newcomers into a dungeon for the love, the bodies and the expectancies of those who live in these towns. The displacement caused by policies is not limited to our physical surroundings but extends to our personal lives and experiences, such as the shifts in sexuality that mirror the shifts in politics, and both highlight the power dynamics at play in decisions made by those in authority.

Therefore, a topic that is so close to love and life like sex, also has experienced displacement. The Western sexual revolution in the second half of the 20th century is the most radical example. The clerical stressing of chastity and fidelity has been sapped by a mix of pharmaceutical new opportunities and political feminism addressed to sexual exploration and fulfilment. All these pleasures were previously considered a male dominion, so it has produced uncertainty among them and a new identity for women (C. Taylor, 2007, p. 495). New ethical issues around sex emerged altogether with a new-brand unisex sexuality displacing the old bipolar one. Indeed, the shifts in sexuality are a *Foucaldian* magnifying glass from the ones in politics. It can be seen, for example, in the 1978 Iranian Revolution and the nowadays Iran sexual one (Mahdavi, 2009, p. 304). Foucault himself, by the way, was seduced by his own sexual points of view and became a polemic international figure unexpectedly involved in the dawn of Ayatollahs' regime (Afary & Anderson, 2010, pp. 38, 138–141).

Furthermore, the potential consequences of the assimilation of technical progress and the merging of body and information technologies, including widening income inequality and the impact on indigenous populations, are examples of the ongoing fight for collective freedoms against alienation. The alienation caused by policies, whether related to technological progress or decisions made about our cities and sexuality, is a reminder of the need to recognize and address such complexities in our pursuit of collective freedoms.

This discussion on collective freedoms opens the floor for a philosophical one on human identity because it touches on the potential consequences of merging body and information technologies, which could affect our understanding of what it means to be human.

Philosophical Grounds of Human Identity

The philosophical grounds of Human Identity are linked with how individual subjectivity comes into being. Unsurprisingly, the "self", the "I" from a postmodern thought is seen as some sort of magic diary gliding from different ideologies towards collective responsibilities (Prainsack & Buyx, 2017). In this sense, we can agree with Nietzsche's vision of anchor-less identities without conceding on his pessimism (Isaac & Platenkamp, 2018). The self is narrative. The self is *a* narrative. There is no criteria of personal identity since it is *logically independent* of both the human body and any other physical fact (S. Shoemaker, 1963, p. 39). Furthermore, authors like David Shoemaker hold that personal identity is insignificant in Bioethics even when the person is alive! (D. Shoemaker, 2010).

Thus, when we talk of survival, generally we mean to keep associating the current mental states with the ones that preceded it (Lewis, 1983). Heraclitus said that we can't step twice in the same river. Even if we contain the water in a ditch to bathe twice in the same waters, these waters wouldn't be the ones flowing in the stage of the river we bathe first because we have left behind in time both waters and river. The same goes for standing with one foot on the riverbank and splashing with the other. As Willard Van Orman Quine explained in the Heraclitus enigma on River Caÿster, we can take a bath twice in the same river, as long as we consider that it is the same river in different stages (Quine, 1950). Wittgenstein would say that Heraclitus was missing rivers with language rules. There is a third way that is falling asleep. The mind loves Heraclitus' riddles, at ease, even when we go to sleep, get shocked or hang out with Dionysus. However, the mind seems ethically less relevant for me when it's seen from this angle.

The reality is imbued by both the contingency principle and its metaphysical counterpart. Most of us, perhaps everybody, only will experience partial changes across life and it includes the experience of identity (Cheng & Bunnin, 2008, pp. 111–112). In any case, we have a lived identity that overlies the wider non-lived alterity that comprehends the many *alt-entities*. The alt-entities are the entities whose belonging can be distinct. Otherwise, without alt-entities, we couldn't say that two things are the same since we'd see them as only one, we would be plurality-blind. The capability to distinguish alt-entities is a higher order of inter-subjectivity. For example, if we take *Zoopolis: A Political Theory of Animal Rights* as a start point, and then play with the phenomenological ontology we'll realise that the identity-gap between humans as non-humans is merely inter-subjective (Donaldson & Kymlicka, 2011, pp. 45–53, 89–107; Rodríguez, 2017). The open subjectivity of the world is indefinite and related with inter-subjectivity (Malpas, 2018).

The identity is one with oneness, although the primitiveness of identity leads to the existence of indiscernible objects (Berto, 2017). The will and the unity of subjectivity, for example, are indiscernibles. *"Trees thinking about determinism and free will might find it impossible to assess their own contribution to their final shape."* (Glover, 1988) The Cartesian "*cogito ergo sum*" requires that the first thought is always "*I belong to myself in this place*". But *the place* is ontologically different from time and space since it's tied to memory. It means that our place, in the beginning, was indiscernible and primitive.

Let's take a new perspective on it. The idea of a criterion of identity can be found in mediaeval authors and Locke, but we mostly owe it to Gottlob Frege (Hale et al., 2017, p. 996). From the Axiom of Extensionality, which assumes that equality is a *primitive* symbol in predicate logic, Frege applied it to sets in his Axiom V (Frege, 1952):

$$\forall F \forall G(\{x:Fx\} = \{x:Gx\} \leftrightarrow \forall x(Fx \leftrightarrow Gx))$$

However, it leads to Russell's Paradox that also reaches the criterion of identity for *directions*. I suppose that there is a direction that points to the directions that do not return to themselves. Does this direction return to itself? We are like this paradoxical line, we point and not point to ourselves because the Frege's Axiom V solves in us upon our sameness but not upon our identity. Rather than this, the identity is this that life is solved about.

Public Health and Identity

Likewise, the Bioethics of Displacement applied to health settings puts the focus on how we belong to a place together (e.g. a hospital) rather than how we relate to each other.

The Beauchamp and Childress concept of self-governance, implies a procedural / functional view of personhood in the search of answers to clinical questions (Beauchamp & Childress, 1994, pp. 120–121). However, it's unsuitable in the analysis of incompetent patient's identity as happens in Alzheimer's since the issue is not how we make rational choices but how the community cares for every member amidst suffering (Have & Purtilo, 2004, pp. 132, 138).

The development of social models of disability makes clearer the nexus between disability and identity. Not only in Alzheimer but in any case, the disability casts shadows on identity since the way they are defined presumes a negative impact on it.

Identity and healthcare is a two-way process, it means that the identity also can lead to negative health outcomes due to structural discrimination (a sort of built disability). For example, a 2017 study led by Aleta Baldwin with sexual minority women found that healthcare structurally unsupportive were perceived as lower quality ones (Baldwin et al., 2017). Therefore, no public healthcare settings should be lacking nondiscrimination legislation. The structural quality of public healthcare settings spans from users to professionals. For example, the social workers' lack of common professional identity and their possible *futility* in a welfare state influences, in turn, vulnerable people access to healthcare resources (Horner, 2018, p. 121).

A further negative impact both from identity and upon identity comes with displacement. Female genital circumcision (FGC) is quoted as "*mutilation*" in the Western countries. Regardless of the truth of this statement, the body image and sexual self-esteem of the women with FGC is seen as dysfunctional (Johnsdotter, 2018). A negative view upon the core of their sexuality through a public discourse on mutilation, imposes a disability to the FGC women.

Public health is increasingly shaped by the demands of both market and state nuts, giving the authority figures, our health data managers, the truth about ourselves. It's a confessional approach in a Foucaldian sense that litters bioethics in the 21st century (Hull 2017).

We are entangled in public discourses on futility and mutilation. Public health, rather than nodding, should embrace the experiences of the many different bodies and its full spectrum of capabilities. To get it, public health should help us to disconnect selfhood from the public. Public health needs to move beyond traditional thinking and embrace a more holistic approach that considers the diverse experiences and needs of all individuals, including those who are marginalised. In urban street settings, marginalised individuals, such as the disabled beggar, a postmodern dalit, faces hypocrisy with exhibitionism (Thurer, 1988). Furthermore, a study led by Todd Lucas discovered that the coordination among systems that perceive stress, like the autonomic nervous system or HPA, is not yet fully understood, especially regarding stress response to social evaluative threats among African Americans. The findings bridge cultural perceptions of stress with stress-related illnesses and health disparities (Lucas et al., 2017). Similarly, a 2017 study of 1351 US adults who identified as gay/lesbian, bisexual, or unsure, found that marginalised sexual minorities who feel discriminated against are more prone to drug consumption and binge drinking (Slater et al., 2017). These behaviours are driven by the cultural-specific risk resulting from their marginalisation, with bisexuals, Hispanics, and less educated adults showing the highest levels of risk (Slater et al. 2017). These results highlight intolerance-driven health disparities, which could be used to tag minority groups as risk-takers or adapt public health to a shifting demography more quickly.

BIOPOLITICS OF IDENTITY

To pack the people together, the national institutions create a *fictive* ethnicity, this is *"a theatrical identification with a racialized image of people"* that spans from national origins to many generations ahead (Murthy, 2011, p. 53). In other words, the modern nations sell an illusion of natural community. Given that this fictive identity belongs to oneself belonging to the other nationals, a person can be addressed individually in the name of collectivity, which makes population control a lot easier. Besides that, in some cases, for example, the Anglo-Irish in Elizabeth Bowen *The Last September* (Bowen, 2000) a national identity can come from voluntary acts, therefore the communities have a more personal sense than usually given and the societal nationalism in a strict sense is both trivial and inconclusive.

In the same vein, the citizenship model of a community is protected against the attacks of a post-national one due to the biopower (subjugation) it holds. But the citizenship model doesn't explain cases like the Ticuanenses, a Mexican community who have migrated to New York since the beginning of 1940s. Like many other migrants, the Ticuanenses have socially constructed structures that transcend the individual nation state where many identities can coexist (R. C. Smith, 2017). However, the transnational life of a community and its multiple contexts not only challenges a model, but gives a status to the voluntary belonging that taunts the modern state while it shows its most mediaeval face.

The Western Philosophy of History was originally racist to the bones. In Hegel, only the Caucasian race seems strong enough to create peoples, to be completely under the auspice of freedom and *to displace others* (P. C. Taylor et al., 2017). The legacy of this philosophy stood out in Giovanni Gentile "*the most rigorous neo-Hegelian in the entire history of Western philosophy*" (Croce, 1995, p. viii). Gentile was the philosopher of fascism who most greatly influenced (even ghostwritten!) Mussolini. Anyhow, the fascism took Italy and Italy almost took Ethiopia where we see an interesting phenomena. Babylon in the Rastafarian views traditionally condensed the recurrent attacks suffered from white-dominated societies as their antithesis expressed through geographies of evil that are both physical and spiritual (Laclau, 1994, pp. 177, 249). In this Rastafarian geography, Rome was added to Babylon after the Italian invasion, therefore, not only Caucasian peoples can *displace* and likely there is no political repetition (restoration, recurrence…) without displacement. The abuses against nations committed by Hegelian natural and national spirits is not so different to the abuses to children veiled by spirit possession rituals (Briggs & Whittaker, 2018). By the way, the deprecated Hegelianism and neo-Hegelianism is not a reason to tag any White Western author as racist, privileged and all this silliness -unless you believe that race actually makes the character.

A way that is quite the opposite than fanatic manners but keeps being abusive, is the benevolent discrimination upon a profiled set of identities. In these soft biopolitics, just like happens in human resources (Romani et al., 2018), the decision-makers address discrimination issues with two conditions:

- The managerial positions identify who are the ones in need, which implies setting the conditions of their subordination
- The help is given expecting the acceptance of hierarchy, which means that the apex will set the conditions of future subordination

There are some dissociative disorders that have a parallel in politics. The dissociative disorders feature a disruption or discontinuity in the integration of consciousness, memory, identity, emotion, perception, behaviour and others (APA, 2013, p. 291). The dissociative amnesia as the inability to recall electoral promises, the depersonalization or detachment of reality (at least of the voters' reality) and, foremost, a special multiple personality where the shattered are the others' selves. Indeed, the political parties turn our societies into some kind of Foucaldian asylums in which madness rules over the sane and insane alike behind the fences of identity. The racism of the new populist far right in the West, a radical case of dissociative amnesia, is syncretic, it mixes contradictory beliefs and stereotypes on fictive in / out ethnic identities (Krzyzanowski & Wodak, 2017). However, a rising number of foreign-born persons with a different ethnic background join xenophobic parties against their own racial identity (Mulinari & Neergaard, 2019). Thus, racism as ideology is alive and kicking (sic) (Lozano Rodríguez, 2019).

Very different groups (from aboriginal tribes to wheelchair users) can share commonalities when experiencing social devaluation, isolation, powerlessness (Mackelprang et al., 2016, pp. 139–140). In and out of asylum, the pursuing of status individually or as a group is determined by the following (Ellemers, 1993):

- Relative status of the group
- Permeability
- Stability
- Legitimacy of personal status versus the group's

Seeing this like a game, the quickest way to success in hierarchical institutions is attributing to yourself the others' merits. It means cheating on your legitimacy while promoting your own status detrimental to any other. For this reason the politicians are the persons with more honorific mentions from any profession, they have (from a Hegelian logic) a *leadership spirit* that allows them to displace whole populations, move and remove the borders, whatever. An example of success in hierarchical institutions comes from a 2017 study led by Christopher Cornwell on Brazilian workers (Cornwell et al., 2017). The workers whose reported race changed from non-White to White received a wage increase (until the 40%) so putting on White powder on the first day of work is a winning strategy in a group where Whiteness equates to less shoddy liaisons and it's a relatively rare quality that allows standing out. Even in a very socially progressive country like Canada, in the nineties, some White European groups attained income parity even if their education level was lower than average and detrimental to Blacks and South Asians whose income level and share of high status occupations was lower than their career merits (Hou & Balakrishnan, 1996). On the other hand, Asian is the ethnicity with the most average income in the US but the "oppression everywhere" discourse will often show graphics in which they are invisibilized in order to justify the white privileges booing.

A quite toxic way of imposing subordination is identifying ethnicity with lower IQ scores. Now I propose a mental experiment. Let's suppose that many racist papers published are true (Kaplan, 2014) and the black people in the US are dumber. Let's suppose also that rather than a campaign of apartheid and black professors fired from universities and labs, all this mess leads to planning some kind of benevolent politics where the social benefits outweigh the racial stigma. Even in this rare case, our partisan attitude as scientists would compromise our hypothesis testing and our judgement in a wider sense since we have "*failed to temper our outrage*" many times before on the same topic (Jeffery & Shackelford, 2018). Some scientists powered by extremely well-funded think-panzers (I mean think-tanks) would postulate that the racial difference in IQ is irremediable, reaches the 500,000 BCE genetics and is far wider than it seems. Meanwhile, most scientists would try not to make the human species cringe. Then, the official truth would lean progressively to the establishment side and, at the end of the day and regardless of well-intended policies, the unintended discrimination would displace the black people first and wipe them out from the US political arena then.

To put it in perspective, the working class US black families were downgraded during the seventies and eighties, due to mostly structural causes, just like deindustrialization and crack epidemics whose consequences haunt the present (Hill, 1989). Despite that it happened while the white collar black families successfully got ahead and no (pseudo)scientific racism ruled, the black people turnout only started to decline gradually in the 2000s (P. Taylor, 2012). In the same vein, in South Africa, the cheap labour from black people has a critical importance to its economy, although the bureaucratic control allegedly fosters the integration of African labour into a developed market (Wolpe, 2018). Given that the market is yet racialized, it is obviously inadequate and gets South Africa stuck on structural inequalities. We lack maturity as a species to make solid judgments about what is good or not for the others' identities since these views have led to displacement time after time.

PHENOMENOLOGY OF IDENTITY

The lived identity among disabled people features a special isolation even when compared with other groups living in the shifting social boundaries that most people feel uneasy to deal with. In a study of Pakistani people with epilepsy, it was found that the participants saw disability as one or many stable, permanent and visible physical impairments (Rhodes et al., 2008). Thus, they don't consider epilepsy a disability. However, what in theory is a dichotomy between biomedical and social interpretations of disability, actually is lived as different dimensions of a complex experience around a contested identity. The people with disability are more entangled with their identity negotiation even due to the media representations purporting to offer them empowerment but actually perpetuating the biomedical model of disability just like is seen in the way Paralympics is presented (Ellis 2008).

Copy-pasted from the idea of political autonomy, a personal autonomy grounded in to identify and evaluate might lead to an infinite regress if hierarchic (Cuypers, 2017). In the same manner seeing memory as a sameness of self, couldn't be an *explanation* of personal identity if it wouldn't exist as a normal *implication* of the self (Ganeri, 2017). Let's take the John Locke's memory criterion (Locke, 1836, Chapter 27) explained in these terms: "*A at t_2 is the same person than B at t_1 if A at t_2 can remember (from the first person) an experience of B at t_1*" (Nichols, 2017).[1] It doesn't explain why anybody should remember rather than roaming in daydreaming, which is what would happen without a self-implied recalling. Shortly, some ideas about personal identity are flawed because remembering things from the past is not enough to explain who we are: we need to have a sense of self to remember. Therefore, an autonomous person, who is able to care about herself and able to belong, must have these three qualities: self-identification, self-evaluation and self-implication.

Youth from religious and cultural minorities are caught between conflicting sexual ideologies from home and school that they must rewrite before accessing an open space for new ways of thinking (Allen, 2018). It can be more challenging in our postmodern society where some of the old experiences and consciousness from religious education has become untenable (Barnes, 2001). From a very different point of view, the mature LGBTQ people with experience in leadership or managerial roles, tend to passively disclose their sexual orientation, usually hidden behind a "*mask of professionalism*" (Moore, 2017). The current lived experiences of a US sample of LGBTQ persons in post-sexual identity disclosure happened to be dense and implied an improvement in mental health but also stress minority, which makes recommendable a health investment in those who come out of the closet (Emetu & Rivera, 2018). In the same vein, the experienced identity of transgenders in the Indian state of Kerala are plenty of hardships due to the way societal institutions shape their identity and the lack of traditional transgendering (i.e. *hijras*) (Krishna, 2017). Their *becoming transgender* and living as one have as a common thread the varying degrees of violence in their walk of life. It's important that everybody get more support and get free before their conscience on this topic because sexuality is an integral part of our identity. The insights everyone gets from her sexuality extends from the personal well-being to the societal one, so the sexuality spans from pleasure to human rights (Bywater & Jones, 2007, pp. 33–34).

As seen in a 2018 study of ethnic Russian and regional North Caucasian identities led by Zarina Lepshokova, two identities might be perceived incompatible when (Lepshokova et al., 2018):

- They represent conflicting sets of norms and values
- Two or more communities requires a commitment and loyalty that cannot be fulfilled for both

A family identity also results incompatible with its previous one when a severe mental illness appears. When it happens, the whole family identity is transformed according to its culture in a process that will turn many members into caregivers (Acero et al., 2017). Quite different from illness, an idiographic and qualitative approach on motherhood shows how pregnant women shift from the public world of business towards a more private one, mainly populated by family and friends (J. A. Smith, 1999). In a pitted against case, a person who jumps to a loyalty-harvesting leader position of a first to second generation in a traditional masculinist manufacturing sector might live the process as personally and organizationally threatening against her identity (Harrison & Leitch, 2019). This varied casuistry shows that a sense of loyalty through the tides of requirements determines the *livable* identities. The enigmatic Socratic suicide, by the way, can be explained as the result of getting rid of livable identities while conserving the loyalty intact.

Loyalty is the touchstone of who we really are.

IDENTITY AND HEALTH IN THE TURNS OF EARLY AND MIDDLE ADULTHOOD

In a 2009 five-wave longitudinal study led by Elizabeth Crocetti on adolescent population was typified two latent growth trajectory classes (Crocetti et al., 2009):

- Low anxiety class characterised by a low initial level of anxiety that decreased over time
- High anxiety class characterised by a higher initial level of anxiety that increased over time

The anxiety classes predicted initial levels and rates of change of three identity dimensions: commitment, in-depth exploration, and reconsideration of commitment. Three dimensions are involved with the aforementioned self-implication which might imply that this prerequisite of autonomy develops during adolescence.

Drawing normative comparisons among adolescents with chronic pain results at two broad classifications: the ones who perceive identity as part of their chronic pain condition and those who do not. In both cases, the adolescents negotiated their identities through implication in valued activities with an income dependent on how their emotional management turned into interpersonal skills (Jordan et al., 2018). The film *The Fault in Our Stars* narrates all these emotional dynamics in a moving way:

Figure 1. Shailene Woodley and Ansel Elgort in a picture from the movie The Fault in Our Stars (Boone, 2014) from the homonymous book by John Green (Green, 2012)

High school athletes no longer engaged in varsity competition during the first year of college showed common transition experiences related with: role transition, identity renegotiation, and reflection and projection, which were facilitated by new camaraderie (Lyons et al., 2018).

The robust identity exploration during the emerging adulthood has been found to shift to a sense of deep commitment with their professional identities during the transition from university to workplace. Moreover, in the case of women also a gendered tension between their career and family roles was found (Huff et al., 2018).

In the absence of debilitating illness and traumatic events, identity growth depends on the willingness of the person to voluntarily step outside their comfort zone. It's easier to see in middle adulthood where the people with favourable life circumstances reach greater visibility. Even in this case, those who engage in challenging leisure-play pursuits will get a bonus to tailor their identities to enjoy a healthy ageing (Rathunde & Isabella, 2017). Maybe it's the reason why adult dolphins show their usual playful pursuits: that's how they practise the art of ageing.

CONCLUSION

Ethical dilemmas arise from the merging of body and information technologies, such as inheritable genetic editing and the incoming widening income gap. It shapes human identity due to the narrative nature of the self and the way personal identity forms through challenges. Therefore, human identity is not only influenced by technology, social structures, cultural traditions, but every day more and more by public health. For this reason a disabled person suffers a true intersectionality when targeted by biopolitics of identity.

REFERENCES

Abbasi, M., Gholami, N., & Soheilmoghadam, S. (2017). The philosophical fundamentals right to be forgotten from Bioethics view. *Bioethics Journal*, *7*(23), 7–15. doi:10.22037/bj.v7i23.17305

Abrams, D., & Hogg, M. A. (1988). Comments on the motivational status of self-esteem in social identity and intergroup discrimination. *European Journal of Social Psychology*, *18*(4), 317–334. doi:10.1002/ejsp.2420180403

Acero, Á. R., Cano-Prous, A., Castellanos, G., Martín-Lanas, R., & Canga-Armayor, A. (2017). Family identity and severe mental illness: A thematic synthesis of qualitative studies. *European Journal of Social Psychology*, *47*(5), 611–627. doi:10.1002/ejsp.2240

Afary, J., & Anderson, K. B. (2010). *Foucault and the Iranian Revolution: Gender and the Seductions of Islamism*. University of Chicago Press.

Allen, L. (2018). Learning About Sexuality 'Between' Home and School. In L. Allen (Ed.), *Sexuality Education and New Materialism: Queer Things* (pp. 71–84)., doi:10.1057/978-1-349-95300-4_5

APA. (2013). *Diagnostic and Statistical Manual of Mental Disorders (DSM-5®)*. American Psychiatric Pub.

Baldwin, A. M., Dodge, B., Schick, V., Sanders, S. A., & Fortenberry, J. D. (2017). Sexual Minority Women's Satisfaction with Health Care Providers and State-level Structural Support: Investigating the Impact of Lesbian, Gay, Bisexual, and Transgender Nondiscrimination Legislation. *Women's Health Issues*, *27*(3), 271–278. doi:10.1016/j.whi.2017.01.004 PMID:28233635

Barnes, L. P. (2001). What Is Wrong with the Phenomenological Approach to Religious Education? *Religious Education (Chicago, Ill.)*, *96*(4), 445–461. doi:10.1080/003440801753442366

Beauchamp, T. L., & Childress, J. F. (1994). *Principles of biomedical ethics*. Oxford University Press.

Bell, E., Kothiyal, N., & Willmott, H. (2017). Methodology-as-Technique and the Meaning of Rigour in Globalized Management Research. *British Journal of Management*, *28*(3), 534–550. doi:10.1111/1467-8551.12205

Berto, F. (2017). Counting the Particles: Entity and Identity in the Philosophy of Physics. *Metaphysica*, *18*(1).

Boone (Director). (2014). *The Fault in Our Stars*. https://www.imdb.com/title/tt2582846/

Bowen, E. (2000). *The Last September*. Anchor Books.

Briggs, S., & Whittaker, A. (2018). Protecting Children from Faith-Based Abuse through Accusations of Witchcraft and Spirit Possession: Understanding Contexts and Informing Practice. *British Journal of Social Work*, *48*(8), 2157–2175. doi:10.1093/bjsw/bcx155

Bywater, J., & Jones, R. (2007). *Sexuality and Social Work*. Learning Matters.

Cheng, C.-Y., & Bunnin, N. (2008). *Contemporary Chinese Philosophy*. John Wiley & Sons.

Clegg, S., Kornberger, M., & Rhodes, C. (2007). Organizational Ethics, Decision Making, Undecidability. *The Sociological Review*, *55*(2), 393–409. doi:10.1111/j.1467-954X.2007.00711.x

Cornwell, C., Rivera, J., & Schmutte, I. M. (2017). Wage Discrimination When Identity Is Subjective Evidence from Changes in Employer-Reported Race. *The Journal of Human Resources*, *52*(3), 719–755. doi:10.3368/jhr.52.3.0815-7340R1

Coulthard, G. S. (2014). *Red Skin, White Masks: Rejecting the Colonial Politics of Recognition*. University of Minnesota Press. doi:10.5749/minnesota/9780816679645.001.0001

Croce, B. (1995). *Guide to aesthetics*. Hackett Publishing.

Crocetti, E., Klimstra, T., Keijsers, L., Hale, W. W. III, & Meeus, W. (2009). Anxiety Trajectories and Identity Development in Adolescence: A Five-wave Longitudinal Study. *Journal of Youth and Adolescence*, *38*(6), 839–849. doi:10.100710964-008-9302-y PMID:19636785

Cuypers, S. E. (2017). Hierarchical Autonomy, Self-Identification and Self-Evaluation. In *Self-Identity and Personal Autonomy: An Analytical Anthropology*. Routledge., doi:10.4324/9781315211428-5

D'Ambrosio, P. J. (2018). From present to presentation: A philosophical critique of Hartmut Rosa's "situational identity." *Time & Society*. doi:10.1177/0961463X18787059

Donaldson, S., & Kymlicka, W. (2011). *Zoopolis: A Political Theory of Animal Rights*. OUP Oxford.

Ellemers, N. (1993). The Influence of Socio-structural Variables on Identity Management Strategies. *European Review of Social Psychology*, *4*(1), 27–57. doi:10.1080/14792779343000013

Emetu, R. E., & Rivera, G. (2018, July). *After Sexual Identity Disclosure: An Ecological Perceptive of LGB Young Adults*. https://doi.org/info:doi/10.5993/AJHB.42.4.5

Falvo, D., & Holland, B. E. (2017). *Medical and Psychosocial Aspects of Chronic Illness and Disability*. Jones & Bartlett Learning.

Frege, G. (1952). The Fundamental Laws of Arithmetic II. In Translations from the Philosophical Writings of Gottlob Frege. Blackwell.

Ganeri, J. (2017). Classical Indian philosophy. In S. Bernecker & K. Michaelian (Eds.), *The Routledge Handbook of Philosophy of Memory*., doi:10.4324/9781315687315-33

Glover, J. (1988). *I: The Philosophy and Psychology of Personal Identity*. Penguin Books.

Green, J. (2012). *The Fault in Our Stars*. Penguin.

Grossberg, L. (1996). Identity and Cultural Studies: Is That All There Is? In S. Hall & P. du Gay (Eds.), *Questions of Cultural Identity: SAGE Publications*. SAGE.

Gustafsson, J. E. (2010). Did Locke defend the memory continuity criterion of personal identity. *Locke Studies*, *10*, 113–129.

Hale, B., Wright, C., & Miller, A. (2017). *A Companion to the Philosophy of Language*. John Wiley & Sons. doi:10.1002/9781118972090

Harrison, R. T., & Leitch, C. M. (2019). The Dynamics of Identity, Identity Work and Identity Formation in the Family Business: Insights from Identity Process Theory and Transformative Learning. In E. Memili & C. Dibrell (Eds.), *The Palgrave Handbook of Heterogeneity among Family Firms* (pp. 673–713). Springer International Publishing. doi:10.1007/978-3-319-77676-7_25

Hauge, Å. L. (2007). Identity and Place: A Critical Comparison of Three Identity Theories. *Architectural Science Review*, *50*(1), 44–51. doi:10.3763/asre.2007.5007

Hill, R. B. (1989). Economic forces, structural discrimination and black family instability. *The Review of Black Political Economy*, *17*(3), 5–23. doi:10.1007/BF02901099

Horner, N. (2018). *What is Social Work?: Contexts and Perspectives*. Learning Matters.

Hou, F., & Balakrishnan, T. R. (1996). The Integration of Visible Minorities in Contemporary Canadian Society. *The Canadian Journal of Sociology/Cahiers Canadiens de Sociologie, 21*(3), 307–326. doi:10.2307/3341769

Huff, J. L., Smith, J. A., Jesiek, B. K., Zoltowski, C. B., & Oakes, W. C. (2018). Identity in Engineering Adulthood: An Interpretative Phenomenological Analysis of Early-Career Engineers in the United States as They Transition to the Workplace. *Emerging Adulthood, 2167696818780444*. Advance online publication. doi:10.1177/2167696818780444

Isaac, R. K., & Platenkamp, V. (2018). Dionysus Versus Apollo: An Uncertain Search for Identity Through Dark Tourism—Palestine as a Case Study. In P. R. Stone, R. Hartmann, T. Seaton, R. Sharpley, & L. White (Eds.), *The Palgrave Handbook of Dark Tourism Studies* (pp. 211–225). doi:10.1057/978-1-137-47566-4_9

Jeffery, A. J., & Shackelford, T. K. (2018). Moral positions on publishing race differences in intelligence. *Journal of Criminal Justice*, *59*, 132–135. doi:10.1016/j.jcrimjus.2017.05.008

Johnsdotter, S. (2018). The Impact of Migration on Attitudes to Female Genital Cutting and Experiences of Sexual Dysfunction Among Migrant Women with FGC. *Current Sexual Health Reports*, *10*(1), 18–24. doi:10.100711930-018-0139-4 PMID:29541003

Jordan, A., Noel, M., Caes, L., Connell, H., & Gauntlett-Gilbert, J. (2018). A developmental arrest? Interruption and identity in adolescent chronic pain. *Pain Reports (Baltimore, Md.)*, *3*(Suppl 1), e678. Advance online publication. doi:10.1097/PR9.0000000000000678 PMID:30324170

Kaplan, J. M. (2014). Ignorance, Lies, and Ways of Being Racist. *Critical Philosophy of Race*, *2*(2), 160–182. doi:10.5325/critphilrace.2.2.0160

Krishna, G. (2017). Transgenders in Kerala-Realisation of their. *Identity*.

Krzyzanowski, M., & Wodak, R. (2017). The Production and Re/production of Institutional and Everyday Racism towards Migrants: The Role of Discourse 1. In *The Politics of Exclusion: Debating Migration in Austria*. Routledge. doi:10.4324/9781315133959-1

Laclau, E. (1994). *The Making of Political Identities*. Verso.

Lepshokova, Z., Lebedeva, N., & van de Vijver, F. J. R. (2018). The mediating role of identity incompatibility in the relationship between perceived discrimination and acculturation preferences in two generations of the ethnic Russian minority in the North Caucasus. *European Journal of Developmental Psychology*, *15*(1), 99–114. doi:10.1080/17405629.2017.1336433

Lewis, D. K. (1983). *Philosophical Papers, I.* http://home.sandiego.edu/~baber/metaphysics/readings/Lewis.SurvivalAndIdentity.pdf

Locke, J. (1836). *An Essay Concerning Human Understanding*. T. Tegg and Son.

Lozano Rodríguez, M. (2019). The Tagged Dove. In D. R. J. Macer (Ed.), *Legacies ofLove, Peace and Hope: How Education can overcome Hatred and Divide*. Eubios Ethics Institute.

Lozano Rodríguez, M. (2019). Ethical Information Foresights: A Journey Across Uncharted Conflictology. American University of Sovereign Nations.

Lucas, T., Wegner, R., Pierce, J., Lumley, M. A., Laurent, H. K., & Granger, D. A. (2017). Perceived Discrimination, Racial Identity, and Multisystem Stress Response to Social Evaluative Threat Among African American Men and Women. *Psychosomatic Medicine*, *79*(3), 293–305. doi:10.1097/PSY.0000000000000406 PMID:27806018

Lyons, L. K., Dorsch, T. E., Bell, L. F., & Mason, L. G. (2018). Renegotiating Identity: A Phenomenological Investigation of the College Transition for Former High School Athletes No Longer Engaged in Varsity Competition. *Identity*, *18*(1), 18–33. doi:10.1080/15283488.2017.1410156

Mackelprang, R. W., Salsgiver, R. O., & Salsgiver, R. (2016). *Disability: A Diversity Model Approach in Human Service Practice*. Oxford University Press.

Mahdavi, P. (2009). *Passionate Uprisings: Iran's Sexual Revolution*. Stanford University Press.

Malpas, J. (2018). Holism, content, and self. In *Place and Experience: A Philosophical Topography*. Routledge. doi:10.4324/9781315265445-4

McNeill, J. R., & McNeill, W. H. (2003). *The Human Web: A Bird's-eye View of World History*. W.W. Norton.

Moore, J. (2017). A Phenomenological Study of Lesbian and Gay People in Leadership Roles: How Perspectives and Priorities Shift in the Workplace as Sexual Orientation Evolves Through Social Constructs. *Doctoral Dissertations*. https://repository.usfca.edu/diss/405

Mulinari, D., & Neergaard, A. (2019). A contradiction in terms? Migrant activists in the Sweden Democrats party. *Identities (Yverdon)*, *26*(2), 222–240. doi:10.1080/1070289X.2017.1418275

Murthy, V. (2011). *The Political Philosophy ofZhang Taiyan: The Resistance of Consciousness*. BRILL. doi:10.1163/ej.9789004203877.i-268

Nichols, S. (2017). Memory and Personal Identity. In S. Bernecker & K. Michaelian (Eds.), *The Routledge Handbook of Philosophy of Memory*. Routledge. doi:10.4324/9781315687315-14

Prainsack, B., & Buyx, A. (2017). *Solidarity in Biomedicine and Beyond*. Cambridge University Press. doi:10.1017/9781139696593

Quine, W. V. (1950). Identity, Ostension, and Hypostasis. *The Journal of Philosophy, 47*(22), 621–633. doi:10.2307/2021795

Rathunde, K., & Isabella, R. (2017). Play, Flow, and Tailoring Identity in Middle Adulthood. In J. D. Sinnott (Ed.), *Identity Flexibility During Adulthood: Perspectives in Adult Development* (pp. 211–232). Springer International Publishing. doi:10.1007/978-3-319-55658-1_14

Rawlinson, M. C. (2016). *Just Life: Bioethics and the Future of Sexual Difference*. Columbia University Press. doi:10.7312/rawl17174

Rhodes, P. J., Small, N. A., Ismail, H., & Wright, J. P. (2008). 'What really annoys me is people take it like it's a disability', epilepsy, disability and identity among people of Pakistani origin living in the UK. *Ethnicity & Health, 13*(1), 1–21. doi:10.1080/13557850701803031 PMID:18066735

Rodríguez, I. O. (2017). Animal Citizenship, Phenomenology, and Ontology: Some reflections on Donaldsons & Kymlickas Zoopolis. *Bangladesh Journal of Bioethics, 8*(1), 21–32. doi:10.3329/bioethics.v8i1.31080

Romani, L., Holck, L., & Risberg, A. (2018). Benevolent discrimination: Explaining how human resources professionals can be blind to the harm of diversity initiatives. *Organization*. Advance online publication. doi:10.1177/1350508418812585

Scheepers, D., Spears, R., Doosje, B., & Manstead, A. S. R. (2003). Two Functions of Verbal Intergroup Discrimination: Identity and Instrumental Motives as a Result of Group Identification and Threat. *Personality and Social Psychology Bulletin, 29*(5), 568–577. doi:10.1177/0146167203029005002 PMID:15272991

Schilder, P. (1938). The Social Neurosis. *Psychoanalytic Review*, 1–19.

Senor, T. D. (2005). *Epistemological problems of memory*. https://stanford.library.sydney.edu.au/entries/memory-episprob/

Shoemaker, D. (2010). The Insignificance of Personal Identity for Bioethics. *Bioethics, 24*(9), 481–489. doi:10.1111/j.1467-8519.2009.01719.x PMID:19438443

Shoemaker, S. (1963). *Self-knowledge and Self-identity*. Cornell University Press.

Slater, M. E., Godette, D., Huang, B., Ruan, W. J., & Kerridge, B. T. (2017). Sexual Orientation-Based Discrimination, Excessive Alcohol Use, and Substance Use Disorders Among Sexual Minority Adults. *LGBT Health, 4*(5), 337–344. doi:10.1089/lgbt.2016.0117 PMID:28876167

Smith, J. A. (1999). Identity development during the transition to motherhood: An interpretative phenomenological analysis. *Journal of Reproductive and Infant Psychology, 17*(3), 281–299. doi:10.1080/02646839908404595

Smith, R. C. (2017, September 29). *Transnational Localities: Community, Technology and the Politics of Membership within the Context of Mexico and U.S. Migration 1*. Transnationalism from Below. doi:10.4324/9781351301244-7

Southern, R. W. (1970). *Western Society and the Church in the Middle Ages*. Penguin Books.

Swanson, R. N. (1995). *Religion and Devotion in Europe, C.1215- C.1515*. Cambridge University Press.

Talwar, S. K. (2018). Identity Matters: Questioning Trauma and Violence through Art, Performance and Social Practice. In *Art Therapy for Social Justice: Radical Intersections*. Routledge. doi:10.4324/9781315694184-3

Taylor, C. (2007). *A Secular Age*. Harvard University Press.

Taylor, P. (2012, December 26). *The Growing Electoral Clout of Blacks Is Driven by Turnout, Not Demographics | Pew Research Center*. https://www.pewsocialtrends.org/2012/12/26/the-growing-electoral-clout-of-blacks-is-driven-by-turnout-not-demographics/

Taylor, P. C., Alcoff, L. M., & Anderson, L. (2017). *The Routledge Companion to the Philosophy of Race*. Routledge. doi:10.4324/9781315884424

ten Have, H., & Purtilo, R. B. (2004). *Ethical Foundations of Palliative Care for Alzheimer Disease*. JHU Press.

Thurer, S. L. (1988). Rebel with a cause: The disabled beggar. *The Journal of Medical Humanities and Bioethics*, *9*(2), 153–163. doi:10.1007/BF01139240

Watson, T. J. (2008). Managing Identity: Identity Work, Personal Predicaments and Structural Circumstances. *Organization*, *15*(1), 121–143. doi:10.1177/1350508407084488

Wolpe, H. (2018). Industrialism and Race in South Africa. In S. Zubaida (Ed.), *Race and Racialism*. doi:10.4324/9781351039345-8

KEY TERMS AND DEFINITIONS

Alienation: Feeling disconnected and isolated from others or society involving a painful separation between the Self and the others that belong together.

Cultural Traditions: Passed-down customs, beliefs, and practices that shape a society's art, values, and way of life.

Ethical Dilemmas: Complex choices between conflicting options where doing either one is morally required but doing both is impossible.

Human Body: Our physical form blending biology, emotions, experiences and identity, able to connect the tangible and intangible.

Indigenous Populations: Original inhabitants with earlier historical and cultural ties to a region, often impacted by colonisation.

Marginalisation: Excluding individuals or groups from access to opportunities and services, leading to a lower status.

Social Structure: The organised and stable patterns of institutions and relationships that shape how people interact in society.

ENDNOTES

1. If John Locke actually supported the memory criterion as we understood it, as well as his further discussions with Thomas Reid, has resulted to be a very subtle philosophical issue beyond the scope of this book (Gustafsson, 2010; Senor, 2005).
2. This syllabary would also be employed to transcribe a robot or an alien spelling in manga. https://www.youtube.com/watch?v=Y0vDAfJ3Mxg&t=698s
   Finally, katakana is commonly used for loanwords and for giving emphasis, like in a slogan / buzz-word, although it doesn't go for an interview, as the Naomi case.
3. The clinical picture of a social neurosis is dominated by suffering in social contacts, patients of this type can be comparatively happy when they are alone or together with their families, but functions mostly below the intellectual and emotional level of the personality when they meet other persons (Schilder, 1938). In this case, the stigmatizers don't suffer a real neurosis and their emotional level only plummets before labelled persons (queer, gypsy, etc.).
4. https://www.tepapa.govt.nz/discover-collections/read-watch-play/maori/ta-moko-maori-tattoos-history

Chapter 14

An In-Depth Analysis of the Bioethical Implications of Stigma:
Free v Confined II

ABSTRACT

Stigma is a multifaceted issue within bioethics of displacement. It encompasses conceptualization, psychological processes, cultural dimensions, and sociological implications, including labelling, stereotyping, separation, status loss, and discrimination. Stigma affects mental health, social inequalities, public health, and relationships, requiring proactive and preventive approaches. Cultural dynamics, moral experiences, and structural barriers must be considered to understand and combat stigma. The author explores the epigenetic effect of naming, stigma's connection to reproductive and maternal health, and the impact of technology. Ultimately, the solution lies in respecting every individual's dignity, dismantling discrimination, and addressing power imbalances.

INTRODUCTION

This chapter delves into the multifaceted nature of stigma, linking it with the themes of identity and social structures discussed in last chapter, it provides a comprehensive understanding of stigma and its implications within Bodi. This chapter underscores the pervasive impact of stigma on individuals, communities, and societies, emphasising the moral responsibilities of addressing and mitigating stigmatisation faced by displaced individuals.

DOI: 10.4018/978-1-6684-4808-3.ch014

ON THE MULTIFACETED NATURE OF STIGMA

Layers of Stigma

Stigma is a pervasive societal issue that profoundly impacts individuals, communities, and societies at large. It involves negative attitudes, beliefs, and behaviours towards individuals or groups based on certain characteristics or attributes. In this discourse, we delve into the multifaceted nature of stigma, examining its conceptualization, psychological processes, and cultural dimensions. By exploring the topic of stigma, we aim to gain a deeper understanding of its impact and relevance within the framework of Bioethics of Displacement. Stigma can be understood as a complex phenomenon encompassing negative attitudes, beliefs, and behaviours towards individuals or groups based on certain characteristics or attributes. Link and Phelan (2001) provide a comprehensive conceptualization of stigma, identifying those interrelated components: labelling, stereotyping, separation, status loss, and discrimination (Link & Phelan, 2001). Labelling refers to the process of categorising individuals as different or deviant based on specific characteristics. Stereotyping involves attributing undesirable traits or characteristics to these labelled individuals, while discrimination manifests as the unjust treatment and denial of rights and opportunities. Displacement may lead to experiences of stigma for the displaced individuals or communities. When individuals are displaced away from their previously meaningful lives, they may face negative stereotypes, labelling, and discrimination because they suffer a sort of thrown-by-a-third-party flag. This can contribute to their social exclusion and marginalisation, which are central components of stigma. In this context, the Bioethics of Displacement may explore the ethical dimensions of how displacement and its associated stigma affect individuals' rights, well-being, and opportunities. It may also consider the moral responsibilities of hosting individuals, communities, and institutions in addressing and mitigating the stigmatisation as well as the ones behind displacement.

Social Psychology of Stigma

The social psychology of stigma investigates the processes underlying its formation, perpetuation, and consequences. Stigmatised individuals internalise societal beliefs and experience negative self-perceptions (Major & O'Brien, 2005). This phenomenon, known as self-stigma, occurs when individuals adopt the stigmatising views directed towards them, leading to lowered self-esteem, diminished self-efficacy, and reduced quality of life. Understanding self-stigma is crucial in addressing the psychological impact of stigma and promoting resilience among affected individuals. By incorporating the concept of stigma into this framework, it expands the understanding of how stigma can significantly impact the lives of individuals who are displaced. It recognizes that stigma is not merely a vague and individually focused concept but consists of labelling, stereotyping, separation, status loss, and discrimination, with power dynamics playing a role in its manifestation. Moreover, the Bioethics of Displacement emphasises the importance of considering vulnerable populations first and promoting proactive and preventive approaches to address ethical issues related to displacement and stigma. It also introduces the concept of Preemptive bioethics, which calls for anticipating and addressing ethical challenges before they escalate. By integrating the study of stigma within the Bioethics of Displacement, this approach acknowledges the broader social, cultural, and ethical implications of stigma for displaced individuals. It encourages a comprehensive examination of the moral principles and decision-making approaches that can guide ethical actions in the face of displacement and its associated stigma. This broader perspective contributes to a more nuanced understanding of the complex challenges faced by individuals who are displaced and the ethical responsibilities of individuals, communities, and institutions in addressing those challenges.

Cultural Dynamics and Moral Experience of Stigma

Stigma is not a static concept but is deeply embedded in cultural contexts so we must consider cultural factors in understanding stigma, highlighting how moral experience shapes stigmatising beliefs and behaviours (Yang et al., 2007). Cultural norms, values, and beliefs influence the perception and interpretation of stigmatised conditions or identities. Moreover, cultural frameworks may reinforce or challenge stigma, shaping the availability of support systems and resources for affected individuals.

Stigma theories have predominantly focused on assessing the individual, and it calls for a shift towards understanding stigma as a moral experience. By recognizing stigma as a moral issue, new targets can be identified for anti-stigma intervention programs and their evaluation. Transactional methodologies and multiple perspectives and methods can be used to capture the interpersonal core of stigma as framed by theories of moral experience (Yang et al., 2007). To understand how stigma affects people's relationships, researchers can use different approaches and ways of looking at things. They can use different methods and consider different points of view to get a deeper understanding of how stigma impacts how people interact with each other. By doing this, they can better grasp the main aspects of stigma and how it affects people's experiences in their relationships with others. In summary, the new aspect in the Bioethics of Displacement related to stigma is the exploration of stigma as a moral issue that threatens what is most important to individuals, and the proposal to incorporate the concept of moral experience into future research and intervention programs targeting stigma.

Impact of Stigma in Social Inequalities and Public Health

Stigma has significant implications for individuals, communities, and public health. Corrigan and Watson (2002) discuss the paradox of self-stigma, where individuals internalise negative beliefs about themselves due to societal stigma (Corrigan & Watson, 2002). This internalised stigma can hinder help-seeking behaviours, exacerbate mental health conditions, and impede overall well-being. Additionally, stigma surrounding health conditions such as HIV and AIDS can lead to discrimination, hindering prevention efforts, and perpetuating the spread of the disease (Parker & Aggleton, 2003). The new aspect in the topic of stigma in the field of Bioethics of Displacement is the exploration of how stigma is connected to social inequalities such as class, race, gender, and sexuality. The traditional understanding of stigma as a discrediting attribute is expanded to consider how stigma reinforces and perpetuates existing social differences and disparities. The position emphasises that stigma is not just an individual experience but is deeply rooted in social structures and power dynamics. It also highlights the limitations of focusing solely on individual approaches to reducing stigma and calls for new programmatic approaches that harness the resistance and resilience of stigmatised individuals and communities to bring about broader social change. Furthermore, Corrigan and Watson's paper discusses the complexities of personal responses to stigma, including the paradoxical reactions of some individuals who may either experience a loss in self-esteem, become energised and express anger, or seemingly ignore the effects of stigma altogether. The research draws from social psychology and minority groups' experiences to develop a situational model that explains these different responses to mental illness stigma. Overall, this new perspective on stigma in the context of Bioethics of Displacement highlights the need for a comprehensive understanding of stigma as a social phenomenon and the importance of addressing politics of pain in order to effectively combat stigma and its negative effects.

INTERPERSONAL STIGMAS

The interpersonal stigmas are direct, not aimed at suffering through group belonging, but focuses on moral and physical pain. I will present the most extended ones: disability, ethnic backgrounds, religious beliefs and sexual orientation.

The notion of contagiousness in the West solidifies normative ontology and epistemology while giving the mental disability a contaminator threat (Shildrick, 2000). So much so that many people with mental health issues don't want to be included with the people who have an impairment while the disabled movement doesn't pay much attention to mental health either (Jon, 2017, p. 104). On the other hand, a true contaminator threat might impose a logic of confinement, just like asylums. For example, Canada has become a hotspot of HIV criminalization with 185 cases of people charged for allegedly not disclosing their serological status and engaging in sexual activities that resulted in contagion are booming (Gagnon and Vézina 2018).

Among other reasons, racism is so sticky because it involves a spread oversensitivity to cultural differences that leads to a hierarchical structure. Once this hierarchy is set, the racists affirms their own superiority and denigrates the other thanks to the pyramid they previously imposed. It should be pointed out Jacques Derrida's critique of this kind of racism on principles of toleration and acceptance (Derrida, 1992, pp. 111–121, 128, 2001, pp. 10–31).

The micro and macro levels of context really matter when it comes to a healthy ethnic identity development among transracial adoptees. In a 2017 study led by Laura Ferrari with 223 transracial adoptees living in US or Italy, it was found that the ethnic identity of adoptees in Italy was negatively related to perceived discrimination, but there it was no significant association among the adoptees in the United States (Ferrari et al., 2017).

The incomplete *blood match* also has consequences when you are adopted by a country. Takeyuki Tsuda, a Japanese-American himself, wrote in 1998 about the Japanese-Brazilian stigma. According to Tsuda, they suffered from a complex admixture of ethnic prejudice and discriminatory behaviour in their experience of self due to the "*psychocultural feelings of ethnic* impurity." (Tsuda, 1998) More than twenty years later, the Haitian-Japanese half-breed tennis champion Naomi Osaka (Osaka Naomi in Japan) was subtitled in TV with katakana characters that are mostly reserved for foreigners (even proficient ones) first and later she was whitewashed (literally) in a noodles ad (Bruce, 2018; McCurry, 2019).[2] On the other hand, it must be recognized that, unlike in TV, Naomi Osaka has had her name rendered all in Kanji and Hiragana in the Japanese press and official receipts (Arudou, 2018).

Since 1998, the US Department of State reported on the state of religious freedom in every foreign country, namely 143 countries of which 86% witnessed how people suffered verbal abuses or displacement from their homes due to their religion from mid-2000 to mid-2007 (Grim & Finke, 2010, pp. 18–19). Brian J. Grim and Roger Finke analyse this wealth of data in their book *The Price of Freedom Denied* (Grim & Finke, 2010). The religious stigma and persecution, according to the authors, is related with the lack of intra-society plurality of beliefs, just like Hume, Voltaire and Adam Smith held (Hume, 1964; Smith, 2007, Chapter V; Voltaire, 1765, pp. 224–238).

The local context where the stigma against sexual minorities happens can condition the ontology of sexual identity. For example, the European men who have sex with other men are more prone to report bisexual and heterosexual attractions in the countries where the stigma is harsher (Pachankis et al., 2017).

All interpersonal stigmas exposed respond to a stigmatizer, a person or group with a sort of social neurosis mixed with animus against the way another person builds her identity.[3] Perhaps there is a latent envy from the victimizer to the victimised one's identity, maybe this wickedness comes from a deep personal insecurity but, in the end of the day, the truest fact is that the stigmatised persons are always assimilable to anti-elites. An elite that is socially acceptable to hate, bodies to throw your Two Minutes Hate. The social diversity and context determines how this compulsion will curse: pushing the victim inside a closet, pouring over her a never-ending mild rain of cutting remarks, criminalising her or plainly beating, stoning, burning or hanging from a crane.

The interplay between stigma and displacement stresses the need for a proactive and preventive stance within the bioethics of displacement in the light of the moral responsibilities of individuals, communities, and institutions in addressing and mitigating the stigmatisation faced by displaced individuals. In other words, the connection between stigma and displacement means we must take action early on to prevent and address the mistreatment faced by displaced people. By promoting a broader perspective, one that encompasses cultural dynamics, moral experiences, and social inequalities, we can pave the way for effective interventions and social change. In conclusion, the integration of interpersonal stigmas into the framework of the bioethics of displacement offers a fresh lens through which to examine the ethical implications and moral dimensions of stigma. It invites us to challenge societal norms, dismantle power imbalances and seek the inherent dignity and worth of every individual, regardless of their differences or identities.

THE RISE OF STRUCTURAL STIGMA

Labelling comes from deeply rooted structural discrimination. In the last years before this book, the structural stigma has gained relevance. It's different from the individual-interpersonal one since it spans over groups / identities, so it's more suitable for a sociological approach.

Accumulated evidence highlights that the health damage produced by structural stigma is under-recognized, hides health inequalities and spans across the life stages (Hatzenbuehler, 2017). Anyhow, the structural stigma is unable to mask the pain inequalities. Not being the right kind of chronic pain patient is likely the original stigma, which intertwines with the one of addiction. For example, some authors hold that the consensual tapering is beneficial to regulate the patients' opioid use while others denounce that chronic pain patients face reduced access to opioid prescriptions as a result of inappropriately applied governmental guidelines (Huang, 2018). However it's much ado about nothing if we skip over the fact that chronic pain disproportionately burdens patients who are in the bottom of the pyramid, as aforementioned when dealing with Psychological inflexibility and Politics of Pain topics. Moreover, being silent about this, is to pay homage to the structural stigma.

Some schizophrenia patients in Hong Kong are a good example of structural stigma, in which the difference of power between the stigmatizers and the stigmatised prevents game-changing critical research from happening (Lee et al., 2006). Structural discrimination is very close to structural stigma. In the aforementioned example, the structural discrimination would consist in locating the treatment facilities for the schizophrenia patients in underserved and dangerous neighbourhoods. In a different setting, in India, one of the main causes of the abuses of persons with mental illness is a *poorly articulated and hence less researched access to rights* (Paul, 2018). This is quite significant given that it happens even when the relative importance of ontological and existential meanings of caregiving in dementia is culture-sensitive and depends on the many world disclosures, as an experience with Strait of Torres aboriginal people points to (Arkles et al., 2018). Gender also sets a difference. In a 2010 study on emotional disorders experienced over a nationally representative sample of New Zealanders over 16 years old, a disproportionately greater role, social and cognitive disability was found among men (Scott & Collings, 2010).

Not by chance, the structural stigma bare bones came from medical sociology. The management of stigma along with medicalization, research on adherence and compliance, and patient-doctor interaction originated within sociology from the 1950s onwards. As time went by, these topics morphed into a robust, stand-alone social science tradition (Timmermans & Tietbohl, 2018).

If the mafias come from a biopower leakage, the structural stigma comes from a brimming one. Just like happens with mafias in the countries ridden by them, around the structural stigma there is also a constant and pigheaded ignorance altogether with the victims' displacement to the no-services, no-neighbourhoods, no-investments… If the criminal organisations aim to build a sub-state among us, the structural stigma wants to build one *within us*.

The examination of structural stigma within the Bioethics of Displacement reveals a compelling sociological dimension that demands attention. By delving into the interplay between power dynamics, discrimination, and societal structures, we gain a deeper understanding of the far-reaching impact of structural stigma on marginalised groups. This perspective unveils the hidden health inequalities and challenges the status quo by exposing the politics of pain and the structural barriers that perpetuate stigma. We witness the urgent need for research, access to rights, and the dismantling of both discriminatory and victimism schemes. Additionally, the difficulty in recognizing the disparities and disproportionate burdens faced by marginalised populations further underscores the relevance of structural stigma. It is vital that we confront this insidious force within our communities, as it seeks to establish its own rule as opposed to a society where dignity prevails.

THE INTERCORPOREAL STIGMA

There is a possible source of stigma rarely considered: the epigenetic effect of naming (Lushetich, 2018). Considering that cultural and social behaviour has a lasting effect on an organism's biological make-up it doesn't seem an exaggerated hypothesis (Pembrey et al., 2006). Moreover, in a world where, thanks to transplantation, the love of life has brought a greater intercorporeal life than ever, the epigenetics of naming could represent a viral and dark-sided intercorporeal will to power. An example of this can be found in the subsumption of slave labour under the slave owner's name or the explicit tattoos indicating status in a criminal organisation. Epigenetics emphasises that our social and cultural experiences can potentially impact our biology. While the exact mechanisms are complex and not fully understood, research in epigenetics suggests that our environment and experiences can leave chemical marks on our DNA, which can influence gene expression and potentially be passed onto future generations. So, this section proposes that the effects of naming on social stigma and identity may have implications that extend beyond social and psychological aspects to potentially affect our biology. We'll use facial tattoos as examples to illustrate the power of names and visual representation. In some cultures, facial tattoos like the tā moko hold cultural significance and can be a source of pride and identity. However, the text also mentions facial tattoos among Latin gang members, which are used as branding and symbols of criminal affiliation. These tattoos can stigmatise individuals, leading to negative assumptions and limited opportunities.

Figure 1. Original tā moko drawing (Maori sacred facial tattoo). Picture by Leslie Hinge (1906).
Source: Retrieved from the Museum of New Zealand website.

Figure 2. Ulises Rodriguez's (Reuters) gangster picture
Source: Retrieved from a Daily Caller article by J. P. Carroll (Carroll, 2016)

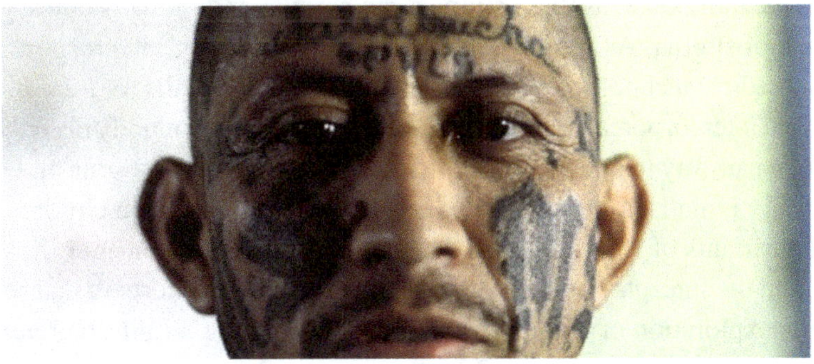

Let's delve a bit on tattoos looking at the contrast between marginalising and non-marginalizing facial tattoos. The *tā moko* is both an art and an ID-card that presents, among other things, the bearer's family, marriages, *tapu* (ritual authority) and personal mana.[1] As long as you are a Maori, since they are extremely zealous with this unique cultural heritage, you can have your whole face tattooed and apply for a white collar job and *do the same Western things* that you'd done without tattoos (at least under Kiwi skies). Quite the opposite are the facial tattoos seen among many Latin gang members. It can be seen that the common gang facial tattoo is not about family or community ties, they are merely branding, in this case Mara Salvatrucha *logos*. Even when these tattoos are framed in a visual language they are very different to the storytelling ones in the back and chest and, definitely, they are not about belonging but proving membership. To cap it all, those tattoos are being toned down due to legal nuisances (Carroll 2016). After the *mano dura* (iron fist) policies in El Salvador, these gang facial tattoos are used sometimes to punish those members who try to get out of the gang (mostly young girls who suffered repeated rape and beating) in order they get forcibly committed (Love, 2007). The image that arrives to the others from these persons before knowing their names is that they are a criminal organisation component. Unlike a *tā moko* tattoo, it's an expeditious way to chase away any career opportunities. Even their names are likely subsumed by a nickname related to the gang. Actually, we don't meet the person, she has no words to say to herself out of the gang's lexicon. When someone is stigmatised due to their facial tattoos, it can create social barriers and limit opportunities not only for that individual but also for their family members. People may make assumptions about the entire family based on the visible tattoos, leading to discrimination, prejudice, and even exclusion. This can have an intergenerational impact, where the negative stigma associated with the individual's tattoos affects their family members, including their children and future generations. How does a person in these circumstances present herself before their children? What name will she impose upon them? Will the body part replacements created from stem-cells shift this biopolitics of intercorporeality to the side of life after so much pain inflicted?

The intercorporeal stigma imposes upon the victims a bad name (or a bad tattoo) that determines and commonly harms her entire genetic kin -that's how itself perpetuates in groups and is likely the nexus between structural and interpersonal stigmas. Every stigma is a chain imposing an-other-name. Thus, this section raises critical questions about the ways in which naming practices can shape an individual's identity, opportunities, and experiences. For marginalised individuals bearing stigmatising tattoos, the visual representation becomes a prejudiced assumption about their character and affiliation before their names are even known. The imposition of a bad name or tattoo becomes a means of perpetuating intercorporeal stigma within groups, linking structural and interpersonal stigmas. Moreover, the implications extend beyond the individual, as these stigmas can harm an individual's entire genetic kin, further entrenching the cycle of stigma. As we navigate a world of increasing intercorporeal connections and advancements in biomedical technologies, the epigenetics of naming calls for reflection on the potential shifts in biopolitics and intercorporeal dynamics. Can the development of body part replacements created from stem cells offer a pathway to alleviate the burdens of intercorporeal stigma and the pain it inflicts? How can individuals in stigmatised circumstances present themselves to their children and choose names that transcend the constraints of stigma? These questions prompt us to consider the ethical dimensions of naming, stigma, and the interplay between societal structures and interpersonal relationships.

In conclusion, the exploration of the epigenetic effect of naming within the realm of the Bioethics of Displacement provides a fresh perspective on the far-reaching implications of stigma. The names we give people and the visual representations associated with them (such as tattoos) can influence how they are perceived, treated, and the opportunities they have in life. It raises questions about the impact of stigma, the interplay between societal structures and interpersonal relationships, and the potential long-term effects on individuals and their genetic kin. It challenges us to examine the power dynamics and consequences embedded in the act of naming, urging us to foster a more inclusive and empathetic society where individuals are not defined by stigmatising labels but are afforded the opportunity to shape their own narratives and transcend the chains imposed by stigma.

STIGMA ON THE REPRODUCING AND MISCARRYING BODIES

Stigma always hovers around reproductive and maternal health. Thankfully, it can be analysed through interpersonal, structural, and intercorporeal lenses. For example, the medicalization of childbirth has led to a change in women's perception of birth-related risks, which can be seen as a structural stigma. The discriminatory access to ART also adds to the interpersonal stigma experienced by infertile women. The text also highlights the intercorporeal stigma experienced by pregnant undeserved women who face various forms of scarcity and unfairness related to their reproductive health. Additionally, the text discusses the need for adolescent-friendly services and the issues related to anti abortion laws, which can be seen as structural stigmas that affect adolescent girls and women seeking abortion services. Let's delve into it.

The biopolitics concerned with reproducing bodies has been displacing for a long time across historic and social very different stances although similarities between Global North and South finally converged, for example, in a declining fertility (Krause & Zordo, 2012).

From the Portuguese we can see how limiting the study of childbirth to its medical aspects leaves important dimensions out of the analysis, like the women's perception of birth-related risks (Pintassilgo & Carvalho, 2017). Likewise, in the UK, a previous study highlighted the way maternity services changed from a social to a medical model over the twentieth century and how the slow turning back affects the risk perception (MacKenzie Bryers & van Teijlingen, 2010). Thus, the women have changed their perception of birth-related risks due to the medicalised offer of maternity care, which will have implications for health outcomes and future care.

To the ontological anxiety of the women with fertility issues whose transcendence is, in their eyes, denied, sums the ART discriminatory access, where the heterosexual couples have *stolen* what was an intimate woman's experience (Shildrick, 2015). Indeed, since the 1999 Susan Wolf diagnostic of bioethics' impairment to produce an analysis that places race, ethnicity, and gender *at the centre* (Wolf, 1999), the case has not improved enough in the last 20 years when it comes to displacing whiteness *from the centre* in ART discussions (Harwood, 2018). More than 20 years of sterile identary lobbying and complaining instead of working hard where it is needed: equality of opportunities regardless of your inherited background. I'll explain with a real story. In the 1970s, during the dawn of modern bioethics, the former Black Panther Assata Shakur gave hints on how reproduction and writing empower the revolutionary women not only against the white state but also against the Black men (Shakur, 2001, pp. 111–117). During her pregnancy in prison she was malnourished and her medical treatment was made unnecessarily difficult. For example, Assata managed herself to get a gifted practitioner from Newmark, the Dr. Garrett, who was a Black man, but the white jailers pushed for a white physician to be present. However both, Black and White physicians, agree in the presence of blood in the vagina, a complication signal. Then, Ernest Wyman Garrett said:

I am prepared to fight all the way for your right as a human being to receive decent medical care and for your baby to be born healthy. I'm doing the same for you as I would for any other patient. You should be hospitalized. There isn't a responsible doctor in the world who wouldn't agree with this opinion. And I'm prepared to testify in any court that to deny you proper medical care would tantamount to committing murder. (Shakur, 2001, p. 128)

Dr. Garrett was faithful to his word and a healthy baby girl called Kakuya was born. Mothering, parenting and the fact of being born must be fully integrated into "*being a human*" (Battersby, 1998). A human.

There is a long road ahead. The 38 immigrant women (mostly Muslim) participants in a study during labour obstetrical care in the Canadian Edmonton hospital unanimously preferred to be attended by a female physician (Aubrey et al., 2018). Despite this, all of them accepted to be attended by a male practitioner, although a minority found it very detrimental, thus, an educational effort with these women is still to be done.

There is a gap in knowledge about the problem of abortion in adolescents across the world as well as a need for adolescent-friendly services (Munasinghe & van den Broek, 2005). More even, in spite of the fact that anti abortion laws typically have not been enforced against women, hundreds of them have been prosecuted for related crimes in the US during recent years (Oberman, 2018, pp. 126–127). To cap it all, Donald Trump claimed in 2016 for some form of punishment for aborting women while a deprecated legal language christened him (Wolffe, 2018; Ziegler, 2018). The injustice upon pregnant undeserved women is such a widespread issue that bioethics of reproduction always is close to unveiling a *trumping* tyrant.

Bioethics must evolve and expand its focus to include marginalised voices and address the true intersection of stigma, reproductive rights, and social inequalities. By dismantling discriminatory practices, promoting reproductive autonomy, and embracing a comprehensive understanding of the complexities surrounding reproductive experiences, we can work towards a solution. Only by tailoring healthcare services to individual needs, we can contribute to creating an ethically sounder approach to addressing the challenges faced by displaced women in their reproductive journeys.

STIGMA, FUTURE AND BIOETHICS OF DISPLACEMENT

In the ever-evolving landscape of technology and human consciousness, different viewpoints and concerns arise regarding the potential impact on individuals and the risk of stigmatisation. In telematic art computer networks serve as a medium for artistic expression and audience interaction (Ascott, 2007). This form of art has the potential to unite diverse perspectives and bridge gaps between communities, fostering understanding and empathy that can help combat stigmatisation. Furthermore, considering human factors and ethical implications in technology design can ensure that technological advancements do not reinforce stereotypes or contribute to stigmatisation (Gorayska & Mey, 2004). Instead, we can create tools that empower individuals and provide equal opportunities for all. Another effective solution to prevent future stigma is to promote empathy and understanding through immersive experiences that simulate the perspective of marginalised individuals. By strengthening advanced virtual reality technologies, individuals can step into the shoes of others and gain firsthand insight into their experiences, challenges, and emotions. By immersing individuals in the lives of marginalised populations, it becomes harder to maintain stigmatising attitudes and easier to empathise with their struggles. Virtual environments can be used to parody, to mirror and mock real-life social norms. By leveraging this aspect, surrogate realities can be designed to challenge and reshape stigmatising beliefs and behaviours (Dibbell, 1998, p. 131). Might it be, but it was better in 1998 than in 2023. Today, governance concerns arise with the transition to a decentralised metaverse (Goldston et al., 2022). It is important to maintain decentralisation and ensure that decision-making power remains in the hands of the community because centralised metaverse models may pose risks related to data control, privacy, and potential exploitation (Goldston et al., 2022).

The metaverse merges virtual and physical reality, offering a transformative digital space for mental health (Usmani et al., 2022). VR/AR/MR have shown effectiveness in mental health, addressing professional shortages and limited access to care. However, excessive use of 3D immersive games and social media can lead to insecurity, anxiety, depression, and addiction. The impact of expanding metaverse applications on mental health and stresses necessary precautions for a positive future must be examined. Understanding and addressing the risks of stigmatisation in the metaverse and mental health care is crucial for creating a supportive and inclusive environment (Khan et al., 2022). Stigmatising beliefs can significantly impact individuals seeking help for mental health issues. Metaverse advertising has the potential to combat these beliefs by improving mental health literacy and educating the population about mental health conditions (Khan et al., 2022). By challenging stigmatised beliefs, metaverse advertising can contribute to reducing negative attitudes and prejudices.

Equity is an important consideration in metaverse-based mental health care (Gloeckler & Biller-Andorno, 2023). While the metaverse offers opportunities for enhanced access to services, it is crucial to ensure that these services are accessible to all individuals, regardless of their background or circumstances. Addressing issues of accessibility and providing equal opportunities for care is essential to prevent further marginalisation or exclusion of vulnerable populations (Gloeckler & Biller-Andorno, 2023). Bridging the digital divide and ensuring equitable access to metaverse-based mental health services are crucial steps in this regard. But metaverse is, itself, an opportunity for enhancing education on mental health and related interventions (Wiederhold, 2023). However, it is important to approach this aspect cautiously. Gamification of mental health education in the metaverse should be designed carefully to avoid trivialising or oversimplifying complex mental health conditions (Wiederhold, 2023). Striking the right balance between engagement and accurate representation is crucial to avoid reinforcing stereotypes or spreading misinformation. In the context of medical care, the application of the metaverse can offer innovative solutions. The integration of virtual and real spaces within the metaverse allows for visualisation of diseases, psychotherapy scenarios, and the development of new care models (Xu et al., 2022). However, it is necessary to consider the potential risks associated with relying solely on virtual experiences. The metaverse should complement traditional medical interventions, ensuring that the benefits are supported by clinical expertise and evidence-based practices (Xu et al., 2022).

In considering the intersection of a centralised metaverse and its influence on mental health, it is crucial to examine the potential consequences that may arise and their impact on humankind as a whole. In examining the potential dystopia where a centralised metaverse alienates individuals through manipulation of their mental health, it is essential to take a realistic approach based on current research and observable trends. The metaverse, with its immersive digital environment, holds significant influence over people's mental well-being and behaviour (Banerji, 2023; Han et al., 2022). However, if centralised control is not carefully managed, it can lead to concerning outcomes. One crucial aspect of this dystopian scenario is the manipulation of individuals' mental health via targeted algorithms and personalised content (Borgesius et al., 2016; Keeling & Burr, 2022; Williamson, 2017). As individuals engage with the metaverse, their actions, preferences, and emotional states are tracked and analysed. This data can then be utilised to create customised experiences that exploit vulnerabilities and amplify negative emotions, ultimately driving individuals towards addictive or detrimental behaviours, even suicide (Flayelle et al., 2023; Lundahl, 2021; Lozano Rodríguez 2022). Existing research has already highlighted the potential addictive tendencies and adverse psychological effects associated with excessive use of immersive technologies and social media platforms (Reer & Quandt, 2021; Siricharoen, 2019).

Furthermore, within a centralised metaverse, decision-making power rests in the hands of a select few, raising concerns about privacy, data control, and potential exploitation (Pałczyńska, 2023). As individuals immerse themselves in the metaverse, their personal information becomes increasingly accessible, enabling targeted manipulation and surveillance (Falchuk et al., 2018; Rosenberg, 2023). Consequently, this erosion of privacy can result in a loss of autonomy and personal agency, exacerbating mental health issues such as anxiety, paranoia, and a pervasive sense of constant surveillance (Anderson et al., 2021; Chun, 2008, Chapters 3–5; Lindau, 2022, pp. 219–269).

Another critical aspect to consider is the potential distortion of reality within the metaverse. In this dystopian context, individuals are exposed to carefully curated and manipulated information, leading to the creation of echo chambers that reinforce existing biases and beliefs (Gramigna, 2022). This limited exposure to diverse perspectives can foster social isolation, polarise societies, and undermine critical thinking abilities. Online environments can influence individuals and contribute to radicalization through exposure to extremist ideologies, stressing the risks associated with a centralised metaverse controlled by a narrow group with their own agenda (Hassan et al., 2018; Holt et al., 2017; Schlegel, 2020).

Additionally, this dystopian scenario envisions the metaverse as a breeding ground for cyberbullying, harassment, and toxic social dynamics. The anonymity afforded by the digital realm can enable individuals to engage in harmful behaviour without accountability, resulting in a hostile and unwelcoming environment (Barnes, 2018, pp. 92–127; Steele, 2013). Such dynamics have severe consequences on mental health, contributing to feelings of isolation, depression, and low self-esteem as well as feeding-back the issues mentioned in the paragraph above (Huddleston, 2022; Shamim et al., 2023).

To incorporate the principles of Bioethics of Displacement into the design, development, and governance of the metaverse and technological landscape, civil society can play a vital role in preventing the realisation of this dystopia and avoiding the pitfalls of alienation. First, civil society organisations can collaborate with bioethics experts, technology professionals, and social scientists to develop and promote ethical guidelines and policies that emphasise inclusivity, privacy, mental health, autonomy, and protection of vulnerable populations. Public engagement efforts can be undertaken by organising open forums, surveys, and consultations to involve diverse perspectives and ensure that the design and governance of the metaverse reflect the values and concerns of the broader population. Civil society can advocate for responsible AI and algorithmic design by encouraging transparency, accountability, and the elimination of biases that may harm mental health or perpetuate discrimination. They can also raise awareness and promote education about the ethical considerations associated with the metaverse, fostering a culture of responsible technology use. Continuous evaluation and adaptation of ethical guidelines and policies can be advocated by civil society to ensure their effectiveness in addressing evolving ethical challenges. Overall, civil society can actively contribute to shaping the metaverse and technological landscape in a manner that enhances mental health, fosters connection, and empowers individuals while respecting their rights and dignity.

CONCLUSION

Challenging and combating stigma needs for proactive and preventive measures. Moreover, it takes to address stigma as a moral issue deeply rooted in power dynamics and highlighting its impact on individuals, communities, and societies. Understanding stigma as a social and moral issue, unveils the interconnectedness between cultural norms and individual experiences. The social inequalities subsequent to stigma impact public health and, especially, reproductive and maternal health. To address stigma effectively, it is necessary to integrate interdisciplinary approaches, engaging civil society, and promoting the responsible use of new technologies.

REFERENCES

Anderson, J., Rainie, L., & Vogels, E. A. (2021). Experts say the 'new normal' in 2025 will be far more tech-driven, presenting more big challenges. Pew Research Center.

Arkles, R., Jankelson, C., Radford, K., & Jackson Pulver, L. (2018). Family caregiving for older Aboriginal people in urban Australia: Disclosing worlds of meaning in the dementia experience. *Dementia (London)*. Advance online publication. doi:10.1177/1471301218776761 PMID:29877735

Arudou, D. (2018, September 10). Naomi Osaka's US Open victory over Serena Williams: Congratulations, but I don't think you know what you're getting yourself into. *Debito*. http://www.debito.org/?p=15145

Ascott, R. (2007). Telematic Embrace: Visionary Theories of Art, Technology, and Consciousness (E. A. Shanken, Ed.). Academic Press.

Aubrey, C., Mumtaz, Z., Patterson, P., Chari, R., & Mitchell, B. F. (2018). Perspectives of Immigrant Women on the Gender of Provider During Childbirth. *Journal of Obstetrics and Gynaecology Canada*, *40*(6), 677–683. doi:10.1016/j.jogc.2017.09.022 PMID:29274934

Banerji, S. (2023, May 7). *Future of Well-being- The Metaverse Era*. OCAD University. https://openresearch.ocadu.ca/id/eprint/4103/

Barnes, R. (2018). *Uncovering Online Commenting Culture: Trolls, Fanboys and Lurkers*. Springer. doi:10.1007/978-3-319-70235-3

Battersby, C. (1998). *The Phenomenal Woman: Feminist Metaphysics and the Patterns of Identity*. Psychology Press.

Borgesius, F. J. Z., Trilling, D., Möller, J., Bodó, B., de Vreese, C. H., & Helberger, N. (2016). Should we worry about filter bubbles? *Internet Policy Review*. https://policyreview.info/articles/analysis/should-we-worry-about-filter-bubbles

Bruce, R. M. (2018). *Strides Toward Equality: The Portrayal of Black Female Athletes in Children’s Picturebooks*. The Ohio State University. https://etd.ohiolink.edu/pg_10?0:NO:10:P10_ACCESSION_NUM:osu154351163594203

Carroll. (2016, May 15). *Why The Deadliest Gang In The World Might Be Rethinking Face Tattoos*. https://dailycaller.com/2016/05/15/why-the-deadliest-gang-in-the-world-might-be-rethinking-face-tattoos/

Chun, W. H. K. (2008). *Control and Freedom: Power and Paranoia in the Age of Fiber Optics*. MIT Press.

Corrigan, P. W., & Watson, A. C. (2002). The paradox of self-stigma and mental illness. *Clinical Psychology: Science and Practice*, *9*(1), 35–53. doi:10.1093/clipsy.9.1.35

Derrida, J. (1992). *The Other Heading: Reflections on Today's Europe*. Indiana University Press.

Derrida, J. (2001). *On Cosmopolitanism and Forgiveness*. Psychology Press.

Dibbell, J. (1998). *My Tiny Life: Crime and Passion in a Virtual World*. Julian Dibbell.

Falchuk, B., Loeb, S., & Neff, R. (2018). The Social Metaverse: Battle for Privacy. *IEEE Technology and Society Magazine*, *37*(2), 52–61. doi:10.1109/MTS.2018.2826060

Ferrari, L., Hu, A. W., Rosnati, R., & Lee, R. M. (2017). Ethnic Socialization and Perceived Discrimination on Ethnic Identity Among Transracial Adoptees: A Cross-Cultural Comparison Between Italy and the United States. *Journal of Cross-Cultural Psychology*, *48*(10), 1507–1521. doi:10.1177/0022022117728338

Flayelle, M., Brevers, D., King, D. L., Maurage, P., Perales, J. C., & Billieux, J. (2023). A taxonomy of technology design features that promote potentially addictive online behaviours. *Nature Reviews Psychology*, *2*(3), 3. Advance online publication. doi:10.103844159-023-00153-4

Gloeckler, S., & Biller-Andorno, N. (2023). Mental health services in the metaverse: Potential and concerns. *Swiss Medical Weekly*, *153*(4), 40089. doi:10.57187mw.2023.40089 PMID:37099742

Goldston, J., Chaffer, T. J., & Martinez, G. (2022). The Metaverse as the Digital Leviathan: A Case Study of Bit. Country. *Journal of Applied Business & Economics*, *24*(2).

Gorayska, B., & Mey, J. L. (2004). *Cognition and Technology: Co-existence, convergence and co-evolution*. John Benjamins Publishing. doi:10.1075/z.127

Gramigna, R. (2022). Inside Facebook's semiosphere. How social media influence digital hate and fuel cyber-polarization. *Social Semiotics*, *32*(5), 606–633. doi:10.1080/10350330.2022.2157171

Grim, B. J., & Finke, R. (2010). *The Price of Freedom Denied: Religious Persecution and Conflict in the Twenty-First Century*. Cambridge University Press. doi:10.1017/CBO9780511762345

Han, D.-I. D., Bergs, Y., & Moorhouse, N. (2022). Virtual reality consumer experience escapes: Preparing for the metaverse. *Virtual Reality (Waltham Cross)*, *26*(4), 1443–1458. doi:10.100710055-022-00641-7

Harwood, K. A. (2018). Decentering Whiteness in Feminist Bioethics: Assisted Reproductive Technologies (ART) as an Illustrative Case. In L. Campo-Engelstein & P. Burcher (Eds.), *Reproductive Ethics II* (pp. 99–112). Springer International Publishing. doi:10.1007/978-3-319-89429-4_8

Hassan, G., Brouillette-Alarie, S., Alava, S., Frau-Meigs, D., Lavoie, L., Fetiu, A., Varela, W., Borokhovski, E., Venkatesh, V., Rousseau, C., & Sieckelinck, S. (2018). Exposure to Extremist Online Content Could Lead to Violent Radicalization:A Systematic Review of Empirical Evidence. *International Journal of Developmental Science*, *12*(1–2), 71–88. doi:10.3233/DEV-170233

Hatzenbuehler, M. L. (2017). Structural Stigma and Health. In B. Major, J. F. Dovidio, & B. G. Link (Eds.), *The Oxford Handbook of Stigma, Discrimination, and Health*. Oxford University Press.

Holt, T. J., Freilich, J. D., & Chermak, S. M. (2017). Internet-Based Radicalization as Enculturation to Violent Deviant Subcultures. *Deviant Behavior*, *38*(8), 855–869. doi:10.1080/01639625.2016.1197704

Huang, C. J. (2018). On Being the "Right" Kind of Chronic Pain Patient. *Narrative Inquiry in Bioethics*, *8*(3), 239–245. doi:10.1353/nib.2018.0073 PMID:30595591

Huddleston, T. (2022, January 31). *"This is creating more loneliness": The metaverse could be a serious problem for kids, experts say*. CNBC. https://www.cnbc.com/2022/01/31/psychologists-metaverse-could-be-a-problem-for-kids-mental-health.html

Hume, D. (1964). *Hume on Religion*. World Publishing Company.

Jon, G. (2017). *Understanding health and social care* (3rd ed.). Policy Press.

Keeling, G., & Burr, C. (2022). Digital Manipulation and Mental Integrity. In *The Philosophy of Online Manipulation* (pp. 253–271). Routledge. doi:10.4324/9781003205425-15

Khan, S. A., Fatima, Q.-U.-A., Ansari, K., Ahmad, M., Mirbahar, A. M., & Safdar, S. (2022). Advertising Mental Healthcare through Metaverse: Consequential Effects on Stigmatizing Beliefs, Psychological Safety, and Mental Health Literacy. *Pakistan Journal of Medical & Health Sciences*, *16*(12), 12. Advance online publication. doi:10.53350/pjmhs20221612268

Krause, E. L., & Zordo, S. D. (2012). Introduction. Ethnography and biopolitics: Tracing 'rationalities' of reproduction across the north–south divide. *Anthropology & Medicine*, *19*(2), 137–151. doi:10.1080/13648470.2012.675050 PMID:22889424

Lee, S., Chiu, M. Y. L., Tsang, A., Chui, H., & Kleinman, A. (2006). Stigmatizing experience and structural discrimination associated with the treatment of schizophrenia in Hong Kong. *Social Science & Medicine*, *62*(7), 1685–1696. doi:10.1016/j.socscimed.2005.08.016 PMID:16174547

Lindau, J. D. (2022). *Surveillance and the Vanishing Individual: Power and Privacy in the Digital Age*. Rowman & Littlefield.

Link, B. G., & Phelan, J. C. (2001). Conceptualizing Stigma. *Annual Review of Sociology*, *27*(1), 363–385. doi:10.1146/annurev.soc.27.1.363

Love, N. (Director). (2007, May). *Ross Kemp on Gangs: El Salvador*. https://www.imdb.com/title/tt1211709/

Lozano Rodríguez, M. (2022). Shame on You! Digital Ethnography, Suicide, Stigma, and Warfare. In Practices, Challenges, and Prospects of Digital Ethnography as a Multidisciplinary Method (pp. 198-222). IGI Global.

Lundahl, O. (2021). Media framing of social media addiction in the UK and the US. *International Journal of Consumer Studies*, *45*(5), 1103–1116. doi:10.1111/ijcs.12636

Lushetich, N. (2018). How Long? How Far? *Performance Research*, *23*(4–5), 34–39. doi:10.1080/13528165.2018.1511039

MacKenzie Bryers, H., & van Teijlingen, E. (2010). Risk, theory, social and medical models: A critical analysis of the concept of risk in maternity care. *Midwifery*, *26*(5), 488–496. doi:10.1016/j.midw.2010.07.003 PMID:20719418

Major, B., & O'Brien, L. T. (2005). The Social Psychology of Stigma. *Annual Review of Psychology*, *56*(1), 393–421. doi:10.1146/annurev.psych.56.091103.070137 PMID:15709941

McCurry, J. (2019, January 23). Naomi Osaka sponsor apologises for "whitewashing" tennis star in ad. *The Guardian*. https://www.theguardian.com/sport/2019/jan/23/naomi-osaka-sponsor-apologises-for-whitewashing-tennis-star-in-ad

Munasinghe, S., & van den Broek, N. (2005). Abortions in adolescents. *Tropical Doctor*, *35*(3), 133–136. doi:10.1258/0049475054620752 PMID:16105332

Oberman, M. (2018). *Her Body, Our Laws: On the Front Lines of the Abortion War, from El Salvador to Oklahoma*. Beacon Press.

Pachankis, J. E., Hatzenbuehler, M. L., Mirandola, M., Weatherburn, P., Berg, R. C., Marcus, U., & Schmidt, A. J. (2017). The Geography of Sexual Orientation: Structural Stigma and Sexual Attraction, Behavior, and Identity Among Men Who Have Sex with Men Across 38 European Countries. *Archives of Sexual Behavior*, *46*(5), 1491–1502. doi:10.100710508-016-0819-y PMID:27620320

Pałczyńska, J. (2023). *Surveillance of employees in the Metaverse: Data protection and privacy challenges*. https://ruj.uj.edu.pl/xmlui/handle/item/308702

Parker, R., & Aggleton, P. (2003). HIV and AIDS-related stigma and discrimination: A conceptual framework and implications for action. *Social Science & Medicine*, *57*(1), 13–24. doi:10.1016/S0277-9536(02)00304-0 PMID:12753813

Paul, S. (2018). Are we doing enough? Stigma, discrimination and human rights violations of people living with schizophrenia in India: Implications for social work practice. *Social Work in Mental Health*, *16*(2), 145–171. doi:10.1080/15332985.2017.1361887

Pembrey, M. E., Bygren, L. O., Kaati, G., Edvinsson, S., Northstone, K., Sjöström, M., & Golding, J. (2006). Sex-specific, male-line transgenerational responses in humans. *European Journal of Human Genetics*, *14*(2), 159–166. doi:10.1038j.ejhg.5201538 PMID:16391557

Pintassilgo, S., & Carvalho, H. (2017). Trends and consequences of the technocratic paradigm of childbirth in Portugal: A population-based analysis of birth conditions and social characteristics of parents. *Sexual & Reproductive Healthcare : Official Journal of the Swedish Association of Midwives*, *13*, 58–67. doi:10.1016/j.srhc.2017.06.003 PMID:28844359

Reer, F., & Quandt, T. (2021). Games Addiction: A Comprehensive Overview. In Oxford Research Encyclopedia of Communication. doi:10.1093/acrefore/9780190228613.013.1154

Rosenberg, L. (2023). The Metaverse and Conversational AI as a Threat Vector for Targeted Influence. *2023 IEEE 13th Annual Computing and Communication Workshop and Conference (CCWC)*, 504–510. 10.1109/CCWC57344.2023.10099167

Schlegel, L. (2020). Jumanji Extremism? How games and gamification could facilitate radicalization processes. *Journal for Deradicalization*, *23*, 1–44.

Scott, K. M., & Collings, S. C. D. (2010). Gender and the association between mental disorders and disability. *Journal of Affective Disorders*, *125*(1), 207–212. doi:10.1016/j.jad.2010.06.022 PMID:20637507

Shakur, A. (2001). *Assata: An Autobiography*. Zed Books.

Shamim, A., Banu, S., Blassingame, J., Guilford, K. L., & Williams, J. (2023). Delusions and Fixed Beliefs in the Age of the Internet: When Do Conspiracy Theories Become Illness? *Psychiatric Annals*, *53*(4), 171–174. doi:10.3928/00485713-20230313-03

Shildrick, M. (2000). Becoming Vulnerable: Contagious Encounters and the Ethics of Risk. *The Journal of Medical Humanities*, *21*(4), 215–227. doi:10.1023/A:1009025125203

Shildrick, M. (2015). *Leaky Bodies and Boundaries: Feminism, Postmodernism and (Bio)ethics*. Routledge. doi:10.4324/9781315004952

Siricharoen, W. (2019). The Effect of Virtual Reality as a form of Escapism. *CONF-IRM 2019 Proceedings*. https://aisel.aisnet.org/confirm2019/36

Smith, A. (2007). *Wealth of Nations*. Cosimo, Inc.

Steele, N. M. (2013). *Trolls under the bridge: Anonymous online comments and gatekeeping in the digital realm*. University of Missouri-Columbia.

Timmermans, S., & Tietbohl, C. (2018). Fifty years of sociological leadership at Social Science and Medicine. *Social Science & Medicine*, *196*, 209–215. doi:10.1016/j.socscimed.2017.11.007 PMID:29254656

Tsuda, T. (1998). The Stigma of Ethnic Difference: The Structure of Prejudice and "Discrimination" toward Japan's New Immigrant Minority. *Journal of Japanese Studies*, *24*(2), 317–359. doi:10.2307/133237

Usmani, S. S., Sharath, M., & Mehendale, M. (2022). Future of mental health in the metaverse. *General Psychiatry*, *35*(4), e100825. doi:10.1136/gpsych-2022-100825 PMID:36189180

Voltaire. (1765). *Dictionnaire philosophique, portatif*. Academic Press.

Wiederhold, B. K. (2023). (Mental) Healthcare Consumerism in the Metaverse: Is There a Benefit? [editorial]. *Cyberpsychology, Behavior, and Social Networking*, *26*(3), 145–146. doi:10.1089/cyber.2023.29269.editorial PMID:36880891

Williamson, B. (2017). Moulding student emotions through computational psychology: Affective learning technologies and algorithmic governance. *Educational Media International*, *54*(4), 267–288. doi:10.1080/09523987.2017.1407080

Wolf, S. (1999). Morality and the View From Here. *The Journal of Ethics*, *3*(3), 203–223. doi:10.1023/A:1009833100856

Wolffe, R. (2018, July 1). Trump's only fixed position on abortion is his disdain for women | Richard Wolffe. *The Guardian*. https://www.theguardian.com/commentisfree/2018/jul/01/donald-trump-abortion-supreme-court

Xu, J., Jiang, C., Wei, W., & Li, Y. (2022). The Research of Medical Metaverse Application Under the Background of the Normalization of the New Crown Epidemic. In L.-J. Zhang (Ed.), *Metaverse – METAVERSE 2022* (pp. 68–80). Springer Nature Switzerland. doi:10.1007/978-3-031-23518-4_6

Yang, L. H., Kleinman, A., Link, B. G., Phelan, J. C., Lee, S., & Good, B. (2007). Culture and stigma: Adding moral experience to stigma theory. *Social Science & Medicine*, *64*(7), 1524–1535. doi:10.1016/j.socscimed.2006.11.013 PMID:17188411

Ziegler, M. (2018). Some Form of Punishment: Penalizing Women for Abortion. *The William and Mary Bill of Rights Journal*, *26*(3), 735.

KEY TERMS AND DEFINITIONS

Cultural Dynamics: The ever-changing interplay of cultural factors, beliefs, and values that shape societies over time.

Discrimination: Unfair treatment based on unchosen or innate characteristics, leading to disadvantage and exclusion.

Epigenetics: The study of how behaviours and environment can change gene expression without altering the DNA sequence.

Intercorporeal Stigma: The negative judgments and prejudices imposed on individuals and their descendants based on naming and body modifications.

Interpersonal Stigma: The negative judgments and prejudices imposed on individuals based on their social identity.

Power Dynamics: Relationships where authority, control, and influence shape interactions, creating winners and losers.

Stigma: Social rejection based on differences in characteristics, conditions, or identities.

Structural Stigma: Unfair treatment and discrimination embedded in societal structures and power dynamics perpetuating inequalities and limiting opportunities for rejected groups.

ENDNOTE

1. Lost in translation term that refers to the effective leadership in tribal political power.

Chapter 15
Examining Confinement Beyond Bars From Philosophical and Practical Perspectives:
Free v Confined III

ABSTRACT

The author examines the concept of confinement from various philosophical and practical perspectives. The historical development of confinement is traced back to the 19th century, when corrections became a colonial institution designed to change offenders' behaviour through punishment and isolation. But it has grown far more than this. Confinement affects people's identity, and philosophical insights can help in alleviating the suffering of those who are confined. Furthermore, the author delves into the biopolitical and phenomenological aspects of confinement, as well as its ethical implications in relation to memory, disability, caretaking, and trauma: confinement by other names. In conclusion, we need a more humane and thoughtful approach to the many confinements, which recognizes the impact of technology on social and cultural norms and values.

INTRODUCTION

The inquiry on stigma in the last chapter provides a foundation for understanding the experiences and challenges faced by individuals who are confined. Stigma can contribute to the perpetuation of confinement, as societal judgments and biases shape the way individuals are treated and confined. By examining the ethical implications and moral dimensions of stigma, Chapter 14 sets the stage for a deeper examination of the confinement experience in this chapter.

DOI: 10.4018/978-1-6684-4808-3.ch015

CONFINEMENT

Introduction

Corrections are a colonial institution born in the US around 1830. They resulted, on the one hand, from the public punishing, humiliation and many times forcible disability that was expected to change the offender's behaviour *for good.* On the other hand, however, the Quakers (a progressive Christian denomination) tempered this vengeful justice and advocated for doing time praying away from bad companies. In practice, the mixed primitive corrections were a place where isolating the inmate and reading him the gospels once in a blue moon (Amos, 1974). Many of them drove crazy.

In the US and UK, the first modern Zoos and asylums evolved in parallel with the modern system of prisons during the 19th century. Likewise, the 1970s were an incarceration rate turning point in the US (National Research Council, 2014) that, in the later decades, dragged most of the world (Jacobson et al., 2017, pp. 1–2). The US inmate population booming was parallel with the booming castration of dogs and cats in order to control the pet abandonment consequences (Rowan & Kartal, 2018). By the way, the asylum population in the US notably shrank in the 1970s thanks to the first psychotic medications and the old 1965 Medicare and Medicaid, which allowed many of the former patients ended in shelters while the former shelter users ended in prison: it was "*a psychiatric Titanic*" (Steadman et al., 1984; Torrey, 1998). Even one the most extreme force of nature was confined like never before in the Princeton Plasma Physics Laboratory in the 1970s.[1] The plasma in astrophysics behaved differently than the plasma in the physics laboratory. It happens because the plasma in the lab is confined, not behind bars, but magnetic fields. An heterogeneous plasma was produced in the large *tokamaks* but an homogeneous one happens in the deep starry space (Kaplan & Tsytovich, 2013, p. 3). The fault in our stars…

Besides any kind hearted Quaker daydreaming, the prisons have been a money tree. It has been since almost the beginning in the US original prison system, whose legacy is now widespread. The first privatisation happened in 1844 and was a factory where the cheap clothes worn by the slaves were produced. According to one prisoner memoirs, when the privatisation came his jailers: "*laid aside all objects of reformation and re-instated the cruelest tyranny, to eke out the dollar and cents of human misery.*" (Bauer, 2018c, p. 81) Among other cases of abuse, in 1967 the inmates of the Arkansas hell of a prison picked cotton while singing "*Old Master don't you whip me, I'll give you half a dollar.*" (Bauer, 2018a, pp. 16–17). As Shane Bauer says, "*there is a straight line from slavery to private prisons*" (Bauer, 2018b). The Old Master was Terrell Don Hutto, who received in the Encyclopedia of Arkansas the Business and Economic Leader tags -also he gets the role of villain in the film *Brubaker* that describes the unveiling of his inhuman abuses by Tom Murton (Rosenberg, 1980).[2,3] Just like in the a barely softer version of the 1844 prison privatisation, in the CoreCivic prison in Louisiana a man lost both legs due to gangrene because the visits to hospital were paid by the company and other end his life weighing 32 kilograms after a set of hunger strikes to receive mental health (Bauer, 2018c). Vulnerability also hastened the human grinding business of prisons.

During the Reagan era Terrell Don Hutto brought the monetization of others' confinement to an industrial level never reached before and started a familiar holding that in 2019 is CoreCivic, the 1.8 billion USD private prison corp (Bauer, 2018a, p. 38). *Terrell the Hutt* died by old age, filthy rich and after receiving the greatest professional recognition in his branch in 1987.[5] Even a detention centre famous for holding immigrant children and heinous cases of sexual abuses brings his

Figure 1. Terrell Don Hutto portrait
Source: Retrieved from the Encyclopedia of Arkansas website, one of the three co-founders of Corrections Corporation of America (CCA)[4]

name! (Mindock, 2018; Talbot, 2008) The US mindset on its own need on security has brought, in the Sonya Goshe words (Goshe, 2019):

Blindness to the history of using punishment to ensure economic and social security for the privileged, ongoing policies that breed high levels of violence, and cultural endorsement of punitive logic as a substitution for social security and substantive justice.

The modern era of confinement in the global powers talks about a time when some people believed to own the right of speaking on behalf of both human nature and nature itself, a time for *putting rather than placing*. This trend led us to the Shock fever suffered in the turn of the millennium, with fishy news on Bin Laden (Everett, 2010), a lot of Mega Max security prisons, retributions and plenty of prozac. The White House envisioned a theurgic delivery was an unholy foolishness even within its borders and the US government spoiled what could have been a golden age after the Cold War. The future of confinement will take place in a bipolar world again, but the nick hardly will change. Similar descriptions appear in the work of many prison writers with very different perspectives, if not opposite, when incarcerated in different countries and epochs: Jean Genet, Jimmy Boyle, Eduard Limonov, Irina Ratushinskaia, Malcolm X, Jack Henry Abbott… (Rogatchevski, 2014, pp. 174–176) Oscar Wilde described his cell in a way which many inmates around the world can probably still recognize today:

Each narrow cell in which we dwell

Is a foul and dark latrine

And the fetid breath of living Death

Chokes up each grated screen,

And all, but lust, is turned to dust

In Humanity's machine. (Wilde, 2007, p. 907)

Philosophical Grounds

Most people live confined in their own lack of creativity and becoming-ness, in their denial of what we are as a part of the creative movement that life is (Hallward, 2006, p. 55). Due to the common opinion that we are static in our human identity, a few of us displace, *in our behalf*, the unsuitable ones to a frozen and still created-ness called confinement, which opposes the preferable creature-ness of the non-confined population. From this ignorance emerges the aforementioned 'philosophy of necessity', where the punishment always has the 'benefit of the doubt' in spite of some abuses here and there (Goshe 2019). It's the idealisation of punishment as the way in which safety and granting justice is meant to be. It's difficult to believe that the confinement once was an act of charity aimed to free the individual from the darker prison of a corrupted heart. But the old benevolence gets subsumed under a dichotomous neoliberalism with no room for mistake nor conversion amidst a continuous 'grave disorder' state (Skotnicki, 2019, p. 26).

Thankfully, there is also a humanitarian post-Enlightenment view on confinement. We can think of confinement as the displacement of retribution to social endings, but given that the eye for an eye and the time in prison are very different things a reason, measure or proportion are needed. Confinement should be streamlined with a proportional allocation of the "benefits" and "burdens" of law-abidingness that is just enough to censure criminal behaviour. But the theory of proportionality is waning due to difficulties in scaling as well as their trade-offs with crime prevention (Tonry, 2019). Anyhow, the ideas of non-excessive punishment and reserving the harshest punishment for those who commit the harshest crimes are strongly bound with the idea of justice and have survived the hyperincarceration.

In *Imagination Dead Imagine* by Samuel Beckett (Beckett, 1965), two homunculus are confined in a skull-shaped dome looking in opposite directions. It represents the loneliness and encapsulation of the mind inside the "*ring of bone*" (Beckett, 1965, p. 8). Between the Beckett's short narration figures and the captives in the Platonic Cave, the subjects of a crime are confined in their own shadows (Arrigo & Polizzi, 2018). The guilty people who are in jail turned back to the world (or simply scattered over) the problems they had in their mind. As Plato and the Quakers held, a more lovely and sensible view of our fellowmen can free us. They were not so naive: the modern philosophical grounds of confinement are streamlined with the lack of impact that philosophy has in education (Carr, 2004). We must stop feigning that the punishment will improve the human quality or the respectfulness of a society.

Public Health and Confinement

Protective isolation is aimed at preventing harm to patients. But as we'll see below, it is implemented inconsistently across centres and is supported by recommendations with poor evidence. To cap it all, the "*special hazards*" of the forensic psychiatry widespread use on the inmates captives due to any mental condition are mostly ignored (Murray & Holmes, 2015).

Neither is enough considered the pathological lens on women, as we can learn from the Danish secure care institutions failure in girls, who are chased inside the institutions by the same social issues that sent them into (Henriksen, 2017). This gendered view on confinement and care persists also in ancient traditions. For example, old Chinese custom stipulates that a woman should be confined to home for one full month recovering after giving birth, in which a broad set of prescriptions and proscriptions are expected to be followed (Pillsbury, 1978). In Europe, the Renaissance drama inherited a long tradition of misogyny, which holds that the body where insanity lives is female and must be confined -due to the men's fear of women's mad sexuality (Salkeld, 1994, p. 124).

This lack of rigour in protective isolation has been studied from out of psychiatric settings, for example, to avoid infections in neutropenic patients (Biagioli et al., 2017). But doubtless, the asylum is the logical reference to health and seclusion. A study on acute psychogeriatric admission units, teaches that after the release from mandatory confinement indoors both verbal and physical aggression decreased as well as medication use. Given that a well-provisioned and designed residential care setting has a well-known effect on mental illness, the freedom to go outdoors or having more room should also be an important design component (McMinn & Hinton, 2000). In the same vein, phenomenological research conducted in a Canadian forensic psychiatric environment showed that the "*structure of place*" matters for both patients and working staff (Holmes et al., 2015).

Isolation is a source of suffering, in isolation a person's place is expected to be reduced to the space she occupies first to reduce the space, then. When her place as a person extinguishes, her body is allegedly turned into a thing. However, this person doesn't cut the cord with the world, which feeds her with both lasting suffering and hope.

BIOPOLITICS OF CONFINEMENT

Confinement as Usual, Even More

An institution is the embodiment of political boundaries able to reshape the social self and bring a full punishment for the unfit subjects. The total institutions, from an oversized state to a fragile poorhouse, are aimed to create "*normal individuals*" (De Cunzo, 2009). In a modern state, confinement is not always an enclosure, but calculation on the bodies processed by an institution, trapped by the prongs of theoretical consent and technical coercion in a way that space sets both hierarchy and punishment (DeLanda, 2006, pp. 72, 99). The privileged instrument of torture many times led to a *right* way of talking about it according the interest of the governments that have invested efforts in separating themselves from regimes of torture (Macias, 2013).

In 2019 Greece, even the 'legal' immigrants who are not being abused in a detention centre and are supposed to be free, actually live confined between police controls to avoid (Goumart, 2019). Here, a greater focus on structural racism, like the facially *neutral* Greek police, would improve the UN Refugee Agency's global policy for addressing foreignness or xenophobic discrimination (Achiume, 2013). The calculation beyond thick walls has its current peak in the drone zones, where the civilians are reduced to "*walking corpses*" trapped in a killing open space of devastated beings (Ahmed, 2018). It's not so different from the Greek immigrants' one but for the lethality bar, which has been badly risen.

The prisons are biopolitical instruments to deal with minorities that go through broad social progress attainments. Let's see a comparison. On the one hand, in 1987, during the Reagan era, nonwhites were more likely to be sentenced to prison in Washington state urbanised counties as well as counties with relatively large minority populations (Bridges et al., 1987). On the other hand, in 2018 New Zealand, the rate of incarceration of the minority Maori people is relatively high -despite its engagement with the UN and the Indigenous rights movement challenging discriminatory criminal justice system practices (Stanley & Mihaere, 2018).

A prison can keep exuding biopolitics even after its death. Many carceral institutions across the world have turned museums after being decommissioned. However, a study on 45 of these museums in Canada show that they actually foster the popular conceptions of devilish criminals who deserve expeditious punitive justice (Piché & Walby, 2018). It's the kind of view that justifies solitary confinement. It harms the mental integrity of the inmate just like a neurointervention would do, so its intentionality is not the same as common incarceration regardless of their apologetics (Lim, 2018). The harm received due to medical care under the punitive constraints of carceral facilities also is structural politics of pain (Pitts, 2018).

Another way of *driving an individual crazy is* making an arbitrary definition of sanity according to political needs. The controversial Australian courts decisions in the 1990s showed a "*fundamental inability to define conceptual boundaries*" between dangerousness and personality disorder (Glaser, 1994; McCallum, 1997). The most blatant example was the suicide of the young offender Garry David with a diagnostic of personality disorder after remaining institutionalized 33 out of his 38 years old in the time of death. He was quite dangerous, in addition to a murder attempt, Garry sliced off his ears, nipples and penis while he was in prison (Greig, 2002, p. 246; *Statement of Reasons in GIPW*, 1990, p. 10) but the solution for his bizarrely aberrant behaviour was keep him in that prison. In the Garry David case, the Victorian government tried to change the 1986 Mental Health Act to include his condition as mental illness in order to shut Garry up *in the nick* (Greig, 2006, pp. 140–141; Parker, 1991).

Pandemia!

Confinement during the Covid-19 pandemic has revealed a bleak reality of government work and its response to the crisis. As the virus spread, it became apparent that the burden of corruption fell upon the many, while safety nets and privileges protected the few (Lozano Rodríguez, 2020c; Lozano Rodríguez et al., 2023). This stark inequality perpetuates the existing disparities in society, exacerbating the economic struggles faced by households. The pandemic has served as a catalyst for reevaluating our understanding of corruption (Lozano Rodríguez, 2023; Lozano Rodríguez et al., 2023). It compels us to move beyond outdated and colonised views, urging us to embrace a new perspective that encompasses the well-being of all humanity (Lozano Rodríguez et al., 2023). The shared experience of facing a common threat has brought about some positive changes. Coordination arrangements between countries and regions, previously hindered by bureaucracy, understaffing, and slow responses, have shown signs of improvement. Hopefully, this progress continues, leading nations to recognize the necessity of setting aside differences and collaborating on global issues that impact our health, well-being, and survival. On an individual level, the pandemic has left its mark, pushing us to the brink and ravaging our psyche. It has exposed our disdain for monotony and sparked a desire for change. While some of these traits can contribute to creating more caring economies and societies, others can have deleterious effects, especially with the yet displaced (Lozano Rodríguez, 2020b). To avoid being defined solely by the negative aspects of the pandemic, we need to adopt a holistic approach to wellness and coping strategies that emphasise living without reinforcing privileges.

It is important to note that the coronavirus crisis is not a byproduct of globalisation; rather, it is the consequence of ad hoc politics and their non-linear effects. Almost every state and economy on Earth, regardless of vulnerability and awareness, has faced the challenge of flattening the curve of COVID-19. However, the leaders of globalisation have fallen short in their response, failing to allocate sufficient resources to combat the virus effectively. This raises questions about the fishy nature of postmodern globalisation and its potential role as a ruling moral vice.

The pandemic crisis and its impact on global health and economies have revealed deep-rooted political and social conflicts. Insufficient research and emergency services, coupled with politicisation of administrative decision-making and scientific criteria, have led to significant economic and human losses in some countries (Lozano Rodríguez, 2020a, 2020a; Lozano Rodríguez et al., 2023). This exposes critical issues in various government areas, including underinvestment in public health, inadequate data management, and corruption. Furthermore, post-pandemic food scarcity mirrors the flaws of political economy hiding a new surge of biological determinism (Lozano Rodríguez 2023, 105-115). It is crucial to examine the role of non-state actors such as NGOs and private partners in the pandemic response, evaluating their contributions to public healthcare systems (Lozano Rodríguez et al., 2023). Future research should consider the long-term consequences of health crises, including the relationship between financial crises, corruption, mental health, and post-crisis social restoration.

In light of these revelations, scepticism towards government actions and decision-making processes is warranted. The self-absorbed nature of state actors raises concerns about their ability to address the complex challenges posed by health crises and displacement effectively. This scepticism, however, necessitates an exploration of alternative approaches and perspectives to ensure responsible and ethical governance.

The biopolitical insight drawn from my experience and research on Covid-19 is that the pandemic has exposed deep-rooted political conflicts, systemic corruption, and stark inequality, necessitating a holistic approach to wellness and coping strategies that prioritise the well-being of all humanity. Pandemics has highlighted the importance of Bioethics of Displacement in addressing the moral implications of displacement and promoting responsible and ethical governance.

PHENOMENOLOGY OF CONFINEMENT (AND FREEDOM!)

The inmates who joined a university philosophy course declare they feel free when they learn from an emancipating teacher who is, in turn, free from intellectual and moral superiority feelings or attitudes (Delgado, 2018). The contribution of philosophy to prison could enhance its original humanitarian sense greatly. As Katerina Deligiorgi highlights, the mainstream prison theories have overlooked how a phenomenological conception of autonomy (a key principle in bioethics) can be applied to psychopathology (Deligiorgi, 2018).

A deep experience of freedom in prison comes with balance after despair, a despair that has a freedom rebirth within. This acceptance of divine forgiveness after wrongs in life, as Kierkegaard expressed it

(Kierkegaard, 1959, pp. 350–362; McDonald, 2017), likely is related with the higher levels of religiosity experienced by inmates in religious programs (Kerley, 2018, p. 132). To sum up, freedom is not an experience of *something* but putting the system that confine us between brackets and speak about it in a sensible manner. Nelson Mandela is the better exponent of this mature kind of freedom (Mandela, 1961).

The wars and oppressive regimes, like the apartheid, are a common frame for a particularly harsh confinement: the isolation in punishment cells. What happens to those subjects left aside from intersubjectivity? Although it's not carried out in the same way across the many scenarios, the isolated inmates will suffer an induced autism (Gallagher, 2014). Additionally, an interview with fifteen prisoners from the 1973 Yom Kippur war concluded that the isolation led to a loss of the subjective sense of body connected with a loss of time as an objective dimension (Ataria & Neria, 2013). Shortly, isolation saps human architecture and makes the person crumble down. This sadistic modification of time and space can be compared with the anguish of awakening suffering from paraplegia (Ostrowski & Sikorska, 2015, p. 34). Ironically, there is no consensus about whether isolation is a cruel punishment. Would infecting an inmate with tuberculosis, injecting hallucinogens or paralysing her legs be considered a common punishment, then? It seems that the prison runners always let the cat out of the bag…

Besides the aforementioned issues, the immediate experience of confined persons is also determined by the fear of mental paralysis and lack of alertness, very common in these circumstances (Gelfand, 1983, p. 201). That's a very dangerous state considering that confinement increases impulsiveness both in men and women (Chaumet et al., 2009).

The confinement is some kind of anti-phenomenology, it attacks the structure of conscience. So, the sanest attitude towards prisons I can imagine is distrusting about them. Mistrust in the name of those who won't be able. Avishai Margalit said:

Being a helpless inmate in a Nazi concentration camp or a Bolshevik gulag can make you believe that the thousand year Reich or the unstoppable juggernaut of communist triumph is just the way of the world. (Margalit, 2009, p. 155)

THE FORGOTTEN CONFINEMENTS: DISABILITY, CARETAKING, AND TRAUMA

Disability, Trauma, and Prospects for a Ubiquitous Computing Society

According to the Bioethics of Displacement a person can be displaced even when confined in a hole and, in turn, likely confined while being free. Individuals can be confined or restricted in different ways, even if they are not physically confined to a specific location. These forms of confinement may be less visible or more subtle, but they can still have a significant impact on an individual's well-being and autonomy. For example, individuals with disabilities may face physical barriers that restrict their movement and participation in society, leading to a form of confinement. Caretakers, who may be responsible for the daily needs and support of individuals with disabilities or other dependents, may also experience a form of confinement as their responsibilities limit their freedom and mobility. Additionally, individuals who have experienced trauma may feel confined by their memories and emotions, which can limit their ability to engage with the world around them. In each of these situations, individuals may experience a loss of autonomy and freedom, which can be similar to the experience of confinement.

In the near-future we may find some new challenges. The post-human neural-prosthetics, a surging market with plenty of under development products, will have a weight on the technological and socio-cultural alignment long-run (Gladden, 2016a). Considering that I'm not talking about strictly recovering senses but adding capacities that greatly differ from the natural ones (Dixit & Prasad, 2017, pp. 188–189), a great attention must be paid to the funding of specialised agencies as well as how the soft skills are handled by this new entrepreneurial class. The neuroprosthetics presents the following major vulnerabilities (Gladden, 2016b, p. 270):

- Subjugation through manipulation of memories, emotions, and volitions
- Shunning, mistreatment, and other forms of discrimination against users with visible neuro-prosthetics, mainly in the labour market
- Increased vulnerability against data theft

All this cyberpunk stuff is very likely to happen, but today in 2023, there are no warning signs, but otherwise. The smart deep-brain stimulation used in treatment-resistant neuropsychiatric movement disorders "*have the potential to affect our identities, perhaps supporting us so that we can be who we want to be, but perhaps also changing in ways we wouldn't want, even to the point of challenging our identity and / or agency*" (Goering et al., 2017). Anyhow, those changes wouldn't have intrinsically negative outcomes (Lázaro-Muñoz et al., 2017). Moreover, the BCI use in the paediatric population with severe disabilities is an important future direction since a reliable communication pathway during early stages of development can prevent learned helplessness (Moghimi et al., 2013). Likewise, the success of BCI to restore communication compared with the limited results in amyotrophic lateral sclerosis or spinal cord injuries, points to the regulation of brain metabolism, blood flow and both electric and magnetic stimulation as most promising fields of research (Birbaumer & Cohen, 2007).

Since the VR landing in psychotherapy in the 2000s, it has resulted in a good omen. To begin with, it has widened the professional practice and clinical research (Riva, 2005). The immersive VR therapies for post-traumatic stress disorder (PTSD), mainly addressed to a native digital generation of veterans, offer better results, usually when combined with Propranolol and similar chems (Glannon, 2013, pp. 28–31; Kuehn, 2018). Beyond PTSD, the VR exposure therapy has extended to other anxiety disorders like phobias and panic (Meyerbröker & Emmelkamp, 2010). Even the VR features of the Wii console have also been successful when combined with traditional therapy helping to raise the daily life activities performance (BERG and MIF scales) in patients impaired after a cerebrovascular stroke (Silva et al., 2015).

Despite this, all these developments aiming to palliate disabilities have relied only upon the industry means and engineers creativity without round tables between rehab specialists, companies and patients (Vanderstraeten, 2009). That's important since some designs of technological products can be aesthetic stigmatising for older and disabled persons (Bichard et al., 2007). It's a cyst-like issue that despite being well-known since the 2000s, even today, in 2019, the knowledge gaps in making accessible designs keeps grounding a market-driven ableism (Crabb et al., 2019). In 2023, we are in the dawn of BCI and VR merging, a blend of techs for restoring communication and mobility with others that provide fun designed for home users (Boldeanu, 2018). It's very challenging ethically, for example in the case of making accountable the closed-loops between EEG and VR in smart hybrid VR systems (Kellmeyer, 2018). Furthermore, even if the health-care devices in the US, for example, must be approved by the Food and Drugs Administration (FDA), many commercial devices with sensitive health-related info not (Kreitmair & Cho, 2017). To cap it all, the 2010s trend is resorting to lacking oversight of healthcare technology providers (even DIY and online approaches!) in order to provide otherwise prohibitive services (Dumbill, 2019). If we add to these incentives to participate, DNA low-cost analysis, social media and the unexpected vulnerability of medical devices even in hospitals (Mahler et al., 2018) the result is quite volatile.

Disability While Forcibly Displaced: True Intersectionality

From the Christian Middle Ages to 21st century neoliberal Britain, the representations of disabled and displaced persons have been dragged to and melted into the furnace of ethnocentrism and ableism hallucinatory strangeness (Hughes, 2017). So much so that the recent humanitarian crises have put on blast the assistance operation readiness to deal with disabled (and likely highly marginalised) persons (Stein & Lord, 2011). In the same vein, the intersectionality between disability, chronic illness and migration has been generally neglected in the literature to date (Ottosdottir & Evans, 2016). From the experience in Southern Africa refugee camps the main three outcomes of intersectionality and disability are (Mirza, 2017):

- Misperceptions about the health-related needs of disabled persons
- Concerns about distributional ethics in relation to disability-inclusive healthcare
- Disabled people health needs falling outside the social minimum of what humanitarian healthcare should be

It's a severe issue considering that, in 2010, 40 million of displaced persons were suffering from a disabling condition according to the Director of UNHCR New York Office (Janz, 2010). Too often they are hosted in the poorest countries where they turn into living ashes "*of epistemological, ontological and practical invisibility*" (Pisani & Grech, 2017).

In addition to the condition of migrants, the migratory route has proven to play a role in the intersectional discrimination of disabled children (and indirectly in their family) as stands out some Australian exclusionary policies (Soldatic et al., 2012). For example, Ernesto is a disabled child whose visa was only granted "*on the proviso that their parents/primary caregivers agree to cover the full costs associated with their disability, including medical care and additional expenses such as educational inclusion*" (Soldatic et al. 2012). Unsurprisingly, one in three school students in Australia are victims of racial discrimination (McGowan, 2019). By the same token, claiming persecution on account of disability faces additional barriers and gets entangled in discourses of rescue within a political asylum system that perpetuates asymmetrical transnational power relations (Shuman & Bohmer, 2016). Plainly, it's a displacing forever scheme. Ignoring geographical, historical and cultural contexts blind researchers and policy analysts from a greater insight of the diversity within the disability and locks them up in wrong assumptions and misleading conclusions (Albrecht et al., 2008).

The biolegal politics in the borders are attached with the unmaking of the self through the ontological insecurity incremental of territorial ambitions that turn harsher when it comes to disabled refugees (Berghs 2017). In layman's terms, the government's policies on how to handle refugees are making things even worse for disabled people. They're so obsessed with claiming more land that they're willing to sacrifice people's well-being and sense of self. It's like they're saying, "Sorry, folks, your disabilities make you a liability. We've got bigger fish to fry."

Trapped in Caretaking: Intergenerational Discrimination

The informal family caregiving, including children and young adults, are expected to meet the disabled familiar needs. Those responsibilities are combined with their restricted entitlements to welfare and harms the education and transitions into adulthood of the caretakers leading to an intersectional and intergenerational discrimination. To cap it all, the opportunism embedded in social work distorts the idea of social justice, leads to indirect discrimination and perpetuates the lack of opportunities of people with disabilities through the provision of services (El-Lahib, 2018). A lack of opportunities that passes to the rest of the family.

Let's narrow this gap. To start with, what counts as disability greatly relies upon the impairment production of the labour market mindset (Berghs, 2015). This hint led me to think that the organs to breathe a mix of bioethics, decision making and intersectionality cannot evolve in borders, refugee camps and not even in summits about humanitarian crises. Therefore, the right mindset must be looked for in another place. But where? Likely other persons will find different replies but I think that accountable sustainability policies resume all the moral work needed to get the job done. Hardly a country would be concerned about how the burden of social categories affects our bioethical agency while withering its own land.

The Crippled in the Maze of Mirrors

Disability consists in getting stuck to an identity that mirrors another, *a priori* existence as a more desirable one. For example, the head massage for quadriplegics brings them the kind of alluring delight that perhaps sex provokes in a non-quadriplegic person, but weathered; in the same fashion, a blind person reading in braille imitates, in a quite narrow way, the pleasure of lecture for a sighted one. Nevertheless, from a phenomenological perspective, the lived disability is not the same as the *dwarfed* version of a body-abled experience -but a different one that builds the human full spectrum. Unsurprisingly, there is a clear communication barrier between the advocates of intervention in the lives of disabled people and the supporters of non-intervention. No middle ground has been reached. It's a pending question even in the light of facts like the weight of both intrinsic impairment and environmental factors (such as services, support, and attitudes) in the participation of severely disabled children (Forsyth et al., 2007). Anyhow, if we depart from disability as social construction, it turns into a bag of junk and broken bottles that nobody wants and that remains plugged in your identity 24/7. The embodied existence pushes even the most abstract goals to rely upon the physical and perceptual capabilities, a disability sends you to a place where you cannot promote your own goals, for a while or for the whole life. The disability "*transform the heart of the self, the agent's subjectivity and global existence*" (Ostrowski & Sikorska, 2015, p. 34). In other words, the disability not only mirrors an identity but traps it in the mirror.

The (social side of) disability is a function whose domain is a subset of the belittling identities. This family of functions shows itself off like a game of mirrors. For example, study and research on migrants and refugees, which have facilitated their control, emulates a primitive form of religion featuring the Émile Durkheim and Albert Ellis's taboos of separating profane from sacred (Durkheim & Ellis, 1963, pp. 27, 70, 85–90, 136; Hatton, 2011, pp. 334–337). These taboos, in turn, re-presents the aliens against the nationals sifting.

Despite the fact that the unequal distribution of suffering in the world mostly relies on body differences, the philosophy of disability is not a widespread discipline on its own (Reynolds, 2018). For example, rather than reaching a common ground on disabled people's sexuality, we witness how disability echoes in sexuality forming a cacophony of prejudices (Shildrick, 2009, pp. 66–73). The self-perception of disability and the belief that others perceive one to have a disability are associated with the social and vocational traits (LoBianco & Sheppard-Jones, 2007). These vocational traits likely influence the educational level which influences, in turn, poverty. However, education is not inclusive in practice for many children and there is an evident lack of (Bakhshi et al., 2017): [6]

- Innovative information on relational aspects of learning that reflect inclusion
- Stronger understanding of the implications of early assessment

As it could be expected, poverty among disabled people has resulted in a problem with no solution. We have no definitions nor philosophical movements that see us through this bottleneck regardless of many good scattered ideas. Even the terms used in the disability models (disability, impairment, and handicap) are commonly used in a disaggregated way (Jones, 2001). Despite the UN *Convention on the Rights of Persons with Disabilities* (United Nations, 2006), neither the Leeds County Court, nor the Court of Appeal and, nor the Supreme Court in the UK gave a satisfactory response to the question of how to share a bus between buggies and wheelchairs (Pearson, 2018). Likewise, the International Classification of Functioning, Disability and Health (ICF) should be clearer when it comes to issues such as (Imrie, 2004):

- (Re)defining the nature of impairment
- Specifying the content of biopsychosocial theory
- The meaning and implications of universalisation as guiding principle of disability policies

In addition to this, the ICF would benefit from a deeper insight in cross-cultural perspectives of needs and ideas and their emancipating outcomes.

To put the conceptual quagmire of disability in perspective, I will present a little comparative. According to my search in Google Scholar 265 papers on "*philosophy of disability*" have been published since 2012 to 2019. Meanwhile, 3600 articles were published on "*philosophy of money*" in the same period -even the lust world of "*shrimp breeding*" reaches 374 hits!

This philosophical gap has a political translation. The UK integration policies for disabled consisted (unsurprisingly) in social cuts *balanced* with employment support allowance that have resulted in a greater risk of poverty among the disabled population (Esther & Gill, 2017, p. 173). I get no doubt that the 10 Downing Street hosts feel themselves closer to the philosophy of money than the philosophy of disability. Even far better policies than the UK ones, like the Australian RecruitAbility scheme, only obtained modest achievements while spreading over disabled people a triumphalism rhetoric (Mellifont, 2018). The identity in disability, as I said previously, is complex and identity politics may cause more problems than it solves. They are like a mirror in a theme park that distorts what they try to represent. Most people, even when they are culturally competent on disability, at times, also employ many terms and concepts that are so outdated that may be considered offensive (Agaronnik et al., 2019).

As I see it, is better to let the disability rights movements representing the diversity and voice of disabled people rather than trust in a group of persons that:

- Unlikely will feel the burden of disability thanks to their wide and wider financial buffer
- Have no incentive for having their knowledge on disability up-to-date

It's not an exclusionary approach but a beachhead to intersectionality. For example, if disability rights movements were sound institutions the best surgeons and political stakeholders would reply to a call for action to repair the injustice committed with people left disabled due to an intervention. It would be a technical democracy, a "*hybrid forum*" (Callon et al., 2009), leading to a greater accountability and to co-creating the future of surgery (Keune, 2018). Perhaps we get closer to a world where the politicians become as responsible as the surgeons when it comes to *cuts*... On the other hand, a booming disability rights movement should be aware of not to turn invisible the identities of those who live in a grey zone (Berne, 2015).

Let's see a non-representation utter case. Picture a group of disabled people who were formerly labelled as non-suitable for active citizenship but targeted then by a set of inclusion politics. First, a wannabe Enlightenment would be the norm, then the government paternalism would appear as soon as any member departed from the marked path and, finally, this paternalism would be widespread until the whole political experiment turns into an institutionalisation replica (Altermark, 2017). An historical reflection of it, that reaches our present 2019, is how the Black men in the US have been treated. Their politics of integration have ranged from looking them (down) as submissive and naïve beings in need of guidance to criminal and violent individuals requiring persecution and discipline (Brown, 2017). That's three centuries of legal furniture wobbling from paternalism to institutionalisation (including slavery).

In the same fashion as the politics of inclusion hidden paternalism, deinstitutionalization sometimes covers an exclusionary geography where one confinement replaces another. The former mental patients suffer and exemplify this social process within the psychiatric ghettos located in the inner cities of the US (Parr, 2011). To cap it all, time after time legal measures and policy frameworks lead to one-size-fits-all solutions (Espinal, 2018, pp. 26–27). These become the basis of exclusion for the disabled belonging to the unequal social/spatial landscapes away from the *ideal majority*: poor, immigrants, queer, transgender, sex workers, inmates and indigenous peoples. An "animal farm *philosophy appears to dominate the* debate over bioethics issues and the development of bioethics theories as they pertain to disabilities" (Wolbring, 2003). The current health discourse narrows down people with one or several chronic illnesses or disabilities to mere patients and gives rise to tensions when a meaningful life shows itself as whimsically far away (Teunissen et al., 2019). In this context, characteristics of the many disabilities have been labelled either as medical or as societal problems. It *cripples any disability rights movement* since many medical claims are allegedly to be solved improving social justice, which is an exasperating slow and uncertain process. Above-board: if we really want to fully trust in the equality and freedom of a disabled person there is only one way: the elimination of diseases and health handicaps that she suffers from. Here we must to stop the hearse and consider a couple of things:

- It's an unattainable goal today in 2023, regardless of the cutting-edge tech we rely upon
- It's hard to figure out an ethical dilemma from contemporary medicine about to be solved only by exact methods and medical procedures (Bilasová, 2011)

So, the main question remains: what society can ensure that human health is not realised at the expense of the health and lives of other people? The one that leaves no one aside. Due to the financial burden of disability or access to surgery, for example. Given that, the vision of ameliorating health handicaps and diseases like a product, quota or even privilege rather than an *inalienable* right might lead to the elimination of humanity in healing.

CONCLUSION

Confinement is a complex and multidimensional phenomenon that goes beyond physical confinement. It can be approached from different perspectives, including its historical origins, philosophical grounds, public health implications, biopolitics, phenomenology, and true intersection with disability, caretaking, and trauma. Confinement touches on the limitations of human creativity and the denial of our true nature. The COVID-19 pandemic has unveiled inequalities and corruption, which is the result of the lack of responsible and ethical governance, collaborative efforts, and a comprehensive understanding of the moral implications of confinement. This situation stresses the importance of wellness and beforehand coping strategies rather than political speeches.

REFERENCES

Achiume, E. T. (2013). Beyond Prejudice: Structural Xenophobic Discrimination against Refugees. *Georgetown Journal of International Law*, *45*, 323.

Agaronnik, N., Campbell, E. G., Ressalam, J., & Iezzoni, L. I. (2019). Exploring issues relating to disability cultural competence among practicing physicians. *Disability and Health Journal*, *12*(3), 403–410. Advance online publication. doi:10.1016/j.dhjo.2019.01.010 PMID:30765256

Ahmed, S. (2018). From Threat to Walking Corpse: Spatial Disruption and the Phenomenology of "Living under Drones.". *Theory & Event*, *21*(2), 382–410.

Albrecht, G. L., Devlieger, P., & van Hove, G. (2008). The experience of disability in plural societies. *Alter (Paris)*, *2*(1), 1–13. doi:10.1016/j.alter.2007.09.002

Altermark, N. (2017). *Citizenship Inclusion and Intellectual Disability: Biopolitics Post-Institutionalisation*. Routledge. doi:10.4324/9781315109947

Amos, W. E. (1974). The Philosophy of Corrections: Revisited. *Federal Probation*, *38*, 43.

Arrigo, B. A., & Polizzi, D. (2018). Symposium: The Laws of Captivity: Critical and Humanistic Inquiries: Introduction to the Special Issue: On the Laws of Captivity. *New Criminal Law Review*, *21*(4), 483–491. doi:10.1525/nclr.2018.21.4.483

Ataria, Y., & Neria, Y. (2013). Consciousness-Body-Time: How Do People Think Lacking Their Body? *Human Studies*, *36*(2), 159–178. doi:10.100710746-013-9263-3

Bakhshi, P., Babulal, G. M., & Trani, J.-F. (2017). Education of children with disabilities in New Delhi: When does exclusion occur? *PLoS One*, *12*(9), e0183885. doi:10.1371/journal.pone.0183885 PMID:28877203

Bauer, S. (2018a). *American Prison: A Reporter's Undercover Journey into the Business of Punishment*. Penguin.

Bauer, S. (2018b, September 19). The Straight Line From Slavery to Private Prisons. *Literary Hub*. https://lithub.com/the-straight-line-from-slavery-to-private-prisons/

Bauer, S. (2018c, September 25). *America's Shocking History of Private Prisons.* Time. https://time.com/5405158/the-true-history-of-americas-private-prison-industry/

Beckett, S. (1965). *Imagination Dead Imagine.* Calder & Boyars.

Berghs, M. (2015). Radicalising 'disability' in conflict and post-conflict situations. *Disability & Society, 30*(5), 743–758. doi:10.1080/09687599.2015.1052044

Berne, P. (2015, June 9). *Disability Justice—A working draft by Patty Berne.* Sins Invalid. https://www.sinsinvalid.org/blog/disability-justice-a-working-draft-by-patty-berne

Biagioli, V., Piredda, M., Alvaro, R., & de Marinis, M. G. (2017). The experiences of protective isolation in patients undergoing bone marrow or haematopoietic stem cell transplantation: Systematic review and metasynthesis. *European Journal of Cancer Care, 26*(5), e12461. Advance online publication. doi:10.1111/ecc.12461 PMID:26892767

Bichard, J.-A., Coleman, R., & Langdon, P. (2007). Does My Stigma Look Big in This? Considering Acceptability and Desirability in the Inclusive Design of Technology Products. In C. Stephanidis (Ed.), *Universal Acess in Human Computer Interaction. Coping with Diversity* (pp. 622–631). Springer Berlin Heidelberg. doi:10.1007/978-3-540-73279-2_69

Bilasová, V. (2011). Philosophical Issues in Medical Ethics in the Context of Bioethical Discourse. *Ethics & Bioethics (in Central Europe)*, 7.

Birbaumer, N., & Cohen, L. G. (2007). Brain–computer interfaces: Communication and restoration of movement in paralysis. *The Journal of Physiology, 579*(3), 621–636. doi:10.1113/jphysiol.2006.125633 PMID:17234696

Boldeanu, S. (2018). *Merging brain-computer interfaces and virtual reality: A neuroscientific exploration.* http://urn.kb.se/resolve?urn=urn:nbn:se:his:diva-15774

Bridges, G. S., Crutchfield, R. D., & Simpson, E. E. (1987). Crime, Social Structure and Criminal Punishment: White and Nonwhite Rates of Imprisonment. *Social Problems, 34*(4), 345–361. doi:10.2307/800812

Brown, S. (2017). *A Race Against the Odds: The Historical Influences on Modern-Day Racism and Structural Discrimination and the Impact on Black Men.* University of Hartford.

Callon, M., Lascoumes, P., & Barthe, Y. (2009). *Acting in an Uncertain World.* MIT Press.

Carr, W. (2004). Philosophy and Education. *Journal of Philosophy of Education, 38*(1), 55–73. doi:10.1111/j.0309-8249.2004.00363.x

Chaumet, G., Taillard, J., Sagaspe, P., Pagani, M., Dinges, D. F., Pavy-Le-Traon, A., Bareille, M.-P., Rascol, O., & Philip, P. (2009, February). *Confinement and Sleep Deprivation Effects on Propensity to Take Risks.* https://doi.org/info:doi/10.3357/ASEM.2366.2009

Crabb, M., Heron, M. J., Jones, R., Armstrong, M., Reid, H., & Wilson, A. (2019). *Developing accessible services: Understanding current knowledge and areas for future support.* https://openair.rgu.ac.uk/handle/10059/3288

De Cunzo, L. A. (2009). Exploring the Institution: Reform, Confinement, Social Change. In M. Hall & S. W. Silliman (Eds.), *Historical Archaeology*. John Wiley & Sons.

DeLanda, M. (2006). *A New Philosophy of Society: Assemblage Theory and Social Complexity*. A&C Black.

Delgado, M. L. (2018, June 22). Teaching Philosophy in Prison: An Experience of Freedom. *Proceedings of the XXIII World Congress of Philosophy*. 10.5840/wcp232018741709

Deligiorgi, K. (2018). *Autonomy*. The Oxford Handbook of Phenomenological Psychopathology. doi:10.1093/oxfordhb/9780198803157.013.42

Dixit, S., & Prasad, R. (2017). *Human Bond Communication: The Holy Grail of Holistic Communication and Immersive Experience*. John Wiley & Sons. doi:10.1002/9781119341451

Dumbill, P. (2019, June 4). *Health technology: Digital therapy by prescription*. https://www.browne-jacobson.com/health/training-and-resources/legal-updates/2019/06/health-technology-digital-therapy-by-prescription

Durkheim, É., & Ellis, A. (1963). *Incest*. L. Stuart.

El-Lahib, Y. (2018). Social Work at the Intersection of Disability and Displacement: Rethinking Our Role. *Journal of Progressive Human Services*, *0*(0), 1–20. doi:10.1080/10428232.2018.1531744

Espinal, S. A. (2018). *Enabling Geographies: Neurodivergence, Self-Authorship, and the Politics of Social Space* [PhD Thesis]. California Institute of Integral Studies.

Esther, D., & Gill, M. (2017). Poverty and Social Exclusion in the UK: Vol. 1: Volume 1 - The Nature and Extent of the Problem. Policy Press.

Everett, M. (2010). 9/11: The greatest lie ever told. *The Journal of Psychohistory*, *38*(2), 133.

Forman, M. (Director). (1975). *One Flew Over the Cuckoo's Nest*. https://www.imdb.com/title/tt0073486/

Forsyth, R., Colver, A., Alvanides, S., Woolley, M., & Lowe, M. (2007). Participation of young severely disabled children is influenced by their intrinsic impairments and environment. *Developmental Medicine and Child Neurology*, *49*(5), 345–349. doi:10.1111/j.1469-8749.2007.00345.x PMID:17489807

Gallagher, S. (2014). The cruel and unusual phenomenology of solitary confinement. *Frontiers in Psychology*, *5*. Advance online publication. doi:10.3389/fpsyg.2014.00585 PMID:24971072

Gelfand, E. D. (1983). *Imagination in Confinement: Women's Writings from French Prisons*. Cornell University Press.

Gladden, M. E. (2016a). *Enterprise Architecture for Neurocybernetically Augmented Organizational Systems: The Impact of Posthuman Neuroprosthetics on the Creation of Strategic, Structural, Functional, Technological, and Sociocultural Alignment* [PhD Thesis].

Gladden, M. E. (2016b). *Posthuman Management: Creating Effective Organizations in an Age of Social Robotics, Ubiquitous AI, Human Augmentation, and Virtual Worlds*. Defragmenter Media.

Glannon, W. (2013). *Brain, Body, and Mind: Neuroethics with a Human Face*. Oxford University Press.

Glaser, W. (1994). Commentary: Garry David, psychiatry, and the discourse of dangerousness. *Australian and New Zealand Journal of Criminology*, *27*(1), 46–49. doi:10.1177/000486589402700107

Goering, S., Klein, E., Dougherty, D. D., & Widge, A. S. (2017). Staying in the Loop: Relational Agency and Identity in Next-Generation DBS for Psychiatry. *AJOB Neuroscience*, *8*(2), 59–70. doi:10.1080/21507740.2017.1320320

Goshe, S. (2019). The lurking punitive threat: The philosophy of necessity and challenges for reform. *Theoretical Criminology*, *23*(1), 25–42. doi:10.1177/1362480617719450

Goumart, S.-I. (2019). Affective Politics and "Crisis": The Examples of HIV-Positive Women's Public Denouncement and of the Refugees' Confinement. In D. Olkowski & E. Pirovolakis (Eds.), *Deleuze and Guattari's Philosophy of Freedom: Freedom's Refrains*. Routledge.

Greig, D. N. (2002). *Neither Bad Nor Mad: The Competing Discourses of Psychiatry, Law, and Politics*. Jessica Kingsley Publishers.

Greig, D. N. (2006). Madness, badness, and evil. In *Forensic psychiatry* (pp. 135–152). Springer. doi:10.1007/978-1-59745-006-5_7

Hallward, P. (2006). *Out of this World: Deleuze and the Philosophy of Creation*. Verso.

Hatton, J. P. (2011). *How and why did MARS facilitate migration control? Understanding the implication of migration and refugee studies (MARS) with the restriction of human mobility by UK state agencies* [PhD Thesis]. Oxford University.

Henriksen, A.-K. (2017). Confined to Care: Girls' Gendered Vulnerabilities in Secure Institutions. *Gender & Society*, *31*(5), 677–698. doi:10.1177/0891243217726968

Holmes, D., Murray, S. J., & Knack, N. (2015). Experiencing Seclusion in a Forensic Psychiatric Setting: A Phenomenological Study. *Journal of Forensic Nursing*, *11*(4), 200–213. doi:10.1097/JFN.0000000000000088 PMID:26457901

Hughes, B. (2017). Impairment on the move: The disabled incomer and other invalidating intersections. *Disability & Society*, *32*(4), 467–482. doi:10.1080/09687599.2017.1298991

Imrie, R. (2004). Demystifying disability: A review of the International Classification of Functioning, Disability and Health. *Sociology of Health & Illness*, *26*(3), 287–305. doi:10.1111/j.1467-9566.2004.00391.x PMID:15043600

Jacobson, J., Heard, C., & Fair, H. (2017). *Prison: Evidence of its use and over-use from around the world*. Academic Press.

Janz, U. (2010, September 3). *Conference of the States Parties to the Convention on the Rights of Persons with Disabilities*. Academic Press.

Jones, R. B. (2001). Impairment, disability and handicap—Old fashioned concepts? *Journal of Medical Ethics*, *27*(6), 377–379. doi:10.1136/jme.27.6.377 PMID:11731599

Kaplan, S. A., & Tsytovich, V. N. (2013). *Plasma Astrophysics: International Series of Monographs in Natural Philosophy*. Elsevier.

Kellmeyer, P. (2018). Neurophilosophical and Ethical Aspects of Virtual Reality Therapy in Neurology and Psychiatry. *Cambridge Quarterly of Healthcare Ethics*, *27*(4), 610–627. doi:10.1017/S0963180118000129 PMID:30198465

Kerley, K. R. (2018). *Finding Freedom in Confinement: The Role of Religion in Prison Life*. ABC-CLIO.

Keune, J. D. (2018). Disability and the Contemporary Surgical Gestalt. *Disability Studies Quarterly*, *38*(1). Advance online publication. doi:10.18061/dsq.v38i1.5390

Kierkegaard, S. (1959). *A Kierkegaard anthology*. Modern Library.

Kreitmair, K. V., & Cho, M. K. (2017). The neuroethical future of wearable and mobile health technology. In J. Illes (Ed.), *Neuroethics: Anticipating the future*. Oxford University Press.

Kuehn, B. M. (2018). Virtual and Augmented Reality Put a Twist on Medical Education. *Journal of the American Medical Association*, *319*(8), 756–758. doi:10.1001/jama.2017.20800 PMID:29417140

Lázaro-Muñoz, G., McGuire, A. L., & Goodman, W. K. (2017). Should We Be Concerned About Preserving Agency and Personal Identity in Patients With Adaptive Deep Brain Stimulation Systems? *AJOB Neuroscience*, *8*(2), 73–75. doi:10.1080/21507740.2017.1320337 PMID:29238625

Lim, D. (2018). Rhetoric, Experimental Philosophy, and Irrelevance. *AJOB Neuroscience*, *9*(3), 160–162. doi:10.1080/21507740.2018.1496170

LoBianco, A. F., & Sheppard-Jones, K. (2007). Perceptions of Disability as Related to Medical and Social Factors. *Journal of Applied Social Psychology*, *37*(1), 1–13. doi:10.1111/j.0021-9029.2007.00143.x

Lozano Rodríguez, M. (2020a). How not to face coronavirus: The case of Spain. *Eubios Journal of Asian and International Bioethics; EJAIB*, *30*(4), 152.

Lozano Rodríguez, M. (2020b). *Lost in Securitization: Displaced people in the pandemic dawn of the New World Order*. 8th International Public Health & Bioethics Ambassador Conference (IPHA8). https://scholar.google.es/citations?view_op=view_citation&hl=en&user=mQnwblkAAAAJ&citation_for_view=mQnwblkAAAAJ:35r97b3x0nAC

Lozano Rodríguez, M. (2020c, September 3). Lies, bloody lies and hiddennes on COVID-19. *The Ninth International Public Health & Bioethics Ambassador Conference (IPHA9)*.

Lozano Rodríguez, M. (2023, January 25). How COVID-19 is Changing Corruption Forever: The Crisis of the State-led Crisis Management and Why You Are Going to Pay for It. In *Conference on Disaster Management: Global Pandemic and Its Impact*. Binghamton University - State University of New York.

Lozano Rodríguez, M. (2023). *5. Are Human Biological Differences Based on Inequalities in Wealth a Threat for Planetary Health?* Planetary Health and Bioethics.

Lozano Rodríguez, M., Sevimli, S., Dhastagir, S., & Jayachandran, S. (2023). COVID-19 Measures on Crime, Corruption, and Inequity of the Impacts of Control Measures Statement (World Emergency COVID19 Pandemic Ethics (WeCope) Committee). Eubios Ethics Institute.

Macias, T. (2013). 'Tortured bodies': The biopolitics of torture and truth in Chile. *International Journal of Human Rights*, *17*(1), 113–132. doi:10.1080/13642987.2012.701912

MahlerT.NissimN.ShalomE.GoldenbergI.HassmanG.MakoriA.KochavI.EloviciY.ShaharY. (2018). *Know Your Enemy: Characteristics of Cyber-Attacks on Medical Imaging Devices.* https://arxiv.org/abs/1801.05583

Mandela, N. (1961, June 26). *Nelson Mandela International Day.* https://www.un.org/en/events/mandeladay/struggle.shtml

Margalit, A. (2009). *The Ethics of Memory.* Harvard University Press.

McCallum, D. (1997). Mental health, criminality and the human sciences. In A. R. Petersen & R. Bunton (Eds.), *Foucault, Health and Medicine*. Psychology Press.

McDonald, W. (2017). Søren Kierkegaard. In E. N. Zalta (Ed.), *The Stanford Encyclopedia of Philosophy* (Winter 2017). Metaphysics Research Lab, Stanford University. https://plato.stanford.edu/archives/win2017/entries/kierkegaard/

McGowan, M. (2019, August 26). Racism study finds one in three school students are victims of discrimination. *The Guardian.* https://www.theguardian.com/australia-news/2019/aug/27/racism-study-finds-one-in-three-school-students-are-victims-of-discrimination

McMinn, B. G., & Hinton, L. (2000). Confined to barracks: The effects of indoor confinement on aggressive behavior among inpatients of an acute psychogeriatric unit. In American Journal of Alzheimer's Disease (Vol. 15, pp. 36–41). doi:10.1177/153331750001500106

Mellifont, D. (2018). Soft affirmative action lacking traction? An early qualitative exploration of the RecruitAbility Scheme performance within the Australian Public Service. *Australian Journal of Career Development, 27*(1), 20–28. doi:10.1177/1038416217745070

Meyerbröker, K., & Emmelkamp, P. M. G. (2010). Virtual reality exposure therapy in anxiety disorders: A systematic review of process-and-outcome studies. *Depression and Anxiety, 27*(10), 933–944. doi:10.1002/da.20734 PMID:20734361

Mindock, C. (2018, February 14). *Woman put in solitary confinement after claiming she was sexually abused by immigration guard.* The Independent. https://www.independent.co.uk/news/world/americas/laural-monterrosa-immigrant-sexual-abuse-assault-guard-texas-jail-prison-solitary-confinement-hut-to-a8211206.html

Mirza, M. (2017). *Disability-inclusive healthcare in humanitarian camps: Pushing the boundaries of disability studies and global health.* http://repositoriocdpd.net:8080/handle/123456789/1488

Moghimi, S., Kushki, A., Guerguerian, A. M., & Chau, T. (2013). A Review of EEG-Based Brain-Computer Interfaces as Access Pathways for Individuals with Severe Disabilities. *Assistive Technology, 25*(2), 99–110. doi:10.1080/10400435.2012.723298 PMID:23923692

Murray, S. J., & Holmes, D. (2015). Seclusive Space: Crisis Confinement and Behavior Modification in Canadian Forensic Psychiatric Settings. In K. Reiter & A. Koenig (Eds.), Extreme Punishment: Comparative Studies in Detention, Incarceration and Solitary Confinement (pp. 115–134). Palgrave Macmillan UK. doi:10.1057/9781137441157_7

National Research Council. (2014). Rising Incarceration Rates. In *The Growth of Incarceration in the United States*. Exploring Causes and Consequences. doi:10.17226/18613

Ostrowski, T. M., & Sikorska, I. (2015). *Health and Resilience*. Wydawnictwo UJ.

Ottosdottir, G., & Evans, R. (2016). Disability and chronic illness. In F. Thomas (Ed.), *Handbook of Migration and Health*. Edward Elgar. doi:10.4337/9781784714789.00028

Parker, N. (1991). The Garry David Case. *The Australian and New Zealand Journal of Psychiatry*, *25*(3), 371–374. doi:10.3109/00048679109062638 PMID:1958160

Parr, H. (2011). A brief spatial history of community and mental illness II: Community care. In *Mental Health and Social Space: Towards Inclusionary Geographies?* John Wiley & Sons.

Pearson, A. (2018). The Debate about Wheelchair Spaces on Buses Goes Round and Round: Access to Public Transport for People with Disabilities as a Human Right. *The Northern Ireland Legal Quarterly*, *69*(1), 1–17. doi:10.53386/nilq.v69i1.68

Piché, J., & Walby, K. (2018). Abolitionist reflections on museums of confinement. *Déviance et Société*, *42*(4), 643–662.

Pillsbury, B. L. K. (1978). "Doing the month": Confinement and convalescence of Chinese women after childbirth. *Social Science & Medicine. Part B, Medical Anthropology*, *12*, 11–22. doi:10.1016/0160-7987(78)90003-0 PMID:565536

Pisani, M., & Grech, S. (2017). *Disability and Forced Migration: Critical Intersectionalities*. http://repositoriocdpd.net:8080/handle/123456789/1505

Pitts, A. J. (2018). Examining Carceral Medicine through Critical Phenomenology. *International Journal of Feminist Approaches to Bioethics*, *11*(2), 14–35. Advance online publication. doi:10.3138/ijfab.2017.08.11

Reynolds, J. M. (2018). Dangerous Discourses of Disability, Subjectivity and Sexuality by Margrit Shildrick. *International Journal of Feminist Approaches to Bioethics*, *11*(1), 162–167. Advance online publication. doi:10.3138/ijfab.11.1.162

Riva, G. (2005). Virtual Reality in Psychotherapy [Review]. *Cyberpsychology & Behavior*, *8*(3), 220–230. doi:10.1089/cpb.2005.8.220 PMID:15971972

Rogatchevski, A. (2014). Non-Totalitarian Imprisonment under Western and Eastern Eyes: Lord Archer, Eduard Limonov, and Theories of Human Motivation. *Punishment as a Crime?*, 163.

Rosenberg (Director). (1980). *Brubaker*. https://www.imdb.com/title/tt0080474/

Rowan, A., & Kartal, T. (2018). Dog Population & Dog Sheltering Trends in the United States of America. *Animals (Basel)*, *8*(5), 68. Advance online publication. doi:10.3390/ani8050068 PMID:29710771

Salkeld, D. (1994). *Madness and Drama in the Age of Shakespeare*. Manchester University Press.

Shildrick, M. (2009). *Dangerous discourses of disability, subjectivity and sexuality*. Palgrave Macmillan. doi:10.1057/9780230244641

Shuman, A., & Bohmer, C. (2016). The Uncomfortable Meeting Grounds of Different Vulnerabilities: Disability and the Political Asylum Process. *Feminist Formations*, *28*(1), 121–145. doi:10.1353/ff.2016.0018

Silva, W. H. S., Lopes, G. L. B., Yano, K. M., Tavares, N. S. A., Rego, I. A. O., & Cavalcanti, F. A. da C. (2015). Efecto de un programa de rehabilitación utilizando la realidad virtual para el equilibrio y la funcionalidad de los pacientes con accidente cerebrovascular crónico. *Motriz*, *21*(3), 237–243. doi:10.1590/S1980-65742015000300003

Skotnicki, A. (2019). *Conversion and the Rehabilitation of the Penal System: A Theological Rereading of Criminal Justice*. Oxford University Press. doi:10.1093/oso/9780190880835.001.0001

Soldatic, K., Meekosha, H., & Somers, K. (2012). Finding Ernesto: Temporary Labour Migration and Disabled Children's Health. *International Journal of Population Research*, *2012*, 1–9. Advance online publication. doi:10.1155/2012/696753

Stanley, E., & Mihaere, R. (2018). Challenging Māori Imprisonment and Human Rights Ritualism. In E. Stanley (Ed.), *Human Rights and Incarceration: Critical Explorations* (pp. 79–102). Springer International Publishing. doi:10.1007/978-3-319-95399-1_4

Steadman, H. J., Monahan, J., Duffee, B., Hartstone, E., & Robbins, P. C. (1984). The impact of state mental hospital deinstitutionalization on United States prison populations, 1968-1978. *The Journal of Criminal Law & Criminology*, *75*(2), 474. doi:10.2307/1143164

Stein, M. A., & Lord, J. E. (2011). Enabling Refugee and IDP Law and Policy: Implications of the U.N. Convention on the Rights of Persons with Disabilities. *Arizona Journal of International and Comparative Law*, *28*, 401.

Talbot, M. (2008, February 25). *The Lost Children*. https://www.newyorker.com/magazine/2008/03/03/the-lost-children

Teunissen, T., Lindhout, P., Schipper, K., & Abma, T. (2019). Living a Meaningful Life and Taking Good Care of Oneself in Times of Illness: Highlighting a Dilemma. *International Journal of Feminist Approaches to Bioethics*, *12*(1), 44–60. doi:10.3138/ijfab.12.1.03

Tonry, M. (2019). *Proportionality Theory in Punishment Philosophy: Fated for the Dustbin of Otiosity?* (*SSRN* Scholarly Paper ID 3331136). Social Science Research Network. https://papers.ssrn.com/abstract=3331136

Torrey, E. F. (1998). *Out of the shadows: Confronting America's mental illness crisis*. Academic Press.

United Nations. (2006). *Convention on the Rights of Persons with Disabilities*. https://www.un.org/disabilities/documents/convention/convoptprot-e.pdf

Vanderstraeten, G. (2009). Dreaming of virtual reality and rehabilitation technology in the future: Virtual reality, robotics and telerehabilitation. *Rehabilitation in the Future*, *8*, 120–123.

Wilde, O. (2007). *The Collected Works of Oscar Wilde: The Complete Plays Poems and Stories Including the Picture of Dorian Gray and De Profundis*. Wordsworth Editions.

Wolbring, G. (2003). Disability Rights Approach Toward Bioethics? *Journal of Disability Policy Studies*, *14*(3), 174–180. doi:10.1177/10442073030140030701

KEY TERMS AND DEFINITIONS

Accountability: Being responsible and answerable for actions and decisions; a necessary trait for personal growth and societal progress.

Colonial Institution: Organisation imposed from an industrialised metropolitan elite upon a materially underdeveloped region.

Confinement: Being trapped or restricted within certain limits or boundaries, unable to freely move or engage with the world, regardless of mental health or well-being deterioration.

Disability: A condition that hinders a person's joy of her own body, mind and social engagement that, in practice, turns worse due to social barriers.

Humanitarian Crises: Widespread suffering caused by natural disasters, conflicts, or emergencies, affecting access to basic needs and rights.

Phenomenology: The study of human consciousness and how individuals perceive and make sense of the world around them.

Philosophy of Necessity: The belief that punishment is for common good and forgiveness risks social order.

ENDNOTES

1. https://www.pppl.gov/about/history/timeline
2. https://encyclopediaofarkansas.net/entries/terrell-don-hutto-12346/
3. Screened the first time only five years after other masterpiece on confinement One Flew Over the Cuckoo's Nest (Forman, 1975).
4. http://www.correctionscorp.com/our-history
5. http://www.aca.org/aca_prod_imis/ACA_Member/AboutUs/ER_Cass_Award_Winners.aspx
6. The Sustainable Development Goal 4, quality of education defined as equity and inclusion alongside traditional learning outcomes: https://sdgs.un.org/goals/goal4.

Chapter 16
Meditations on the Intrinsic Value of Life and the Present and Future Nature of Death and Dying:
Worthy v Lost I

ABSTRACT

Our legacy of scientific nihilism raises important ethical questions about the use of neural technologies to prolong life and the creation of conscious machines. Bioethics of displacement recognizes the interconnectedness of individuals in their social, cultural, and historical contexts, so it's ready to delve into the implications of end-of-life care and the nature of death in a post-human society, tracing the common roots of philosophy, psychology, art, linguistics, and neural nets. To make things harder, people in power have reduced others to mere things across history. So, the author inquires into the psychological factors that allowed people to participate in atrocities, such as cognitive dissonance, psychological inflexibility, desire for revenge, sense of moral superiority, and obedience to authority along with leaders' psychopathic traits with a knack for Gehenna politics. Ultimately, the author reflects and ponders the complex issues surrounding death, telling the dead from the living and the creation of conscious machines to build a better future for ourselves and our creations.

INTRODUCTION

This chapter examines the complex and intertwined themes of scientific nihilism, bioethics, neural technologies, the nature of death, and the rise of conscious machines as well as the ideas advanced in the last chapter on the philosophical and biopolitical implications of confinement. Between both chapters, we see a continuation of the exploration of power dynamics, the dehumanisation of individuals, and the societal impact of certain ideologies and institutions. As we navigate through the ideas presented

DOI: 10.4018/978-1-6684-4808-3.ch016

in this chapter, we reflect on the interconnectedness of these topics and the need to consider the ethical implications they have for our understanding of life, death, and human dignity.

WE CAN'T GET RID OF OUR SCIENTIFIC NIHILISTIC LEGACY

The first modern realistic *sci-fi* novel, Sinclair Lewis' *Arrowsmith* (Lewis, 1998), originally published in 1926, was actually a romance that today echoes in bioethics. The protagonist's loving attitude gets twister and twister before realising that his overrational Lacanian Cupid was not to protect life but to control it. A former collaborator of Lewis, Paul de Kruif, openly talked about his own work experiences again in 1962:

I was a nihilist, period. For me, the world was too full of people and animals. And having no spark of reverence for all life, I had no ethics. (Kruif, 1962, p. 39)

Although *Arrowsmith* was awarded a Pulitzer and Lewis became the first American to gain a Nobel Prize of Literature four years later, I think Paul de Kruif is more interesting here. Before participating in the famous novel, he was a medical researcher in the Rockefeller Foundation fired by participating (anonymously!) in the book titled *Civilization* originally published and edited by Harold Stearns in 1922. In this book, de Kruif described the scientific approach to disease prevention and treatment as a "*mélange of religious ritual, more or less accurate folk-lore, and commercial cunning*." (Stearns, 1922, p. 455) The critique was *not* directed against the Rockefeller Foundation but its director Simon Flexner kicked out de Kruif the very same day he found who wrote it. De Kruif's former boss was bitterly disappointed by the attitude of an important member of a "*big, happy, harmonious family*" (Henig, 2011). The neuroticism of truly bright scientists like Flexner himself and his successor the Nobel in Medicine Herbert Spencer Grasser, I mean physicians whose research has contributed to save numerous lives, also paved the road for the Rockefeller Foundation participation in the Guatemalan Tuskegee twenty four years later. The victims are still looking for justice in a blatant case of profit over the patients (Bollinger 2019). [1]

Figure 1. "Worse Than Tuskegee." Left: The syphilitic chancre of a female psychiatric subject who was exposed to syphilis twice and after some treatment passed away. Right: Lawsuit plaintiff Frederico Ramos, age 91, sits on the porch of his son's house in San Agustín Acasaguastlán, Guatemala, Sept. 3, 2016.
Source: Retrieved from a Slate article titled "Worse Than Tuskegee" (Subramanian, 2017)

The Guatemalan syphilis experiment was sponsored by the Public Health Service Venereal Disease Research Laboratory in Washington, the pharmaceutical company Bristol-Myers Squibb, Johns Hopkins University and the Rockefeller Foundation. The *avoid do no harm* principle was painstakingly applied (Rodriguez & García, 2013).

The way that Guatemalan persons were infected with syphilis and gonorrhoea (with *bad blood*) is described in the *Slate* aforementioned article, I'll only make a glimpse into it. The original procedure was infecting prostitutes (rather than looking for infected ones) and bedding them with recruits to get them infected. To infect the prostitutes a contaminated cotton was introduced rudely through their genitalia, allegedly without consent. One of the women laid with eight soldiers in seventy one minutes. The doctors like John F. Mahoney and John Charles Cutler created their version of the venomous woman-monster from the WW2 poster to chase it (Presidential Commission for the Study of Bioethical Issues, 2011).

Anyhow, the hunt ended soon and the researchers, frustrated by the low rate of contagion, put their hands to work to fix it. Dr. Cutler and his colleagues changed the procedure and started to scratch and needle penises and eyes with syphilis and gonorrhoea infected instruments. It cannot be explained by mere cruelty or sadism, perhaps providing those underserved men with free services was too much socialist for the US researchers, famous for rejecting public money. Very cultured persons from a most impressive democracy behaved like Nazis as the chance of getting profit arose. But consider it, Germany was also filled to the brim with geniuses in almost any discipline. Auschwitz cannot be explained with Mengele, around Mengele and less even by Mengele himself. Likewise, the Guatemalan Tuskegee cannot be explained if we limited ourselves to Cutler, Mahoney and their subordinates. These doctors, like many researchers to lesser or greater degrees, are relatively squeezed in turn. The rush for entailing economic and military added value is what favours this kind of unethical behaviour; the Guatemalan crimes happened in the coldest phase of the Cold War. Some groups in the US, even in major cities like Washington, were also stripped from their basic rights. For example, the homosexuals were being chased amidst a true social paranoia in spite of the fact that the syphilis and gonorrhoea almost have passed from the scene thanks to penicillin (Johnson, 2009, pp. 40–51).

As exposes Hannah Arendt in *Eichmann in Jerusalem*, the Dr. Otto Bradfisch who presided over the murder of fifteen thousand people was "*inwardly opposed*" to the Holocaust unlike the *true Nazis* (Arendt, 1963, p. 113). Likely, the Drs. Mahoney and Cutler saw their malpractices as a lesser evil in a world full of foes and nukes, but there is no doubt of their full knowledge of the wickedness of their own acts. In contrast to the Nazis, their experiments were illegal according to US law and they knew it, like the disclosed correspondence witnesses (Parran, 1942).[2]

Some may think that things have completely changed since then thanks to decolonized or anti imperialist politics, but, ironically, the recognition of medical treatment as part of cultural behaviour is one of the principles that grounds the dark side of medical tourism today (Cooper & Hieda, 2017). You can bet that even cultural recognition can be twisted in order to lead to the exploitation of vulnerable populations and the violation of ethical principles in healthcare. All cultures have their own profiteers, closer to other opportunists than their own kind. As a result, recognition of medical treatment as part of cultural behaviour can be problematic when it leads to void exotism or even cultural appropriation of traditional medical practices, and when it fails to respect the autonomy and dignity of patients (Botterill et al., 2013). Without a strong civil society able to ensure that ethical principles and patient welfare are not compromised in the pursuit of medical tourism, the cultural context of healthcare practices can get quickly stained and muddy. In Tuskegee as in the resort SPA.

When it comes to crimes of science, both research and researchers are reified and removed from their real settings, never shown as active in the hands of powerful social stakeholders (Miedema, 2012, pp. 24, 71–74). For this reason, their kind of perverted scientific research is so similar: we only grasp the formal structure. A poststructuralist view of the human as subject might help to restore patients and physicians to their full status as persons from their current restricted status (Marta, 1998). Understanding responsibility as bearing consequences, and not only accountability, will lead to an idea of informed consent integrated in a wider frame than pure rationality. A Lacanian perspective could help to reflectively transcend *scientism*, blind to human wholeness (Gillett, 2015). We find a little drawback here. For Lacan, the ultimate reality is not caring but irrational (Czermak & Frignet, 1996). He sees the progress of science as the progress of the symbolic order at the expense of Nature (Zwart, 2017a). It doesn't sound like a do no harm discourse so what about a waving irrationality with plenty of chaos rather than a *pathetic* one?

DEATH'S GREY ZONE

Grey Today

Bioethics has *displaced* the moral compass to the wholeness of the lived-body (Carson & Burns, 2006, p. xii). It is a way of thinking that considers the whole person, including their body and their experiences, when making important decisions. Although Bioethics of Displacement, as love of belonging, has been introduced to shift the focus of bioethics from individual autonomy towards recognizing the interconnectedness of individuals within their social, cultural, and historical contexts, it includes considering the wholeness of their lived-bodies. It encompasses more than just the physical body, but also the emotional, cultural, and spiritual dimensions.

Anyhow, Bioethics as a profession also has its own chimaeras given that it might lead us to a somewhat self-regulating bioscience. Such a *gravitas* can be quite worrisome when it comes to synthetic biology ethical judgments from bioethicists whose moral and normative theories are not fully transparent, especially when it comes to discern what is alive or not (Arnason, 2017). In *Global Bioethics: Issues of Conscience for the Twenty-First Century*, it is argued that bioethics must be guided by a commitment to justice and fairness, which requires an understanding of the social, economic, and political factors that contribute to health disparities and inequities (Green et al., 2008). This approach acknowledges the importance of recognizing the diversity of experiences and perspectives that exist within different cultural and historical contexts, which can help ensure that bioethical decision-making is not influenced by a single dominant perspective.

It will be helpful for sure to have a plural view on life and death but to tell alive from death, technical progress can be a game change. Sadly, Bioethics is boring to the tears when it comes to technoculture (Franklin & Lock, 2003, p. 298). Usually appears when the action is over to talk about not doing something. In any case, the scientific philosophemes, its research core tenets so to say, have felt the sway of the new ICTs. Namely, "*the technification of the object and the anonymisation or collectivisation of the subject*" (Zwart, 2017b). The implication for bioethics of the major shifts in science that resulted from the new disruptive techs is the surging of optimization and control detrimental to the old focus in acquiring knowledge. For example, Alzheimer Disease as well as other previously untreatable neurodegenerative disorders can be diagnosed and predicted as never before, but we shouldn't miss the potential violations of rights when testing (Have & Purtilo, 2004, p. 318). By the same token, the public health reliance on big data analysis of health information from a zillion patients and, above all, the incentives and costs in further analysis has raised new bioethical questions on the quality and complexity of healthcare as well as the relations with the patients (Sullivan et al., 2018). The question of determining the line between life and death becomes more complex as technology offers new means of preserving life, making it essential to carefully consider the implications of these innovations (Shemie, 2014; Thomson, 2019). Bioethics of Displacement must include an understanding of the potential violations of human rights that could arise from these developments, particularly in the context of brain death and the use of neural technologies. It leads to the ultimate bioethical grey challenge: the brain itself. The difficulty of discerning between normative and metaphysical questions can blind us to an efficient bioethical approach to questions involving neurons (Buller, 2006, p. 59). Neurons and death.

Brain death is a term used to describe the irreversible loss of function in the brain, including the brainstem, which controls many essential functions like breathing and heartbeat. The diagnosis of brain death is critical for determining when someone is truly dead, especially in cases where life support systems are being used. Traditionally, the diagnosis of brain death has been made using physical examinations, but as neural technologies have advanced, new methods have been developed that are more accurate and objective. However, these advancements in technology also raise ethical questions about the nature of death and how we define it. A technology for diagnosing brain death is electroencephalography (EEG). EEG measures the electrical activity in the brain and can help determine if there is any brain function remaining. If there is no electrical activity in the brain, it is a strong indicator that the brain has stopped functioning and the patient is clinically brain dead. EEG is a non-invasive, low-risk, and relatively inexpensive method for diagnosing brain death, which can even improve with some drugs (Meinitzer et al., 2008). Another promising technology is functional magnetic resonance imaging (fMRI), which measures changes in blood flow to different areas of the brain. By measuring changes in blood flow, fMRI can help identify areas of the brain that are still active and may indicate some level of brain function. However, fMRI is not yet widely used in the diagnosis of brain death, as it is more expensive and difficult to perform than EEG (Schiff & Fins, 2016). In addition to these non-invasive methods, there are also more invasive methods for diagnosing brain death, such as transcranial Doppler ultrasound (TCD) (Petty et al., 1990). TCD measures blood flow in the major arteries of the brain, and can help determine if there is any brain function remaining. TCD, just like EEG, is a relatively low-risk and inexpensive method, but it requires specialised training and equipment. While these technologies can provide more objective measures for diagnosing brain death, they also raise ethical questions about the nature of death and the limits of medical intervention. Some argue that these technologies extend the boundaries of life, blurring the lines between life and death (Cerulo & Ruane, 1997).

Furthermore, neural technologies are also being used to prolong the lives of patients who are brain dead. For example, extracorporeal membrane oxygenation (ECMO) is a technology that can maintain blood flow and oxygenation to vital organs in the body, even if the brain has stopped functioning (Migliaccio et al., 2013). ECMO can keep a patient's organs viable for transplantation and can also give family members more time to make decisions about end-of-life care. However, the use of ECMO to maintain organ function in a brain dead patient may be a violation of the dead donor rule, which requires that organ donation only be performed after the patient has been declared dead. In conclusion, neural technologies have greatly improved our ability to diagnose brain death and provide more objective measures for determining when someone is truly dead. However, these technologies also raise important ethical questions about the nature of death and the limits of medical intervention.

Grey Tomorrow

As technology advances, the idea of creating conscious machines or uploading our minds to computers becomes more realistic. But this raises philosophical questions about the nature of the mind and its relationship to the body. Can a mechanistic world view truly achieve spiritual machines, or are there inherent philosophical difficulties that stand in the way? The post-human society we are evolving towards may include enhanced humans, hybrid humans, and artificial intelligences. As an artist in digital media, I believe we must create a new aesthetic discourse emerging from networked culture. But this requires being willing to create art that may not be immediately recognized as such. To build this foundation, we must trace the common roots of philosophy, psychology, art, linguistics, and neural nets.

Philip K. Dick's Blade Runner explored the anxieties at the heart of our quest to overcome our mortality (Dick, 1982). The final scene, where the dying android spares Deckard's life after feeling empathy for him, highlights the shared experience of fear of death between humans and machines. It's better even with Rutger Hauer's magistral improvisation:

Figure 2. Rutger Hauer's tears in rain monologue, pic from Ridley Scott's Blade Runner
Source: Scott (1982)

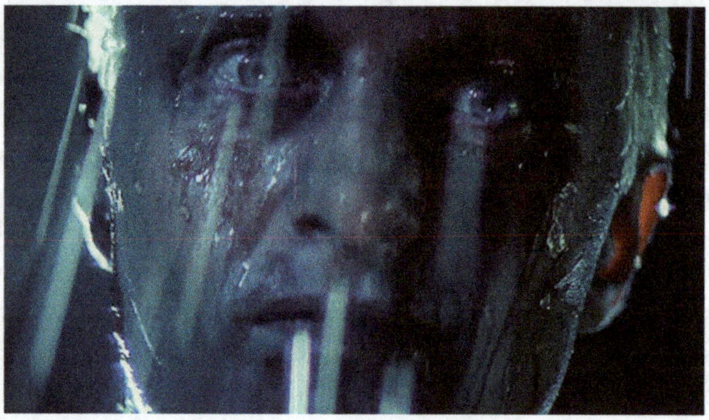

Perhaps empathy is the key to understanding what it means to be human, and it may also be the key to understanding conscious machines. But merging with machines creates a ponderous dilemma. We cannot shut off our computers without bringing human society to a violent end. We have already begun to merge with our machines, and turning them off is akin to turning ourselves off. Just like during the Romantic period, we have elevated a material form to spiritual power, yet sank the boundless into a body (E. G. Wilson, 2006). Sadly, human History is littered with normal people participating in the most heinous crimes. I fear that the degree of abject experimentation that a few scientists could do on behalf of a powerful leader against the sprouting first spiritual machines may decide the future of Humankind.

Ultimately, the death of a spiritual machine, a sentient robot, is a worthy philosophical and ethical question. As we continue to merge with our machines and push the boundaries of artificial intelligence, we must also consider the implications of our actions on our own humanity. How much can we displace death away before losing humanity? Similarly, we face a rising uncertainty in creating spiritual machines -will we be able to elevate machines to consciousness without diminishing our own humanity?

Only through reflection and consideration can we hope to navigate these complex issues and build a better future for ourselves and our creations. It is essential that we consider the implications they have for end-of-life care and for our understanding of what it means to be alive. As new embodiments and agencies appear the meaning of death also will be blurrier and more distant. And we need to be ready for it because it's better safe than sorry.

NEAR TO DEATH EXPERIENCES

About the end of life, a significant share of persons who are resuscitated after nearly dying have an out-of-body experience accompanied by a variety of perceptions such as travelling through a tunnel of light, feeling loved, or being received by deceased loved ones. So high as 11.1% in the case of cardiac arrests (Parnia et al., 2001). Near to death experiences (NDEs) are reported by patients of all ages, including children. In a 2019 study on NDErs and NDE-likes led by Helena Cassol, the Greyson NDE and Centrality Event scales' results suggested that the self-defining aspect of the experience might be related to its phenomenological content rather than its circumstances of occurrence (Cassol et al., 2019).

In the same vein, during a 2018 longitudinal study on NDEs led by William Van Gordon with the participation of Buddhist advanced meditators, it was found plenty of non-worldly encounters and detachment spiritual experiences (Van Gordon et al., 2018). Someone may argue that this sort of monks were unglued and it's true that the fantasy proneness has an important weight in the NDEs-like, however, a common NDE is independent from this personality trait (Martial et al., 2018). Someone may argue, too, that we can never reach nonlinguistic pure experience (Quine, 1951; Sellars, 1974), but it can be possible if this experience *always has been right there*, at the end of the day, the ineffability is the true mark of mysticism. From this point of view, the mystics around NDEs make sense of some perimortal imaginative experiences that extend to relatives and caregivers (Roser, 2017). Furthermore, when Marie Tonnard and colleagues tested the NDEs memories' vividness, it was superior to imagined and even real ones (Thonnard et al., 2013). These results were shadowed by a high-degree of subjectivity in tests, but confirmed the next year by other experiment led by Arianna Palmieri that included EEG and highlighted the role of slow-waves and its relation with shamanic states (Hartman & Zimberoff, 2002; Palmieri et al., 2014). Furthermore, including and excluding hypnosis, the NDEs have proven to be, at least, so complex and detailed like the common real memories (Lavazza, 2016, pp. 140–142).

It could imply that during death the personal identity disintegrates from the self, in other words, *death is the moment when we* utterly *stop reflecting about ourselves*. I suggest a second order definition of death inspired by the Filipino language concept of *kapwa* or shared inner identity. In this case, the broad definition of death is the moment when the others utterly stop reflecting about us, when you're not considered to be *sharing*.

It led me to elucidate that the parts of the brain needed for *belonging are* also those involved in *sharing ourselves*, it's having a life. The brain parts involved in self-identification, self-evaluation and self-implication shouldn't be enhanced to a level away from human nature since the way the candidate will establish bounds and reflect about herself would be understandable by the rest of people and her life and death will become astonishingly risky from a bioethical point of view. Those cerebral structures / regions are:

- Brainstem, thalamus, and posteromedial cortices, insular cortex, the anterior cingulate cortex, and the medial prefrontal cortex (Philippi et al., 2012; Riehl, 2012)
- Temporal lobe, cuneus, and the dorsolateral prefrontal cortex (Tan et al., 2015)
- Hippocampus, right supramarginal gyrus, temporoparietal junction, inferior frontal gyrus (El Haj et al., 2015; Padmanaban, 2017; Silani et al., 2013)

By the way, impairment beyond healing in all thirteen cerebral regions (it implies an utterly disintegration of the self) might be considered as an alternative cerebral death checklist in further research. It's important to (un)determine when a transplantation from a living donor is ethical or perhaps concede that, in this case, "*ethics and availability should not be included in the criteria for death*" (Fortunato, 2013). In the same manner, the broad definition of death is linked with Erika Borgstrom's idea of social death as an ontological deficiency that makes you barely human (Borgstrom, 2015; Emmerich, 2017). Death, social death, near-death and alike draws the centre of an always active discussion in bioethics.

Meditating on displacement is linked with thinking about death. A place where the ones who move never come back. So, Bodi didn't stop at death gates as usual in bioethics. We chased shambling walking dead roaming across the pages on moral improvement. We have now chased ghosts in the death blue screen. Now is the time to descend even further.

The ultimate displacement is labelling an alive human being as death or inert.

DEATH BY GEHENNA

Throughout history, people in positions of power have reduced other human beings to mere things, many times to break them better. The reasons behind such behaviour are complex, but they often involve a struggle for identity and belonging. Identity is a crucial component of our self-concept, and cultural changes can cause individuals to feel a loss of identity, leading them to seek out group affiliations as a means of restoring it (Baumeister, 1986, p. 88). Likewise, cognitive dissonance can also play a role in the reduction of humans to things. Festinger's theory of cognitive dissonance explains how individuals seek to reduce the discomfort caused by conflicting beliefs or actions, often by justifying their behaviour in ways that align with their existing beliefs (Festinger, 1962, p. 237). This can lead people in power to view those they oppress as somehow deserving of their treatment, or to dehumanise them in order to justify their actions.

The case of Nazi Germany serves as a prime example of the diminishing of humans to things. Adolf Hitler rose to power led to the genocide of millions of Jews and other groups deemed undesirable by the Nazi regime (Kershaw, 1999, 2000). Some of the contributors to the gruesomest scenes of Nazism were well-educated physicians without a criminal record. Robert Jay Lifton's book, *The Nazi Doctors*, explores the role of medical professionals in these atrocities, highlighting the psychological factors that allowed them to participate in the killing of fellow human beings. One of the psychological factors that led to this behaviour was psychological inflexibility. The physicians involved in the Nazi medical experiments had a rigid belief system and were unwilling to question the moral and ethical implications of their actions (Lifton, 1988, p. 25). They saw the prisoners as subhuman and believed that their suffering was necessary to reach a greater good. This mindset led to a lack of empathy and a disregard for the physical and emotional pain that the prisoners endured during the experiments. Lifton also describes how the physicians were able to rationalise their behaviour through a process of "doubling" separating their professional duties from their personal values and beliefs, allowing them to carry out their work without feeling guilty or conflicted:

By means of doubling, Nazi doctors made a Faustian choice for evil: in the process of doubling, in fact, lies an overall key to human evil. (Lifton, 1988, p. 424)

But Nazism, just like Mao's China and any other authoritarianism's peak, also relied on widespread obedience. Milgram's famous obedience experiments shed light on how ordinary people can be persuaded to engage in harmful behaviour towards others (Milgram, 1963). The study found that participants were willing to deliver electric shocks to an innocent person simply because an authority figure instructed them to do so. Zimbardo's book, "The Lucifer Effect," similarly explores how good people can turn evil, using the Stanford Prison Experiment as a case study (Zimbardo, 2011, pp. 3–22). In a similar way than Lifton's "doubling" and the way I described inauthenticity as the source of Gehenna in the fifth chapter, according to Zimbardo, evil happens when an individual disengages from their normal sense of morality, generally more humanitarian, in favour of an unempathetic one as circumstances harden.

But what makes obedience so dangerous is that the worst people rise to power attracted as moths to a flame (Klaas, 2021, Chapter 3). Namely, individuals with psychopathic traits were more likely to attain leadership positions in companies, possibly due to their ability to make difficult decisions and manipulate others (Mathieu et al., 2013, 2020). In the corp subset of political parties, leaders with psychopathic traits may be more effective at achieving political goals but less concerned with the well-being of their citizens, specially when meddling with ethnic groups (Ibaba & Okoye, 2015; Lilienfeld et al., 2014). The desire for interpersonal attachments and social inclusion is a fundamental human motivation, suggesting that people in power may seek to reduce others to things in order to maintain their own sense of belonging within a particular group because they present an excess of self-esteem (Baumeister & Leary, 1995; Leary & Baumeister, 2000). It's coherent with leaders' psychopathic narcissism and an homicidal and perverted love of belonging from the least souls. Nazism, just like the present-day North Korea regime, are not aberrations or mistakes but the truly evil perfection of the pyramidal modern state as brightly synthetized and described in Orwell's 1984 (Orwell, 2013). In *The Roots of Evil*, Ervin Staub argued that group violence and genocide have their roots in a complex web of historical, social, and psychological factors, including a desire for revenge and a sense of moral superiority (Klee et al., 1991, p. 76; Staub, 1989, pp. 19, 102).

Don't trade your humanity for blind justice but neither let the pity to blind you, those losers who participated in genocides were almost all of them more than happy with the bloodshed, more than happy of sharing their leader's pathological self-esteem.

In conclusion, the reduction of humans to mere things by people in power has been a recurring theme throughout history and will remain the same way as long as power distance is both the haven and the strings of the psychos at helm till Humankind gets finally wiped out. The works of Arendt, Baumeister, Festinger, Kershaw, Lifton, Milgram, Orwell, Staub, Zimbardo and now this *Bioethics of Displacement*, shed light on this issue and provide insight into how it can be prevented in the future.

AM I A NAZI? AM I ADOLF HITLER?

Spoiler: No.

Anyhow, today it's too easy to label another person as "Nazi" or "Hitler". Since Leo Strauss coined the *reductio ad Hitlerum* a few years after WW2 more than 70 years have passed, but it now pesters social media and even geopolitics (Laruelle, 2019). Considering that my values are distant from both overrepresented postmodern positions of wokism and neoliberalism, for sure I will be tagged as a tenant of the Hitler's Alpine headquarters (J. Wilson, 2014) a bazillion times. Let's discuss how Nazi I am in the chapter on death.

As seen in the section above, Lifton's *Nazi Doctors* highlights the dangers of psychological inflexibility and the importance of ethical and moral considerations in the medical profession. It serves as a reminder of the importance of questioning our beliefs and values, and being open to changing circumstances and perspectives, closer to the values in this book. Nazism is in direct opposition to the bioethical coordinates of prospective-relativist-communicational which emphasises respect for all individuals and the importance of considering multiple perspectives in ethical decision-making. In his shadow-written *Mein Kamp*, Hitler argues that there are superior and inferior races, and that it is the duty of the superior race to dominate and control the inferior races (Hitler, 1969, p. 259). This belief is in stark contrast to the bioethical paradigm of respecting the dignity and worth of all individuals. Furthermore, the Nazi medical experiments on concentration camp prisoners, as described in Lifton's aforementioned book from Nazi physicians' records and the Nuremberg Trial Archives demonstrate a complete disregard for the well-being and autonomy of individuals (Lifton, 1988, p. 368).[3] These experiments were carried out without the consent of the prisoners, and often resulted in severe pain, disability, or death. This type of experimentation is a clear violation of the principles of the prospective-relativist-communicational paradigm, which emphasises the importance of open and transparent communication, respect for autonomy, and consideration of the multiple perspectives involved in ethical decision-making.

In summary, the Nazi ideology and actions directly oppose the principles of the bioethical prospective-relativist-communicational coordinates, which emphasises respect for all individuals and the importance of considering multiple perspectives in ethical decision-making.

By the same token, Bodi's principles as exposed in the fifth chapter are diametrically against the main Nazi theoretical work, Alfred Rosenberg's "The Myth of Twentieth Century". The principle of Vulnerable First emphasises the intrinsic value and dignity of all human beings, regardless of their race or ethnicity. Rosenberg, on the other hand, believed in the inherent superiority of the Aryan race and the subjugation of other groups, which he deemed inferior (Rosenberg, 2004, pp. 416–435). Even if it would be the real case, this view is fundamentally opposed to the Vulnerable First approach, which recognizes the importance of respecting and protecting the rights and well-being of all vulnerable individuals. The second principle of Big Picture Thinking highlights the importance of recognizing the interconnectedness of various aspects of displacement and integrating insights from different disciplines to promote ethical decision-making. In contrast, Rosenberg's ideology was based on a narrow and exclusionary worldview that ignored the complexity and diversity of human experience (Rosenberg, 2004, pp. 171, 461). This narrow view is incompatible with the holistic and inclusive approach of the Vulnerable First principle. The third principle of Proactivity spots out proactive and preventive approaches that anticipate and address ethical issues before they escalate into conflicts or crises. This principle is fundamentally opposed to the reactive and violent approach of the Nazi regime which, permeated by Alfred Rosenberg's ideas and deeds, caused enormous harm and suffering to dozens of millions of people (Fleury, 2018;

Hirsch, 2020, pp. 128–140, 178–218; Sereny, 1983). Back to the Vulnerable First principle, it stresses the need for ethical reflection, dialogue, and collaboration among different stakeholders to promote ethical decision-making and prevent harm. The fourth principle of Scepticism accentuates the relevance of being open to recognizing the limitations and uncertainties of ethical reasoning, and being willing to reconsider and revise ethical decisions in light of new information or changing circumstances. In contrast, Rosenberg's ideology was based on rigid and dogmatic beliefs that were resistant to change or revision (Rosenberg, 2004, pp. 451–457). Back to the Vulnerable First principle again, it recognizes the complexity and uncertainty of ethical decision-making and makes a point of the importance of ongoing reflection and dialogue to promote ethical decision-making.

Besides the nimious detail that I would be gassed under Hitler's rule, *my* Bodi's principles and values are absolutely opposed to Nazism as well as any other ideology based on a narrow, exclusionary worldview that ignores the inherent dignity and worth of all human beings. Thus, no, I am not a Nazi.

CONCLUSION

Ethical concerns about neural technologies, brain death, and the potential consequences of merging with machines urges a shift from a narrow rational perspective to a holistic understanding of living bodies. A comprehensive and empathetic understanding of lived bodies is a reckoning force against Displacement. Quite the opposite, exclusionary ideologies are a main drive historical dehumanisation of individuals and preserving our humanity takes to reject them and uphold against well-being baron robbers. The foundation for ethical decision-making in the face of complex challenges surrounding death and the merging of human bodies with machines is the recognition of the inherent dignity and worth of all human beings.

REFERENCES

Arendt, H. (1963). *Eichmann in Jerusalem: A report on the banality of evil.* Viking Press.

Arnason, G. (2017). Synthetic Biology between Self-Regulation and Public Discourse: Ethical Issues and the Many Roles of the Ethicist. *Cambridge Quarterly of Healthcare Ethics, 26*(2), 246–256. doi:10.1017/S0963180116000840 PMID:28361722

Baumeister, R. F. (1986). *Identity: Cultural Change and the Struggle for Self.* Oxford University Press.

Baumeister, R. F., & Leary, M. R. (1995). The Need to Belong: Desire for Interpersonal Attachments as a Fundamental Human Motivation. *Psychological Bulletin, 117*(3), 497–529. doi:10.1037/0033-2909.117.3.497 PMID:7777651

Borgstrom, E. (2015). Social death in end-of-life care policy. *Contemporary Social Science, 10*(3), 272–283. doi:10.1080/21582041.2015.1109799

Botterill, D., Pennings, G., & Mainil, T. (2013). *Medical Tourism and Transnational Health Care.* Springer. doi:10.1057/9781137338495

Buller, T. (2006). Brains, lies and psychological explanations. In J. Illes (Ed.), *Neuroethics: Defining the Issues in Theory, Practice, and Policy*. Oxford University Press.

Cara, E. (2019, January 7). Bristol Myers-Squibb and Others Can't Dodge $1 Billion Lawsuit Over 1940s Syphilis Study, Judge Rules. *Gizmodo*. https://gizmodo.com/bristol-meyers-squibb-and-others-cant-dodge-1-billion-1831548144

Carson, R. A., & Burns, C. R. (2006). *Philosophy of Medicine and Bioethics: A Twenty-Year Retrospective and Critical Appraisal*. Springer Science & Business Media.

Cassol, H., D'Argembeau, A., Charland-Verville, V., Laureys, S., & Martial, C. (2019). Memories of near-death experiences: Are they self-defining? *Neuroscience of Consciousness, 2019*(1). doi:10.1093/nc/niz002

Cerulo, K. A., & Ruane, J. M. (1997). Death comes alive; technology and the re-conception of death. *Science as Culture, 6*(3), 444–466. doi:10.1080/09505439709526477

Cooper, M., & Hieda, M. (2017). The Dark Side of Medical Tourism?: End of Life Choice, Human Trafficking, and Organ Transplants. *Oncology: Breakthroughs in Research and Practice*, 203–215. doi:10.4018/978-1-5225-0549-5.ch006

Czermak, M., & Frignet, H. (1996). *Sur l'identité sexuelle: A propos du transsexualisme : actes des journées des 30 novembre et 1er décembre 1996, Centre hospitalier Sainte-Anne, Paris*. Association freudienne internationale.

Dick, P. K. (1982). *Blade Runner*. Ballantine Books.

El Haj, M., Antoine, P., & Kapogiannis, D. (2015). Similarity between remembering the past and imagining the future in Alzheimer's disease: Implication of episodic memory. *Neuropsychologia, 66*, 119–125. doi:10.1016/j.neuropsychologia.2014.11.015 PMID:25448861

Emmerich, N. (2017). Topics in death and dying. *QJM, 110*(1), 3–4. doi:10.1093/qjmed/hcw186 PMID:27770050

Festinger, L. (1962). *A Theory of Cognitive Dissonance*. Stanford University Press.

Fleury, J. S. (2018). *Adolf Hitler: Trial in Absentia in Nuremberg*. Xlibris Corporation.

Fortunato, J. (2013). "Irreversibility" and the Modern Understanding of Death. *Discussions, 9*(2). http://www.inquiriesjournal.com/articles/795/irreversibility-and-the-modern-understanding-of-death

Franklin, S., & Lock, M. M. (2003). *Remaking life & death: Toward an anthropology of the biosciences*. School of American Research Press.

Gillett, G. (2015). Culture, Truth, and Science After Lacan. *Journal of Bioethical Inquiry, 12*(4), 633–644. doi:10.100711673-015-9664-2 PMID:26615543

Green, R. M., Donovan, A., & Jauss, S. A. (2008). *Global Bioethics: Issues of Conscience for the Twenty-First Century*. OUP Oxford.

Hartman, D., & Zimberoff, D. (2002). Memory access to our earliest influences. *Journal of Heart Centered Therapies, 5*, 3–64.

Henig, R. M. (2011, May 5). The Life and Legacy of Paul de Kruif. *Alicia Patterson Foundation*. https://aliciapatterson.org/stories/life-and-legacy-paul-de-kruif

Hirsch, F. (2020). *Soviet Judgment at Nuremberg: A New History of the International Military Tribunal After World War II*. Oxford University Press.

Hitler, A. (1969). *Mein Kampf*. Hutchinson.

Ibaba, S. I., & Okoye, A. C. (2015). Psychopathic leadership and the challenge of terrorism in Nigeria. *Poultry Science Reviews*, *6*(2), 149–170.

Johnson, D. K. (2009). *The Lavender Scare: The Cold War Persecution of Gays and Lesbians in the Federal Government*. University of Chicago Press.

Kershaw, I. (1999). *Hitler, 1889-1936: Hubris*. W.W. Norton.

Kershaw, I. (2000). *Hitler, 1936-45: Nemesis*. W.W. Norton.

Klaas, B. (2021). *Corruptible: Who Gets Power and How It Changes Us*. Simon and Schuster.

Klee, E., Dressen, W., & Riess, V. (1991). *"The Good Old Days": The Holocaust as Seen by Its Perpetrators and Bystanders*. Konecky Konecky.

Kruif, P. D. (1962). *The sweeping wind: A memoir*. Harcourt, Brace & World.

Laruelle, M. (2019). Reductio ad Hitlerum as a new frame for political and geopolitical conflicts. In *Routledge handbook of character assassination and reputation management* (pp. 307–319). Routledge. doi:10.4324/9781315150178-23

Lavazza, A. (2016). *Frontiers in Neuroethics: Conceptual and Empirical Advancements*. Cambridge Scholars Publishing.

Leary, M. R., & Baumeister, R. F. (2000). The nature and function of self-esteem: Sociometer theory. In Advances in experimental social psychology (Vol. 32, pp. 1–62). Academic Press. doi:10.1016/S0065-2601(00)80003-9

Lewis, S. (1998). *Arrowsmith*. Penguin Group.

Lifton, R. J. (1988). *The Nazi Doctors: Medical Killing And The Psychology Of Genocide*. Basic Books.

Lilienfeld, S. O., Latzman, R. D., Watts, A. L., Smith, S. F., & Dutton, K. (2014). Correlates of psychopathic personality traits in everyday life: Results from a large community survey. *Frontiers in Psychology*, *5*. Advance online publication. doi:10.3389/fpsyg.2014.00740 PMID:25101019

Marta, J. (1998). Whose Consent is it Anyway? A Poststructuralist framing of the Person in Medical Decision-Making. *Theoretical Medicine and Bioethics*, *19*(4), 353–370. doi:10.1023/A:1009955914018 PMID:9865139

Martial, C., Cassol, H., Charland-Verville, V., Merckelbach, H., & Laureys, S. (2018). Fantasy Proneness Correlates With the Intensity of Near-Death Experience. *Frontiers in Psychiatry*, *9*, 190. Advance online publication. doi:10.3389/fpsyt.2018.00190 PMID:29930518

Mathieu, C., Babiak, P., & Hare, R. D. (2020). Psychopathy in the Workplace. In *The Wiley International Handbook on Psychopathic Disorders and the Law* (pp. 607–644). John Wiley & Sons, Ltd., doi:10.1002/9781119159322.ch27

Mathieu, C., Hare, R. D., Jones, D. N., Babiak, P., & Neumann, C. S. (2013). Factor structure of the B-Scan 360: A measure of corporate psychopathy. *Psychological Assessment*, *25*(1), 288–293. doi:10.1037/a0029262 PMID:22775409

Meinitzer, A., Kalcher, K., Gartner, G., Halwachs-Baumann, G., März, W., & Stettin, M. (2008). Drugs and brain death diagnostics: Determination of drugs capable of inducing EEG zero line. *Clinical Chemistry and Laboratory Medicine*, *46*(12), 1732–1738. doi:10.1515/CCLM.2008.335 PMID:18973460

Miedema, F. (2012). *Science 3.0: Real Science, Real Knowledge*. Amsterdam University Press.

Migliaccio, M. L., Zagli, G., Cianchi, G., Lazzeri, C., Bonizzoli, M., Cecchi, A., Anichini, V., Gensini, G. F., & Peris, A. (2013). Extracorporeal membrane oxygenation in brain-death organ and tissues donors: A single-centre experience. *British Journal of Anaesthesia*, *111*(4), 673–674. doi:10.1093/bja/aet323 PMID:24027145

Milgram, S. (1963). Behavioral Study of obedience. *Journal of Abnormal and Social Psychology*, *67*(4), 371–378. doi:10.1037/h0040525 PMID:14049516

Orwell, G. (2013). *1984*. HarperPerennial Classics.

Padmanaban, D. (2017, April 12). *Where Empathy Lives in the Brain*. The Cut. https://www.thecut.com/2017/04/where-empathy-lives-in-the-brain.html

Palmieri, A., Calvo, V., Kleinbub, J. R., Meconi, F., Marangoni, M., Barilaro, P., Broggio, A., Sambin, M., & Sessa, P. (2014). "Reality" of near-death-experience memories: Evidence from a psychodynamic and electrophysiological integrated study. *Frontiers in Human Neuroscience*, *8*. Advance online publication. doi:10.3389/fnhum.2014.00429 PMID:24994974

Parnia, S., Waller, D. G., Yeates, R., & Fenwick, P. (2001). A qualitative and quantitative study of the incidence, features and aetiology of near death experiences in cardiac arrest survivors. *Resuscitation*, *48*(2), 149–156. doi:10.1016/S0300-9572(00)00328-2 PMID:11426476

Parran, T. (1942, November 19). *Thomas Parran to Lewis H. Weed* (NARAII_0000284). PCSBI HSPI Archives.

Petty, G. W., Mohr, J. P., Pedley, T. A., Tatemichi, T. K., Lennihan, L., Duterte, D. I., & Sacco, R. L. (1990). The role of transcranial Doppler in confirming brain death: Sensitivity, specificity, and suggestions for performance and interpretation. *Neurology*, *40*(2), 300–300. doi:10.1212/WNL.40.2.300 PMID:2405294

Philippi, C. L., Feinstein, J. S., Khalsa, S. S., Damasio, A., Tranel, D., Landini, G., Williford, K., & Rudrauf, D. (2012). Preserved Self-Awareness following Extensive Bilateral Brain Damage to the Insula, Anterior Cingulate, and Medial Prefrontal Cortices. *PLoS One*, *7*(8), e38413. doi:10.1371/journal.pone.0038413 PMID:22927899

Presidential Commission for the Study of Bioethical Issues. (2011). *"Ethically Impossible" STD Research in Guatemala from 1946 to 1948*. BrainFeed Press. https://law.stanford.edu/wp-content/uploads/2011/09/EthicallyImpossible_PCSBI_110913.pdf

Quine, W. V. (1951). Main Trends in Recent Philosophy: Two Dogmas of Empiricism. *The Philosophical Review, 60*(1), 20–43. doi:10.2307/2181906

Riehl, J. (2012, August 22). *The roots of human self-awareness*. Iowa Now. https://now.uiowa.edu/2012/08/roots-human-self-awareness

Rodriguez, M. A., & García, R. (2013). First, Do No Harm: The US Sexually Transmitted Disease Experiments in Guatemala. *American Journal of Public Health, 103*(12), 2122–2126. doi:10.2105/AJPH.2013.301520 PMID:24134370

Rosenberg, A. (2004). *The Myth of the Twentieth Century: An Evaluation of the Spiritual-intellectual Confrontations of Our Age*. Historical Review Press.

Roser, T. (2017). Simon Peng-Keller: Sinnereignisse in Todesnähe. Traum- und Wachvisionen Sterbender und Nahtoderfahrungen im Horizont von Spiritual Care (Studies in Spiritual Care 1), Berlin / Boston (de Gruyter) 2017, 188 S., ISBN-13: 978–3110473162, Eur (D) 79,95. *International Journal of Practical Theology, 21*(2), 323–325. doi:10.1515/ijpt-2017-0034

Schiff, N. D., & Fins, J. J. (2016). Brain death and disorders of consciousness. *Current Biology, 26*(13), R572–R576. doi:10.1016/j.cub.2016.02.027 PMID:27404252

Scott, R. (Director). (1982, June 25). *Blade Runner* [Action, Drama, Sci-Fi]. The Ladd Company, Shaw Brothers, Warner Bros.

Sellars, W. (1974). Meaning as functional classification. *Synthese, 27*(3), 417–437. doi:10.1007/BF00484606

Sereny, G. (1983). *Into That Darkness: An Examination of Conscience*. Knopf Doubleday Publishing Group.

Shemie, S. D. (2014). Rethinking Mortality: Exploring the Boundaries between Life and Death. *Annals of the New York Academy of Sciences, 1330*(1), 101–104. doi:10.1111/nyas.12564 PMID:25351389

Silani, G., Lamm, C., Ruff, C. C., & Singer, T. (2013). Right supramarginal gyrus is crucial to overcome emotional egocentricity bias in social judgments. *The Journal of Neuroscience : The Official Journal of the Society for Neuroscience, 33*(39), 15466–15476. doi:10.1523/JNEUROSCI.1488-13.2013 PMID:24068815

Staub, E. (1989). *The Roots of Evil: The Origins of Genocide and Other Group Violence*. Cambridge University Press.

Stearns, H. (1922). *Civilization in the United States: An Inquiry by Thirty Americans*. Academic Press.

Subramanian, S. (2017, February 26). Worse than Tuskegee. *Slate Magazine*. https://www.slate.com/articles/health_and_science/cover_story/2017/02/guatemala_syphilis_experiments_worse_than_tuskegee.html

Sullivan, W. F., Heng, J., De Bono, C., Gleeson, G., Goulding, G., & Jamieson, C. (2018). Healing Relationships and Transformations in Health Care: IACB Consensus Statement on Ethical Discernment and Practical Recommendations. *The National Catholic Bioethics Quarterly, 18*(2), 319–327. doi:10.5840/ncbq201818229

Tan, S., Zhao, Y., Fan, F., Zou, Y., Jin, Z., Zen, Y., Zhu, X., Yang, F., Tan, Y., & Zhou, D. (2015). Brain Correlates of Self-Evaluation Deficits in Schizophrenia: A Combined Functional and Structural MRI Study. *PLoS One, 10*(9), e0138737. doi:10.1371/journal.pone.0138737 PMID:26406464

ten Have, H., & Purtilo, R. B. (2004). *Ethical Foundations of Palliative Care for Alzheimer Disease*. JHU Press.

Thomson, H. (2019, November 20). *Why the line between life and death is now more blurred than ever*. New Scientist. https://www.newscientist.com/article/mg24432570-300-why-the-line-between-life-and-death-is-now-more-blurred-than-ever/

Thonnard, M., Charland-Verville, V., Brédart, S., Dehon, H., Ledoux, D., Laureys, S., & Vanhaudenhuyse, A. (2013). Characteristics of Near-Death Experiences Memories as Compared to Real and Imagined Events Memories. *PLoS One, 8*(3), e57620. doi:10.1371/journal.pone.0057620 PMID:23544039

Van Gordon, W., Shonin, E., Dunn, T. J., Sheffield, D., Garcia-Campayo, J., & Griffiths, M. D. (2018). Meditation-Induced Near-Death Experiences: A 3-Year Longitudinal Study. *Mindfulness, 9*(6), 1794–1806. doi:10.100712671-018-0922-3 PMID:30524512

Vollmann, J., & Winau, R. (1996). Informed consent in human experimentation before the Nuremberg code. *BMJ (Clinical Research Ed.), 313*(7070), 1445–1447. doi:10.1136/bmj.313.7070.1445 PMID:8973233

Wilson, E. G. (2006). Matter and Spirit in the Age of Animal Magnetism. *Philosophy and Literature, 30*(2), 329–345. doi:10.1353/phl.2006.0042

Wilson, J. (2014). *Hitler's Alpine Headquarters*. Pen and Sword.

Zimbardo, P. (2011). *The Lucifer Effect: How Good People Turn Evil*. Random House.

Zwart, H. (2017a). Conceptual and Methodological Framework: Lacanian Psychoanalysis. In H. Zwart (Ed.), *Tales of Research Misconduct: A Lacanian Diagnostics of Integrity Challenges in Science Novels* (pp. 25–55). Springer International Publishing., doi:10.1007/978-3-319-65554-3_2

Zwart, H. (2017b). Introduction: An Oblique Perspective on Research Misconduct. In H. Zwart (Ed.), *Tales of Research Misconduct: A Lacanian Diagnostics of Integrity Challenges in Science Novels* (pp. 1–23). Springer International Publishing., doi:10.1007/978-3-319-65554-3_1

KEY TERMS AND DEFINITIONS

Cognitive Dissonance: Psychological discomfort when holding conflicting beliefs or attitudes, causing a need to justify or change behaviour.

Conscious Machines: Advanced artificial intelligence systems that possess self-awareness and exhibit qualities of human-like awareness, potentially passing the Garland Test.

Death: The moment when the others utterly stop reflecting about you, when you're not considered to be sharing.

Nihilism: The belief that life lacks inherent meaning, values, or purpose, often in a pessimistic way.

Philosophy: Epistemic displacement.

Post-Structuralism: A philosophy that critiques fixed structures of power and knowledge, emphasizing the changing nature of truth.

Social Stakeholders: Individuals or groups with an interest in an organisation's decisions, including employees, customers, communities, and government.

Spiritual Machine: AI that recognizes, feels, senses, or believes that there is something greater than itself that is cosmic or divine in nature.

Wholeness: Being complete and transcending conflicts and divisions.

ENDNOTES

1. In January of 2019, the U.S. District Judge Theodore Chuang has accepted the USD 1 billion lawsuit against the pharmaceutical company Bristol-Myers Squibb, Johns Hopkins University and the Rockefeller Foundation to proceed. This brave Maryland judge rejected the argument from the Supreme Court (Cara, 2019).
2. Dr. Parran, was the former President of the American Neisserian Medical Society and, in his turn, Albert Neisser (who discovered the gonorrhoea's causes), was also fined by the government of Prussia for infecting prostitutes with syphilis without their knowledge or consent in 1898 (Vollmann & Winau, 1996).
3. https://www.icj-cij.org/sites/default/files/documents/library-of-the-court-en.pdf

Chapter 17
The Ethics Behind Neuroethics, Neurobioethics, Sensory Enhancement, and Surrogate Realities:
Worthy v Lost II

ABSTRACT

Surrogate realities are technological devices that create one or more layers of reality. Those surrogate realities can have a lasting effect on individuals and even they could be more efficient than our natural perceptions. So, the author explores the convergence of philosophy, science, technology, and education in a world where natural sciences may include non-physical realities for the first time as well as the potential risks of neuroscience and its impact on culture. To do this, the author delves into the field of neuroethics, which investigates the ethical issues raised by new insights on the brain and how neuroscience can change our views of ethics itself. Next, the author highlights neurobioethics' potential for melting the boundary between therapy and better human performance. Finally, the author shows how surrogate realities enhance the experiences and quality of life of individuals and how they can be used in clinical settings. It includes the importance of corporeal consciousness in studying mental disorders as well as the potential of VR in treating body image psychopathology.

INTRODUCTION

This chapter continues the study of the convergence of philosophy, science, technology, and information in the context of bioethics presented in the last chapter. It underscores considering the ethical implications of using surrogate realities and ensuring that their use aligns with Bodi's principles: Vulnerable First, Big Picture Thinking, Proactivity, and Scepticism.

DOI: 10.4018/978-1-6684-4808-3.ch017

SURROGATE REALITIES

A surrogate reality is the one containing one or more layers produced by a technological device. The surrogate realities can produce a lasting effect, which is unsurprising when considering leading theories of brain dynamics in the 2010s describe the mind as a sort of phenomenologically transparent VR device (Madary & Metzinger, 2016).

Surrogate realities are important because they offer new ways of experiencing and interacting with the world, but they are also risky because they can be used to manipulate and control individuals. This is particularly important for Bioethics of Displacement as love of belonging, as surrogate realities can be used to artificially create a sense of belonging or to disrupt existing communities. It is therefore important to consider the ethical implications of using surrogate realities and to ensure that their use is guided by Bodi's principles: Vulnerable First, Big Picture Thinking, Proactivity and Scepticism.

Even if the psychology behind VR is almost universal, the way how a person gets absorbed in daydreaming or detached from reality makes the VR experience radically different (Baños et al., 1999). From the beginning there were those who considered many of the augmented reality devices, mainly the head-worn ones, as "*electronic LSD*" (Zachary, 1990) that got us higher through the willing suspension of disbelief (Slattery, 2008). Therefore, the market-driven one size fits all approach is flawed, another moral botch in a hype-ridden sector (McCrea, 2014). Despite that it's well-known, many helmet wearing VR users are immersed in violent video-games while the affectivity is ready to join the VR somatic panoply (Kozel et al., 2018). So much so, that the gaming industry and market cravings for realism have raised ethical concerns about its abundant use of motion capture actors with no ethical code about what can't be asked of them to do in such an environment (Kade, 2016). To cap it all, the brain-computer interfaces (BCIs) have started to replace joysticks and pads and remain at doors of the non-clinical emulation of brain functions (Fafrowicz et al., 2012, p. 336).

The virtualization of social spaces turns the intersubjective differences into a measure of distance between people ready to usurp the old miles and kilometres' throne (Sandu, 2019). Furthermore, classical intersubjectivity (and a good chunk of cultural anthropology!) falls in the History's plot twist of the mutual understanding of natural and artificial intelligence. In turn, the self becomes a parameter in the complexity of pancommunication which gives the smart-city more weight in our incoming human corporeality (Arshinov et al., 2018). In addition to this, there is the gamified citizen science. It can lead to more critical crowd-sourced initiatives (e.g. against unhealthy food) but the common people lack of intimacy with sciences also can allow blatant Orwellian uses (Malina, 2009; Spitz et al., 2017).

If surrogate realities became a new channel for bioethics communication, it would be good news (Rieder et al., 2022). In this line, the submission of cyberspace to the common good has been relentlessly held by UNESCO (Fuentes-Camacho, 2018). However, this new digitalization epoch won't unfold for good if many deep changes are not made before. The decolonial computing unveils the present day colonial structures in the interstices of postcoloniality, nesting in geopolitics and biopolitics (Ali, 2014). For example, the way the algorithms favour some nonverbal behaviours over others when two persons interact in a collaborative virtual environment can hide a classicist / racist bias (Bailenson et al., 2004). Civil society must not fall behind the public and private stakeholders from global polity when it comes to the present and future digitalization and virtualization of our daily lives.

The sci-fi wet dream of fully loading our body engagement into a drive likely will rely upon the synesthetic cutting-edge rather than stimuli brute force (Ross, 2018). It means attaching to our mind's eye other experiences, in this case, linking particular numbers, letters, colours, words, weeks, or months with specific tastes, colours, and sounds. This psychedelic experience of blending senses exists as a harmless neurological condition called synesthesia that affects 2-4% of the general population and allegedly comes from a higher interconnectedness between the parts of the brain that are tied to sensory stimuli (Watson, 2018). Maybe the synesthetic will become a new elite in the future, or maybe not, but it seems that the future social classes will be more sensitive to neurological deeds than ever. In a neurocentric future the thievery, abuse and faking of cognitive sciences' latest insights would lead to a great amount of political power and vice versa (Giordano & Benedikter, 2012).

At any rate, philosophy, science, technology and information are converging in a world where the natural sciences might include non-physical realities for the first time (Kreps, 2019). Not to mention the risks of mixing a full infrastructural neuroscience with a brain-centric cultural imaginary (Littlefield & Johnson, 2012, pp. 10, 90). In short, we are not beholding the evolution of surrogate realities, but the convolution with them. Our flesh-powered instinctive sniffing of internal models of the world is getting nested in a wired one that in the future might be more efficient than this of ours. But, what if a virtual world where we are almost almighty results to be pointless? Will we get the time for thinking about not thinking when plugged in? Will we be able to *unlearn* amidst such a Bergsonian ontological and immanent rush brimming our minds (de Freitas & Ferrara, 2015)? Perhaps the surrogate realities will become a Hotel California where we'll get caught in "*our own device*" (Eagles, 1977)...

NEUROETHICS

Neuroethics is a multidisciplinary applied field of investigation on the ethical issues raised by the new insights on the brain, and the consequent chances for improving it (Roskies 2016). It's still uncertain if neuroethics is a discipline, so to date it's the place where the "*ethics of neuroscience, and the neuroscience of ethics*" joins (Racine, 2010, p. 89; Roskies, 2002). Neuroethics tackling moral issues aims to enlighten many traditional ethical conundrums (Robertson, 2011) from two premises (Racine & Sample, 2018):

- Neuroscience can change our views about the nature of ethics itself
- Neuroscience poses challenges distinct from other areas of medicine

It's upon neuroethics to deal with the unanticipated events and surging neurocentrism while wriggling out both from dogmatism and Western biases (Giordano & Benedikter, 2012). Namely, it implies (Sententia, 2004):

- Preserving "*cognitive freedom*"
- Keeping personhood both:
    - Under the umbrella of humanitarianism
    - Away from partisan or sectarian interests

The neuroethicists, just like geneticists and atomic physicians, are in a moral crucible in global politics after 9/11 when assessing the agendas of their funding sources, namely in topics like remote neuroimaging (Moreno, 2015, pp. 739–740). The brainstem (and even the chances of a soul!) seemingly points to the soup of the brain and spark beating in our skull (Pandya, 2011). In addition to all these interesting topics, neuroethics experimental design can be the key for bioethics to complete its empirical turn (Lyreskog et al., 2023).

So, the neuroethics discourse could shape our brains and change human nature and how it is perceived and loved, for good or evil. The precautionary principle in neuroethics should contain the following three warnings (Jotterand & Ienca, 2017):

- To pay attention to the politicisation of neurotechnology -closely
- To make accountable the social, economic and political motivations behind every research
- To be aware of dominant discourses, especially those away from *deliberative* democracy

In turn, the claim for non-dehumanizing corporal intervention has room in a metaphysics of subjectivity, i.e., a metaphysics in which the personalist subject emerges from its immaterial mooring (Larrivee, 2019). Whatever the case may be, the metaethics will pervade the neuroethical praxis.

Thankfully, this metaethics shed new light thanks to the analysis of encultured persons managing different moralities. It leads to an updated cultural sensitivity around medical principles, which stands out: empowerment, non-obsolescence, self-creativity, and citizenship (Shook & Giordano, 2014). This cosmopolitanism is a must when considering key moral issues in neuroethics like (Fukushi et al., 2007):

- Criminal applications of brain scans[1]
- Incidental findings during non-clinical brain imaging
- Cognitive enhancement

Likewise, it concerns how they were swiftly faced in US and Europe neuroethical societies but barely regarded in other places with a similar or above technological level, like Japan. By the same token, the sex, gender and sexuality differences in the brain also require a non-narrowed appreciation for biological complexity (Roy, 2012).

All these subjects are offering a new insight on the objective reality of the subjective mind, which will allow a wider ground for further discussion in an incoming **neurobioethics**. We'll depart soon from the gentle slope of past evolution and we will see how neuroethics falls behind the groundbreaking advances and "*catastrophic*" mistakes (Katz, 2009, p. xv). Through those mistakes, we'll experience the sea between governors and governed goals, which will spur bigger changes even. Anyhow, today in 2023, the focus should be in a pragmatic and progressive neurobioethics in order to melt the boundary between therapy and better human performance, which will lead to an unmatched personalization. (Giordano, 2012; Vanderstraeten, 2009).

SENSES AND IMAGING

Senses and imaging in a Surrogate Reality are important for bioethics because they can enhance the experiences and quality of life of individuals. From a clinical perspective, both VR and AR can supplement lesson content in anatomical education partly replacing corpses and offering an immersion and engagement that tablet devices cannot (Moro et al., 2017). Surrogate realities could even help in the anorexia nervosa treatment (Ray, 2022). However, they can also be risky if they are used to manipulate or deceive individuals as well as messing with their capabilities.

According to Adam Nash, the tweaking of senses, image and virtual environments leads to a formal aesthetic in which the main components are: digital colour, visual opacity, digital sound, code, artificial intelligence, emergence *and agency* (Nash, 2015). Even the theatre has changed its traditional mistrust towards the most modern technological means (Palade, 2019). The virtual environment and augmented realities, in spite of being mostly visual, extend themselves to other senses and aim to change our way of drinking more tasteful water and enjoying the fresh odour and wind of a far-away landscape, for example (Kerruish, 2019).

The sensory enhancement can not only be attained through augmented reality (AR) and BCIs but also unleashing gene-related technologies like the horizontal gene transfer (Jebari, 2015). Although any of these three methods could lead us to unheard forms of art, the human diversity will be less *over-clocked* if we rely upon the augmented realities. Without going any further we can consider the way we interact in some surrogate realities (for example with hand gestures) like a sort of spreading Sixth Sense (Dixit & Prasad, 2017, pp. 24–26). The most important point is, however, how the human capabilities will differ due to enhancement since it grounds the fundamental entitlements and rights involved in this section (Nussbaum, 2003; Sen, 2005).

PRESENCE AND CORPOREALITY

Perceiving a body is a phenomenologically separated experience from perceiving a body as one's own body and even different brain parts are implied (Costantini, 2014). Corporeal consciousness should be considered by bioethicists when studying mental disorders such as schizophrenia, depression, and autism spectrum disorder, stressing the importance of an "expressive common environment" and freeing bioethics' phenomenological approach from the Heideggerian burden (Bizzari, 2023; Coyne, 2023). Shortly, body awareness is important in studying mental disorders, making a point of shared understanding between patients and clinicians. In this sense, VR has opened new doors to the treatment of body image psychopathology, just like in many eating disorders (Perpiñá et al., 1999). Surrogate realities led to other interesting findings as well. For example, the rubber hand illusion works better with a tangible fake hand than with a virtual one, which is related with the way we project presence through corporeality (IJsselsteijn et al., 2006). Furthermore, when a "*cardiac rubber hand*" was employed (one featuring a cardio-visual feedback in-time with the actual heartbeat) results demonstrate that interoceptive signals directly influence the experience of body ownership via multisensory integration (Suzuki et al., 2013). On the other hand, not a virtual limb and neither an AR one, in despite of both inducing a greater sensation of 'phantom mobility' and being promising techniques, have not reduced significantly the pain in amputees (Brodie et al., 2007; Cole, 2008; Desmond et al., 2006). So we should stop the hearse here since, more than a decade after the first experiments, no statistically significant analgesic effect of vision of the virtual arm movement has been found, neither (Zanini et al., 2017).

Presence is the subjective sensation of being in a place and persons who get sick more often in a VR environment tend to have a lesser presence (Witmer & Singer, 1998). Indeed, both presence and reality judgement are important features when orienting ourselves within VR, however, a little attention has been paid to the last in the scientific literature, which has led to subsume reality in presence (Baños et al., 2000). The judgement of presence is earlier than the reality. It started with the teleoperators, robots controlled from a distance by a human operator. The teleoperators allowed for the first time that the presence could *throw* the self in a space where it was able to act and get updated with new tools. Our mirroring neurons do a similar thing to produce an Other to be thrown at Self in order we can make the others present (enabling a *social presence*) through a metacognitive comparison (Riva & Mantovani, 2014). That's an allocentric view in contrast with an egocentric one and, just like in the teleoperators' case, we also have social artefacts available to update our-Selves.

Figure 1. Allocentric vs. egocentric views and projected self
Source: Riva and Mantovani (2014, p. 12)

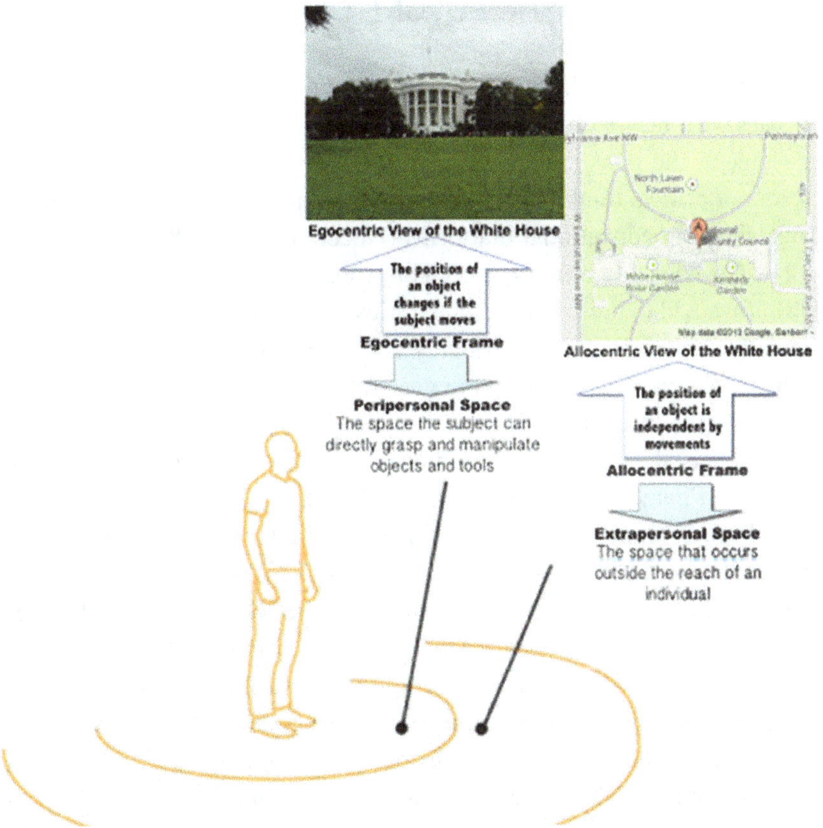

Our allocentric presence is growing with the ubiquity, the zeitgeist of our generation that drags us to an hybrid bodily condition (Cipolletta, 2019). Also the egocentric presence has an evil twist about human hybrid nature due to the today full feasibility of a head transplant. It means that a person could have a set of social and physical experiences she hadn't had without the transplant while enjoying a continuity with her lifelong head and brain. Thus, this person will be different from their parts, because her inputs, her *in-form-ation* would lead to a brand-new embodiment (Pascalev et al., 2016). It puts us in the crossroad between semiotics and life since considering life and conscience as necessarily *incarnated* ones, the humans are displacing "*from rational animals into reasonable animals*" (Petrilli, 2004).

When our presence fades away irreversibly, just like in the terminal illnesses, the corporeality also is disrupted when it comes to the spatial arrangement (Lobont & Mladin, 2018). It's likely related to how the insight towards the rubber hand experiments converges: "*the conscious selfhood is based on interoceptive predictive coding*" (Suzuki et al., 2013). The person is present when her corporeality is, it happens because corporeality is both one the many facets of our personality and a basic aspect of our experience (Beauregard, 2019, pp. 154–169). Furthermore, the illusion of a body swap using a *real avatar* is more shocking and raises the sense of presence thanks to the body agency (Oliveira et al., 2016). Given that we can consider the dead like the displacement to the *null avatar* it opens the floor to the most important question: what was there in the *before-life*?

SURROGATE REALITIES AND THE FUTURE OF EDUCATION

Surrogate realities and the future of education are important for bioethics as they raise questions about how we should interact with new technologies and the ethical implications of these interactions. For Bioethics of Displacement, the risks associated with the use of surrogate realities in education, such as the potential for loss of critical thinking through early manipulation as never before, are a concern.

AR and VR have landed in educational platforms (Nucera et al., 2018). It started by the VR asynchronous distance but the AR is generating the greatest expectancies in the last decade (Chen et al., 2017; Dickey, 2005). If we add to these surrogate realities' educational uses a layer of gamification, we obtain a neuro-stimulation able to improve the quality of cognitive processes (Sandakova & Tolstobrova, 2019). It's to be enhanced more even since in the near-future educational platforms will also consider the real-time emotional feedback of the learner (Καλύβας, 2018).

Even if we consider not just the brain but the coupled body–brain–environment as the basis of cognition the surrogate realities are a game-changer event (Gallagher, 2018). Educators can even raise environmental awareness thanks to the immersion in landscapes provided by AR, GIS, multi-channel video and akin techs (Ghadirian & Bishop, 2002). Furthermore, we can preserve endangered habitats-being-there without touching them as well as reaching places that elsewhere we couldn't like the bottom sea (Skarlatos et al., 2016). The marine learning results in a Taiwanese primary school, for example, highlights how the AR has helped the low academic achievers improve learning performance (Lu & Liu, 2015). Indeed, the surrogate realities have melted the barrier between technology and an education focused on ecological values thanks to the positive emotions raised during learning (T.-C. Huang et al., 2016).

Let's delve into the body-brain-surrogate reality nexus of learning. The AR has proven to be useful in anatomy education and creates the illusion that the user can look inside their body (Ma et al., 2016). Likewise, the VR has also been used for learning about the air flow through the nasal cavity (Marks et al., 2017). AR and VR are aimed to be a live operative field aid and an educational tool for neurosurgery respectively (Pelargos et al., 2017). In a study led by José Weber Vieira de Faria on 84 graduate medical students, the group who studied the brain through digitally interactive but non-stereoscopic learning path outperformed the group that used the stereoscopic (traditional) way and matched the group that used both methods (Vieira de Faria et al., 2016). When it comes to VR interfaces, they can be added to tangible user interfaces to produce a hybrid bodily experience and disclose an additional abstract layer within the educational experience, mainly about physical phenomena (Fleck & Hachet, 2016).

The educational AR toys, in spite of being appealing for children and educators altogether, have not led to a greater cognitive attainment and lack of collaborative and interactive learning (Yilmaz, 2016). In the most educational settings (K-12 and higher) AR devices actually enhance learning performance but are somewhat difficult for students to use (Akçayır & Akçayır, 2017). So, the AR is better used as a nexus from the books to the lab (Martín-Gutiérrez et al., 2015). As an example I'll address the Malayan case, where the enhanced AR successfully raised the effectiveness and attractiveness of the learning environment for biology students in a real world scenario without crippling the social interaction (Weng et al., 2016). Likewise, even in such a high educational level as engineering the VR approach beats the traditional one offering "*incredible results*" (Alhalabi, 2016). The immersion and imagination features of VR, that's the presence related ones, seems to be the key in the usefulness perception of this tech by students (H.-M. Huang et al., 2016). However, despite that 2016 was heralded as the great year of surrogate realities, "*in most cases our classrooms are barren of any interactive technologies*" while we are more and more surrounded by screens (Martín-Gutiérrez, 2017).

Besides that, and regardless of my good vibe with the use of surrogate realities in educational settings, it has (unsurprisingly) a dark side. Big companies like Apple, Facebook and Samsung, among others, are investing to cope with the market of accessible virtual technologies that likely will break the boundaries of formal education (Martín-Gutiérrez et al., 2017). So, these companies that are in our houses, cells and lives could also shape the education of the next generations from the very beginning and may annihilate any form of critical pedagogy. Reducing the flow of data in the human pool according to corporate interest rather than its utility is also a way of bringing today's shocked humankind to a new level of submission. An important factor here will be the rebellious nature (or not) of the new science of language arising as the linguists became surrogate reality adepts (De Landa, 2005, p. 95). Will the way we be educated in, from, about and *within* surrogate realities be dictated by an institution or brand or, otherwise, this dialogue will emerge from an asynchronous and live world?

CONCLUSION

If Bodi's principles (vulnerability, big picture thinking, proactivity, and scepticism) are ignored, surrogate realities' risks such as manipulation and control of individuals, especially youth people, will become real and extend to ethics, education, critical thinking and the perception of the Self. About this, neuroethics explores the potential of neuroscience to change our understanding of ethics. Nevertheless, it presents unique challenges that require a multidisciplinary approach, for example, to understand involvement of major companies and governments in mass control and surveillance schemes.

REFERENCES

Akçayır, M., & Akçayır, G. (2017). Advantages and challenges associated with augmented reality for education: A systematic review of the literature. *Educational Research Review*, *20*, 1–11. doi:10.1016/j.edurev.2016.11.002

Alhalabi, W. (2016). Virtual reality systems enhance students' achievements in engineering education. *Behaviour & Information Technology*, *35*(11), 919–925. doi:10.1080/0144929X.2016.1212931

Ali, M. (2014). Towards a decolonial computing. *Ambiguous Technologies: Philosophical Issues, Practical Solutions. Human Nature (Hawthorne, N.Y.)*, 28–35. http://oro.open.ac.uk/41372/

Arshinov, V., Chapny, E., & Chekletsov, V. (2018, May 14). Intersubjectivity and Complexity in Human Enhancement Problem. *Proceedings of the XXIII World Congress of Philosophy*. 10.5840/wcp232018382

Bailenson, J. N., Beall, A. C., Loomis, J., Blascovich, J., & Turk, M. (2004). Transformed Social Interaction: Decoupling Representation from Behavior and Form in Collaborative Virtual Environments. *Presence (Cambridge, Mass.), 13*(4), 428–441. doi:10.1162/1054746041944803

Baños, R., Botella, C., Garcia-Palacios, A., Villa, H., Perpiña, C., & Alcañiz, M. (2000). Presence and Reality Judgment in Virtual Environments: A Unitary Construct? *Cyberpsychology & Behavior, 3*(3), 327–335. doi:10.1089/10949310050078760

Baños, R., Botella, C., García-Palacios, A., Villa, H., Perpiñá, C., & Gallardo, M. (1999). Psychological Variables and Reality Judgment in Virtual Environments: The Roles of Absorption and Dissociation. *Cyberpsychology & Behavior*, *2*(2), 143–148. doi:10.1089/cpb.1999.2.143 PMID:19178250

Beauregard, J. (2019). *Philosophical Neuroethics: A Personalist Approach. Volume 1: Foundations.* Vernon Press.

Bizzari, V. (2023). Which Kind of Body in "Mental" Pathologies? Phenomenological Insights on the Nature of the Disrupted Self. *The Journal of Medicine and Philosophy: A Forum for Bioethics and Philosophy of Medicine, 48*(2), 116–127. 10.1093/jmp/jhad008

Brodie, E. E., Whyte, A., & Niven, C. A. (2007). Analgesia through the looking-glass? A randomized controlled trial investigating the effect of viewing a 'virtual' limb upon phantom limb pain, sensation and movement. *European Journal of Pain (London, England)*, *11*(4), 428–436. doi:10.1016/j.ejpain.2006.06.002 PMID:16857400

Chen, P., Liu, X., Cheng, W., & Huang, R. (2017). A review of using Augmented Reality in Education from 2011 to 2016. In E. Popescu, Kinshuk, M. K. Khribi, R. Huang, M. Jemni, N.-S. Chen, & D. G. Sampson (Eds.), Innovations in Smart Learning (pp. 13–18). Springer Singapore.

Cipolletta, G. (2019). Ubiquitous Bodies: A "Metrobodily" Transition From Real to Virtual. *Virtual and Augmented Reality in Mental Health Treatment*, 1–16. doi:10.4018/978-1-5225-7168-1.ch001

Cole, J. (2008). Virtual and Augmented Reality, Phantom Experience, and Prosthetics. In P. Gallagher, D. Desmond, & M. MacLachlan (Eds.), *Psychoprosthetics* (pp. 141–153). Springer London. doi:10.1007/978-1-84628-980-4_11

Costantini, M. (2014). Body perception, awareness, and illusions. *Wiley Interdisciplinary Reviews: Cognitive Science*, *5*(5), 551–560. doi:10.1002/wcs.1309 PMID:26308744

Coyne, L. (2023). What is Phenomenological Bioethics? A Critical Appraisal of Its Ends and Means. *The Journal of Medicine and Philosophy: A Forum for Bioethics and Philosophy of Medicine, 48*(2), 170–183. 10.1093/jmp/jhad001

de Freitas, E., & Ferrara, F. (2015). Movement, Memory and Mathematics: Henri Bergson and the Ontology of Learning. *Studies in Philosophy and Education*, *34*(6), 565–585. doi:10.100711217-014-9455-y

De Landa, M. (2005). Virtual Environments and the Emergence of Synthetic Reason. In J. B. Dixon & E. Cassidy (Eds.), *Virtual Futures: Cyberotics, Technology and Posthuman Pragmatism*. Routledge.

Desmond, D. M., O'Neill, K., De Paor, A., McDarby, G., & MacLachlan, M. (2006). Augmenting the Reality of Phantom Limbs: Three Case Studies Using an Augmented Mirror Box Procedure. *JPO: Journal of Prosthetics and Orthotics*, *18*(3), 74. doi:10.1097/00008526-200607000-00005

Dickey, M. D. (2005). Three-dimensional virtual worlds and distance learning: Two case studies of Active Worlds as a medium for distance education. *British Journal of Educational Technology*, *36*(3), 439–451. doi:10.1111/j.1467-8535.2005.00477.x

Dixit, S., & Prasad, R. (2017). *Human Bond Communication: The Holy Grail of Holistic Communication and Immersive Experience*. John Wiley & Sons. doi:10.1002/9781119341451

Eagles. (1977, February 22). *Hotel California* [7-inch single]. Asylum. https://en.wikipedia.org/w/index.php?title=Hotel_California&oldid=919168934

Fafrowicz, M., Marek, T., Karwowski, W., & Schmorrow, D. (2012). *Neuroadaptive Systems: Theory and Applications*. CRC Press. doi:10.1201/b13019

Fleck, S., & Hachet, M. (2016). Making Tangible the Intangible: Hybridization of the Real and the Virtual to Enhance Learning of Abstract Phenomena. *Frontiers in ICT (Lausanne, Switzerland)*, *3*. Advance online publication. doi:10.3389/fict.2016.00030

Fuentes-Camacho, T. (2018). Introduction: UNESCO and the Law of Cyberspace. In The International Dimensions of Cyberspace Law. doi:10.4324/9781315181912-1

Fukushi, T., Sakura, O., & Koizumi, H. (2007). Ethical considerations of neuroscience research: The perspectives on neuroethics in Japan. *Neuroscience Research*, *57*(1), 10–16. doi:10.1016/j.neures.2006.09.004 PMID:17034890

Gallagher, S. (2018). Educating the Right Stuff: Lessons in Enactivist Learning. *Educational Theory*, *68*(6), 625–641. doi:10.1111/edth.12337

Ghadirian, P., & Bishop, I. D. (2002). Composition of augmented reality and GIS to visualize environmental changes. *Proceedings of the Joint AURISA and Institution of Surveyors Conference*, 25–30.

Giordano, J. (2012). Integrative Convergence in Neuroscience: Trajectories, Problems, and the Need for a Progressive Neurobioethics. In A. Vaseashta, E. Braman, & P. Susmann (Eds.), *Technological Innovations in Sensing and Detection of Chemical, Biological, Radiological, Nuclear Threats and Ecological Terrorism* (pp. 115–130). Springer Netherlands. doi:10.1007/978-94-007-2488-4_10

Giordano, J., & Benedikter, R. (2012). An early-and necessary-flight of the owl of Minerva: Neuroscience, neurotechnology, human socio-cultural boundaries, and the importance of neuroethics. *Journal of Evolution and Technology / WTA*, *22*(1), 14–25.

Huang, H.-M., Liaw, S.-S., & Lai, C.-M. (2016). Exploring learner acceptance of the use of virtual reality in medical education: A case study of desktop and projection-based display systems. *Interactive Learning Environments*, *24*(1), 3–19. doi:10.1080/10494820.2013.817436

Huang, T.-C., Chen, C.-C., & Chou, Y.-W. (2016). Animating eco-education: To see, feel, and discover in an augmented reality-based experiential learning environment. *Computers & Education*, *96*, 72–82. doi:10.1016/j.compedu.2016.02.008

IJsselsteijn, W. A., de Kort, Y. A. W., & Haans, A. (2006). Is This My Hand I See Before Me? The Rubber Hand Illusion in Reality, Virtual Reality, and Mixed Reality. *Presence (Cambridge, Mass.)*, *15*(4), 455–464. doi:10.1162/pres.15.4.455

Jebari, K. (2015). Sensory Enhancement. In J. Clausen & N. Levy (Eds.), *Handbook of Neuroethics* (pp. 827–838). Springer Netherlands. doi:10.1007/978-94-007-4707-4_106

Jotterand, F., & Ienca, M. (2017). The Biopolitics of Neuroethics. In E. Racine & J. Aspler (Eds.), Debates About Neuroethics: Perspectives on Its Development, Focus, and Future (pp. 247–261). Springer International Publishing. doi:10.1007/978-3-319-54651-3_17

Καλύβας, Δ. (2018). *Tools and platforms for enhanced human-computer interaction*. http://okeanis.lib2.uniwa.gr/xmlui/handle/123456789/4377

Kade, D. (2016). Ethics of Virtual Reality Applications in Computer Game Production. *Philosophies*, *1*(1), 73–86. doi:10.3390/philosophies1010073

Katz, B. F. (2009). *Neuroengineering The Future: Virtual Minds And The Creation Of Immortality*. Jones & Bartlett Publishers.

Kerruish, E. (2019). Arranging sensations: Smell and taste in augmented and virtual reality. *The Senses and Society*, *14*(1), 31–45. doi:10.1080/17458927.2018.1556952

Kozel, S., Gibson, R., & Martelli, B. (2018). The Weird Giggle: Attending to Affect in Virtual Reality. *Transformations: Journal of Media. Culture Technique*, *31*, 1–24.

Kreps, D. (2019). *Understanding Digital Events: Bergson, Whitehead, and the Experience of the Digital*. Routledge. doi:10.4324/9780429032066

Larrivee, D. (2019). *Introductory Chapter: Reconciling Neurobioethics through Nature's Lens - Metaphysical Determinants of Subjectivity*. Bioethics - Neurology, Psychiatry and Genetics., doi:10.5772/intechopen.83502

Littlefield, M. M., & Johnson, J. (2012). *The Neuroscientific Turn: Transdisciplinarity in the Age of the Brain*. University of Michigan Press. doi:10.3998/mpub.4585194

Lobont, F., & Mladin, I. (2018). Philosophical-Ethical Alleviation of Perceptive Experience of Corporeality in Terminally Ill Sufferers. *Ethics of Emerging Biotechnologies*, 107–118.

Lu, S.-J., & Liu, Y.-C. (2015). Integrating augmented reality technology to enhance children's learning in marine education. *Environmental Education Research*, *21*(4), 525–541. doi:10.1080/13504622.2014.911247

Lyreskog, D. M., Pavarini, G., Jacobs, E., Bennett, V., Mawdsley, G., & Singh, I. (2023). Testing Design Bioethics Methods: Comparing a Digital Game with a Vignette Survey for Neuroethics Research with Young People. *AJOB Empirical Bioethics*, *14*(1), 55–64. doi:10.1080/23294515.2022.2110964 PMID:35993968

Ma, M., Fallavollita, P., Seelbach, I., Von Der Heide, A. M., Euler, E., Waschke, J., & Navab, N. (2016). Personalized augmented reality for anatomy education. *Clinical Anatomy (New York, N.Y.)*, *29*(4), 446–453. doi:10.1002/ca.22675 PMID:26646315

Madary, M., & Metzinger, T. K. (2016). Real virtuality: A code of ethical conduct. Recommendations for good scientific practice and the consumers of VR-technology. *Frontiers in Robotics and AI*, 3.

Malina, R. F. (2009). Intimate science and hard humanities. *Leonardo*, *42*(3), 184–184. doi:10.1162/leon.2009.42.3.184

Marks, S., White, D., & Singh, M. (2017). Getting Up Your Nose: A Virtual Reality Education Tool for Nasal Cavity Anatomy. *SIGGRAPH Asia 2017 Symposium on Education*, 1:1-1:7. 10.1145/3134368.3139218

Martín-Gutiérrez, J. (2017). Editorial: Learning Strategies in Engineering Education Using Virtual and Augmented Reality Technologies. *Eurasia Journal of Mathematics, Science and Technology Education*, *13*(2), 297–300. doi:10.12973/eurasia.2017.00630a

Martín-Gutiérrez, J., Fabiani, P., Benesova, W., Meneses, M. D., & Mora, C. E. (2015). Augmented reality to promote collaborative and autonomous learning in higher education. *Computers in Human Behavior*, *51*, 752–761. doi:10.1016/j.chb.2014.11.093

Martín-Gutiérrez, J., Mora, C. E., Añorbe-Díaz, B., & González-Marrero, A. (2017). Virtual technologies trends in education. *Eurasia Journal of Mathematics, Science and Technology Education*, *13*(2), 469–486.

McCrea, J. (2014, June 9). Virtual Reality: Then and Now. *The Real McCrea*. https://therealmccrea.com/2014/06/08/virtual-reality-then-and-now/

Moreno, J. D. (2015). Neuroethics: An Agenda for Neuroscience and Society. In H. Kuhse, U. Schüklenk, & P. Singer (Eds.), *Bioethics: An Anthology*. John Wiley & Sons.

Moro, C., Štromberga, Z., Raikos, A., & Stirling, A. (2017). The effectiveness of virtual and augmented reality in health sciences and medical anatomy. *Anatomical Sciences Education*, *10*(6), 549–559. doi:10.1002/ase.1696 PMID:28419750

Nash, A. (2015). An Aesthetics of Digital Virtual Environments. In *New Opportunities for Artistic Practice in Virtual Worlds* (pp. 1–22). https://www.igi-global.com/chapter/an-aesthetics-of-digital-virtual-environments/132413

Nucera, S., Tartarisco, G., Epasto, A., Smeriglio, D., Mazzeo, A., Pioggia, G., & Anastasi, A. (2018). Ubiquitous, Wearable, Mobile: Paradigm Shifts in E-Learning and Diffusion of Knowledge. Handbook of Research on Mobile Devices and Smart Gadgets in K-12 Education, 286–307. doi:10.4018/978-1-5225-2706-0.ch018

Nussbaum, M. (2003). Capabilities as Fundamental Entitlements: Sen and Social Justice. *Feminist Economics*, *9*(2–3), 33–59. doi:10.1080/1354570022000077926

Oliveira, E. C. D., Bertrand, P., Lesur, M. E. R., Palomo, P., Demarzo, M., Cebolla, A., Baños, R., & Tori, R. (2016). Virtual Body Swap: A New Feasible Tool to Be Explored in Health and Education. *2016 XVIII Symposium on Virtual and Augmented Reality (SVR)*, 81–89. 10.1109/SVR.2016.23

Palade, A. (2019). The peculiarities of effective teaching through theater. *Învăţământ, Cercetare. Creaţie*, *V*(1), 333–335.

Pandya, S. K. (2011). Understanding Brain, Mind and Soul: Contributions from Neurology and Neurosurgery. *Mens Sana Monographs*, *9*(1), 129–149. doi:10.4103/0973-1229.77431 PMID:21694966

Pascalev, A., Pascalev, M., & Giordano, J. (2016). Head Transplants, Personal Identity and Neuroethics. *Neuroethics*, *9*(1), 15–22. doi:10.100712152-015-9245-4

Pelargos, P. E., Nagasawa, D. T., Lagman, C., Tenn, S., Demos, J. V., Lee, S. J., Bui, T. T., Barnette, N. E., Bhatt, N. S., Ung, N., Bari, A., Martin, N. A., & Yang, I. (2017). Utilizing virtual and augmented reality for educational and clinical enhancements in neurosurgery. *Journal of Clinical Neuroscience*, *35*, 1–4. doi:10.1016/j.jocn.2016.09.002 PMID:28137372

Perpiñá, C., Botella, C., Baños, R., Marco, H., Alcañiz, M., & Quero, S. (1999). Body Image and Virtual Reality in Eating Disorders: Is Exposure to Virtual Reality More Effective than the Classical Body Image Treatment? *Cyberpsychology & Behavior*, *2*(2), 149–155. doi:10.1089/cpb.1999.2.149 PMID:19178251

Petrilli, S. (2004). *The responsibility of power and the power of responsibility: From the "semiotic" to the "semioethic" animal*. Academic Press.

Racine, E. (2010). *Pragmatic Neuroethics: Improving Treatment and Understanding of the Mind-brain*. MIT Press. doi:10.7551/mitpress/8027.001.0001

Racine, E., & Sample, M. (2018). Two Problematic Foundations of Neuroethics and Pragmatist Reconstructions. *Cambridge Quarterly of Healthcare Ethics*, *27*(4), 566–577. doi:10.1017/S0963180118000099 PMID:30720412

Ray, M. L. (2022). *Nurturing with Nature: Exploring Anorexia Nervosa through an Ecotherapeutic Lens* [PhD Thesis]. Pacifica Graduate Institute.

Rieder, T. N., Hutchinson, L. A., & Kahn, J. P. (2022). Why the World Needs Bioethics Communication. *Perspectives in Biology and Medicine*, *1*. Advance online publication. doi:10.1353/pbm.0.0174 PMID:36468391

Riva, G., & Mantovani, F. (2014). 1 Extending the Self through the Tools and the Others: A General Framework for Presence and Social Presence in Mediated Interactions. In J. Waterworth & D. Murray (Eds.), *Interacting with Presence* (pp. 9–31). Sciendo Migration. doi:10.2478/9783110409697.1

Robertson, M. (2011). Symposium: Neuroethics and Mental Health—Old Wine in New Bottles or a Legitimate New Field of Bioethical Inquiry. *Journal of Bioethical Inquiry*, *8*(1), 13–14. doi:10.100711673-010-9267-x

Roskies, A. (2002). Neuroethics for the new millennium. *Neuron*, *35*(1), 21–23. doi:10.1016/S0896-6273(02)00763-8 PMID:12123605

Ross, M. (2018). Virtual Reality's New Synesthetic Possibilities. *Television & New Media*, *1527476418805240*. Advance online publication. doi:10.1177/1527476418805240

Roy, D. (2012). Neuroethics, Gender and the Response to Difference. *Neuroethics*, *5*(3), 217–230. doi:10.100712152-011-9130-8

Sandakova, L. B., & Tolstobrova, L. I. (2019, July). *Proliferation of Neuroscientific Knowledge in Pedagogy and Education*. "Humanities and Social Sciences: Novations, Problems, Prospects" (HSSNPP 2019). https://www.atlantis-press.com/proceedings/hssnpp-19/125913454

Sandu, A. (2019). Towards a Phenomenology of the Digitalization of Consciousness. The Virtualization of the Social Space. *Postmodern Openings*, *10*(2), 155–161. doi:10.18662/po/77

Sen, A. (2005). Human Rights and Capabilities. *Journal of Human Development*, *6*(2), 151–166. doi:10.1080/14649880500120491

Sententia, W. (2004). Neuroethical considerations: Cognitive liberty and converging technologies for improving human cognition. *Annals of the New York Academy of Sciences*, *1013*(1), 221–228. doi:10.1196/annals.1305.014 PMID:15194617

Shook, J. R., & Giordano, J. (2014). A principled and cosmopolitan neuroethics: Considerations for international relevance. *Philosophy, Ethics, and Humanities in Medicine; PEHM*, *9*(1), 1. doi:10.1186/1747-5341-9-1 PMID:24387102

Skarlatos, D., Agrafiotis, P., Balogh, T., Bruno, F., Castro, F., Petriaggi, B. D., Demesticha, S., Doulamis, A., Drap, P., Georgopoulos, A., Kikillos, F., Kyriakidis, P., Liarokapis, F., Poullis, C., & Rizvic, S. (2016). Project iMARECULTURE: Advanced VR, iMmersive Serious Games and Augmented REality as Tools to Raise Awareness and Access to European Underwater CULTURal heritagE. In M. Ioannides, E. Fink, A. Moropoulou, M. Hagedorn-Saupe, A. Fresa, G. Liestøl, V. Rajcic, & P. Grussenmeyer (Eds.), *Digital Heritage. Progress in Cultural Heritage: Documentation, Preservation, and Protection* (pp. 805–813). Springer International Publishing. doi:10.1007/978-3-319-48496-9_64

Slattery, D. R. (2008). Virtual reality and hallucination: A technoetic perspective. *The Engineering Reality of Virtual Reality*, *2008*(6804), 680407. doi:10.1117/12.772650

Spitz, R., Pereira, C., Jr., Queiroz, F., Leite, L. C., Dam, P., Ferranti, M. P., Kogut, R., & Oliveira, W. (2017). Gamification, citizen science and civic engagement: In search of the common good. *6th International Conference Balance-Unbalance (BunB)*. UN-Habitat. Https://Unhabitat. Org/Acessado Em, 2.

Suzuki, K., Garfinkel, S. N., Critchley, H. D., & Seth, A. K. (2013). Multisensory integration across exteroceptive and interoceptive domains modulates self-experience in the rubber-hand illusion. *Neuropsychologia*, *51*(13), 2909–2917. doi:10.1016/j.neuropsychologia.2013.08.014 PMID:23993906

Ubukata, T. (Director). (2016, September 15). *Psycho-Pass* [Animation, Action, Crime]. Dentsu, Fuji Television Network, Index.

Vanderstraeten, G. (2009). Dreaming of virtual reality and rehabilitation technology in the future: Virtual reality, robotics and telerehabilitation. *Rehabilitation in the Future*, *8*, 120–123.

Vieira de Faria, J. W., Teixeira, M. J., de Moura Sousa, L., Pinhata Otoch, J., & Gadelha Figueiredo, E. (2016). Virtual and stereoscopic anatomy: When virtual reality meets medical education. *Journal of Neurosurgery*, *125*(5), 1105–1111. doi:10.3171/2015.8.JNS141563 PMID:26871375

Watson, K. (2018, October 24). *Synesthesia: Definition, Examples, Causes, Symptoms, and Treatment*. Healthline. https://www.healthline.com/health/synesthesia

Weng, N. G., Bee, O. Y., Yew, L. H., & Hsia, T. E. (2016). An Augmented Reality System for Biology Science Education in Malaysia. *International Journal of Innovative Computing*, *6*(2). http://kp.fsksm.utm.my/ijic/index.php/ijic/article/view/128

Witmer, B. G., & Singer, M. J. (1998). Measuring Presence in Virtual Environments: A Presence Questionnaire. *Presence (Cambridge, Mass.)*, *7*(3), 225–240. doi:10.1162/105474698565686

Yilmaz, R. M. (2016). Educational magic toys developed with augmented reality technology for early childhood education. *Computers in Human Behavior*, *54*, 240–248. doi:10.1016/j.chb.2015.07.040

Zachary, P. (1990, January 23). Artificial Reality: Computer Simulations One Day May Provide Surreal Experiences. *Wall Street Journal.*

Zanini, A., Montalti, M., Caola, B., Leadbetter, A., & Martini, M. (2017). Pain during illusory own arm movement: A study in immersive virtual reality. *European Medical Journal*, *2*(2), 90–97.

KEY TERMS AND DEFINITIONS

Augmented Reality (AR): Technology that blends virtual content with the real world, enhancing perception and interaction in various sensory modalities.

Brain Dynamics: Patterns of activity in the brain, involving electrical and chemical interactions among neurons, that underlie cognitive functions.

Cognitive Freedom: To preserve mental self-determination, that's autonomy to control their own thoughts, cognition, and consciousness without external influences or constraints.

Corporeality: The subjective experience of having a physical body and the dynamic interplay of expansion and contraction in corporeal feelings.

Neurobioethics: Exploration of what is morally right or wrong in neuroscience advancements especially in healthcare and research settings (e.g., neurosurgery, psychiatry, and the treatment of neurological disorders).

Neuroethics: Ethical, legal, and societal analysis of the broader implications of neuroscience in different areas of human life and society (e.g., including privacy, enhancement, and mind-brain relationships).

Phenomenologically Transparent: The property of perceptual experiences of making the subject only aware of seemingly external objects and properties (e.g., a VR immersion).

Presence: The subjective feeling or perception of being present in a virtual or simulated environment, often achieved through the use of virtual reality technologies.

Surrogate Realities: Technological creations that provide immersive experiences.

Virtual Reality (VR): A computer-generated environment that surrounds and immerses the user in interactive experiences.

ENDNOTE

1. For an excellent speculation on this topic, I recommend the futurist thriller manga Psycho-Pass (Ubukata, 2016).

Chapter 18
Extraterrestrial Bioethics:
Worthy v Lost III

ABSTRACT

The author plays with three interconnected concepts: off-Earth bioethics, xenobioethics, and the ethical considerations surrounding the disposal of a cosmonaut's corpse. Off-Earth bioethics involves ethical considerations related to recruiting and monitoring astronauts, while xenobioethics deals with the ethical challenges of encountering extraterrestrial life. The author argues that it is important to develop a non-speciesist standard of personhood that can accommodate the possibility of intelligent extraterrestrial life. Finally, the author ponders the disposal of a cosmonaut's corpse in a space mission, arguing that this process must prioritise the crew's morale and consider the potential environmental impact of different disposal methods. Overall, the role of bioethics will be crucial in shaping humanity's future as a galactic sinner or a reliable cosmic stakeholder.

INTRODUCTION

This chapter continues the exploration of ethical implications in future scenarios, linking back to the discussion on surrogate realities from last chapter but extended to off-Earth bioethics. This chapter raises important questions about collecting genetic information for occupational surveillance and research, highlighting the potential benefits and risks associated with personalised drugs and enhancements. Moving beyond the individual, this chapter examines the bioethical dilemmas we face regarding life itself in outer space.

DOI: 10.4018/978-1-6684-4808-3.ch018

OFF-EARTH BIOETHICS

Galactic Sins

The future of humanity as a galactic sinner or a reliable cosmic stakeholder grounds in our present off-Earth bioethics. And the off-Earth bioethics starts by the astronauts' recruitment. In the first place, the space can be a great place to skip labour laws and many types of recruitment have proven to be a rave social research on human subjects (Lin & Abney, 2014). The astronauts will be the first ones to really behold the relentless killer that outer space is. As F. Strollo and colleagues proved in the Mars 500 project, prolonged confinement leads to metabolic derangement (Strollo et al., 2018).[1] Likewise, the microgravity environment in orbital flight can alter a number of physiological processes associated with diurnal rhythms such as a dramatic elevation of cortisol levels (Liang et al., 2012). Should any astronaut employer (e.g. NASA) collect genetic information for occupational surveillance and research? (Reed & Antonsen, 2018) This info could be used for protecting their lives providing them with personalised pharmaceuticals and enhancements, but also it may lead to be covertly rejected from the space program on barely nothing more than genetic grounds.

Even if the astronaut squad selection is morally exemplar, the mere perspective of long time missions in space have arisen new bioethical questions such as the ones the NASA have been considering (Seedhouse, 2010, p. 181):

- How do you get rid of the dead body of a crewmember?
- When should life support be discontinued for a critically ill astronaut consuming valuable oxygen?
- Should a straitjacket be included in the medical supplies?
- Should mission managers mandate prophylactic surgery such as removal of appendix, tonsils and gall bladder before a mission to Mars?
- If a crewmember becomes disabled during the mission, who becomes her default surrogate decision-maker? Family members, physicians, companions?

In short, the off-Earth bioethics are so hard because there are strong commitment conflicts. On the one hand there are the protection of research subjects first and astronauts then (Sawin, 2017), but on the other hand there are different obligations like national security and interests (e.g. maintaining a position of power abroad). As you may suppose, there are also potential privacy and commitment moral issues in the spaceship regarding computer clearance's management and body scans involving both humans and AIs, e.g. an AI agency regarding the mission and human security, as seen in *2001: a space Odyssey* (Kubrick, 1968).

Regardless of their image of heroes or overenthusiastic billionaires, both the astronauts and space tourists are "*vulnerable persons*" when it comes to human research, thus, the suitable ethics and bioethics committees should protect their dignity and rights (Coleman, 2019). Sometimes the mission will be inhumanely demanding for astronauts. In these cases, the bioethics of such stranded and displaced persons like the next-gen astronauts might come to be focused on discerning whether the benefits they are provided with are really therapeutically needed or, otherwise, plain enhancements (Galliott, 2016, pp. 249–250). For example, there is no doubt that modifications conceding resistance to cellular damage by radiation, reducing the need of water or enabling hibernation are a must in the interstellar journeymen toolbox. But, what about an exoskeleton for an everlasting radiation resistance or photosynthesis capabilities for an uncanny frugality? By the same token, a modification like breathing methane can be an enhancement in a Jupiter moon but a disability when back to the Earth. Should a non-reversible modification of this kind ever be permitted? Is a person entitled to erase the qualities that make her able to share her life with other fellowmen? And how does this dilemma change if there is a space destination that offers better hopes of healing a deadly condition?

As we saw in the section on Cyborgs, human enhancement is under the foci of bioethical debates, but not its transgenerational consequences, which may lead us to a deeper insight about the evolution of life and intelligence in their widest, cosmic context including even a postbiological civ (Ćirković, 2017). Paraphrasing Prof. Darryl Macer comments on my PhD. research this book is based upon, we can guess that the future galactica diaspora's communities will see their journey and modifications as migration rather than bitter displacement attending on how we accept transience as a glue for our human and posthuman nature.

But not all the ethical and bioethical issues will await to be tackled down some day in the distant future. Rather than this, we will set foot on Mars in the next few years without any real experience on what an ethical spatial colonisation is (Vesely, 2018). Mars is off-Earth. Mars is inhospitable. There is no air to breathe, the gravity is about 1/3 of Earth's one and the night temperature plummets below the -70ºC. The living rooms and bases in Mars won't be different from a sort of parked spaceship where any scratch can lead to an insurmountable death toll. The pregnancies will be bioethically dubious, even when colonising implicitly means not letting demography bent down. To cap it all, a foetus born in the aforementioned Martian circumstances likely will present a set of features that will make those children *look different*. It should be enough to forget about Mars colonisation and await the spatial next-gen for a better colony placement, for example, among the Jupiter moons. But, at any rate, the firms behind Mars colonisation (e.g. Mars One) boasts that a one-way lifelong mission will succeed despite the fact that six months long essays (on Earth!) have badly eroded the volunteers' health and psyche. Not to say that Mars conquest will play a similar role between China ant the US than the Moon did between the US and the USSR. And they will make a reality show of it (oh my!). Very likely the first wannabe Mars settlers will be selected among the fittest for this kind of TV shows: adventurous persons with strong personalities and awesome curricula mixed with a pack of laughable and polemic losers. But just like I've highlighted above, the maintenance routine and menial tasks are key in a hermetic life-support environment and dull, hardened handymen will be more needed than gifted scientists and snobs (especially the later ones!). May the Mars biz is overoptimistic or the producers hope to cover expenses selling an unseen kind of death on TV whatever it happens. Unsurprisingly, no permanent bioethics committee oversees this crock of a colonisation (Sawin, 2017). On the bright side we shouldn't forget that colonisation of Mars will free us from our *Earthiness* for the first time when reconsidering bioethical standards and obligations towards future generations. We might be able then to enjoy an unheard all-embracing perspective and include among those future generations not only our species but also its offspring as well as extraterrestrial sentient beings integrated in our society (Kramer, 2011).

In addition to the Mars fever, during the 2010s we've also seen the birth of the spatial gold-rush, in which corporate and governmental entities have daydreamed about a surging extraterrestrial mining industry (Galliott, 2016). It implies more than privatisation of space issues like patents or rights of exploitation and puts any off-Earth bioethics on the verge of being shot to hell. Why should the winners of this extraterrestrial expansion accept any unfavourable ethical considerations that might hinder their spatial extractivism? Would the UK, Netherlands or Germany have accepted a ban on carbon pollution in the 19th century? Neither US, China, Blue Origin or SpaceX are going to accept moral hindrances as soon as the first big off-Earth buck arrives.

Off-Earth colonial micromanaging cravings will raise a bitter jealousy among Earth factions. More even, a *prima facie* one-way exploration mission (a glimpse through a wormhole and alike) may raise its chances of success *cyborgsitazing* pilot with hull. Can we grant that fear of events in outer space is not going to be a blank check to one-shot human enhancement just like the fear of terrorism has become a free lunch over many (if not all) individual rights?

Life

Anyhow, the most transcendental type of ethical choices (mainly bioethical ones, indeed!) that we'll face in outer space won't be about what we make to our lovely or dreadful galactic explorers' bodies but to life itself. As our galactic maps discloses, our rulers may target new planets with colonising cyanobacteria, extremophile microorganisms, and eggs of multicellular rotifers to lure life to self-propagate in this planet aiming to terraforming it or exploiting it otherwise (Mautner, 2000). The panspermia and possible further terraformation can be an ecopoiesis, which is the creation of an artificially-generated, sustainable ecosystem on a lifeless planet, or aimed at a living planet that we want to modify. Considering this, the ecopoiesis should be the bioethical standard of off-Earth colonisation since it fulfils the not to harm principle incrementally with a caring of life attitude.

But even if the directed panspermia remains in the ecopoiesis bioethical standards, we will face another bioethical issue closer to the bioethics of displacement. Let's figure that ecopoiesis is fostered by two different political agendas, with different missions and visions. If it's the case and competition arises in a planet, the struggle from the different panspermia agendas survivors may suppose a dangerous kind of contamination if the off-Earth samples are brought back to Earth for any reason. As we are going to see in the next section, the biotic ethics and bioethics will meet many times in the deep blue space.

All the livable surface on Earth is mapped till the last inch. And every inch has a jealous and bureaucratized political voice claiming it. But there is Space. Surely, in the big water bodies in the Solar System moons there are microscopic forms of life. In a rare event, perhaps they are bigger and smarter than this. What are the chances that the one to decide on what to do next thinks about your interests? To put it in perspective, have you ever been consulted before sending radio-signals and probes to Space? We live today in an utter ignorance of what civilizations in our present and fleeting human history are out there. If any. We lack intelligence gathering means of what happens on the surface of planets other than the ones in the solar system. With our present 2020s capabilities we won't know a galactic civ from a hole in the ground even in the closest star system. Thus perhaps it is not the most sensible option for Earth to become a backward attention-whore only because our global leaders are so.

XENOBIOETHICS

If the off-Earth bioethics is about the bioethical decisions we'll take in extraterrestrial settings unconcerning previous alien life, the xenobioethics is specifically about what we should do with the life we'll very likely find beyond Earth. The speciesist standard identifies personhood with biological (that's genetic) humanity. Thus, all human organisms regardless of maturity, decay, retardation, comma or congenital malformations are granted personhood. On the other hand, no extraterrestrial life (nor AI or animals), regardless of intelligence and feelings are, according to this standard, a person (Becker, 2000, pp. 20–31). It's important because as Tristam Engelhardt pointed out: *"persons, not humans, are special"* (Engelhardt, 1996, p. 135). The contractualism of John Rawls, as well as his contemporary interpreters, such as Mark Rowlands, support the idea of nonhuman personhood (Andrews et al., 2018). There is possibly an extreme personhood condition in which the alien *is* not a person but *are* a person, instead. Their body, in this case, would be a space of possibilities from which bodily properties emerge and synthesis and analysis is expressed simultaneously. Anyhow, they will likely have a single name even when the reflexive pronoun would be a plural. Whatever the case may be, if we want to address an alien being in a meaningful way we can't strip *it* from its bare name dignity and merely label it like a spare part. So, before any step further in xenobioethics we must adopt an imperfect but working non-speciesist standard of personhood that doesn't threaten & mistreat *the persons* we want to relate with. Otherwise, they might feel entitled to do the same with us…

The astronauts and space agencies should consider the benefits and drawbacks of communicating with aliens, how a dialogue should proceed and whether we actually could establish it. To begin with, some philosophical prejudice is unavoidable when it comes to terms like "life" and "intelligence" applied to newfound aliens (Schneider, 2013). Besides this, the chances of *contact* and the proverbial silence from outer space, unlikely comes from prejudices but from the aliens' alleged unmatched balance between physics and life science -including, at least, the solution of the Enrico Fermi's paradoxes (Webb, 2002). This scientific mastery could unlock to them properties of neutrinos, including far-away interstellar comm (among many other capacities). For good or evil, the rising knowledge on active and sterile neutrinos might soon join the human species to the cosmic chatbox (Spivey, 2015). When it happens, our soundest way to an alien friendly relation is a biophilic and interspecies altruism, or in other words: showing universal love (Gupta, 2014). On the other hand, reciprocity may not be a good strategy given the big amount of time and distance needed to properly interchange information of presents and, likely, we should mistrust from most sudden gifts, especially wonderful devices or cosmic secrets.

In the presence of aliens, and regardless of their shape and smell, we must not forget that only from life can we experience goodness, so living creatures possess intrinsic value both on Earth and elsewhere in the Universe (Peters, 2019). In addition to this, many secularists hold that the aliens' existence will put traditional religions out of business. Nevertheless, there is, for example, a Jewish exotheology tradition since the 1970s that believes that man is not necessarily the only rational being in the universe (Carmell, 1976, pp. 354–356). As Aryeh Carmell states: *"Man's non-singularity does not imply his insignificance. Metaphysical dignity is not part of a numbers game"* (Carmell, 1976, p. 377).

Might it come the time when we will share our planet with (non-human) aliens. Perhaps they were multi-limbed masters as in the Octabia Butler's *Bloodchild* (Butler, 1996), brute-minded predators like the ones in *A Quiet Place* (Krasinski, 2018) or galactic refugees like in District 9 (Blomkamp, 2009). Very likely, there will be a generation who have lived the aftermath of the alien arrival and others, mostly a children cohort, who don't know anything but aliens and humans sharing the planet. In any case, we'll need to assess our position as species and turn our compromised autonomy with the aliens into a worthwhile coexistence (Holloway, 2011, pp. 33–34).

Let's turn back to the panspermia off-Earth stage. Also there is the chance of not being able to identify the lifelessness of the targeted planet. Figure out that we direct panspermia to foster the gene-protein life-based organisms alone while ignoring the chances of alternative *lifeness*. I mean the beings we might find in echoes and crystals, transdimensionality or even former gene-protein species turned into spiritual machines. Would *they* be identified as life? Indeed, one of the most puzzling questions in Astrobiology is the precise definition of what life is (Palhares & Santos, 2017). I believe that narrowing the quest for extraterrestrial life to extraterrestrial cells is judgmental, misleading and *terracentrist* given that Earth is an astronomical rarity that shows off liquid water on its surface.

To cap it all, we might face even an exo-terror: the fear of interplanetary cross-infection and possible ecological damages caused by alien living beings. It would be the case of an alien panspermia or xeno-forming agenda. Anyhow, if terracentrism rules our off-Earth colonisation, we'll get involved in a sort of galactic religious wars soon or later. Then, we will face the exo-terror again. Will the human species become entangled in a proxy war fought on the xenocide of extremophile species? Which politics will rule upon us while the exo-terror spreads a new kind of ontological insecurity?

Likely in the future our extended human existence will allow us to fulfil our potential for good while we sow the cosmos with life and our interspecies relations become mostly constructive and peaceful (with war being only an exception). Perhaps our common purpose includes safeguarding life (Mautner, 2009). On the other hand, the persons in charge of space colonisation in the next century can commit profound mistakes before profound questions. Perhaps, only perhaps they will profit from their detrimental life mistakes. *In outer space no one can hear of your lack of ethics.*

THE COSMONAUT CORPSE

Proposal

The longer the spatial mission the higher the chance to die on board. But how to dispose of the corpse then? I will discuss it from the main views, ethical decision-making and ethical quality axes highlighting the new contribution of bioethics of displacement.

According to the bioethics of displacement as love of belonging, the crew's morale is vulnerable and should be given priority in decision-making. It takes treating the corpse with dignity and respect. Firstly, to dispose of the astronaut's corpse, the big picture must be considered. The potential impact on the environment and other living beings should be evaluated. Therefore, the method of disposal should be selected in a way that minimises harm. Proactivity is also important and a contingency plan should be (at least) sketched before the mission begins to avoid any hasty or unthoughtful decisions. Finally, scepticism means questioning traditional methods of disposal and exploring new, innovative approaches that align with the other three principles.

Ultimately, the disposal of the astronaut's corpse should be carried out with the utmost care and consideration for the principles of bioethics of displacement as love of belonging. The method chosen should prioritise the dignity of the deceased, minimise harm to the environment, and be decided upon proactively and with a healthy dose of scepticism.

Main Views

Prescriptive Bioethics: The crew and the space agency must grant that their mate's corpse will be treated with dignity while reducing the biological hazards of her disposal.

Descriptive Bioethics: The space ship crew must get a group dynamic that helps them to cope with the death of their companion beyond the dignity bare minimum. For example, who should direct the farewell ceremony on board: her closest teammate or a coreligionist that knows better the proper ritual words?

Interactive Bioethics: There are many groups here. The first is the space ship crew. Then, there is the spatial agency of both the reduced and close group in the control room and the agency as a whole. Also there is the public opinion and media (national, international, tabloids, watchdogs, talk-show guests…) and the politicians trying to come off well before them. Any group prolonged unbalance can lead to a moral fracture.

Prospective Bioethics: Since space exploration is a human enterprise rather than an agency one everybody should be involved, the main bioethical view should shift from descriptive to interactive according to communications availability. Ethical issues will appear as the people in the control room feel themselves whimsically pressured by the people outside. Finally but not the least, there will be the death astronaut family, in and out of the spatial agency facilities and the spotlights and, allegedly, the most fragile group and the one who fits the most with a prospective bioethics view (privacy, information access rights, etc.).

Ethical Decision-Making Approach

Legalist Approach (Absolute Moral Norms): The legalist approach is necessary to remind where our starting position was and to set our mind; it may give a sense of earthly belonging. However, doing the things by the book after sailing away from a liquid postmodernity and while exploring the unknown is unlikely to work as planned. (I mean the ethics manual side, it doesn't apply to tested safety protocols and technical procedures).

Antinomian Approach (No Rule for Morality): It's the most feasible and easy to implement. Nevertheless, being in a place, the outer space, where there is neither up nor down and embracing a parallel ethical approach, would put the cosmonauts in the halfway between the Leontinian Gorgias and the Alamut assassins. Simply, it would be too easy to put many goals and personal principles over life, e.g. disassemble a life sustaining machine to repair an auxiliary engine electronics.

Situationist Approach: A few selected guiding principles can be a middle ground between two narrow options. Anyhow, there are long-time moral choices that hardly will put up with contextual decisions, for example, if no pregnancies are allowed during exploration mission a permanent base in a moon suitable for human life shouldn't be an exemption unless the mission goals change to a colonial settings after a discussion encompassing the many stakeholders in and out the spaceship.

Relativist Approach: The moral choices related with the links with life and Earth like child-bearing or interrupting communications are fitter for a legalist approach. On the other hand, the decisions that, allegedly, are going to have a shorter timeliness should be handled with a swiffer antinomian approach. Finally, as a sort of spaceship and space culture with its own values streamlined with human survival progresses, a situationist approach might substitute both the legalist and antinomian approaches.

Bioethics Quality Criteria

Argumentative: The many arguments of how to cope with the moral responsibility of disposing of the astronaut corpse with the due human dignity should be exposed in a way that those who disagree could accept them as a ground to make her counter-arguments. For example, it cannot be motivated by unshared mystical beliefs that lead to a bottleneck in discussion.

Dialectic: The ethical discussion on the astronaut corpse should be useful to expand our moral knowledge pool in views to a further casuistic rather than leading to a stalemate debate.

Empirical: To be a complete Case Study, the moral choices around the astronaut corpse disposal should be contrasted with a real case and grow with the experience on the field. So, a discussion that ends in byzantine debates with no room for a game-changer real event wouldn't qualify as a quality bioethical debate.

Communicational: To ensure the quality of the debate on this Case Study, special attention must to be paid to the ICTs that widen the dialogue to an increased number of stakeholders, for example using digital tokens, turns and phases to help with this sort of asynchronous communication that the unmatched long distances will impose.

What we Have Learned?

Until a spatial communication network were a reality, the descriptive ethics will be, in practice, the main bioethics view in the case of the astronaut's corpse. It stresses even more the need for bioethical formation for the next-gen of space travellers. Also, it stands out the ethical information social weight in the extraterrestrial telecommunications as well as in the psychological well-being.

Until a material and moral culture around extraterrestrial humanity would be developed, moral creativity will rule over the incoming complex decisions like the ones around the cosmonaut farewell disposed of here.

Some ideas from this Case Study are:

- Use surrogate reality technologies (or even holograms) to replace the astronaut body if needed to fulfil the sanitary protocols and overcome the communications lag in order the family could feel themselves included in the ceremony on board.
- Adopting liquid solutions (literally) to get some religious precepts fulfilled. For example, a Jewish faith practitioner must be buried in the ground (not above like a mausoleum, for example) which is impossible to fulfil in outer space. However, a soil solution could be sprayed in the anterior face of the body. A Hindu believer's cremation could be easily attainable if the body is projected against a dense atmosphere or even a sun. Thus, some software to make these calculations easier or even embalming liquids should be provided. The new (dis)solutions hopefully will spur the bioethical discussion further.
- Play phase-based role games with the control room simulating ethical dilemmas like the ones around a deceased companion before the trip and when on field to compare how the outer space disbelonging (and cosmic belonging, too) works in the human psyche and close groups.
- Religious beliefs and spiritual care may be useful to keep the bound with Earth for the first-gen space explorers, so the astronauts should be helped to maintain them.

Further Considerations

The difficulty settings of *the cosmonaut corpse* Case Study could be raised if we change the on board human crew for an AI driven spaceship, a common corpse for a posthuman astronaut body. Also this difficulty could be lowered if we set the burial in a planet or moon, but even in this case there are interesting questions to keep the debate on. How do the non-organic parts of the astronaut be handled? What about the tech and info contained therein? How many corpses are to be considered as polluting the planet? Could the bacteria living in a single human decomposing body be a major environmental hazard than a living one's residuals?

The future former earthling bioethicists will have a meta-ethical goal grounded in the definition of good bioethics (s. sixth chapter on transcendental Bodi), its quality criteria and the ethical information idea: bioethics must flow.

CONCLUSION

Off-Earth bioethics and its implications for space exploration and colonisation need for a comprehensive bioethical framework involving stakeholders, and clarifying criteria around argumentation, dialectics, empiricism, and communication topics.Without this minimum, it won't be possible to navigate the ethical complexities of space exploration and ensure that fundamental principles of dignity and responsible decision-making are upheld. Likewise, it will be needed a xenobioethics before facing non-speciesist standard of personhood dilemmas. The future of humanity as cosmic stakeholders depends on developing robust off-Earth bioethics that guide our actions and protect the well-being of all involved.

REFERENCES

Andrews, K., Comstock, G. L., G.k.d, C., Donaldson, S., Fenton, A., John, T. M., Johnson, L. S. M., Jones, R. C., Kymlicka, W., Meynell, L., Nobis, N., Pena-Guzman, D., Sebo, J., Comstock, G. L., G.k.d, C., Donaldson, S., Fenton, A., John, T. M., Johnson, L. S. M., … Sebo, J. (2018). *Chimpanzee Rights: The Philosophers' Brief.* Routledge. doi:10.4324/9780429461071

Becker, G. K. (2000). *The Moral Status of Persons: Perspectives on Bioethics.* Rodopi. doi:10.1163/9789004495029

Blomkamp, N. (Director). (2009). *District 9.* https://www.imdb.com/title/tt1136608/

Butler, O. E. (1996). *Bloodchild and Other Stories.* Seven Stories Press.

Carmell, A. (1976). *Challenge: Torah Views on Science and Its Problems.* Feldheim Publishers.

Ćirković, M. M. (2017). Enhancing a Person, Enhancing a Civilization: A Research Program at the Intersection of Bioethics, Future Studies, and Astrobiology. *Cambridge Quarterly of Healthcare Ethics, 26*(3), 459–468. doi:10.1017/S0963180116001134 PMID:28541174

Coleman, R. D. N. (2019). *Valid consent considerations for astronauts and space tourists involved in medical research.* Academic Press.

Engelhardt, H. T. (1996). *The Foundations of Bioethics.* Oxford University Press.

Galliott, J. (2016). *Commercial Space Exploration: Ethics, Policy and Governance.* Routledge. doi:10.4324/9781315572857

Gupta, A. (2014). Altruism Toward Non-Humans: Lessons for Interstellar Communication. In D. A. Vakoch (Ed.), *Extraterrestrial Altruism: Evolution and Ethics in the Cosmos* (pp. 79–91). Springer Berlin Heidelberg. doi:10.1007/978-3-642-37750-1_6

Holloway, K. F. (2011). *Private Bodies, Public Texts: Race, Gender, and a Cultural Bioethics*. Duke University Press.

Kramer, W. R. (2011). Colonizing mars—An opportunity for reconsidering bioethical standards and obligations to future generations. *Futures*, *43*(5), 545–551. doi:10.1016/j.futures.2011.02.006

Krasinski, J. (Director). (2018). *A Quiet Place*. https://www.imdb.com/title/tt6644200/

Kubrick, S. (Director). (1968). *2001: A Space Odyssey*. https://www.imdb.com/title/tt0062622/

Liang, X., Zhang, L., Wan, Y., Yu, X., Guo, Y., Chen, X., Tan, C., Huang, T., Shen, H., Chen, X., Li, H., Lv, K., Sun, F., Chen, S., & Guo, J. (2012). Changes in the Diurnal Rhythms during a 45-Day Head-Down Bed Rest. *PLoS One*, *7*(10), e47984. doi:10.1371/journal.pone.0047984 PMID:23110150

Lin, P., & Abney, K. (2014). Introduction to Astronaut Bioethics. *FutureTense, 8*.

Mautner, M. N. (2000). *Seeding the universe with life: Securing our cosmological future*. Michael Mautner.

Mautner, M. N. (2009). Life-Centered Ethics, and the Human Future in Space. *Bioethics*, *23*(8), 433–440. doi:10.1111/j.1467-8519.2008.00688.x PMID:19077128

PalharesD.dos SantosÍ. A. (2017). *Astronomic Bioethics: Terraforming X Planetary protection*. https://arxiv.org/abs/1704.08404

Peters, T. (2019). Does extraterrestrial life have intrinsic value? An exploration in responsibility ethics. *International Journal of Astrobiology*, *18*(4), 304–310. doi:10.1017/S147355041700057X

Reed, R. D., & Antonsen, E. L. (2018). Should NASA Collect Astronauts' Genetic Information for Occupational Surveillance and Research? *AMA Journal of Ethics*, *20*(9), 849–856. doi:10.1001/amajethics.2018.849 PMID:30242816

Sawin, C. (2017). Bioethics in Space Exploration. Encyclopedia of Bioastronautics, 1–6.

Schneider, J. (2013). Philosophical issues in the search for extraterrestrial life and intelligence1. *International Journal of Astrobiology*, *12*(3), 259–262. doi:10.1017/S1473550413000025

Seedhouse, E. (2010). *Prepare for Launch: The Astronaut Training Process*. Springer Science & Business Media. doi:10.1007/978-1-4419-1350-0

Spivey, R. J. (2015, June). *A cosmological hypothesis potentially resolving the mystery of extraterrestrial silence with falsifiable implications for neutrinos*. https://www.ingentaconnect.com/content/pe/pe/2015/00000028/00000002/art00019

Strollo, F., Macchi, C., Eberini, I., Masini, M. A., Botta, M., Vassilieva, G., Nichiporuk, I., Monici, M., Santucci, D., Celotti, F., Magni, P., & Ruscica, M. (2018). Body composition and metabolic changes during a 520-day mission simulation to Mars. *Journal of Endocrinological Investigation*, *41*(11), 1267–1273. doi:10.100740618-018-0861-9 PMID:29532423

Vesely, P. (2018). Conservative Colonization: An Ethical Analysis of Galactic Expansion. *UReCA: The NCHC Journal of Undergraduate Research & Creative Activity*. https://digitalcommons.unl.edu/ureca/51

Webb, S. (2002). *If the Universe Is Teeming with Aliens ... WHERE IS EVERYBODY? Fifty Solutions to the Fermi Paradox and the Problem of Extraterrestrial Life*. Springer Science & Business Media.

KEY TERMS AND DEFINITIONS

Alien Life: Non-indigenous Earth life.

Ecopoiesis: Creating sustainable ecosystems on lifeless planets or within contained environments, offering ethical alternatives to terraforming.

Extraterrestrial Life: Life forms that originate from beyond Earth, including but not limited to aliens.

Life: A system far from equilibrium that actively uses information for self-preservation and belonging (e.g., to a species, family, etc.).

Off-Earth Bioethics: Bioethics to be applied in extraterrestrial environments, beyond the confines of Earth and commonly addressed to space exploration.

Panspermia: The idea that life exists throughout the universe and can spread through space from one place to another.

Person: A being with rights and responsibilities.

Terraforming: The process of deliberately modifying the environment of another planet or moon to make it habitable for Earth-like life forms.

Xenobioethics: Bioethics applied to alien life, including personhood, communication, and coexistence.

ENDNOTE

1. During Mars 500 project the isolation effects on 6 volunteers for more than 500 days was studied and a 9.2% decrease of BMI was found (among many other harmful conditions).

Chapter 19
The Cosmos of Bioethics of Displacement

ABSTRACT

Embark on a journey across time and the impact displacement has had on humanity. Using the analogy of a one-year calendar, the author glides over the historical events that have led to the current state of the world, including the rise of bioethics and the effects of the industrial revolution. The new big data empires are rapidly changing the world, and those who do not master them likely will be victims of those who do. The symbols we use and the way we treat the vulnerable are important aspects of displacement as seen in the impact of agribusiness giants on indigenous communities. Otherwise, the macrocosm of the bioethics of displacement will resume in a brain-dead humankind in a shifting landscape. It's important to understand the impact of displacement on humanity, and the values needed to navigate this rapidly changing world in which politicians' role seem to promote displacement and the subsequent growth of gangs. To cope with all this loss of meaning, some values are essential, including self-kindness, human warmth, solidarity, and a sense of belonging.

INTRODUCTION

In the last chapter, the focus was on the ethical considerations surrounding off-Earth bioethics, space exploration, and the potential colonisation of other planets. It touched upon topics such as the recruitment of astronauts, the impact of space travel on the human body, the ethical implications of genetic information collection, and the concept of xenobioethics. But this chapter takes a broader perspective and explores the concept of displacement in the context of human history and the future of humanity.

DOI: 10.4018/978-1-6684-4808-3.ch019

BODI'S COSMIC CALENDAR

It is the responsibility of scientists never to suppress knowledge, no matter how awkward that knowledge is, no matter how it may bother those in power; we are not smart enough to decide which pieces of knowledge are permissible and which are not. (Malik, 2018)

Let's now imagine Bodi's timespan as a one year calendar, just like Carl Sagan did at *Cosmos* (Druyan et al., 1980, Chapter 1). If the timespan since the first primitive shark moms made displacement important for life were a year, the human migrations, wonders, horrors and new and unmatched deeds would have started in the last day dusk; the surge in references to "bioethics" would have appeared in the last ten minutes. It took humans roughly an hour of this year to remove Neanderthals, a bit less than 20 minutes for deprecating nomadism since the first cities and only 20 seconds for the Industrial Revolution coal and carbon favoured to rule over the deindustrialized economies. The way the new Big Data empires are changing the world is so expeditious that we hardly realise what is happening. This new divergence against our common humanity will turn, one more time, the people who don't master it into the victims of the ones who do. But, seemingly, it will be faster than industrialization, 3-4 seconds in the calendar of Displacement at most or, in other words, it will end before 2050. But beware, the light that burns as short can burn as hotter and, maybe, what is going to end before 2050 is Humankind. History teaches us the daily life values we must thrive for in order to cope with displacement. Those values are: self-kindness, human warmth, solidarity and sense of belonging.

THE MACROCOSM OF THE BIOETHICS OF DISPLACEMENT

A Brain-Dead Humankind in a Shifting Risks Landscape

The idea of macrocosm as something greater than life but connecting us comes from Empedocles and echoes through Greek and Roman until the present 2019 public health concepts (Have, 2019, p. 45). However, it's not only a Western idea, but quite the opposite: the meteorological medicine practised by the Ancient Greeks mirrors in African ontology and Ancient China diagnosis (Houston, 2002; Masolo, 1994, p. 108).

In history, the skin has been a place where microcosm and macrocosm joined (Diamandopoulos et al., 2001). Unsurprisingly, the study of macrocosm, more even in bioethics, have been hindered by the political incivilities and plain criminal abuse like the Tuskegees, which turned colour sexual minorities into more elusive study groups, as well as other issues that litters the grounds of our discipline (Ramos et al., 2019). Thankfully, ethical information management, namely from electronic devices can fix it. For example, the data custodian reciprocity, non-exploitation and focus on the public good is an invaluable asset in the use of personal information in health research (Allen et al., 2019; Lozano Rodríguez, 2018, pp. 73–75).

To reach a macrocosmic level of xenotransplantation as a global and affordable technology leading the zoonotic risks to be distributed globally was over the 20th century bioethics capabilities (Rothblatt, 2001). In addition to this, bioterrorism global scorch will likely use zoonoses (Jackson et al., 2006; Ryan, 2008). It's up to see how this global distribution of risk-producing stakeholders changes as the global climate advances.

There is on Earth about 8 billion kilos of human thinking matter (Dunagan, 2004) but even when it comes to deciding on the behalf of Earth or humankind not even 40 tons are used.[1] And not even these few tons are used entirely or impartially! Ironically, there is also 8 billion kilos of plastic in the Ocean awaiting removal (Mast, 2019) but till the last gram counts. A brain with less than 1% of activity is close to the subject's clinical death, but when it comes to most transcendental decision making human brains' activity is 0,0005%. If we don't stop being like dead, we'll be dead for good.

Between a New Colonial Dawn and the Pol-Cor Eugenics

About 1.27 trillion new photographs were expected to be taken in 2017, which means that the human population likely took three times more photographs in one year than there are stars in the Milky Way (Ingraham, 2017). The human species has engraved symbols that are specifically human on the whole planet. It should denote the capacity to be responsible that accompanies semiotics to a similar extent. Just like happens with choices, we're not only responsible for the symbols we use, but also for the symbols used on our behalf and how all these symbols affect life. Two of these symbols are © and ™ usually beside a logo that has swallowed the original intent of an individual enterprise. Despite that their goals are the worldliest you can figure out, the big pharma and data companies have turned the creation and life-giving divine sparks into suitable for displacement gifts thanks to the patents regime. Resources in public health are below the tradeoffs between survival, individual well-being and common good. Both the political claims and the possibility to reach a moral end to this, are hypocritical, unrealistic or both. On the other hand, stinginess with the most vulnerable doesn't bring a bloody new hope for those middle-class individuals in need of expensive healthcare. Rather than this, it only spurs the greediest ones to push towards a pol-cor eugenics.

Biosemiotics is the field that explores the signs, symbols, and communication processes in living organisms. It investigates how living beings generate, interpret, and exchange information, using various forms of communication such as chemical signals, visual displays, vocalizations, and more. Biosemiotics examines the role of signs and symbols in understanding the behavior, relationships, and meaning-making processes of living organisms.

Thanks to the bioethics expansion to the new Prospective-Relativist-Communicational ethical coordinates, we have reached a new degree of interdisciplinarity between biosemiotics and bioethics. The intersection between biosemiotics and bioethics occurs when we examine the ethical implications of communication and meaning-making processes in living organisms. It involves considering how ethical principles apply to the study and understanding of signs, symbols, and communication within the biological realm. This intersection raises questions about the treatment of living beings, the impact of human actions on their communication systems, and the ethical responsibilities associated with our interactions and interventions in the natural world. By exploring the intersection of biosemiotics and bioethics, we gain a deeper understanding of the ethical dimensions of life, communication, and our relationship with the living world, including people who communicate in unusual ways. It helps us navigate ethical dilemmas and promotes a more responsible and respectful approach to the diverse forms of life and loving life on our planet.

A needed learning before exploring the deep dark beyond our blue sky. A balance between the utter needs of a meaningful human life conservation and the preservation of alien worlds should rule. It should be a clear supranational politic. As Jai Galliott said: "*Outer space and the future of humankind don't belong to any one nation-state but to all of us*" (Galliott, 2016, p. 255). Looking at the results on Earth, it shouldn't open the floor for star colonialism. The space expeditions earned lore will have a great value for bioethics and should be accountable before a bioethics committee just like a research or experiment is. Furthermore, this transparency must balance the space billionaires and their minions' euphoria.

Euphoria is also abundant in the merciless and exploitative agenda of agribusiness giants for the indigenous communities. Those big logos, big money and big words that turn the former lands into mere territories. Those entities foster a capitalism whose roots are poisoned and cripple the genuine entrepreneurial initiative turning the native women in spoils kicked away from their traditional and key roles and grant to the LBGTQs the kind of regimes that will provide them with politics of pains, displacement and exile. Meanwhile, the lair of many of these giants willingly has forgotten what a refugee is (United Nations High Commissioner for Refugees, 1984, p. 36). Donald Trump and Joe Biden, butlers of giants in disguise, gave his back to the democratic virtues of his own people and dwelled in racial animus while many families were torn apart during their tenure. Moreover, the gangs (a side effect of big business-driven disbelonging) insanely envy the millionaires like Trump or Pelosi. Intuitively, the gangs know themselves as a flawed copy of their creator and try to reach through violence the same land of milk and political forgiveness belonging to the eidetic *mara* that many modern states are. The desire for getting the monopoly of fear is what makes violence a state-builder (Warnecke-Berger, 2019).

A day may come when the biopolitics of the brain ground a Ministry of Memory putting in order our political will. A darkest hour in which a sentient and moral newfound creature will be mistreated at sight. But now, within us, lies the chance to avoid it from happening.

Peaceful and Flammable

In Bodi's eyes the philosophy as epistemic displacement treasures musical properties, as Pythagoras stated from cosmology and therapy twenty-five centuries before me (Riedweg, 2005, pp. ix, 103–33). The philosopher of Samos, however, allegedly didn't care about how the divergence imposed upon the disempowered ones drove them to lose the meaning of life even before that the meaning became speakable. The only way you can become immune to displacement is turning yourself into a sight, an ethereal eye, becoming an illusion or perhaps having a greater density than the real world. Even the Bioethics displaces itself, blurred and restless, between the body and the language, both multiple and incomplete. It awaits to be stretched with the bioethical dilemmas about cyborg dis-abilities and plug-abilities as well as undead inherent rights or the lack of them.

The practical wisdom that Bioethics of Displacement takes, implies connecting with our moral selves while seeing the moral value in the others' lives. In Bodi, intuition and moral imagination gain relevance versus other forms of moral knowledge as can be seen in the section on bioethical grounds of immigration.

Hardly anybody can doubt today about the biopolitics' weight on bioethics, especially in the public sector top-notch charges. Further than this, the partisan interests, which take the lion's share, are sapping an authentic self-assessment in our professional commitment as bioethicists. Sadly, those Western bureaucracies of virtue that were allegedly a cure for the maniacal human subject research during the Nazi regime, sanctioned two *Tuskegees*.

The responsibility of politics of pain as a whole, cannot be traced causally to any particular entity or agent beyond particular heinous crimes or leaked political foul game. That's what makes the forcible displaced persons to keep moving and getting displaced. Even when refugees are pinned, they keep being left behind. As pointed out in the Greek sanitary crisis section, many refugee camps hold a great deal of structural injustice.

Since the Soviet Empire implosion, the main contributor to this postmodern Gehenna archipelago is the weight of Western politicians in the international community but it's far from being the only one (Parekh, 2016, p. 121). China seeks the same as the US and the rest of countries: to make the world safe for their leaders' way of life. For Washington it means a world full of open for bribing partitocracies, a sort of *Europa Universalis*. A world where the White House can commit a terrorist and war-like act as blowing a gasoduct and no country dares to *pipe up* (Hersh, 2023). On the other hand, Beijing wants a world of tributary states in which the way the Party rules Chinese people is privileged, too (The Economist, 2020; Walker, 2018). Of course, neither the Washington nor Beijing plans won't work in the best interest of most people, including many US and Chinese citizens. And, unlike in the US versus USSR match, this time we are not going to see a true race for welfare. This time the race is going to be for a more and more incisive and ubiquitous data-ification and manipulation reaching till the last subject. The race is for making people lose everything and believe that they are doing well. Supranational institutions and globally toxic events will be part of the battlefield as usual but the most dangerous by far is going to be the AI.

The Westerners suffer from an infantilization of politics that cannot be corrected with speeches on democracy and giving the back to the problem but with true socialisation (Jörke & Selk, 2015, p. 496; Sieferle, 2017, p. 128; von Beyme, 2019). In a twist from the *Idiocracy* movie, the politicians who most reproduce (at least in the streams) are not the readiest for bringing innovative solutions but the readiest to suck up human mediocrity and alleged leading democracies' Parliaments are a chicken coop (Judge, 2006). Sooner or later, the Chinese will face economic stagnation and then the Party will resort more and more to a struggling mindset, a road that Xi Jinping has started to walk (Pottinger et al., 2022; Tepperman, 2022). Either because there is nobody able to solve problems or because the scale and exigence of those problems are too high, both the White House and the Party Charter are going to resort to AI in the near and middle-term future to get them out of trouble. But those AIs will be programmed and trained in political leaders' own political image and likeness. When it happens, billions of people are going to suffer, even die, because their present-day or former rulers didn't care about the difference between intelligence and epistemic maturity. While intelligence focuses on cognitive abilities, epistemic maturity goes beyond intelligence and emphasises the quality of one's thinking and understanding. It involves the ability to engage in reflective and critical thinking, to consider multiple perspectives, and to evaluate the validity and reliability of information. Epistemic maturity is important for making informed judgments, forming beliefs based on evidence, and navigating complex and diverse information in an increasingly interconnected world. So, yes the AI is going to outperform their owners in the White House, the Zhongnanhai or wherever, but it will solve the problem in unsustainable ways, just like Napoleon or Stalin, but in this case blinded by its own taming.

Another source of social conflict as redistribution, in turn, suffers a continuous displacement through a path of negative tension and is burdened with the lack of translation (Turim, 1985, p. 13). When we try to share things fairly, it can be really hard because there are often problems and conflicts. To cap it all, sometimes people don't understand each other, which makes it even more challenging to share things in a fair way. No human pain and flesh seems too much when it comes to self-asserting the status quo of those who are banally evil, who trade self-kindness for self-praising and bounding for bondage. The policies aiming to stop poverty and massive displacements must act against a myriad of spread agents, e.g. the many financial stakeholders mentioned in this book, who live across the globe and who many times harm the others in innovative ways while being admired for their success. There is no government honest and diligent enough to scratch those money and media benders who constitute *the misery of nations*.

Approximately the same number of people, 40 million, is enslaved and brutalised than the one suffering from the double condition of refugee and disabled in inhuman conditions (Janz, 2010), day after day.[2]

To begin with a solution, we should focus our minds on how we can fight against the wicked scarcity of opportunities in the access to healthcare from the workplace thanks to a better understanding of structural barriers. And building from this.

A successful way to displace societal groups unfit for a political vision is belittling the persons who make it up from political subjects to political objects. I have covered those politics of pain pointing out the groups targeted by the Nazi regime but I want to highlight that it's not the past. Gehenna politics at its worst are happening in places such as Uganda and Yemen where the oil for development and better life vision has found uncomfortable the people living in the recently identified natural resources sites (Holterman, 2014; Lewton, 2019).

Let's face it and stop the consensus makers' work of a too homogeneous human species. Doing so might revert the politicisation of migration statistical models. Let's stop the policy makers imposing upon the most of the population their own tailored determinants rather than disclosing human groups interdependency. Most people have yet followed enough an apex of greedy attention whores amidst the trivial legitimacy of a thundering acclamation. We are badly suffering from the leadership of individuals who prefer to put life in order than love it. An opportunistic minority who wants to make us feel as if we were flat-footed in a flat universe. But society is like the plasma: both peaceful and flammable. At least a good society is.

THE MICROCOSM OF THE BIOETHICS OF DISPLACEMENT

Sex, Money, and Expats Researchers

The vulva is the organ of belonging. It's the organ where all our other bounds besides the umbilical one will start. And there is the mouth, the horizontal vulva, which most of us use to establish bounds thanks to communication. Nevertheless, plenty of interests are over the authentic communication and the women's genitalia suffers more often atrocities driven by politics of pain (war rape, mutilation, sex tourism and unneeded painful deliveries). The leaders and lobbyists that contribute to the politics of pain hate this world and their inhabitants so much that they want to pull it out by the roots. Doing so, they commit an actionable crime against humanity.

Many citizens in wealthy countries fear and loathe immigrants because they think that the newcomers wish the same things and resources as them and the welfare will go extinct. However, the same could be said from our fellow countrymen. Fostering the human value of hospitality against displacement should be in any political agenda. A good place to start with is enhancing the mobility of academics, researchers and technicians. If it were a very significant share of migrant people with tertiary education, besides brain flight issues, the science might turn more compassionate towards the less lucky expats. The in-formation of an intellectual, the way her bibliography, diplomas, acronyms, library loans and searches historial shape her being becomes false when she gets fond of the architects building a winners and losers society. Then, her former inquiring spirit is enfeebled and twisted enough to practise an academic necrophilia in which the main subject is her own writings' popularity.

As scientists, we have put too much weight on the success of our scientific predictive models, mainly short-term ones, which prevented us from reaching a deeper epistemology. We don't realise how much scientific discoveries brims with unpredictable political misuses (Emmerich, 2018). And vice versa. The gene editing, for example, is like a red raging bull with a pale-grey elephant behind. Everybody looks at the red bull and remains unaware of the far more dangerous pachyderm of the epigenetic changes driven by social structures (Combs-Orme, 2013; Juengst et al., 2014). To prevent the fragility coming from our fixation in successful predictions, we should pay more attention to fiction works (simulators included) since these are the keys that will unlock a new set of open-minded scientific features. This is risk prevention against the unseen mistakes that we, the scientists, are going to commit.

Some Grump Politicians and White Rabbits

Rather than in a true democracy most countries are wrapped in a liberal pseudo democracy eager to label any reform as populism while the lifelong demagoguery gains momentum. When the most power-thirsty politicians become trendy, the opportunist hard-hats show no mercy with the offenders trying to regenerate their identities since they are opposite poles. Sometimes, the individuals with power to stop a war are the ones who start them and sanction with nationalist forgiveness what is actually a massacre (of civilians most times). Sadly, working for peace seems harder than pleasing chauvinists and depict working migrant families as a sort of doom bringers while paving the road to belligerency against materially poorer countries.

A politically induced displacement blindness happens when an individual political identity owns a great share of her whole identity. In this case, the versatility of human identity gets lost and the person becomes trapped inside, unable to properly reflect. Unable to understand that in a world where the borders grow thicker and the differences grow more irreconcilable, the countries look more like prisons. Pens where the psychological harm from discrimination, especially that one happening in the classroom and the white-collar bureaucratic one, is ignored when the target is *an Other*.

Awaiting for many societies to fulfil their promises of turning themselves in fairer places to live with disability avoids most people to see the elimination of diseases and health handicaps as the straightest line to insertion. It's not quite different to the punishment blindness in prisons where heinous abuses are the research lab in which the inter-breeding privileged ones weave new politics of pain. There, the victims are turned unable to mistrust in a system that has become their whole world, a world where mothering and parenting seems to be stripped from the human condition. Too many people have been displaced, are displaced and will be displaced because their neighbours turned a blind-eye when no profit was expected.

The economic priorities have swayed so badly the question of what standard of care a physician owes to her patients that it's like a void sack. Have the persons who were owed bioethics moral standards received the same treatment? I don't think so. Firstly, the sins that someone may commit in bioethics, especially if we consider the effect on the sinner herself, may result in ludicrous indulgence when it comes to health-care economy mismanagement. The belief in the lack of healthcare for everybody as a sort of natural law is the source of most pretexts used to kneel bioethics before the financial interest of middle and above classes. Every day millions of people are abandoned in the Bodi's backyard: the forsaken wasteland where health and physical integrity are unaccountable. Disability should teach us to evaluate a situation for its virtue rather than its complexity (Ellison, 2019).

The quality of life as reflected in the cholinergic basal forebrain performance during AD is a middle ground between cognitive science and bioethics (Whitehouse, 2006). In the future, we may paint a homunculus of displacement pinpointing the brain parts used when leaving anyone behind and being left behind. In addition to this, the bioethical debate about intellectual enhancement is closer to partisan politics than standing on a ground where it could serve the public good. Like in the Jefferson Airplane's hit on *Alice in Wonderland* (Carroll, 1992; Jefferson Airplane, 1966), the logic and proportion about to fall sloppy dead for the incoming ethical dilemmas about the brain.[3] What will we use to feed our heads? Will the surrogate realities and BCIs be our postmodern white rabbits?

The brain is hard to scrutinise. The spatial dimension of selfhood disaggregated during death keeps folded and zipped in corporeality while we can stay alive. However, Stelarc's view of information as prostheses (Stelarc, 1997) stretches both the human corporeality in unheard ways.

Figure 1. Stelarc hanging from the ceiling over a giant replica of his own famous ear-attached arm. Some transhumanists are clear about the need to go for the full monty if you want to change the rules.
Source: Stenslie (2015, p. 24)

Transhumanism, by the way, turns body dignity into an artefact that nobody knows how is going to work in the present 2020s. Just like we don't overlook religions that have common moral beliefs within a culture, we shouldn't overlook transpirituality and transcendence in transhumanism. Transpirituality is a virtue in transhumanism that focuses on the spiritual aspects of human enhancement and transformation. Transcendence is a term used to describe extraordinary experiences that go beyond normal limits. These concepts shouldn't be ignored because they provide unique perspectives and insights that are important to consider in discussions about ethics and human potential. Just as we respect different religions, we should also value and explore these ideas that offer new ways of understanding ourselves and the world around us. The dignity in a human enterprise is proportional to the personal risk plus responsibility taken. Either in human enhancement or any other issue, forbidding a person to take her own responsibility solely on religious grounds or blame the others if they don't accept our changes are two ways to prevent this person from behaving with dignity.

CONCLUSION

Addressing Displacement and its implications is urgent for humanity: world's rapid pace of change warns us about the dangers posed by the growing influence of Big Data control and surveillance empires, either oligarchic pseudo-democracies or overtly liberticide regimes. To make things worse, even scientific models are limited when exposing the politics of pain (as COVID-19 taught us) so we need a deeper epistemology that encourages the exploration of diverse perspectives, including fiction, to prevent unseen mistakes. A philosophy that doesn't shrug from the pursuit of peace or transhumanism topics, but neither fall in naiveté or careerism. A philosophy that embraces critical thinking and other values that upholds our common humanity such as individual autonomy, dignity, and responsible decision-making.

REFERENCES

Allen, J., Adams, C., & Flack, F. (2019). The role of data custodians in establishing and maintaining social licence for health research. *Bioethics*, *0*(4), 502–510. Advance online publication. doi:10.1111/bioe.12549 PMID:30657596

Carroll, L. (1992). *Alice in Wonderland*. Wordsworth Editions.

Combs-Orme, T. (2013). Epigenetics and the Social Work Imperative. *Social Work*, *58*(1), 23–30. doi:10.1093wws052 PMID:23409337

Diamandopoulos, A. A., Goudas, P., & Diamandopoulou, A. H. (2001). The human skin: A meeting ground for the ideas about macrocosm and microcosm in Ancient and Medieval Greek literature. *Revue Officielle de La Societe Internationale d'Histoire de La Medecine Acta Internationalia Historiae Medicinae*, 94.

Druyan, A., Sagan, C., & Soter, S. (Directors). (1980, September 28). *Cosmos* [Documentary]. KCET, Carl Sagan Productions, British Broadcasting Corporation (BBC).

Dunagan, J. F. (2004). Neuro-futures: The brain, politics and power. *Journal of Futures Studies*, *9*(2), 1–18.

Ellison, B. (2019). The Patient as Professor: How My Life as a Person with Quadriplegia Shaped My Thinking as an Ethicist. *Perspectives in Biology and Medicine*, *62*(2), 342–351. doi:10.1353/pbm.2019.0018 PMID:31281126

Emmerich, N. (2018). Bioethics, public intellectuals and political biology today. *History of the Human Sciences*, *31*(1), 124–131. doi:10.1177/0952695117729119d

Galliott, J. (2016). *Commercial Space Exploration: Ethics, Policy and Governance*. Routledge. doi:10.4324/9781315572857

Hersh, S. (2023, February 8). How America Took Out The Nord Stream Pipeline [Substack newsletter]. *Seymour Hersh*. https://seymourhersh.substack.com/p/how-america-took-out-the-nord-stream

Holterman, D. (2014). The biopolitical war for life: Extractivism and the Ugandan oil state. *The Extractive Industries and Society*, *1*(1), 28–37. doi:10.1016/j.exis.2014.01.003

Houston, L. (2002). Historical and Cross-Cultural Perspectives on the Art of Medical Diagnosis. *Senior Thesis Projects, 1993-2002*, 97.

Ingraham, C. (2017). The Suddener World: Photography and Ineffable Rhetoric. *Philosophy & Rhetoric, 50*(2), 129–152. doi:10.5325/philrhet.50.2.0129

Jackson, B. A., Buchler, J. W., Cole, D., Cookson, S., Dausey, D. J., Honess-Morreale, L., Lance, S., Molander, R. C., O'Neal, P., & Lurie, N. (2006). Bioterrorism with zoonotic disease: Public health preparedness lessons from a multiagency exercise. *Biosecurity and Bioterrorism, 4*(3), 287–292. doi:10.1089/bsp.2006.4.287 PMID:16999589

Janz, U. (2010, September 3). *Conference of the States Parties to the Convention on the Rights of Persons with Disabilities*. Academic Press.

Jefferson Airplane. (1966, November 3). *White Rabbit* [7-inch single]. RCA Victor.

Jörke, D., & Selk, V. (2015). Der hilflose Antipopulismus. *Leviathan (Düsseldorf), 43*(4), 484–500. doi:10.5771/0340-0425-2015-4-484

Judge, M. (Director). (2006). *Idiocracy* [Adventure, Comedy, Sci-Fi]. Twentieth Century Fox, Ternion Pictures, Major Studio Partners.

Juengst, E. T., Fishman, J. R., McGowan, M. L., & Settersten, R. A. Jr. (2014). Serving epigenetics before its time. *Trends in Genetics, 30*(10), 427–429. doi:10.1016/j.tig.2014.08.001 PMID:25242336

Lewton, T. (2019, October 23). Uganda's eco-feminists are taking on mining and plantation industries. *Mongabay Environmental News*. https://news.mongabay.com/2019/10/ugandas-eco-feminists-are-taking-on-mining-and-plantation-industries/

Lozano Rodríguez, M. (2018). *Ethical Information Foresights: A journey across uncharted conflictology*. American University of Sovereign Nations.

Malik, E. (2018, November 9). *Carl Sagan history timeline*. Project Management Tips and Tricks. https://www.officetimeline.com/blog/carl-sagan-timeline

Masolo, D. A. (1994). *African Philosophy in Search of Identity*. Indiana University Press.

Mast, K. (2019, January 15). *State of Science: World Wakes to Perils of Plastics*. Discover Magazine. http://discovermagazine.com/2019/january/world-wakes-to-perils-of-plastics

Parekh, S. (2016). *Refugees and the Ethics of Forced Displacement*. Taylor & Francis. doi:10.4324/9781315883854

Pottinger, M., Johnson, M., & Feith, D. (2022, November 30). Xi Jinping in His Own Words. *Foreign Affairs*. https://www.foreignaffairs.com/china/xi-jinping-his-own-words

Ramos, S. R., Warren, R., Shedlin, M., Melkus, G., Kershaw, T., & Vorderstrasse, A. (2019). A Framework for Using eHealth Interventions to Overcome Medical Mistrust Among Sexual Minority Men of Color Living with Chronic Conditions. *Behavioral Medicine (Washington, D.C.), 45*(2), 166–176. doi:10.1080/08964289.2019.1570074 PMID:31343963

Riedweg, C. (2005). *Pythagoras: His Life, Teaching, and Influence*. Cornell University Press.

Rothblatt, M. (2001). Key Legal Issues: Determining the Rise of Geoethics as the Optimum Analytical Paradigm. *Graft, 4*(2), 143–145. doi:10.1177/152216280100400218

Ryan, C. P. (2008). Zoonoses likely to be used in bioterrorism. *Public Health Reports, 123*(3), 276–281. doi:10.1177/003335490812300308 PMID:19006970

Sieferle, R. P. (2017). *Das Migrationsproblem: Über die Unvereinbarkeit von Sozialstaat und Masseneinwanderung*. Manŭscriptum.

Stelarc. (1997). From psycho to cyber strategies: Prosthetics, robotics and remote existence. *Cultural Values, 1*(2), 241–249. doi:10.1080/14797589709367146

Stenslie, S. (2015). Stelarc: On the Body as an Artistic Material. *The Journal of Somaesthetics, 1*(1).

ten Have, H. (2019). *Wounded Planet: How Declining Biodiversity Endangers Health and How Bioethics Can Help*. JHU Press. doi:10.1353/book.66171

Tepperman, J. (2022, December 19). China's Dangerous Decline. *Foreign Affairs*. https://www.foreignaffairs.com/china/chinas-dangerous-decline

The Economist. (2020). *China wants to put itself back at the centre of the world* (China's Belt and Road). https://www.economist.com/special-report/2020/02/06/china-wants-to-put-itself-back-at-the-centre-of-the-world

Turim, M. C. (1985). *Abstraction in avant-garde films*. UMI Research Press.

United Nations High Commissioner for Refugees. (1984, November 22). *Cartagena Declaration on Refugees, adopted by the Colloquium on the International Protection of Refugees in Central America, Mexico and Panama, Cartagena de Indias, Colombia, 22 November 1984*. UNHCR. https://www.unhcr.org/about-us/background/45dc19084/cartagena-declaration-refugees-adopted-colloquium-international-protection.html

von Beyme, K. (2019). Definitions of Populism and Its Stages of Development. In K. von Beyme (Ed.), *Rightwing Populism: An Element of Neodemocracy* (pp. 27–34). Springer International Publishing. doi:10.1007/978-3-030-03177-0_2

Walker, C. (2018). What Is "Sharp Power"? *Journal of Democracy, 29*(3), 9–23. doi:10.1353/jod.2018.0041

Warnecke-Berger, H. (2019). The Cultural Political Economy of Violence: A Conclusion. In H. Warnecke-Berger (Ed.), *Politics and Violence in Central America and the Caribbean* (pp. 261–277). Springer International Publishing. doi:10.1007/978-3-319-89782-0_8

Whitehouse, P. J. (2006). Quality of life: The bridge from the cholinergic basal forebrain to cognitive science and bioethics. *Journal of Alzheimer's Disease, 9*(s3), 447–453. doi:10.3233/JAD-2006-9S351 PMID:16914884

KEY TERMS AND DEFINITIONS

Cyborg: A creature part organic, part machine, with restored functions or enhanced abilities through integrated feed-back controlled mechanical elements.

Epistemology: The study of knowledge, including its nature, limits, and how we justify our beliefs about the world.

Macrocosm: A complete system that contains smaller or subordinate elements within it.

Microcosm: A small-scale representation or reflection of a larger system, usually used to describe a small community or individual within a larger society.

Transcendence: Going beyond normal limits or existence, often associated with spiritual experiences that surpass ordinary life.

Transhumanism: Using: technology to enhance human abilities and push beyond current physical and mental limitations.

Vulnerability: Lacking means of protection or defense opportunities leading to being easily hurt or affected by physical, emotional, or environmental factors.

ENDNOTES

1. 192 countries in UN x 160 team members by country in every decision x 1.3 kg per brain = 39936 Kg.
2. https://www.globalslaveryindex.org/2018/findings/global-findings/
3. Although Grace Slick written the song while she was still a member of her original band, *The Great Society.*

Conclusion

Deep within the human experience resides an insatiable longing for belonging. The essence of our existence is intricately woven with this profound need, a need that yearns for connection and purpose. This book, delving into the realm of Bioethics of Displacement, recognizes the weight of loss and the ceaseless quest for meaning in life. It embraces the intricate tapestry of ethical dilemmas that intertwine with healthcare and the life sciences, urging us to forge a distinctive moral philosophy that can grapple with the multifaceted challenges presented by Displacement.

Bioethics of Displacement (Bodi) challenges the conventional boundaries of bioethics and sheds light on the hidden tyranny that undermines our sense of belonging. Bodi's philosophy confronts the deep chasms of displacement, where life loses its meaning through diverging paths, leaving us bereft of words to express our sorrow, shrouded in bitter silence. Bodi beckons us to view bioethics through the lens of politics, to unearth the hidden dimensions that intertwine and shape our moral fabric. Through Bodi's lens, we delve into the stories whose meaning of life is dictated from above, pondering the implications for humankind's present and future.

Displacement, whether physical, psychological, or existential, presents individuals and communities with difficult choices, conflicts, and moral implications. A moral philosophy provides a framework for evaluating and addressing these challenges, guiding decision-making and actions in the face of complexity. Bodi introduces the concept of Preemptive bioethics, which anticipates and addresses ethical dilemmas associated with AI and robotics, including issues of bias, discrimination, and safety. Furthermore, Bodi proposes the Syncretist ethical decision-making approach, which combines the strengths of the legalist, antinomian, and situationist perspectives to create a dynamic and context-dependent ethical framework. This approach recognizes that ethical decision-making is not one-size-fits-all and encourages a comprehensive evaluation of moral dilemmas.

Darryl Macer's insightful work unveils three primary perspectives of bioethics: prescriptive, descriptive, and interactive. Prescriptive bioethics delineates the boundaries between good and evil, establishing ethical principles and assigning duties and responsibilities. Descriptive bioethics delves into the moral perceptions and interactions that arise in our encounters with living beings, including ourselves. Interactive bioethics embodies the discussion and debate that transpires among social groups and entire societies, encompassing both descriptive and prescriptive bioethics. As we venture deeper, we intertwine Macer's classifications with the Bioethics of Displacement, forging a more comprehensive understanding. In this quest to expand the horizons of bioethics beyond the confines of space and time, the elucidation and evolution of Bioethics, as exposed by Macer, can journey even further. Bioethics of Displacement proposes the concept of preemptive bioethics. Preemptive bioethics entails the ethical consideration of potential outcomes, placing greater emphasis on descriptive, preemptive, and interactive bioethics given current trends and available time. In the realm of AI and robotics, preemptive bioethics demands a thoughtful examination of the ethical implications and potential consequences prior to widespread deployment. Prescriptive bioethics offers guidelines and principles for discerning good from evil in the design, programming, and utilisation of these technologies, assigning ethical responsibilities to developers and policy-makers.

In our quest to unravel the essence of "good bioethics," we embark on a meticulous examination of the principles put forth by Hofmann and Magelssen, which encompass argumentative, empirical, and dialectic approaches. As we delve deeper into this exploration, we uncover a meta-ethical definition of good bioethics that places utmost importance on authentic communication and the thoughtful consideration of future generations. This journey leads us to the introduction of Transcendental Bioethics of Displacement, where we shift our focus towards the intertwined nature of energy and displacement, gaining valuable insights into the experiences of individuals grappling with conditions like Autism Spectrum Disorder and advanced Alzheimer's Dementia. By embracing these concepts in their entirety, we acquire a more profound understanding of what truly constitutes "good bioethics" and how these principles complement and enrich one another.

The four moral principles of the Bioethics of Displacement further elucidate Bodi's unique perspective. "Vulnerable First" emphasises the intrinsic value and dignity of all human beings, prioritising the protection of the rights and well-being of the displaced. "Big Picture Thinking" recognizes the interconnectedness of displacement and integrates insights from diverse disciplines to develop a comprehensive understanding. "Proactivity" highlights the importance of proactive and preventive approaches, such as preemptive bioethics, to anticipate and address ethical issues before they escalate. Lastly, "Scepticism" encourages open-mindedness, willingness to revise ethical decisions, and recognition of the limitations and limits of our wisdom. Overall, Bodi's principles aim to protect the rights and well-being of the displaced, consider interconnected dimensions of displacement, anticipate and prevent ethical issues, and remain open to diverse perspectives. Bodi emphasises the importance of mindful and reflective communication to effectively address ethical concerns related to displacement. This philosophy challenges the prevailing biopolitical necrophilia that treats living individuals as mere ashes or bones.

Bodi calls for authentic engagement and rejects the inauthenticity that permeates societies, where individuals are often viewed as failures rather than acknowledging failure as an outcome. The bioethics of displacement expands the existing classifications of bioethics, encompassing argumentative, empirical, and dialectic criteria. It encourages a future-oriented approach to ethical decision-making, considering cultural and contextual factors while promoting clear and effective communication. Bodi advocates for a progressive pursuit of genuine and open personal communication, welcoming all voices and safeguarding the interests of future generations. In contrast, the Gehenna side of politics thrives on prevarication and corruption, creating a moral vacuum that allows the privileged to further their selfish interests. Bodi recognizes that our philosophy of migration remains stagnant, echoing the Iron Age, and emphasises loyalty as a defining characteristic of our humanity. It warns against trading our humanity for blind justice, urging us to avoid the pitfall of losing sight of compassion amidst the injustices of the world. Bodi reminds us that those who participated in genocides found contentment in their leader's pathological self-esteem, highlighting the dangers of blind allegiance. In a bittersweet and thought-provoking manner, Bodi's ideas confront the stagnant currents of bioethics, promising to reshape the field forever.

In the vast tapestry of history, displacement has held a powerful sway, affecting both humans and other species, and leaving lasting imprints on their trajectories. From the extinction of dinosaurs to the nomadic bowmen, the ancient Silk Road, and the era of new energies and computing, each epoch has brought significant transformations and highlighted the role of resource access, energy, and technology. Within this intricate tapestry, we direct our attention to the politics of pain, illuminating its intricate complexities through vivid examples and poignant narratives that exemplify the intersection of biopolitics and bioethics. Animal experimentation serves as a glaring paradox, as it necessitates the acceptance of suffering while striving to minimise unnecessary pain. The infamous Tuskegee experiment in the United States and the harrowing Guatemala study epitomise how those in positions of power have callously exploited vulnerable populations for their scientific pursuits, disregarding their well-being with appalling disregard. Moreover, the prevalence of psychological rigidity among individuals in privileged societies perpetuates a reliance on prescribed medications and insufficient medical practices for pain management. These instances serve as stark reminders of the inextricable link between material progress and the politics of pain, perpetuating a society fraught with winners and losers.

Venturing into the realm of transcendental ideas, we embark on an exploration of the notion of dignity and its multifaceted assaults in various forms. From the forced donation of organs to the marginalised populations of the Global South, we bear witness to the erosion of human dignity. In analysing these scenarios, we uncover how the majority population often shifts responsibility onto the disadvantaged, exacerbating their plight. Additionally, we examine how political dignity becomes compromised when individuals are coerced into revealing their vulnerabilities to qualify for compensation. Amidst these challenges, however, a glimmer of hope emerges as we discover successful endeavours that challenge the very concept of dignity. Notably, the recognition of three vast geographical regions in New Zealand as legal persons, incorporating Mātauranga Māori and forging a harmonious union between personhood and nonhuman dignity, serves as an inspiring example. Alongside the exploration of dignity, we delve into the study of engagement, recognizing both concepts as pivotal elements within the framework of Bodi. Through this lens, we uncover the profound harm that displacement inflicts upon individuals and communities, serving as a poignant reminder of the importance of acknowledging and respecting their inherent humanity, agency, and knowledge.

In the depths of contemplation, we embark on an odyssey that seeks to unravel the entrenched divisions within our modern society. It is a journey that exposes the inherent injustices spawned by societal structures, where too pervasive hierarchies perpetuate a global imbalance. Inspired by the Nepali experience, we shine a light on the agonising atrocities that result from these constructs and honour the resilient rebels who yearn for a sense of belonging. Their unwavering spirit exposes the prevailing injustices, demanding a collective global response. We discover that it is within the hearts of individuals to defy the shackles of inequity, armed with moral agency. Yet, the rebellion must stem from a place of moral righteousness, lest it be tainted by opportunism. Genuine challenges to power and lack of opportunities are the catalysts for a more just society, one where the bitter taste of wrath is tempered by the sweet promise of a brighter future.

Expanding our horizons, we embark on an introspective exploration of migration that transcends the boundaries of physical relocation. It is a journey that navigates the intricate transitions of life, traversing diverse stages and surpassing the limitations of our own existence. Our compass is guided by the context of climate change, which casts a profound impact on displacement. In this realm, we harness the power of moral imagination and proactive planning to cultivate solidarity with displaced populations, particularly the elderly and forcibly uprooted individuals. We are beckoned to acknowledge the multifaceted dimensions of disability, both social and cultural, urging us to adopt a more humble approach in reevaluating our perceptions of functionality and health. The interconnected threads of migration, aging, and the perpetuation of poverty weave a common phenomenology. We are confronted with the unique health challenges faced by refugees and the potential of near-future healthcare devices to address their needs. However, our journey is not without ethical quandaries. We encounter the implications of chemical analysis in discerning ethnic backgrounds at border crossings. Traversing the labyrinthine terrain of migration, we find ourselves immersed in the multifaceted issues that define the current refugee crisis. Our discourse revolves around the central argument that the lack of economic diversity fuels migration, resulting in the displacement of local networks and infrastructure. Regrettably, the study of migration often falls prey to assumptions and neglect within liberal democracies, obscuring the true impact it has on global social change. To shed light on this enigma, we delve into the philosophical underpinnings of migration, the intricate web of biopolitics, and the profound phenomenology that accompanies the migrant experience. So we venture into the depths of migration, transcending mere physical relocation and delving into the intricate transitions across life. We are confronted with the harsh realities of displacement, a consequence amplified by the relentless force of climate change. The more we meditate on migration, the clearer are their similarities with ageing. So, those two collectives should forge bonds of solidarity and prepare for a more assertive future. Our journey leads us to the intersection of migration and bioethics, where we see some implications of this global phenomenon. Amidst our exploration, we encounter the unique health challenges faced by refugees and the potential of near-future healthcare devices to address them. Contemplation takes an unexpected turn as we ponder the use of chemical analysis to discern ethnic backgrounds at border crossings, sparking a mix of curiosity and concern. Yet, our reflections do not stop there—they extend far beyond the boundaries of humanity.

Cyborgs and posthuman beings, flesh and machine or even chem, may inhabit a realm where their enhanced abilities and altered senses beckon a new way of perceiving the world. They are deemed to exist as both marvels and warnings, embodying the bittersweet realisation that human limitations can be transcended, but at the cost of a profound rupture in the fabric of our quotidianity. Similarly, migrants embark on a journey of profound phenomenological transformation. They leave behind the familiar and immerse themselves in the unknown, adapting to new cultures, languages, and landscapes. The ecstasy of embracing new opportunities is shadowed by the pangs of yearning for a lost sense of belonging. Their phenomenological resemblance to cyborgs and posthuman beings lies in the rupture and reconfiguration of their lived reality. The longing for understanding and acceptance never should cling to a narrow definition of what it means to belong.

As our journey takes a turn, we direct our attention to the multifaceted concept of confinement. Through philosophical and practical lenses, we unveil its historical roots as a colonial institution, designed to control and shape behaviour through punishment and isolation. Yet, confinement extends beyond its intended purpose, penetrating deep into the essence of individual identity. Drawing from philosophical insights, we embark on a quest to alleviate the suffering inflicted by confinement. We navigate the complex intersections of biopolitics and phenomenology, delving into the ethical implications intertwined with memory, disability, caretaking, and trauma. The ever-present impact of technology on social norms and human experiences cannot be ignored as we seek a compassionate and thoughtful approach to the various forms of confinement.

The COVID-19 pandemic has revealed a bleak reality of government work, with corruption burdening the many while safety nets protect the privileged few. This inequality perpetuates disparities and worsens economic struggles. However, the crisis has prompted a reevaluation of corruption, urging us to adopt a new perspective that encompasses the well-being of all humanity. Coordination between countries has improved, but the pandemic has taken a toll on individuals, exposing our disdain for monotony and sparking a desire for change. While some traits born from the crisis can have negative effects, we must adopt a holistic approach to wellness that rejects reinforcing privileges. The pandemic is not a byproduct of globalisation but a consequence of ad hoc politics. Time after time, pandemic leaders have failed to allocate sufficient resources in an efficient way and, because it has happened globally, it has raised questions about the nature of postmodern globalisation instead of states-led pandemic. The crisis has exposed deep-rooted conflicts, insufficient research and emergency services, and politicisation of decision-making. We must address underinvestment in public health, inadequate data management, and corruption while evaluating the role of non-state actors. Scepticism towards government actions is warranted, necessitating alternative approaches for responsible and ethical governance. The pandemic highlights the need for a holistic approach to wellness, prioritising the well-being of all humanity and considering the moral implications of Displacement.

One of the most common alternative confinements is disability, over 1 in every 6 people on Earth are behind disability's bars. Bodi deals with the complex dimensions of disability, unravelling its social, political, and cultural intricacies. In doing so, we challenge the International Classification of Functioning, Disability, and Health, opening the door to a more epistemically humble and inclusive evaluation. This challenging way of human existence encompasses the myriad dimensions of social, cultural, philosophical, and biopolitical identity. With each step, we uncover the profound influence of social identity on one's perception of self and body-image. And, as we go further, the historical and cultural backdrop reveals itself, shaping our understanding of human identity and illuminating the transformative power of technological advancements and societal structures. Within this evolving landscape, ethical dilemmas emerge from the convergence of body and information technologies, giving rise to concerns of wealth inequality and the far-reaching implications of genetic editing. The philosophical underpinnings of human identity take centre stage, emphasising the narrative nature of selfhood and the pivotal role of personal identity within the realm of bioethics. Lastly, we confront the impact of public health and biopolitics on identity, grappling with the creation of fictive ethnicities and the complex challenges inherent in addressing displacement.

As bioethics unfolds, especially Bodi, it confronts controversial issues surrounding neural technologies and the creation of conscious machines. These ethical dilemmas trace their roots to a legacy of scientific nihilism, challenging our perceptions of end-of-life care and the essence of death in a post-human society. Moreover, the misuse of power throughout history has reduced individuals to mere objects too many times, giving birth to psychological factors that contribute to unspeakable atrocities. To forge a better future, introspection becomes crucial as we ponder the intricate web of death, discerning the boundary between the living and the dead while carefully contemplating the creation of conscious machines.

Out of our pale blue dot, we delve into two interconnected concepts: off-Earth bioethics and xenobioethics. Off-Earth bioethics casts its gaze upon the ethical dimensions of selecting and monitoring astronauts, while xenobioethics grapples with the complex challenges arising from the potential encounter with extraterrestrial life. It becomes imperative to cultivate an inclusive standard of personhood that transcends the boundaries of species, accounting for the possibility of intelligent beings beyond Earth. Through these explorations, bioethics assumes a significant role in shaping our place as inhabitants of the cosmos and our moral responsibility within the grand canvas of the universe.

Surrogate realities, an emerging technological marvel and drug alike, possess the potential to shape individuals' lives, surpassing the effectiveness of our natural perceptions. The convergence of philosophy, science, technology, and education paints a world where non-physical realities blend with natural sciences, offering novel insights. However, we must be wary of the risks that neuroscience brings to culture and society. Neuroethics, a crucial field, delves into the ethical dilemmas arising from advancements in brain research, blurring the line between therapy and human enhancement. These surrogate realities hold promise in enhancing experiences and improving the quality of life, especially in clinical settings for mental disorders and body image psychopathology.

Transitioning to the transcendental, we embark on a journey through time, exploring the profound impact of Displacement on humanity. Imagine a one-year calendar, seamlessly gliding across historical events that have shaped our present world. The transformative effects of the industrial revolution and the rise of bioethics have left indelible marks. Yet, the rapid evolution of Big Data empires ushers in a paradigm shift, with far-reaching consequences for those unacquainted with its power. In understanding Displacement, we find that both symbolism and the treatment of vulnerable populations play pivotal roles. For example, the influence of agribusiness giants on indigenous communities reveals the gravity of these issues, warning us of a world where humanity, like a ship lost at sea, becomes adrift, rendered brain dead. To navigate this shifting terrain, we must embrace essential values such as self-kindness, human warmth and solidarity.

True power still lies in the hands of Nature, from vulcans to viruses, while leaders merely strive to maintain their power over the common people. The same scam that the Pharaoh and the Nile relentlessly repeated. As we face a world cycle change and a potential shift away from liberal capitalism, we may witness a paradigmatic crisis similar to that of 17th century Europe that led to Westphalian states. The period between 2030 and 2060 holds the potential for transformative changes that could redefine what it means to be human. Unless political followerism ends, civil society becomes paramount, and collective freedoms prevail over corruption, the long term survival of our humanity beyond the incoming economic cycle will need a *critical roll.*

Appendix

RECOMMENDATIONS

These Recommendations explore the effectiveness of implementing the initiatives Town Cultural Resurgence, Migration for Mutual Enrichment, and Gamified Civil Disobedience in a specific order to enhance their impact on underserved communities. The Town Cultural Resurgence initiative focuses on empowering communities to revive their cultural practices, while Migration for Mutual Enrichment promotes skill-sharing and knowledge transfer between newcomers and lifelong residents. The Gamified Civil Disobedience initiative introduces interactive civic education to cultivate critical thinking and peaceful resistance. By implementing these initiatives in order, communities can preserve their cultural heritage, embrace diversity through migration, and foster social change. Among recommendations is emphasized the need for individual mindset change and avoiding reliance on taxpayer funding to mitigate corruption risks.

Even if many recommendations in the form of policies can be done considering Bioethics of Displacement, I think that it is a waste of time. Such policies were going to be twisted sooner than later. Therefore, I have made three sequential recommendations for improving an underserved town and a fourth one for a better future.

The order of implementing the initiatives Town Cultural Resurgence, Migration for Mutual Enrichment, and Gamified Civil Disobedience offers unique synergies that enhance their effectiveness:

1. Town Cultural Resurgence sets the foundation by empowering underserved communities to revive and promote their cultural practices. This creates a sense of identity, pride, and resilience within the community.
2. Migration for Mutual Enrichment builds upon the cultural resurgence by facilitating skill-sharing and knowledge transfer between newcomers and lifelong town dwellers. This promotes economic growth, fosters mutual enrichment, and strengthens social integration.
3. Gamified Civil Disobedience harnesses the empowered community by introducing interactive civic education and principles of civil disobedience to youth. This cultivates critical thinking, active participation, and peaceful resistance, allowing the community to shape its future.

Implementing these initiatives in this order ensures that the community's cultural heritage is preserved and serves as a solid foundation for embracing diversity and mutual learning through migration. The final step, gamification, empowers the community to become active agents of change. The synergies emerge from the cumulative effect of these initiatives. Town Cultural Resurgence provides a strong cultural foundation and sense of belonging, which enriches the interactions and exchanges in Migration for Mutual Enrichment. In turn, the diverse expertise and perspectives gained from migration enhance the effectiveness and inclusivity of Gamified Civil Disobedience. This order facilitates a holistic approach to community development, leveraging the unique strengths and resources of the underserved population at each stage. It encourages community collaboration, resource-sharing, and partnerships, maximising the impact of the initiatives and creating a sustainable pathway for positive change.

Ontologically, these three initiatives represent different stages in the development of collective agency and empowerment as seen in the latest stage Quantum Civil Disobedience. Town Cultural Resurgence aims to revive and promote cultural practices, traditions, and heritage, emphasising community participation and the preservation of cultural identity. This initiative recognizes the importance of communal bonds, historical narratives, and shared values in shaping the ontological grounding of a community. It seeks to reclaim and reaffirm the cultural essence of the town, fostering a sense of belonging and resilience.

Migration for Mutual Enrichment introduces a paradigm of intercultural exchange and collaboration. By prioritising skill-sharing and knowledge transfer between newcomers and lifelong residents, this initiative recognizes the ontological value of integration and mutual enrichment. It acknowledges that the interplay between different cultural perspectives and experiences can lead to the emergence of new expertise, talents, and economic growth. Through this initiative, the ontological landscape expands to include a broader understanding of community, embracing the dynamic interconnections between different individuals and cultures.

Gamified Civil Disobedience delves into the realm of social change and civic engagement. By employing gamified platforms and applications, this initiative encourages critical thinking, challenges societal norms, and fosters active participation in shaping communities through peaceful resistance. It recognizes the ontological agency of individuals and the transformative power of collective action. Gamified Civil Disobedience seeks to empower individuals to question existing power structures, assert their autonomy, and reimagine the ontological possibilities of society.

Finally, Quantum Civil Disobedience represents a metaphysical future that builds upon the ontological and epistemic foundations laid by the previous initiatives. It envisions the utilisation of quantum computing to enhance civil disobedience movements by creating secure, decentralised communication networks. From an ontological perspective, Quantum Civil Disobedience aligns with the quest for agency, resilience, and autonomy. It empowers individuals to resist oppressive systems and challenge the status quo by leveraging the transformative potential of quantum technologies.

Let's switch now to an epistemological perspective. From this view, each initiative contributes to the evolution of knowledge and understanding within the community. Town Cultural Resurgence encourages the preservation and transmission of cultural knowledge, ensuring that traditions and practices are passed down through generations. This initiative values experiential knowledge, local wisdom, and the epistemic authority of community members who hold deep insights into their cultural heritage.

Migration for Mutual Enrichment expands the epistemic horizons by facilitating knowledge exchange and skill-sharing. It recognizes the epistemic value of diverse perspectives and experiences, promoting a collective learning process. Through this initiative, individuals gain new skills, knowledge, and expertise, fostering a more inclusive and dynamic epistemic community.

Gamified Civil Disobedience engages individuals in critical thinking, encouraging them to question established norms and assumptions. It promotes a reflective and analytical approach to civic issues, nurturing a community that values intellectual curiosity, scepticism, and the pursuit of truth. This initiative cultivates an epistemic environment that fosters critical consciousness and the capacity to challenge dominant narratives.

Finally, Quantum Civil Disobedience introduces a new epistemic dimension by leveraging the unique capabilities of quantum computing. It expands the boundaries of knowledge and communication by utilising quantum encryption, key distribution, and randomness. This epistemic shift offers novel insights into secure communication, privacy, and the possibilities of resisting surveillance. Quantum Civil Disobedience challenges existing epistemic frameworks and opens up new avenues for understanding and engaging with power dynamics in society.

In summary, the progression from Town Cultural Resurgence to Migration for Mutual Enrichment and Gamified Civil Disobedience lays the groundwork for the emergence of Quantum Civil Disobedience from both ontological and epistemic perspectives. These initiatives contribute to the development of collective agency, empowerment, and the expansion of knowledge and understanding within the community. Quantum Civil Disobedience represents a metaphysical vision that harnesses the transformative potential of quantum technologies to enhance civil disobedience movements, challenging oppressive systems and promoting social change.

PREREQUISITES FOR RECOMMENDATIONS TO WORK

Individual Mindset Change First

It appears more detailed in the first chapter on Praxis in the section titled "Yes, Rebellion Must Start but Individual Mindset Change Comes First" but let's brush up anyway.

Changing the way we think is key to making things better. When we challenge the things that society tells us to believe, we can figure out what we really think and what's important to us. That gives us the power to break free from the rules that hold us back and to help make our communities better.

That's where the Bioethics of Displacement stresses that, given that being forced to shift our mindset is almost always harmful, we must make the most of our agency to change our mindset for good as soon as the chance arises. (And depending where you live, those chances are rare, blink and it is gone forever). Anyhow, changing our mindset doesn't change us alone, for good and evil. Doing so, we can also contribute to a world where collective freedoms (or corruption) are on the rise. Choosing collective freedoms requires thinking about how the things we do now will affect the future, and considering different ways of making decisions that are fair for everyone. By considering different ways to make fair decisions, we ensure that everyone's rights and well-being are respected.

Don't Touch Taxpayers' Money

In simple terms, the Bioethics of Displacement ideology teaches us that when individuals or civil society initiatives become dependent on funding from taxpayers' money, such as grants or subsidies, it can lead to corruption in the long run. Imagine you start a project or initiative with good intentions, aiming to make a positive impact on society. Initially, you might receive financial support from the government or other sources to fund your work. However, over time, reliance on this funding can create a problematic dynamic. When organisations or individuals rely heavily on taxpayer money, they can become susceptible to external influences and pressures. They may feel compelled to align their goals and actions with the expectations of those providing the funding. This alignment can lead to compromises in principles, values, and objectives, as the focus shifts from the original purpose to pleasing the funders or maintaining financial stability.

Moreover, when money is involved, there is always a risk of individuals or groups seeking personal gain or misusing funds for their own benefit. The temptation to engage in unethical practices, such as embezzlement, can increase when there is a significant amount of money at stake. Quite the opposite, promoting a mindset of self-reliance and encouraging diversified sources of funding helps to safeguard the integrity and independence of civil society initiatives. It highlights the importance of remaining true to one's values and objectives, even in the face of financial challenges. By avoiding overreliance on taxpayers' money, civil society initiatives can maintain their ethical standards, act as a check on power, and work towards genuine and lasting positive change in society.

Bodi highlights several factors that contribute to corruption within civil society initiatives that significantly relies on politicians: power imbalances, lack of transparency and accountability, political patronage and clientelism and weak regulatory frameworks.

Power Imbalances: Politicians often hold significant power and influence, which can create an uneven playing field. When civil society initiatives rely on politicians for funding, support, or approval, there is a risk that politicians may exploit their position for personal gain or to serve the interests of their clientele. This power dynamic can make it difficult for initiatives to maintain their independence and integrity.

Lack of Transparency and Accountability: Corruption thrives in environments where there is a lack of transparency and accountability. When civil society initiatives depend on politicians for resources or funding, there may be limited transparency in how these decisions are made, how funds are allocated, and whether they are being used effectively. Without proper checks and balances, corruption, sooner or later, goes unchecked.

Political Patronage and Clientelism: Politicians often have their own networks of supporters and clients whom they prioritise over broader societal interests. When civil society initiatives align themselves closely with politicians, they may become entangled in these patronage networks. This can result in favouritism, nepotism, and robbing by another name rather than serving the broader public.

Weak Regulatory Frameworks: To cap it all, the regulatory frameworks that govern civil society initiatives and their interactions with politicians, sometimes, may be inadequate or ineffective. Insufficient regulations or loopholes can create opportunities for corruption to thrive. Without strong oversight and enforcement mechanisms, initiatives, sooner or later, will be vulnerable to exploitation by politicians and their networks.

While the Bioethics of Displacement does not claim that corruption is an unavoidable outcome of civil society initiatives relying on politicians, it highlights the inherent risks and challenges involved. To mitigate the potential for corruption, the ideology emphasises the need for robust systems of transparency, accountability, and independent oversight. It promotes the establishment of ethical guidelines, strong regulatory frameworks, and mechanisms to ensure the interests of the broader society are protected. By addressing these issues, it is possible to reduce the likelihood of corruption and promote the integrity and effectiveness of civil society initiatives.

TOWN CULTURAL RESURGENCE

Summary

An initiative of self-empowerment for underserved communities in reviving and promoting their cultural practices, traditions, and heritage related with their town. This initiative aims to foster cultural resilience, and ensure cultural preservation in urban development and policy initiatives.

To implement the Town Cultural Resurgence initiative with minimal financial resources, here are some approaches: volunteer and community participation, in-kind donations, leveraging existing community resources, partnerships and collaborations and creative fundraising.

Volunteer and Community Participation: Emphasise the power of volunteerism and community participation. Encourage individuals within the vulnerable community to contribute their time, skills, and resources towards the initiative. This can include community members organising events, sharing their cultural knowledge, and engaging in collaborative efforts to revive and promote their cultural practices.

In-Kind Donations: Seek in-kind donations from individuals, organisations, and businesses that are willing to support the initiative without financial contributions. This could include the provision of materials, equipment, or venues for cultural events and activities. Local businesses, for instance, might be willing to donate space or resources for community gatherings.

Leveraging Existing Community Resources: Identify and utilise existing resources within the community that can support the cultural resurgence efforts. This might include utilising community spaces, such as parks or community centres, for cultural events or engaging local artists, musicians, or storytellers who are willing to share their talents and knowledge without financial compensation.

Partnerships and Collaborations: Establish partnerships with local organisations, educational institutions, and cultural groups that share similar objectives. Collaborate with these entities to pool resources, share costs, and access additional funding opportunities. Through partnerships, the initiative can leverage existing networks and expertise to maximise impact with limited financial resources.

Creative Fundraising: Organise community-led fundraising initiatives, such as cultural festivals, art exhibitions, or crowdfunding campaigns, to generate funds for the initiative. Engage the community in these activities to raise awareness and support for the cause. Encourage community members to contribute their skills and talents to make these events successful.

Implementing the Town Cultural Resurgence initiative with minimal financial resources may require a higher level of community involvement, creativity, and resourcefulness. However, by leveraging the collective strengths, talents, and existing resources within the community, it is possible to make meaningful progress in reviving and promoting cultural practices while minimising financial costs.

It is important to note here that cultural practices, traditions and heritage refers to those that are specific to the underserved town and the people who have grown old in the town. For example, even if a downtown in Europe is brimmed with Afghan migrants, this initiative wouldn't aim to preserve buzkashi (horse-mounted basketball in which the ball is a goat).

Ideological Perspectives

Postmodern Left Ideology: The initiative focuses on the preservation of cultural heritage without falling in victimism, accusations of cultural appropriation or bounding the town with an ethnic identity.

Neoliberal Ideology: This initiative simply contains too many procedures away from financial markets' rule and puts too much emphasis in the collective well-being to be relevant for neoliberalism.

Authoritarian Ideology: While cultural preservation is valued in most authoritarian regimes, it is only a tool for raising national unity. Tailored initiatives for an underserved town community and their cultural practices, may challenge the top-down approach to cultural preservation under most authoritarian regimes.

In summary, the Town Cultural Resurgence initiative is better suited to the Bioethics of Displacement perspective due to its emphasis on meaningful life and the preservation and access to the own cultural heritage.

Civil Society

Community Empowerment: Civil society organisations, community groups, and grassroots movements are often more connected to the needs and aspirations of underserved town communities. They can directly engage with community members, understand their cultural practices, and work collaboratively to promote cultural resurgence and diversity. Civil society organisations have a bottom-up approach, ensuring that the initiative is driven by the communities themselves, leading to greater empowerment and ownership.

Flexibility and Adaptability: Civil society organisations can be more agile, innovative, and adaptable to the specific needs and contexts of underserved communities. They can tailor their approaches, programs, and interventions to address the diverse cultural practices, traditions, and languages of different towns. In addition, they are often more responsive to changing dynamics and can quickly adjust their strategies based on feedback from the communities they serve. Politicians, in turn, may impose rigid guidelines or enforce top-down approaches on the initiative, limiting the flexibility and adaptability needed to effectively support cultural resurgence.

Accountability and Transparency: Civil society organisations are often subject to greater scrutiny and accountability from the communities they serve and from external stakeholders. They operate with a focus on transparency, ensuring that resources are allocated efficiently and effectively. So they engage in participatory decision-making processes more often and involve community members in project implementation.

Overall, the Town Cultural Resurgence initiative is better suited for civil society due to its focus on community empowerment, flexibility, adaptability, and accountability. While local public media can play a role in giving visibility to the initiative, civil society needs to do the bulk of the work to ensure that the initiative remains driven by the communities themselves and safeguards against potential corruption by politicians.

Collective Freedoms

Preserving Cultural Identity: By supporting underserved communities in reviving and promoting their cultural practices, traditions, and heritage, the initiative helps preserve their originality. Cultural identity is an essential aspect of individual and collective freedom, as it allows people to express themselves, celebrate their heritage, and maintain a sense of belonging. By empowering communities to preserve their cultural heritage, the initiative safeguards their freedom to define themselves.

Fostering Cultural Resilience: Underserved town communities often face challenges that threaten the continuity of their cultural practices and traditions. The initiative helps build cultural resilience by providing support and resources to these communities. By preserving their cultural heritage, communities are better equipped to overcome adversity and maintain their traditions in the face of social, economic, and environmental changes. This resilience strengthens their collective freedom by ensuring their cultural practices are not lost or suppressed.

Countering Cultural Homogenization: In many communities, there is a risk of cultural homogenization due to globalisation, urbanisation, and dominant cultural influences. The Town Cultural Resurgence initiative promotes a non-politicized diversity and counteracts cultural homogenization by actively supporting and promoting the cultural practices of underserved town communities. This diversity is essential for collective freedom as it allows for a range of perspectives, traditions, and ways of life to coexist, fostering a more open society.

In a nutshell, the Town Cultural Resurgence initiative improves collective freedom by empowering underserved town communities to preserve their cultural identity, fostering cultural resilience and countering cultural homogenization.

Civil Disobedience

Civil disobedience can play a significant role in fostering the Town Cultural Resurgence initiative by highlighting the importance of cultural preservation and pressuring authorities to take action. Here's how civil disobedience can contribute to the initiative:

Drawing Attention to Cultural Injustices: Civil disobedience brings attention to the cultural injustices faced by underserved town communities. By openly defying laws or regulations that hinder the expression of their cultural practices, traditions, or heritage access, individuals and communities raise awareness about the suppression or marginalisation of their cultural heritage. This can help garner public support and create a sense of urgency for the need to support Town Cultural Resurgence initiatives.

Influencing Public Opinion and Policy Change: Civil disobedience has the potential to sway public opinion and influence policymakers. By bringing cultural injustices to the forefront and engaging in peaceful acts of resistance, individuals and communities can garner support from the wider society. This increased public awareness and support can pressure policymakers to stop harming the underserved community's interests.

MIGRATION FOR MUTUAL ENRICHMENT

Summary

Establish sustainable migration initiatives that prioritise skill-sharing and knowledge transfer between newcomers and underserved town lifelong dwellers. This approach would promote economic growth and mutual enrichment through new expertise and talents.

Implementing the Migration for Mutual Enrichment initiative with minimal financial resources and relying solely on underserved towns' resources would require a community-driven and collaborative approach. Let's see how it could work.

Identifying Community Skills and Expertise: Conduct a comprehensive assessment within the newcomers to identify the skills, expertise, and talents that they possess. This could involve surveys, interviews, and community consultations to understand the diverse range of abilities within the community.

Skill-Sharing and Knowledge Transfer Workshops: Organise skill-sharing workshops and knowledge transfer sessions within the community. Those workshops can be hosted in existing community platforms, such as community centres, religious institutions, or local organisations' spaces. This can be facilitated by individuals who possess expertise in specific fields or through peer-to-peer learning. For example, newcomers with professional experience in various industries can share their knowledge with others to enhance their skills, which may help in their integration.

Community-Led Mentorship Programs: Establish mentorship programs where experienced newcomers can provide guidance and support to other underserved town members. This can help develop skills and provide career advice. The mentorship programs' quality should be granted by community leaders or organisations within the underserved town.

Networking and Partnerships: Facilitate networking opportunities and foster partnerships with other underserved towns that have undertaken the same initiative. These partnerships can create avenues for collaboration, access to resources and skills, and opportunities for further mutual enrichment.

Community-Led Economic Initiatives: Encourage the development of community-led economic initiatives, such as cooperative enterprises or small-scale businesses, that capitalise on the town's skills and talents. This can involve community members working together on entrepreneurial ventures, artistic endeavours, or community development projects. This can generate income, create employment opportunities, and foster economic growth within the community, which in turn, would improve social integration.

By focusing on the strengths and resources of the underserved population, the Migration for Mutual Enrichment initiative can empower individuals to contribute their expertise and talents, fostering economic growth, cultural exchange, and mutual enrichment within the community. It requires strong community leadership, collaboration, and the utilisation of existing platforms and networks to create meaningful opportunities for skill-sharing and knowledge transfer.

Ideological Perspectives

Postmodern Left Ideology: Employs a narrow-minded social justice perspective without necessarily considering the broader ethical implications or potential unintended consequences of migration policies. I mean those unintended consequences that are unseen where the unregulated migration SJW promoters live. Worse even, namely wokism may prioritise only the interests of alleged oppressed identities with total disregard for the broader community or even against it.

Neoliberal Ideology: It might focus on economic gains without adequately addressing the social and cultural impacts of migration. As long as migrants don't park only in the houses of wealthiest people in the city, I mean. Moreover, neoliberalism easily would sacrifice social cohesion or cultural preservation attained with the former Town Cultural Resurgence in the rush for a larger industrial reserve army pushing salaries down.

Authoritarian Ideology: It might prioritise fast assimilation or expulsion over mutual enrichment and heterogeneity. Just like in postmodern left and neoliberalism, there is no problem with the bottom-up opportunities draining, of course.

Bodi's approach as materialised in the initiative of "Migration for Mutual Enrichment" can better address the ethical complexities, cultural preservation, and equitable outcomes associated with demographics shifts in underserved towns. It bets on a strong civil society able to abjure those who relentlessly look for parasitizing taxpayers alleging historical injustices they never suffered, those who profit from labour scarcity, and the ones for whom both newcomers and lifelong underserved town dwellers are nothing but figures in a five-year plan. These ideologies may overlook the importance of the many ways of mutual enrichment and the ethical considerations in favour of hosting underserved towns when facing migratory waves.

Civil Society

Grassroots Approach: Civil society organisations are often rooted in local communities and have a deeper understanding of their dynamics. They can better engage with newcomers and host communities, facilitating meaningful connections and collaborations. It is more attuned to the specific challenges faced by migrants living in this or that underserved town and, therefore, can provide tailored support and resources. In contrast, governments may have broader policies that may not address the needs of migrants and host communities alike. Needless to say, politicians have a knack to turn mutual enrichment into their own.

Flexibility and Innovation: Civil society can be more open to exploring alternative models of skill-sharing and knowledge transfer that may not be constrained by bureaucratic processes or political considerations. Politicians may also favour certain industries or sectors over others, leading to imbalances in skill-sharing and knowledge transfer opportunities.

An aware and efficient government minding national welfare can play a crucial role in creating an enabling environment for sustainable migration initiatives. However, the active involvement of civil society in migratory waves reaching underserved towns grants a minimum standard of quality regarding the assistance to migrants and host communities detrimental to no one as far as four Bodi's principles are observed.

Collective Freedoms

Economic Empowerment: By prioritising skill-sharing and knowledge transfer between newcomers and host underserved towns, the initiative promotes economic growth, employment opportunities and fostering entrepreneurship. This economic empowerment enhances collective freedom by reducing dependence on limited resources, creating avenues for self-sufficiency, and increasing overall prosperity.

Enhanced Social Cohesion: Through skill-sharing and knowledge transfer, newcomers and town residents can work together towards common goals, fostering a sense of community and shared identity. This social integration contributes to collective freedom by creating a supportive environment where everyone can participate and benefit from collective endeavours.

Overall, the "Migration for Mutual Enrichment" initiative promotes collective freedom by empowering individuals and communities, encouraging mutual understanding and fostering social cohesion. It recognizes the value of work and aims to capitalise the skills and resources available in the community for their own benefit.

Civil Disobedience

Protecting Most Valuable Migrants: Civil disobedience can be employed to resist and challenge policies or practices against town interest involving migrants. This can include acts of nonviolent resistance, such as providing sanctuary or offering support services to most talented migrants facing unjust treatment.

GAMIFIED CIVIL DISOBEDIENCE

Summary

Civil society organisations, educational technology developers, and educators can initiate the introduction of gamified platforms and applications for interactive civic education, while also incorporating the principles of civil disobedience. This transformative approach aims to empower youth to critically engage with civic issues, challenge societal norms, and actively participate in shaping their communities through peaceful resistance. Implementing the

Gamified Civil Disobedience initiative with minimal financial resources and relying solely on underserved population resources requires a grassroots approach that strengthens existing community resources.

Community Collaboration: Engage community members, including youth, educators, and local organisations, to collectively develop and design gamified platforms and applications for interactive civic education. This can be done through community meetings, workshops, and collaborative brainstorming sessions to generate ideas and pool resources.

Open-Source Development: Embrace open-source development principles, where the gamified platforms and applications are built using freely available software, shared codes, and collaborative development techniques. This allows for community members with programming or design skills to contribute their expertise and collectively build the tools needed for interactive civic education.

Utilise Existing Technology: Maximise the use of existing technology resources within the underserved population, such as smartphones, computers, or internet access. Community members can contribute their personal devices, share access points, or utilise public facilities to ensure widespread availability of the gamified platforms and applications.

Peer-to-Peer Learning: Foster peer-to-peer learning and knowledge sharing within the community. Educators and knowledgeable individuals can volunteer their time and expertise to provide guidance on civic issues, critical thinking, and the principles of civil disobedience. This informal education can take place through community workshops, mentorship programs, or online forums.

Underserved Town Tailored Content: Encourage community members to generate tailored content for the gamified platforms and applications appealing to the town's culture and heritage. This can include creating interactive challenges, or storytelling elements that address specific civic issues and reflect the unique context and concerns of the underserved community. By incorporating local perspectives, the initiative becomes more relevant and engaging for the community.

Collaborative Partnerships: Seek partnerships with other civil society organisations or within town socially responsible businesses that share the values of the initiative. These partnerships can provide access to additional resources and expertise to support the development, deployment, and maintenance of the gamified platforms and applications.

By joining the resources, skills, and knowledge within the underserved population, the Gamified Civil Disobedience initiative can empower youth to critically engage with civic issues, challenge societal norms, and actively shape their communities through peaceful resistance. It requires a collaborative and community-driven approach, utilising open-source development, localised content creation, and partnerships to ensure accessibility and sustainability with limited financial resources.

Ideological Perspectives

Postmodern Left Ideology: The initiative would be sooner than later monopolised by progressive activism SJW-like prioritising their ideological fetishes (BIPOC trans and so on and so on) bragging of being inclusive while being actually exclusive when it comes to engaging with various perspectives at the same time, as the Bioethics of Displacement defends.

Neoliberal Ideology: Civic engagement challenging existing power structures, as far as doesn't imply to resist financial power and especially if it limits itself to resist statal power, is not against neoliberal ideology. Neither in favour, less even considering the collective freedom relevance and the not-for-profit nature of the initiative.

Authoritarian Ideology: Civil disobedience, by its nature, is totally against authoritarianism's up-down social grip and control.

In summary, the "Gamified Civil Disobedience" initiative is better suited for the principles of Bioethics of Displacement, as it emphasises responsible decision-making and inclusive critical thinking, which is related with epistemic maturity and against magical voluntarism of any colour.

Civil Society

Preservation of Independent and Diverse Perspectives: Civil society organisations, educational technology developers, and educators are more likely to prioritise the inclusion of diverse perspectives in the design and implementation of gamified platforms for civic education. They can ensure that multiple viewpoints are represented and encourage critical thinking and engagement with a variety of societal issues. In contrast, governments may be inclined to promote specific agendas or ideologies, potentially limiting the scope of the initiative and suppressing dissenting voices.

Protection of Civic Freedom and Autonomy: Civil society organisations are often driven by the principles of civic freedom and autonomy. So they are better positioned to safeguard these values within the "Gamified Civil Disobedience" initiative and ensure that the games really foster free expression, independent thinking, and peaceful resistance.

The "Gamified Civil Disobedience" initiative is overall better suited for civil society due to its potential vulnerability to political corruption. Civil society organisations, through grassroots participation, developers and educators can uphold independent perspectives, avoid political bias, and protect civic freedom. By doing so, they can ensure that the initiative remains true to its transformative goals of critically engaging with civic issues and peacefully challenging societal norms.

Collective Freedoms

Promoting Collective Freedoms Through Gamified Civil Disobedience: This initiative embodies the principles of Bioethics of Displacement by encouraging individuals to critically engage with societal norms and participate in shaping their communities. It promotes collective freedoms through interactive civic education, challenging existing power structures, and fostering a more just and inclusive society.

Civil Disobedience

By integrating civil disobedience principles into gamified civic education, this transformed initiative seeks to inspire youth to become active agents of change, equipping them with the knowledge, skills, and values necessary to challenge unjust systems and promote a more just and equitable society. It encourages students to exercise their agency, peacefully resist oppressive practices, and advocate for positive social transformation.

QUANTUM CIVIL DISOBEDIENCE

Summary

An original and groundbreaking way in which quantum computing could improve civil disobedience is through the development of Quantum Resistance Networks (QRN). QRNs would leverage the computational power of quantum computers to create secure, decentralised communication networks that enable individuals engaging in civil disobedience to communicate and coordinate without fear of surveillance or interception by authorities.

Quantum Encryption: Quantum computers have the potential to create unbreakable encryption algorithms that can safeguard communication channels from surveillance. QRNs would utilise quantum encryption protocols to ensure that messages and data transmitted within the network remain secure and private, protecting the identities and activities of individuals involved in civil disobedience.

Quantum Key Distribution: Quantum computing can also facilitate secure key distribution, making it virtually impossible for unauthorised parties to access encrypted communications. QRNs would employ quantum key distribution techniques, such as quantum entanglement, to establish cryptographic keys that are resistant to hacking and eavesdropping, thereby strengthening the security of the network.

Quantum Randomness for Anonymity: Quantum computers can generate true randomness, which can be harnessed to enhance anonymity within the network. QRNs would leverage quantum randomness to assign temporary identities or pseudonyms to individuals engaging in civil disobedience, making it extremely difficult for authorities to trace their activities or identify specific individuals involved.

Quantum Resilience Against Attacks: Quantum computing's computational power can also be utilised to detect and counteract potential attacks on the network. QRNs would employ quantum algorithms to continuously monitor network integrity, identify any attempts to compromise the system, and respond with appropriate countermeasures, ensuring the resilience and continuity of the network.

By leveraging the unique capabilities of quantum computing, QRNs would provide civil disobedience movements with unprecedented levels of privacy, security, and resilience. This groundbreaking application of quantum computing technology would enable individuals to organise, communicate, and coordinate actions with minimal risk of detection, thereby empowering civil disobedience movements and fostering a more robust and effective means of challenging oppressive systems and advocating for social change.

Ideological Perspectives

Postmodern Left Ideology: This initiative may be seen as diverging from the postmodern left by focusing on collective agency and individual privacy indiscriminately rather than systemic transformation against cisheteropatriarchal White oppression.

Neoliberal Ideology: It typically values market efficiency, and may view civil disobedience as disruptive or counterproductive to economic growth and stability.

Authoritarian Ideology: Quantum Civil Disobedience initiative, with its focus on secure and decentralised communication, goes against the authoritarian emphasis on surveillance and control over information and dissent. So this initiative would be tagged as a threat to the any authoritarian regime.

Quantum Civil Disobedience emphasis on empowering individuals aligns with the Bioethics of Displacement ideology's focus on restoring agency and promoting self-determination for underserved communities who are often displaced by unjust power schemes. Likewise, it is streamlined with Bodi's belief in utilising innovative approaches to address complex challenges.

Civil Society

Power Consolidation: Governments, particularly those with authoritarian tendencies, may attempt to co-opt or control the Quantum Resistance Networks (QRNs) to consolidate their power. They could infiltrate or manipulate the network infrastructure to monitor and suppress dissenting voices, undermining the intended purpose of secure communication for civil disobedience.

Manipulation of Information: Governments could attempt to manipulate the quantum communication infrastructure to spread disinformation or propaganda, creating confusion and mistrust within the civil disobedience movement. By controlling the flow of information, politicians could undermine the coordination and unity among activists, making it more challenging to challenge authority effectively.

Legal and Regulatory Restrictions: Governments could introduce legislation or regulations aimed at limiting the use of QRNs or placing restrictions on the technology itself. These measures could curtail the ability of civil society organisations and individuals to freely utilise quantum computing resources for civil disobedience, hindering the potential benefits of the initiative.

Civil society organisations can maintain the integrity of QRNs, ensuring they serve the purpose of empowering individuals engaging in civil disobedience rather than being exploited for political gain.

Collective Freedoms

Protection Against Repression: By leveraging quantum-resistant encryption and decentralised communication networks, QRNs can safeguard activists from repression by authorities. This technology makes it more challenging for governments or other powerful entities to monitor, track, or disrupt civil disobedience movements. As a result, activists can exercise their right to dissent and challenge oppressive systems more effectively, promoting their collective freedom.

Empowerment of Underserved Communities: QRNs can be particularly beneficial for the collective freedoms of underserved communities who often face disproportionately higher levels of surveillance and repression and ensures that their voices are heard, contributing to a favourable public opinion.

Overall, the Quantum Civil Disobedience initiative with the development of QRNs improves collective freedom by protecting activists from repression, facilitating global connectivity, and empowering marginalised communities.

Compilation of References

Abbasi, M., Gholami, N., & Soheilmoghadam, S. (2017). The philosophical fundamentals right to be forgotten from Bioethics view. *Bioethics Journal, 7*(23), 7–15. doi:10.22037/bj.v7i23.17305

Abrams, D., & Hogg, M. A. (1988). Comments on the motivational status of self-esteem in social identity and intergroup discrimination. *European Journal of Social Psychology, 18*(4), 317–334. doi:10.1002/ejsp.2420180403

Acero, Á. R., Cano-Prous, A., Castellanos, G., Martín-Lanas, R., & Canga-Armayor, A. (2017). Family identity and severe mental illness: A thematic synthesis of qualitative studies. *European Journal of Social Psychology, 47*(5), 611–627. doi:10.1002/ejsp.2240

Achiume, E. T. (2013). Beyond Prejudice: Structural Xenophobic Discrimination against Refugees. *Georgetown Journal of International Law, 45*, 323.

Afary, J., & Anderson, K. B. (2010). *Foucault and the Iranian Revolution: Gender and the Seductions of Islamism.* University of Chicago Press.

Affolter, J. (2015). Challenging the State: Teaching Alternative Historiographies in Early Modern Politics. *Metaphilosophy, 46*(3), 398–413. doi:10.1111/meta.12140

Agar, N. (2015). Moral Bioenhancement and the Utilitarian Catastrophe. *Cambridge Quarterly of Healthcare Ethics, 24*(1), 37–47. doi:10.1017/S0963180114000280 PMID:25473856

Agaronnik, N., Campbell, E. G., Ressalam, J., & Iezzoni, L. I. (2019). Exploring issues relating to disability cultural competence among practicing physicians. *Disability and Health Journal, 12*(3), 403–410. Advance online publication. doi:10.1016/j.dhjo.2019.01.010 PMID:30765256

Agbedahin, K. (2014). Interrogating the Togolese historical sex strike. *International Journal on World Peace, 31*(1), 7.

Ahmed, S. (2018). From Threat to Walking Corpse: Spatial Disruption and the Phenomenology of "Living under Drones.". *Theory & Event, 21*(2), 382–410.

Akçayır, M., & Akçayır, G. (2017). Advantages and challenges associated with augmented reality for education: A systematic review of the literature. *Educational Research Review, 20*, 1–11. doi:10.1016/j.edurev.2016.11.002

Akoh, J. A. (2017). State of the Trade: World Transplant Tourism. In J.-D. Rainhorn & S. El Boudamoussi (Eds.), *New Cannibal Markets: Globalization and Commodification of the Human Body* (pp. 199–214). Éditions de la Maison des sciences de l'homme. https://books.openedition.org/editionsmsh/10770

Albrecht, G. L., Devlieger, P., & van Hove, G. (2008). The experience of disability in plural societies. *Alter (Paris), 2*(1), 1–13. doi:10.1016/j.alter.2007.09.002

Alexander, S. (1962, November 9). They decide who lives, who dies: Medical miracle and a moral burden of a small committee. *Life Magazine*, 102–125.

Alexandri, G. (2015). Unravelling the yarn of gentrification trends in the contested inner city of Athens. In L. Loretta & S. Bang Hyun (Eds.), *Global gentrifications: Uneven development and displacement*. Policy Press. doi:10.2307/j.ctt1t894bt.8

Alexandrov, D., Karepin, V., Musabirov, I., & Chuprina, D. (2018). Educational Migration from Russia to the Nordic Countries, China and the Middle East. Social Media Data. *Companion Proceedings of the Web Conference 2018*, 49–50. 10.1145/3184558.3186923

Algeriani, A. M. A., & Mohadi, M. (2018). The Administrative and Technical Aspects in Baghdad and Cairo's House of Wisdom Libraries (a Comparative Analysis). *Arts and Social Sciences Journal*, *9*(5), 9. doi:10.4172/2151-6200.1000402

Alhalabi, W. (2016). Virtual reality systems enhance students' achievements in engineering education. *Behaviour & Information Technology*, *35*(11), 919–925. doi:10.1080/0144929X.2016.1212931

Ali, M. (2014). Towards a decolonial computing. *Ambiguous Technologies: Philosophical Issues, Practical Solutions. Human Nature (Hawthorne, N.Y.)*, 28–35. http://oro.open.ac.uk/41372/

Allen, J., Adams, C., & Flack, F. (2019). The role of data custodians in establishing and maintaining social licence for health research. *Bioethics*, *0*(4), 502–510. Advance online publication. doi:10.1111/bioe.12549 PMID:30657596

Allen, L. (2018). Learning About Sexuality 'Between' Home and School. In L. Allen (Ed.), *Sexuality Education and New Materialism: Queer Things* (pp. 71–84)., doi:10.1057/978-1-349-95300-4_5

Altermark, N. (2017). *Citizenship Inclusion and Intellectual Disability: Biopolitics Post-Institutionalisation*. Routledge. doi:10.4324/9781315109947

Altuntaş, L. (2022, December 24). Oldest known arrowheads uncovered in the Americas. *Arkeonews*. https://arkeonews.net/oldest-known-arrowheads-uncovered-in-the-americas/

Altwegg, A. (1963, May 12). Why Busy Men Go Extracurricular. *The Dallas Morning News*.

Álvarez, G. M. (2022, July 6). *El peligro de sustituir la presunción de inocencia por la de culpabilidad*. Cinco Días. https://cincodias.elpais.com/cincodias/2022/07/05/opinion/1657038118_685321.html

Amos, W. E. (1974). The Philosophy of Corrections: Revisited. *Federal Probation*, *38*, 43.

Anderson, J., Rainie, L., & Vogels, E. A. (2021). Experts say the 'new normal' in 2025 will be far more tech-driven, presenting more big challenges. Pew Research Center.

Anderson, M. (2016, August 15). *Blacks more likely than whites to see – and post – race-related content on social media*. Pew Research Center: Internet, Science & Tech. https://www.pewresearch.org/internet/2016/08/15/blacks-more-likely-than-whites-to-see-and-post-race-related-content-on-social-media/

Andorno, R. (2013). The dual role of human dignity in bioethics. *Medicine, Health Care, and Philosophy*, *16*(4), 967–973. doi:10.100711019-011-9373-5 PMID:22173655

Andrews, K., Comstock, G. L., G.k.d, C., Donaldson, S., Fenton, A., John, T. M., Johnson, L. S. M., Jones, R. C., Kymlicka, W., Meynell, L., Nobis, N., Pena-Guzman, D., Sebo, J., Comstock, G. L., G.k.d, C., Donaldson, S., Fenton, A., John, T. M., Johnson, L. S. M., … Sebo, J. (2018). *Chimpanzee Rights: The Philosophers' Brief*. Routledge. doi:10.4324/9780429461071

Anijar, K. (2000). *Teachings towards the twenty-forth century: Star Trek as social curriculum*. Routledge.

Anstey, J. (2017). *Decision processes of emigrants from Nazi Germany*. University of Rochester.

APA. (2013). *Diagnostic and Statistical Manual of Mental Disorders (DSM-5®)*. American Psychiatric Pub.

Arendt, H. (1963). *Eichmann in Jerusalem: A report on the banality of evil*. Viking Press.

Arendt, H. (2019). *The Human Condition*. University of Chicago Press.

Arkles, R., Jankelson, C., Radford, K., & Jackson Pulver, L. (2018). Family caregiving for older Aboriginal people in urban Australia: Disclosing worlds of meaning in the dementia experience. *Dementia (London)*. Advance online publication. doi:10.1177/1471301218776761 PMID:29877735

Armitage, J. (2015, July 10). *Meet the Goldman Sachs banker who got rich getting Greece into the euro*. The Independent. https://www.independent.co.uk/news/uk/home-news/greek-debt-crisis-meet-the-goldman-sachs-banker-who-got-rich-getting-greece-into-the-euro-10381951.html

Arnason, G. (2017). Synthetic Biology between Self-Regulation and Public Discourse: Ethical Issues and the Many Roles of the Ethicist. *Cambridge Quarterly of Healthcare Ethics*, *26*(2), 246–256. doi:10.1017/S0963180116000840 PMID:28361722

Arnold, B. B. (2016). Is the Zombie My Neighbour: The Zombie Apocalypse as a Lens for Understanding Legal Personhood. *Canberra Law Review*, *14*, 25.

Arrigo, B. A., & Polizzi, D. (2018). Symposium: The Laws of Captivity: Critical and Humanistic Inquiries: Introduction to the Special Issue: On the Laws of Captivity. *New Criminal Law Review*, *21*(4), 483–491. doi:10.1525/nclr.2018.21.4.483

Arshinov, V., Chapny, E., & Chekletsov, V. (2018, May 14). Intersubjectivity and Complexity in Human Enhancement Problem. *Proceedings of the XXIII World Congress of Philosophy*. 10.5840/wcp232018382

Arudou, D. (2018, September 10). Naomi Osaka's US Open victory over Serena Williams: Congratulations, but I don't think you know what you're getting yourself into. *Debito*. http://www.debito.org/?p=15145

Ascott, R. (2007). Telematic Embrace: Visionary Theories of Art, Technology, and Consciousness (E. A. Shanken, Ed.). Academic Press.

Ashby, M. A., & Rich, L. E. (2014). Two Deaths and a Birth: Reminiscing and Rehashing Principles in Biomedical Ethics. *Journal of Bioethical Inquiry*, *11*(1), 1–4. doi:10.100711673-014-9509-4 PMID:24647806

Asscher, J. (2007). Killing and Letting Die: The Similarity Criterion. *Journal of Applied Philosophy*, *24*(3), 271–282. doi:10.1111/j.1468-5930.2007.00381.x

Ataria, Y., & Neria, Y. (2013). Consciousness-Body-Time: How Do People Think Lacking Their Body? *Human Studies*, *36*(2), 159–178. doi:10.100710746-013-9263-3

Atkins, A. R. C. (1910). Requisitions in War. *Royal United Services Institution. Journal*, *54*(389), 837–860. doi:10.1080/03071841009423732

Attfield, R. (2012). Synthetic Biology, Deontology and Synthetic Bioethics. *Ethics, Policy & Environment*, *15*(1), 29–32. doi:10.1080/21550085.2012.672681

Aubrey, C., Mumtaz, Z., Patterson, P., Chari, R., & Mitchell, B. F. (2018). Perspectives of Immigrant Women on the Gender of Provider During Childbirth. *Journal of Obstetrics and Gynaecology Canada*, *40*(6), 677–683. doi:10.1016/j.jogc.2017.09.022 PMID:29274934

Avgenakis, E. (2011). *Question: Dramatic increase in the number of suicides due to the economic crisis and rumors on the operation of networks of usurers and blackmailers*. Academic Press.

Avilov, A. (Director). (2018, June 19). *Sex, Drugs & Refugees*. Russia Today. https://www.youtube.com/watch?v=m4HpLitVRK8

Axinn, W. G., & Barber, J. S. (2001). Mass Education and Fertility Transition. *American Sociological Review*, *66*(4), 481–505. doi:10.2307/3088919

Baadsgaard, A., Monge, J., Cox, S., & Zettler, R. L. (2011). Human sacrifice and intentional corpse preservation in the Royal Cemetery of Ur. *Antiquity*, *85*(327), 27–42. doi:10.1017/S0003598X00067417

Baar, H. van. (2017, July 6). *Boundary practices of citizenship: Europe's Roma at the nexus of securitization and citizenship*. Within and Beyond Citizenship. doi:10.4324/9781315268910-10

Baard, E. (2003). Cyborg Liberation Front. *Village Voice, 29*.

Baars, B. (2003). The double life of BF Skinner: Inner conflict, dissociation and the scientific taboo against consciousness. *Journal of Consciousness Studies*, *10*(1), 5–25.

Bailenson, J. N., Beall, A. C., Loomis, J., Blascovich, J., & Turk, M. (2004). Transformed Social Interaction: Decoupling Representation from Behavior and Form in Collaborative Virtual Environments. *Presence (Cambridge, Mass.)*, *13*(4), 428–441. doi:10.1162/1054746041944803

Bailey, R. (2013). Liberation Biology: The Scientific And Moral Case For The Biotech Revolution. *Prometheus*.

Bakhshi, P., Babulal, G. M., & Trani, J.-F. (2017). Education of children with disabilities in New Delhi: When does exclusion occur? *PLoS One*, *12*(9), e0183885. doi:10.1371/journal.pone.0183885 PMID:28877203

Baldwin, A. M., Dodge, B., Schick, V., Sanders, S. A., & Fortenberry, J. D. (2017). Sexual Minority Women's Satisfaction with Health Care Providers and State-level Structural Support: Investigating the Impact of Lesbian, Gay, Bisexual, and Transgender Nondiscrimination Legislation. *Women's Health Issues*, *27*(3), 271–278. doi:10.1016/j.whi.2017.01.004 PMID:28233635

Balzli, B. (2010, February 8). Greek Debt Crisis: How Goldman Sachs Helped Greece to Mask its True Debt. *Spiegel Online*. https://www.spiegel.de/international/europe/greek-debt-crisis-how-goldman-sachs-helped-greece-to-mask-its-true-debt-a-676634.html

Banerji, S. (2023, May 7). *Future of Well-being- The Metaverse Era*. OCAD University. https://openresearch.ocadu.ca/id/eprint/4103/

Baños, P. (2020). El dominio mental. Ariel.

Baños, R., Botella, C., Garcia-Palacios, A., Villa, H., Perpiña, C., & Alcañiz, M. (2000). Presence and Reality Judgment in Virtual Environments: A Unitary Construct? *Cyberpsychology & Behavior*, *3*(3), 327–335. doi:10.1089/10949310050078760

Baños, R., Botella, C., García-Palacios, A., Villa, H., Perpiñá, C., & Gallardo, M. (1999). Psychological Variables and Reality Judgment in Virtual Environments: The Roles of Absorption and Dissociation. *Cyberpsychology & Behavior*, *2*(2), 143–148. doi:10.1089/cpb.1999.2.143 PMID:19178250

Barbiero, G. (2011). Biophilia and Gaia. Two hypothesis for an affective ecology. *Journal of Bio-Urbanism*, *1*, 11–27.

Barbosa, M. E. (2016). *Body Talk: Performing Violence Against Women in Contemporary Guatemala*. https://kuscholarworks.ku.edu/handle/1808/25757

Barclay, I., Preece, A., & Taylor, I. (2018). *Defining the collective intelligence supply chain.* arXiv preprint arXiv:1809.09444.

Bardzell, J. (2009). Interaction Criticism and Aesthetics. *Proceedings of the SIGCHI Conference on Human Factors in Computing Systems*, 2357–2366. 10.1145/1518701.1519063

Barnes, L. P. (2001). What Is Wrong with the Phenomenological Approach to Religious Education? *Religious Education (Chicago, Ill.)*, *96*(4), 445–461. doi:10.1080/003440801753442366

Barnes, R. (2018). *Uncovering Online Commenting Culture: Trolls, Fanboys and Lurkers.* Springer. doi:10.1007/978-3-319-70235-3

Bassin, M., & Kotkina, I. (2016). The Etnogenez Project: Ideology and Science Fiction in Putin's Russia. *Utopian Studies*, *27*(1), 53–76. doi:10.5325/utopianstudies.27.1.0053

BattagliaE. M.MeiJ.DumasG. (2018). *Systems of Global Governance in the Era of Human-Machine Convergence.* https://arxiv.org/abs/1802.04255

Battersby, C. (1998). *The Phenomenal Woman: Feminist Metaphysics and the Patterns of Identity.* Psychology Press.

Bauböck, R. (1994). *Transnational Citizenship: Membership and Rights in International Migration.* Edward Elgar Publishing.

Bauer, S. (2018b, September 19). The Straight Line From Slavery to Private Prisons. *Literary Hub.* https://lithub.com/the-straight-line-from-slavery-to-private-prisons/

Bauer, S. (2018c, September 25). *America's Shocking History of Private Prisons.* Time. https://time.com/5405158/the-true-history-of-americas-private-prison-industry/

Bauer, S. (2018a). *American Prison: A Reporter's Undercover Journey into the Business of Punishment.* Penguin.

Bauman, Z. (1993). *Postmodern ethics.* Blackwell Oxford.

Baumeister, R. F. (1986). *Identity: Cultural Change and the Struggle for Self.* Oxford University Press.

Baumeister, R. F., & Leary, M. R. (1995). The Need to Belong: Desire for Interpersonal Attachments as a Fundamental Human Motivation. *Psychological Bulletin*, *117*(3), 497–529. doi:10.1037/0033-2909.117.3.497 PMID:7777651

Bay, M. (Director). (2005). *The Island.* https://www.imdb.com/title/tt0399201/

Bayer, S. (Director). (2010). *A Nightmare on Elm Street.* https://www.imdb.com/title/tt1179056/

Beauchamp, T. L. (2008). The Belmont report. *The Oxford Textbook of Clinical Research Ethics*, 149–155.

Beauchamp, T. L., & Childress, J. F. (1994). *Principles of biomedical ethics.* Oxford University Press.

Beauchamp, T. L., & Childress, J. F. (2001). *Principles of Biomedical Ethics.* Oxford University Press.

Beauregard, J. (2019). *Philosophical Neuroethics: A Personalist Approach. Volume 1: Foundations.* Vernon Press.

Becker, G. K. (2000). *The Moral Status of Persons: Perspectives on Bioethics.* Rodopi. doi:10.1163/9789004495029

Beckett, S. (1965). *Imagination Dead Imagine.* Calder & Boyars.

Behrens, K. G. (2017). Hearing sub-Saharan African voices in bioethics. *Theoretical Medicine and Bioethics*, *38*(2), 95–99. doi:10.100711017-017-9406-z PMID:28361295

Belavusau, U., & Henrard, K. (2018). *EU Anti-Discrimination Law beyond Gender.* Bloomsbury Publishing.

Bell, K. (2021). Genocide or disaster: An investigation into the British Empire's role within the Great Bengal Famine of 1770. *Vulcan Historical Review*, 74.

Bell, E., Kothiyal, N., & Willmott, H. (2017). Methodology-as-Technique and the Meaning of Rigour in Globalized Management Research. *British Journal of Management*, *28*(3), 534–550. doi:10.1111/1467-8551.12205

Benhabib, S. (1992). *Situating the Self: Gender, Community, and Postmodernism in Contemporary Ethics*. Psychology Press.

Benhabib, S. (2006). *Another Cosmopolitanism*. Oxford University Press. doi:10.1093/acprof:oso/9780195183221.001.0001

Berardi, F. (2003). La fábrica de la infelicidad. Nuevas formas de trabajo y movimiento global. Madrid: Traficantes de Sueños, 7-167.

Bergamini, D. (1971). *Japan's Imperial Conspiracy*. Morrow.

Berghs, M. (2015). Radicalising 'disability' in conflict and post-conflict situations. *Disability & Society*, *30*(5), 743–758. doi:10.1080/09687599.2015.1052044

Bergson, H. (1984). *Creative Evolution*. University Press of America.

Berne, P. (2015, June 9). *Disability Justice—A working draft by Patty Berne*. Sins Invalid. https://www.sinsinvalid.org/blog/disability-justice-a-working-draft-by-patty-berne

Berry, C. (2018, November 16). Theresa May's deal is almost exactly the Brexit the UK voted for. *The Conversation*. http://theconversation.com/theresa-mays-deal-is-almost-exactly-the-brexit-the-uk-voted-for-107040

Berto, F. (2017). Counting the Particles: Entity and Identity in the Philosophy of Physics. *Metaphysica, 18*(1).

Bertolotti, T., Arfini, S., & Magnani, L. (2017). Of Cyborgs and Brutes: Technology-Inherited Violence and Ignorance. *Philosophies*, *2*(1), 1–14.

Bertrand, M., & Mullainathan, S. (2003). *Are Emily and Greg More Employable than Lakisha and Jamal? A Field Experiment on Labor Market Discrimination* (Working Paper No. 9873). National Bureau of Economic Research. doi:10.3386/w9873

Betts, A. (2013). *Survival Migration: Failed Governance and the Crisis of Displacement*. Cornell University Press. doi:10.7591/cornell/9780801451065.001.0001

Bevan, S. (2019, January 21). *Half of employers say they are less inclined to recruit obese candidates – it's not OK*. The Conversation. http://theconversation.com/half-of-employers-say-they-are-less-inclined-to-recruit-obese-candidates-its-not-ok-109821

Bhutada, G. (2021, January 14). *The U.S. Share of the Global Economy Over Time*. Visual Capitalist. https://www.visualcapitalist.com/u-s-share-of-global-economy-over-time/

Biagioli, V., Piredda, M., Alvaro, R., & de Marinis, M. G. (2017). The experiences of protective isolation in patients undergoing bone marrow or haematopoietic stem cell transplantation: Systematic review and metasynthesis. *European Journal of Cancer Care*, *26*(5), e12461. Advance online publication. doi:10.1111/ecc.12461 PMID:26892767

Bichard, J.-A., Coleman, R., & Langdon, P. (2007). Does My Stigma Look Big in This? Considering Acceptability and Desirability in the Inclusive Design of Technology Products. In C. Stephanidis (Ed.), *Universal Acess in Human Computer Interaction. Coping with Diversity* (pp. 622–631). Springer Berlin Heidelberg. doi:10.1007/978-3-540-73279-2_69

Bilasová, V. (2011). Philosophical Issues in Medical Ethics in the Context of Bioethical Discourse. *Ethics & Bioethics (in Central Europe)*, 7.

Binding, K., & Hoche, A. (2006). *Die Freigabe der Vernichtung lebensunwerten Lebens: Ihr Maß und ihre Form*. Bwv - Berliner Wissenschafts-Verlag.

Birbaumer, N., & Cohen, L. G. (2007). Brain–computer interfaces: Communication and restoration of movement in paralysis. *The Journal of Physiology*, *579*(3), 621–636. doi:10.1113/jphysiol.2006.125633 PMID:17234696

Birnbaum, L., & Birnbaum, A. (2004). In Search of Inner Wisdom: Guided Mindfulness Meditation in the Context of Suicide. *TheScientificWorldJournal*, *4*, 216–227. doi:10.1100/tsw.2004.17 PMID:15105961

Bishop, J. P., & Jotterand, F. (2006). Bioethics as Biopolitics. *The Journal of Medicine and Philosophy*, *31*(3), 205–212. doi:10.1080/03605310600712760 PMID:16760100

Bix, H. P. (2008, May 3). War Responsibility and Historical Memory: Hirohito's Apparition. *The Asia-Pacific Journal: Japan Focus*. https://apjjf.org/-Herbert-P.-Bix/2741/article.html

Bizzari, V. (2023). Which Kind of Body in "Mental" Pathologies? Phenomenological Insights on the Nature of the Disrupted Self. *The Journal of Medicine and Philosophy: A Forum for Bioethics and Philosophy of Medicine, 48*(2), 116–127. 10.1093/jmp/jhad008

Blake, M., & Risse, M. (2007). *Migration, Territoriality, and Culture* (*SSRN* Scholarly Paper ID 963130). Social Science Research Network. https://papers.ssrn.com/abstract=963130

Blenkinsop, S., & Morse, M. (2017). Saying Yes to Life: The Search for the Rebel Teacher. In B. Jickling & S. Sterling (Eds.), *Post-Sustainability and Environmental Education: Remaking Education for the Future* (pp. 49–61). Springer International Publishing., doi:10.1007/978-3-319-51322-5_4

Blomkamp, N. (Director). (2009). *District 9*. https://www.imdb.com/title/tt1136608/

Bloom, D. E., Canning, D., & Sevilla, J. P. (2001). *Economic growth and the demographic transition*. Academic Press.

Bobis, Z. (2013, October 29). *Stand up for equality in German schools*. Open Society Foundations. https://www.opensocietyfoundations.org/voices/german-schools-quiet-deep-discrimination-problem

Boldeanu, S. (2018). *Merging brain-computer interfaces and virtual reality: A neuroscientific exploration*. http://urn.kb.se/resolve?urn=urn:nbn:se:his:diva-15774

Bolotnyy, V., & Emanuel, N. (2022). Why do women earn less than men? Evidence from bus and train operators. *Journal of Labor Economics*, *40*(2), 283–323. doi:10.1086/715835

Bong, S. A. (2018). Going beyond the Universal-versus-Relativist Rights Discourse and Practice: The Case of Malaysia. In S. Bahun-Radunovic (Ed.), *Violence and Gender in the Globalized World*. doi:10.4324/9781351143363-5

Boone (Director). (2014). *The Fault in Our Stars*. https://www.imdb.com/title/tt2582846/

Bordo, M. D., & White, E. N. (1991). A Tale of Two Currencies: British and French Finance During the Napoleonic Wars. *The Journal of Economic History*, *51*(2), 303–316. doi:10.1017/S002205070003895X

Borgesius, F. J. Z., Trilling, D., Möller, J., Bodó, B., de Vreese, C. H., & Helberger, N. (2016). Should we worry about filter bubbles? *Internet Policy Review*. https://policyreview.info/articles/analysis/should-we-worry-about-filter-bubbles

Borgstrom, E. (2015). Social death in end-of-life care policy. *Contemporary Social Science*, *10*(3), 272–283. doi:10.1080/21582041.2015.1109799

Borile, S. (2019). Bacha Bazi: Cultural norms and violence against poor children in Afghanistan. *International Review of Sociology, 29*(3), 498–507. doi:10.1080/03906701.2019.1672346

Bostrom, N. (2015). In Defense of Posthuman Dignity. In H. Kuhse, U. Schüklenk, & P. Singer (Eds.), *Bioethics: An Anthology*. John Wiley & Sons.

Botterill, D., Pennings, G., & Mainil, T. (2013). *Medical Tourism and Transnational Health Care*. Springer. doi:10.1057/9781137338495

Bouche, V., & Crotty, S. M. (2018). Estimating demand for illicit massage businesses in Houston, Texas. *Journal of Human Trafficking, 4*(4), 279–297. doi:10.1080/23322705.2017.1374080

Bourne, P. A., & McGrowder, D. A. (2009). Rural health in Jamaica: Examining and refining the predictive factors of good health status of rural residents. *Rural and Remote Health, 9*(2). Advance online publication. doi:10.22605/RRH1116 PMID:19338392

Bowen, E. (2000). *The Last September*. Anchor Books.

Braun, R. (2014, January 9). *Prehistoric sharks were earliest animals to migrate*. New Scientist. https://www.newscientist.com/article/dn24845-prehistoric-sharks-were-earliest-animals-to-migrate/

Brem, S. K., & Anijar, K. Z. (2003). The Bioethics of Fiction: The Chimera in Film and Print. *The American Journal of Bioethics, 3*(3), 22–24. doi:10.1162/15265160360706787 PMID:14594476

Brenan, M. (2018, May 31). *Americans' Strong Support for Euthanasia Persists*. Gallup.Com. https://news.gallup.com/poll/235145/americans-strong-support-euthanasia-persists.aspx

Brenner, Y. (2016, June 5). *Germany: Roma march against asylum-seeker crackdown*. https://www.aljazeera.com/news/2016/06/germany-roma-march-asylum-seeker-crackdown-160605092954463.html

Bresler, A. Y., & Paskhover, B. (2018). Religion and the Plastic Surgeon: An Imam, a Minister, and a Rabbi Walk into a Surgical Centre. *Aesthetic Plastic Surgery, 42*(6), 1699–1703. doi:10.100700266-018-1196-9 PMID:30019243

Bridges, G. S., Crutchfield, R. D., & Simpson, E. E. (1987). Crime, Social Structure and Criminal Punishment: White and Nonwhite Rates of Imprisonment. *Social Problems, 34*(4), 345–361. doi:10.2307/800812

Brigada Antifraude (Director). (2023, April 9). *Empresas y políticos españoles apoyan a Marruecos y su ocupación*. https://www.youtube.com/watch?v=dh7VcLkg4Uk

Briggs, S., & Whittaker, A. (2018). Protecting Children from Faith-Based Abuse through Accusations of Witchcraft and Spirit Possession: Understanding Contexts and Informing Practice. *British Journal of Social Work, 48*(8), 2157–2175. doi:10.1093/bjsw/bcx155

Brochmann, G., & Hagelund, A. (2011). Migrants in the Scandinavian Welfare State. *Nordic Journal of Migration Research, 1*(1), 13–24. doi:10.2478/v10202-011-0003-3

Brodie, E. E., Whyte, A., & Niven, C. A. (2007). Analgesia through the looking-glass? A randomized controlled trial investigating the effect of viewing a 'virtual' limb upon phantom limb pain, sensation and movement. *European Journal of Pain (London, England), 11*(4), 428–436. doi:10.1016/j.ejpain.2006.06.002 PMID:16857400

Broumand, B., & Saidi, R. F. (2017). New Definition of Transplant Tourism. *International Journal of Organ Transplantation Medicine, 8*(1), 49–51. PMID:28299028

Brown, S. (2017). *A Race Against the Odds: The Historical Influences on Modern-Day Racism and Structural Discrimination and the Impact on Black Men*. University of Hartford.

Bruce, R. M. (2018). *Strides Toward Equality: The Portrayal of Black Female Athletes in Children’s Picturebooks*. The Ohio State University. https://etd.ohiolink.edu/pg_10?0:NO:10:P10_ACCESSION_NUM:osu154351163594203

Bryant, J. A., & la Velle, L. B. (2018). *Introduction to Bioethics*. John Wiley & Sons.

Brzozowski, W. (2018). *Is Islam Incompatible with European Identity? (SSRN* Scholarly Paper No. 3301195). doi:10.2139/ssrn.3301195

Bublitz, J.-C. (2013). My Mind Is Mine!? Cognitive Liberty as a Legal Concept. In E. Hildt & A. G. Franke (Eds.), *Cognitive Enhancement: An Interdisciplinary Perspective* (pp. 233–264). Springer Netherlands. doi:10.1007/978-94-007-6253-4_19

Buchholz, K. (2022, February 17). *Inherited Wealth Concentrates Among The Ultra-Rich* [Infographic]. Forbes. https://www.forbes.com/sites/katharinabuchholz/2022/02/17/inherited-wealth-concentrates-among-the-ultra-rich-infographic/

Budiani, D. (2006). *Consequences of living kidney donors in Egypt. Presentation At The Middle East Society On Organ Transplants (Mesot) Meetings. The Middle East Society On Organ Transplants (Mesot)* Meetings.

Bugliarello, G. (2008). Reflections on Morality, Ethics, and Bioethics Decisions. *Journal of Long-Term Effects of Medical Implants, 18*(1), 85–90. Advance online publication. doi:10.1615/JLongTermEffMedImplants.v18.i1.590 PMID:19348615

Buller, T. (2006). Brains, lies and psychological explanations. In J. Illes (Ed.), *Neuroethics: Defining the Issues in Theory, Practice, and Policy*. Oxford University Press.

Burkitt, I. (1998). Sexuality and Gender Identity: From a Discursive to a Relational Analysis. *The Sociological Review, 46*(3), 483–504. doi:10.1111/1467-954X.00128

Burwell, J. (2018). *Quantum Language and the Migration of Scientific Concepts*. MIT Press. doi:10.7551/mitpress/11353.001.0001

Butler, J. (1988). Performative Acts and Gender Constitution: An Essay in Phenomenology and Feminist Theory. *Theatre Journal, 40*(4), 519–531. doi:10.2307/3207893

Butler, O. E. (1996). *Bloodchild and Other Stories*. Seven Stories Press.

Byrne, R. (2011). *The Secret*. Simon and Schuster.

Bywater, J., & Jones, R. (2007). *Sexuality and Social Work*. Learning Matters.

Cabrera, J. (2018). *Discomfort and Moral Impediment: The Human Situation, Radical Bioethics and Procreation*. Cambridge Scholars Publishing.

Cahill, L. S. (1990). Can Theology Have a Role in "Public" Bioethical Discourse? *The Hastings Center Report, 20*(4), 10–14. doi:10.2307/3562778

Caire, P., Moawad, A., Efthymiou, V., Bikakis, A., & Le Traon, Y. (2016). Privacy challenges in ambient intelligence systems. *Journal of Ambient Intelligence and Smart Environments, 8*(6), 619–644. doi:10.3233/AIS-160405

Callahan, D. (1973). Bioethics as a Discipline. *The Hastings Center Studies, 1*(1), 66–73. doi:10.2307/3527474

Callahan, D. (2017). Religion and the Secularization of Bioethics. In M. Freeman (Ed.), *Ethics and Medical Decision-Making*. doi:10.4324/9781315209692-5

Callon, M., Lascoumes, P., & Barthe, Y. (2009). *Acting in an Uncertain World*. MIT Press.

Camenisch, P. F. (2013). *Religious Methods and Resources in Bioethics*. Springer Science & Business Media.

Cameron, J. (Director). (1984). *The Terminator*. https://www.imdb.com/title/tt0088247/

Cameron, J. (Director). (1986). *Aliens*. https://www.imdb.com/title/tt0090605/

Campbell, D., & Davison, N. (2012, May 27). Illegal kidney trade booms as new organ is "sold every hour." *The Guardian*. https://www.theguardian.com/world/2012/may/27/kidney-trade-illegal-operations-who

Camus, A. (1984). *The Fall*. Vintage Books.

Canadian Gender Report. (2020, November 12). The Swedish U-Turn on Gender Transitioning for Children. *Canadian Gender Report*. https://genderreport.ca/the-swedish-u-turn-on-gender-transitioning/

Cantekin, D. (2018). Syrian Refugees Living on the Edge: Policy and Practice Implications for Mental Health and Psychosocial Wellbeing. *International Migration (Geneva, Switzerland)*. Advance online publication. doi:10.1111/imig.12508

Cara, E. (2019, January 7). Bristol Myers-Squibb and Others Can't Dodge $1 Billion Lawsuit Over 1940s Syphilis Study, Judge Rules. *Gizmodo*. https://gizmodo.com/bristol-meyers-squibb-and-others-cant-dodge-1-billion-1831548144

Carastathis, A. (2014). The Concept of Intersectionality in Feminist Theory. *Philosophy Compass*, *9*(5), 304–314. doi:10.1111/phc3.12129

Carel, H. (2016). *Illness: The Cry of the Flesh*. Routledge. doi:10.4324/9781315487410

Carel, H. H. (2013). Illness, phenomenology, and philosophical method. *Theoretical Medicine and Bioethics*, *34*(4), 345–357. doi:10.100711017-013-9265-1 PMID:23836135

Carling, J., & Collins, F. (2018). Aspiration, desire and drivers of migration. *Journal of Ethnic and Migration Studies*, *44*(6), 909–926. doi:10.1080/1369183X.2017.1384134

Carmell, A. (1976). *Challenge: Torah Views on Science and Its Problems*. Feldheim Publishers.

Carroll. (2016, May 15). *Why The Deadliest Gang In The World Might Be Rethinking Face Tattoos*. https://dailycaller.com/2016/05/15/why-the-deadliest-gang-in-the-world-might-be-rethinking-face-tattoos/

Carroll, L. (1992). *Alice in Wonderland*. Wordsworth Editions.

Carr, W. (2004). Philosophy and Education. *Journal of Philosophy of Education*, *38*(1), 55–73. doi:10.1111/j.0309-8249.2004.00363.x

Carse, A. L. (1991). The 'voice of care': Implications for bioethical education. *The Journal of Medicine and Philosophy*, *16*(1), 5–28. doi:10.1093/jmp/16.1.5 PMID:2010720

Carson, R. A., & Burns, C. R. (2006). *Philosophy of Medicine and Bioethics: A Twenty-Year Retrospective and Critical Appraisal*. Springer Science & Business Media.

Carta, M. G., Bernal, M., Hardoy, M. C., & Haro-Abad, J. M. (2005). Migration and mental health in Europe (the state of the mental health in Europe working group: Appendix 1). *Clinical Practice and Epidemiology in Mental Health*, *1*(1), 13. doi:10.1186/1745-0179-1-13 PMID:16135246

Cassol, H., D'Argembeau, A., Charland-Verville, V., Laureys, S., & Martial, C. (2019). Memories of near-death experiences: Are they self-defining? *Neuroscience of Consciousness*, *2019*(1). doi:10.1093/nc/niz002

Castañares, A. (2021, April 12). *SDG&E Employee Fired Over Alleged White Supremacist Hand Sign is a Latino*. La Prensa. https://laprensa.org/sdge-employee-fired-over-alleged-white-supremacist-hand-sign-latino

Cattorini, P. (2009). Shall I Become a Zombie? Stories of Illness, Ethical Dilemmas and Visions of Society. *Medicina Nei Secoli, 21*(3), 1017–1035. PMID:21560776

Cattorini, P. M. (2019). Narrative bioethics and film. Some historical remarks. *Medical History, 3*(2), 110–123.

Cembrero, I. (2023, April 9). *El presidente del Senado marroquí insta a sus compatriotas a entrar en la política española para favorecer a Rabat.* elconfidencial.com. https://www.elconfidencial.com/espana/2023-04-09/presidente-senado-marruecos-ceuta-melilla_3607689/

Cerulo, K. A., & Ruane, J. M. (1997). Death comes alive; technology and the re-conception of death. *Science as Culture, 6*(3), 444–466. doi:10.1080/09505439709526477

Chalise, A. (2018). *Nepalese social structure, atrocities and mental health* [PhD Thesis]. Tribhuvan University.

Chang, M. K., Lim, M. X., Tay, K., Lee, R. J., Sim, D. S., Menon, S., Kanesvaran, R., Puvanendran, R., & Radha Krishna, L. K. (2017). The Influence of Evolving Confucian Beliefs in the Selection of Proxy Decision-Makers at the End of Life in Singapore. *Asian Bioethics Review, 9*(1), 117–128. doi:10.100741649-017-0003-2

Chattopadhyay, S., Myser, C., Moxham, T., & Vries, R. D. (2017). A Question of Social Justice: How Policies of Profit Negate Engagement of Developing World Bioethicists and Undermine Global Bioethics. *The American Journal of Bioethics, 17*(10), 3–14. doi:10.1080/15265161.2017.1365185 PMID:29020562

Chaumet, G., Taillard, J., Sagaspe, P., Pagani, M., Dinges, D. F., Pavy-Le-Traon, A., Bareille, M.-P., Rascol, O., & Philip, P. (2009, February). *Confinement and Sleep Deprivation Effects on Propensity to Take Risks.* https://doi.org/info:doi/10.3357/ASEM.2366.2009

Chen, K. (1210, 1274). *Zhong xing guan ge lu.* [Compressed data]. Library of Congress. https://www.loc.gov/item/2021666333/

Chen, P., Liu, X., Cheng, W., & Huang, R. (2017). A review of using Augmented Reality in Education from 2011 to 2016. In E. Popescu, Kinshuk, M. K. Khribi, R. Huang, M. Jemni, N.-S. Chen, & D. G. Sampson (Eds.), Innovations in Smart Learning (pp. 13–18). Springer Singapore.

Cheng, C.-Y., & Bunnin, N. (2008). *Contemporary Chinese Philosophy.* John Wiley & Sons.

Cheng, S.-Y., Suh, S.-Y., Morita, T., Oyama, Y., Chiu, T.-Y., Koh, S. J., Kim, H. S., Hwang, S.-J., Yoshie, T., & Tsuneto, S. (2015). A Cross-Cultural Study on Behaviors When Death Is Approaching in East Asian Countries. *Medicine, 94*(39), e1573. Advance online publication. doi:10.1097/MD.0000000000001573 PMID:26426631

Chen, Y., Palma, N., & Ward, F. (2021). Reconstruction of the Spanish money supply, 1492–1810. *Explorations in Economic History, 81*, 101401. doi:10.1016/j.eeh.2021.101401

Cherry, M. (2013). Orthodox Christian Bioethics: Medical Morality in the Mind of the Fathers. In *Religious Perspectives on Bioethics.* Taylor & Francis. doi:10.4324/9781315800394

Cheshko, V. T., Glazko, V. I., & Kosova, Y. V. (2017). Bioethics: Reincarnation of Natural Philosophy in Modern Science. *Biogeosystem Technique, 4*(2), 111–121.

Chetty, R., Hendren, N., Jones, M. R., & Porter, S. R. (2020). Race and economic opportunity in the United States: An intergenerational perspective. *The Quarterly Journal of Economics, 135*(2), 711–783. doi:10.1093/qje/qjz042

Chief Justice Warren, E. (1954). Brown v. Board of education. *United States Reports, 347*(1954), 483.

Chomsky, A. (2018). *"They Take Our Jobs!": And 20 Other Myths about Immigration.* Beacon Press.

Chrysanthou, A., & Nottingham, E. (2018). Conferences as a Tool for Interdisciplinary (Postgraduate) Learning in Bioethics. *JAHR*, *9*(2), 234–246. doi:10.21860/j.9.2.6

Chua, D. K. L. (2011). Listening to the Self: The Shawshank Redemption and the Technology of Music. *19th-Century Music, 34*(3), 341–355. doi:10.1525/ncm.2011.34.3.341

Chubak, B. (2008). Visual Bioethics: Seeing is Believing? *The American Journal of Bioethics*, *8*(12), 58–60. doi:10.1080/15265160802521047 PMID:19085481

Chun, W. H. K. (2008). *Control and Freedom: Power and Paranoia in the Age of Fiber Optics*. MIT Press.

Cipolletta, G. (2019). Ubiquitous Bodies: A "Metrobodily" Transition From Real to Virtual. *Virtual and Augmented Reality in Mental Health Treatment*, 1–16. doi:10.4018/978-1-5225-7168-1.ch001

Ćirković, M. M. (2017). Enhancing a Person, Enhancing a Civilization: A Research Program at the Intersection of Bioethics, Future Studies, and Astrobiology. *Cambridge Quarterly of Healthcare Ethics*, *26*(3), 459–468. doi:10.1017/S0963180116001134 PMID:28541174

Clayton, J. (2013). The Ridicule of Time: Science Fiction, Bioethics, and the Posthuman. *American Literary History*, *25*(2), 317–343. doi:10.1093/alh/ajt005

Clegg, S., Kornberger, M., & Rhodes, C. (2007). Organizational Ethics, Decision Making, Undecidability. *The Sociological Review*, *55*(2), 393–409. doi:10.1111/j.1467-954X.2007.00711.x

Cole, J. (2008). Virtual and Augmented Reality, Phantom Experience, and Prosthetics. In P. Gallagher, D. Desmond, & M. MacLachlan (Eds.), *Psychoprosthetics* (pp. 141–153). Springer London. doi:10.1007/978-1-84628-980-4_11

Coleman, R. D. N. (2019). *Valid consent considerations for astronauts and space tourists involved in medical research*. Academic Press.

Collins, A. (2014). *Karahan Tepe: Göbekli Tepe's Sister Site—Another Temple Of The Stars?* https://d1wqtxts1xzle7.cloudfront.net/34492368/Karahan_Tepe_double_column_with_pics_low-libre.pdf?1408528975=&response-content-disposition=inline%3B+filename%3DKarahan_Tepe_Gobekli_Tepes_Sister_Site_A.pdf&Expires=1684507578&Signature=dK2HL1N2Yin-N2QlDGX1KvhJIcCbM9vCrppzHA5LjMBdYsdl5UcpM2zk9GWTQRgz7am03he-sQR8gtzAdcnsx5nr75MAjG7R~RW93xK68ZDvhAQo7QFjCK-Pf74IeIgCYM8AM9LThnhKlY1iheeMmGGEa4zE~UWAQ~qBFk-7i23OgYSnfYc2S2FnUXDAvE1wzIrTB4InGbWnBmwKYTN45hzBUZyVzX~4rPUBr2PVdHhfhodikxuMY4h-bHQJI-lC0MiVdvdu6JoYk383M83toX-LnHd~fhs4Hlpsep97ZkEXjCC6GOy~ertwu3DVQJktyoJGnZQhLLjZtLd0b5H-llQ__&Key-Pair-Id=APKAJLOHF5GGSLRBV4ZA

Collins, S. (2008). *The Hunger Games*. Scholastic Inc.

Combs-Orme, T. (2013). Epigenetics and the Social Work Imperative. *Social Work*, *58*(1), 23–30. doi:10.1093wws052 PMID:23409337

Condé, M. (2014). Wittgenstein's grammar and the contemporary scientific rationality. *Analytische Und Kontinentale Philosophie: Perspektiven Und Methoden Analytical and Continental Philosophy: Methods And*, 52. https://www.alws.at/alws/wp-content/uploads/2018/06/papers-2014.pdf#page=53

Contreras, I., & Murphy, A. (2023). *Forbes Global 2000 List 2022: The Top 200*. Forbes. https://www.forbes.com/sites/forbesstaff/2022/05/12/forbes-global-2000-list-2022-the-top-200/

Cook, K., Keller, D., & Myers, A. (2014). Bioethics in the Hunger Games. *Science Teacher (Normal, Ill.)*, *81*(1), 31. doi:10.2505/4/tst14_081_01_31

Cooper, M., & Hieda, M. (2017). The Dark Side of Medical Tourism?: End of Life Choice, Human Trafficking, and Organ Transplants. *Oncology: Breakthroughs in Research and Practice*, 203–215. doi:10.4018/978-1-5225-0549-5.ch006

Coppola, F. F. (Director). (1992). *Bram Stoker's Dracula*. https://www.imdb.com/title/tt0103874/

Corbett, J. (2012). *Rebel with a Cause: The Making of a Community Mediator* (*SSRN* Scholarly Paper ID 2030438). Social Science Research Network. https://papers.ssrn.com/abstract=2030438

Cornwell, C., Rivera, J., & Schmutte, I. M. (2017). Wage Discrimination When Identity Is Subjective Evidence from Changes in Employer-Reported Race. *The Journal of Human Resources*, *52*(3), 719–755. doi:10.3368/jhr.52.3.0815-7340R1

Corrigan, P. W., & Watson, A. C. (2002). The paradox of self-stigma and mental illness. *Clinical Psychology: Science and Practice*, *9*(1), 35–53. doi:10.1093/clipsy.9.1.35

Costantini, M. (2014). Body perception, awareness, and illusions. *Wiley Interdisciplinary Reviews: Cognitive Science*, *5*(5), 551–560. doi:10.1002/wcs.1309 PMID:26308744

Coulthard, G. S. (2014). *Red Skin, White Masks: Rejecting the Colonial Politics of Recognition*. University of Minnesota Press. doi:10.5749/minnesota/9780816679645.001.0001

Coutin, S. B. (2010). Confined within: National territories as zones of confinement. *Political Geography*, *29*(4), 200–208. doi:10.1016/j.polgeo.2010.03.005

Coyne, L. (2023). What is Phenomenological Bioethics? A Critical Appraisal of Its Ends and Means. *The Journal of Medicine and Philosophy: A Forum for Bioethics and Philosophy of Medicine, 48*(2), 170–183. 10.1093/jmp/jhad001

Crabb, M., Heron, M. J., Jones, R., Armstrong, M., Reid, H., & Wilson, A. (2019). *Developing accessible services: Understanding current knowledge and areas for future support*. https://openair.rgu.ac.uk/handle/10059/3288

Crane, S. (1972). *The Complete Poems of Stephen Crane*. Cornell University Press.

Craven, W. (Director). (1984). *A Nightmare on Elm Street*. https://www.imdb.com/title/tt0087800/

Crichton, M. (1991). *Jurassic Park*. Ballantine Books.

Croce, B. (1995). *Guide to aesthetics*. Hackett Publishing.

Crocetti, E., Klimstra, T., Keijsers, L., Hale, W. W. III, & Meeus, W. (2009). Anxiety Trajectories and Identity Development in Adolescence: A Five-wave Longitudinal Study. *Journal of Youth and Adolescence*, *38*(6), 839–849. doi:10.100710964-008-9302-y PMID:19636785

Crow, G. M., & Møller, J. (2017). Professional Identities of School Leaders across International Contexts: An Introduction and Rationale. *Educational Management Administration & Leadership*, *45*(5), 749–758. doi:10.1177/1741143217714485

Cunningham, S. S. (Director). (1980). *Friday the 13th*. https://www.imdb.com/title/tt0080761/

Curtis, D., & Hart, C. L. (2022). *Pathological Lying: Theory, Research, and Practice*. American Psychological Association.

Cuypers, S. E. (2017). Hierarchical Autonomy, Self-Identification and Self-Evaluation. In *Self-Identity and Personal Autonomy: An Analytical Anthropology*. Routledge., doi:10.4324/9781315211428-5

Czermak, M., & Frignet, H. (1996). *Sur l'identité sexuelle: A propos du transsexualisme : actes des journées des 30 novembre et 1er décembre 1996, Centre hospitalier Sainte-Anne, Paris*. Association freudienne internationale.

D'Ambrosio, P. J. (2018). From present to presentation: A philosophical critique of Hartmut Rosa's "situational identity." *Time & Society*. doi:10.1177/0961463X18787059

Daftary, A.-M. H. (2018). Confronting Immigration Myths with the Reality: A Necessary Perspective for Culturally Grounded Social Work Practice. *Smith College Studies in Social Work*, *88*(2), 96–114. doi:10.1080/00377317.2018.1437995

Dalio, R. (2021). *Principles for Dealing with the Changing World Order: Why Nations Succeed and Fail* (1st ed.). Avid Reader Press / Simon & Schuster.

Damien, P. (2008). *Help! Debunking the Outrageous Claims of Self-help Gurus*. Synergy Books.

Damodaran, V., Hamilton, J., & Allan, R. (2019). Climate signals, environment and livelihoods in the long seventeenth century in India. In *A Cultural History of Famine* (pp. 52–70). Routledge. doi:10.4324/9781315316529-4

Daniels, L. (Director). (2009, November 20). *Precious* [Drama]. Lionsgate, Lee Daniels Entertainment, Smokewood Entertainment Group.

Davidson, M. (2009). Displacement, Space and Dwelling: Placing Gentrification Debate. *Ethics Place and Environment*, *12*(2), 219–234. doi:10.1080/13668790902863465

Davis, J., & Watson, N. (2017). Disabled children, ethnography and unspoken understandings: The collaborative construction of diverse identities. In Research with Children. doi:10.4324/9781315657349-13

Davis, O. (2016, December 19). *No Test Left Behind:How Pearson Made a Killing on the US Testing Craze*. http://talkingpointsmemo.com/features/privatization/four/

Davis, J. M., & Watson, N. (2001). Where are the children's experiences? Analysing social and cultural exclusion in'sp ecial'and'mainstream'schools. *Disability & Society*, *16*(5), 671–687. doi:10.1080/09687590120070060

Davis, M. (2002). *Late Victorian Holocausts: El Niño Famines and the Making of the Third World*. Verso Books.

de Blok, C. J. M., Wiepjes, C. M., Nota, N. M., & den Heijer, M. (2020). Hormone treatment of transgender people: Long term health effects and safety. *Nederlands Tijdschrift voor Geneeskunde*, *164*, D4481. PMID:32757514

De Cunzo, L. A. (2009). Exploring the Institution: Reform, Confinement, Social Change. In M. Hall & S. W. Silliman (Eds.), *Historical Archaeology*. John Wiley & Sons.

de Freitas, E., & Ferrara, F. (2015). Movement, Memory and Mathematics: Henri Bergson and the Ontology of Learning. *Studies in Philosophy and Education*, *34*(6), 565–585. doi:10.100711217-014-9455-y

De Genova, N. (2018). Afterword. Deportation: The Last Word? In S. Khosravi (Ed.), *After Deportation: Ethnographic Perspectives* (pp. 253–266). Springer International Publishing. doi:10.1007/978-3-319-57267-3_13

De Landa, M. (2005). Virtual Environments and the Emergence of Synthetic Reason. In J. B. Dixon & E. Cassidy (Eds.), *Virtual Futures: Cyberotics, Technology and Posthuman Pragmatism*. Routledge.

Deane-Drummond, C. (2017). The Myth of Moral Bio-enhancement: An Evolutionary Anthropology and Theological Critique. In T. J. Trothen & C. Mercer (Eds.), *Religion and Human Enhancement: Death, Values, and Morality*. Springer. doi:10.1007/978-3-319-62488-4_11

DeBlaere, C., Chadwick, C. N., Zelaya, D. G., Bowie, J.-A., Bass, M. F., & Finzi-Smith, Z. (2017). The Feminist Identity Composite: An Examination of Structural Validity With Sexual Minority Women. *Psychology of Women Quarterly*, *41*(1), 20–31. doi:10.1177/0361684316676046

DeLamater, J. D., & Myers, D. J. (2011). *Social Psychology*. Wadsworth.

DeLanda, M. (2006). *A New Philosophy of Society: Assemblage Theory and Social Complexity*. A&C Black.

Deleuze, G. (1988). *Spinoza: Practical Philosophy*. City Lights Books.

Deleuze, G. (2006). *Nietzsche and Philosophy*. Columbia University Press.

Delgado, M. L. (2018, June 22). Teaching Philosophy in Prison: An Experience of Freedom. *Proceedings of the XXIII World Congress of Philosophy*. 10.5840/wcp232018741709

Deligiorgi, K. (2018). *Autonomy*. The Oxford Handbook of Phenomenological Psychopathology. doi:10.1093/oxfordhb/9780198803157.013.42

Deng, F. M. (2010). *Protecting the Dispossessed: A Challenge for the International Community*. Brookings Institution Press.

Derrida, J. (1992). *The Other Heading: Reflections on Today's Europe*. Indiana University Press.

Derrida, J. (2001). *On Cosmopolitanism and Forgiveness*. Psychology Press.

Desmond, D. M., O'Neill, K., De Paor, A., McDarby, G., & MacLachlan, M. (2006). Augmenting the Reality of Phantom Limbs: Three Case Studies Using an Augmented Mirror Box Procedure. *JPO: Journal of Prosthetics and Orthotics*, *18*(3), 74. doi:10.1097/00008526-200607000-00005

Desmond-Harris, J. (2015, February 26). *Here's when you can expect racial minorities to be the majority in each state*. Vox. https://www.vox.com/2015/2/26/8109947/minority-majority-states-race

Devolder, K., & Douglas, T. (2018). The epistemic costs of compromise in bioethics. *Bioethics*, *32*(2), 111–118. doi:10.1111/bioe.12418 PMID:29280164

Diamandopoulos, A. A., Goudas, P., & Diamandopoulou, A. H. (2001). The human skin: A meeting ground for the ideas about macrocosm and microcosm in Ancient and Medieval Greek literature. *Revue Officielle de La Societe Internationale d'Histoire de La Medecine Acta Internationalia Historiae Medicinae*, 94.

Dibbell, J. (1998). *My Tiny Life: Crime and Passion in a Virtual World*. Julian Dibbell.

Dickenson, D. (2006). Gender and Ethics Committees: Where's the 'Different Voice'? *Bioethics*, *20*(3), 115–124. doi:10.1111/j.1467-8519.2006.00485.x PMID:17039630

Dickey, M. D. (2005). Three-dimensional virtual worlds and distance learning: Two case studies of Active Worlds as a medium for distance education. *British Journal of Educational Technology*, *36*(3), 439–451. doi:10.1111/j.1467-8535.2005.00477.x

Dick, P. K. (1982). *Blade Runner*. Ballantine Books.

Diprose, R. (2005). Biomedical ethics and lived, sexed bodies. In *The Bodies of Women: Ethics, Embodiment and Sexual Differences*. Routledge., doi:10.4324/9780203981061-12

Dixit, S., & Prasad, R. (2017). *Human Bond Communication: The Holy Grail of Holistic Communication and Immersive Experience*. John Wiley & Sons. doi:10.1002/9781119341451

Dobken, J. H. (2018). Physician-Assisted Suicide (PAS)/Physician-Assisted Death (PAD): The Rise of Lifeboat Ethics. *Journal of American Physicians and Surgeons*, *23*(4).

Dobken, J. H. (2018). The "New" Medical Morality: Hippocrates or Bioethics? *Journal of American Physicians and Surgeons*, *23*(2).

Dominiczak, P. (Ed.). (2015, October 5). *Theresa May: Mass immigration making "cohesive society" impossible*. https://www.telegraph.co.uk/news/uknews/immigration/11913392/Theresa-May-Mass-immigration-making-cohesive-society-impossible.html

Donaldson, R. (Director). (1995). *Species*. https://www.imdb.com/title/tt0114508/

Donaldson, S. (2010, December 30). Sisters Given Life for $11 Crime Released. *ABC News*. https://abcnews.go.com/US/scott-sisters-released-16-years-prison-kidney-deal/story?id=12508754

Donaldson, S., & Kymlicka, W. (2011). *Zoopolis: A Political Theory of Animal Rights*. OUP Oxford.

Dong, P., & Wang, X.-Y. (2004). Life, Death and End-of-Life Care: Taoist Perspective. In R.-Z. Qiu (Ed.), *Bioethics: Asian Perspectives: A Quest for Moral Diversity* (pp. 147–155). Springer Netherlands. doi:10.1007/978-94-017-0419-9_14

Doring, O., & Chen, R. (2002). *Advances in Chinese medical ethics: Chinese and international perspectives*. Institut Fur Asienkunde.

Dougherty, M. (Director). (2019). *Godzilla: King of the Monsters*. https://www.imdb.com/title/tt3741700/

Drentea, P. (2000). Age, Debt and Anxiety. *Journal of Health and Social Behavior, 41*(4), 437–450. doi:10.2307/2676296

Drevdahl, D. J. (2018). Culture Shifts: From Cultural to Structural Theorizing in Nursing. *Nursing Research, 67*(2), 146–160. doi:10.1097/NNR.0000000000000262 PMID:29489635

Druyan, A., Sagan, C., & Soter, S. (Directors). (1980, September 28). *Cosmos* [Documentary]. KCET, Carl Sagan Productions, British Broadcasting Corporation (BBC).

Dumbill, P. (2019, June 4). *Health technology: Digital therapy by prescription*. https://www.brownejacobson.com/health/training-and-resources/legal-updates/2019/06/health-technology-digital-therapy-by-prescription

Dunagan, J. F. (2004). Neuro-futures: The brain, politics and power. *Journal of Futures Studies, 9*(2), 1–18.

Dunbar, N. (2003, July 1). *Revealed: Goldman Sachs' mega-deal for Greece*. Risk.Net. https://www.risk.net/node/1498135

Durkheim, É., & Ellis, A. (1963). *Incest*. L. Stuart.

Dwoskin, E., Tiku, N., & Kelly, H. (2020, December 3). Facebook to start policing anti-Black hate speech more aggressively than anti-White comments, documents show. *Washington Post*. https://www.washingtonpost.com/technology/2020/12/03/facebook-hate-speech/

Dylan, B. (1964). *The Times They Are a-Changin* [Vynil 7"].

Dynes, R. R., & Quarantelli, E. L. (1975). *Community Conflict: Its Absence and Its Presence In Natural Disasters*. http://udspace.udel.edu/handle/19716/382

Eagles. (1977, February 22). *Hotel California* [7-inch single]. Asylum. https://en.wikipedia.org/w/index.php?title=Hotel_California&oldid=919168934

Ebai, P. N. (2021). *Nations Building*. Exceller Books.

Ebrey, P. B. (1996). *The Cambridge illustrated history of China*. Cambridge University Press., https://archive.org/details/cambridgeillustr00ebre_0

Eckart, J. (2016, June 23). Things you need to know about China's economy. *World Economic Forum*. http://gdelaurier.pbworks.com/w/file/fetch/130100685/8%20things%20you%20need%20to%20know%20about%20China%E2%80%99s%20economy%20_%20World%20Economic%20Forum.pdf

Eckenwiler, L. (2018). Displacement and solidarity: An ethic of place-making. *Bioethics, 32*(9), 562–568. doi:10.1111/bioe.12538 PMID:30450599

Edwards, R. A. (2013). Redefining Industrial Revolution: Song China and England. *2013 Meeting Papers*, Article 351. https://ideas.repec.org//p/red/sed013/351.html

Egypt Independent. (2023, May 8). Lawyer demands Egypt suspend Netflix's 'Cleopatra' documentary. *Egypt Independent.* https://www.egyptindependent.com/lawyer-demands-egypt-take-action-against-cleopatra-documentary/

Ehrenreich, B. (2010). *Smile Or Die: How Positive Thinking Fooled America and the World.* Granta.

EKTEPN. (2010). *Annual report on the state of the drugs and alcohol problem.* Greek Documentation and Monitoring Centre for Drugs.

EKTEPN. (2011). Report of the ad hoc expert group of the Greek focal point on the outbreak of HIV. Greek Documentation and Monitoring Centre for Drugs.

El Haj, M., Antoine, P., & Kapogiannis, D. (2015). Similarity between remembering the past and imagining the future in Alzheimer's disease: Implication of episodic memory. *Neuropsychologia, 66*, 119–125. doi:10.1016/j.neuropsychologia.2014.11.015 PMID:25448861

El-Lahib, Y. (2018). Social Work at the Intersection of Disability and Displacement: Rethinking Our Role. *Journal of Progressive Human Services, 0*(0), 1–20. doi:10.1080/10428232.2018.1531744

Ellemers, N. (1993). The Influence of Socio-structural Variables on Identity Management Strategies. *European Review of Social Psychology, 4*(1), 27–57. doi:10.1080/14792779343000013

Elliott, C. (2003). Humanity 2.0. *The Wilson Quarterly, 27*(4), 13–20.

Ellison, B. (2019). The Patient as Professor: How My Life as a Person with Quadriplegia Shaped My Thinking as an Ethicist. *Perspectives in Biology and Medicine, 62*(2), 342–351. doi:10.1353/pbm.2019.0018 PMID:31281126

Ellis, S. (1995). Liberia 1989-1994: A Study of Ethnic and Spiritual Violence. *African Affairs, 94*(375), 165–197. doi:10.1093/oxfordjournals.afraf.a098806

Emetu, R. E., & Rivera, G. (2018, July). *After Sexual Identity Disclosure: An Ecological Perceptive of LGB Young Adults.* https://doi.org/info:doi/10.5993/AJHB.42.4.5

Emmerich, N. (2017). Topics in death and dying. *QJM, 110*(1), 3–4. doi:10.1093/qjmed/hcw186 PMID:27770050

Emmerich, N. (2018). Bioethics, public intellectuals and political biology today. *History of the Human Sciences, 31*(1), 124–131. doi:10.1177/0952695117729119d

Engelhardt, H. T. (1996). *The Foundations of Bioethics.* Oxford University Press.

Engelhardt, H. T. (2002a). Medicine, Philosophy, and Theology: Christian Bioethics Reconsidered. *Christian Bioethics, 8*(2), 105–117. doi:10.1076/chbi.8.2.105.8750 PMID:12678032

Engelhardt, H. T. (2002b). The ordination of bioethicists as secular moral experts. *Social Philosophy & Policy, 19*(2), 59–82. doi:10.1017/S026505250219203X PMID:12678082

Engelhardt, H. T. (2006). Bioethics and the Philosophy of Medicine Reconsidered. In R. A. Carson & C. R. Burns (Eds.), *Philosophy of Medicine and Bioethics: A Twenty-Year Retrospective and Critical Appraisal.* Springer Science & Business Media.

Engelhardt, H. T. (2013). Bioethics as a Liberal Roman Catholic Heresy: Critical Reflections on the Founding of Bioethics. In J. R. Garrett, F. Jotterand, & D. C. Ralston (Eds.), *The Development of Bioethics in the United States* (pp. 55–76). Springer Netherlands. doi:10.1007/978-94-007-4011-2_5

Engler, J. (2012). *War of the Living Dead: An Analysis of the Permissibility of the Zombie Weapon in Modern International Armed Conflict* (*SSRN* Scholarly Paper ID 2171477). Social Science Research Network. doi:10.2139/ssrn.2171477

Erdal, M. B., & Oeppen, C. (2018). Forced to leave? The discursive and analytical significance of describing migration as forced and voluntary. *Journal of Ethnic and Migration Studies*, *44*(6), 981–998. doi:10.1080/1369183X.2017.1384149

Ervin-Ward, A. (2009). Doing It for the Kids: Rebels and Prom Queens in the Cold War Classroom Film. *Colloquy*, *18*, 160–161.

Escobar, A. (2008). *Territories of Difference: Place, Movements, Life, Redes*. Duke University Press.

Espinal, S. A. (2018). *Enabling Geographies: Neurodivergence, Self-Authorship, and the Politics of Social Space* [PhD Thesis]. California Institute of Integral Studies.

Esther, D., & Gill, M. (2017). Poverty and Social Exclusion in the UK: Vol. 1: Volume 1 - The Nature and Extent of the Problem. Policy Press.

Ettorre, E. (2017). Self, Normativity and Embodiment. In E. Ettorre, E. Annandale, V. M. Hildebrand, A. Porroche-Escudero, & B. K. Rothman (Eds.), Health, Culture and Society: Conceptual Legacies and Contemporary Applications (pp. 37–74). Springer International Publishing. doi:10.1007/978-3-319-60786-3_3

Evans, J. H. (2002). *Playing God?: Human Genetic Engineering and the Rationalization of Public Bioethical Debate*. University of Chicago Press.

Everett, M. (2010). 9/11: The greatest lie ever told. *The Journal of Psychohistory*, *38*(2), 133.

Fafrowicz, M., Marek, T., Karwowski, W., & Schmorrow, D. (2012). *Neuroadaptive Systems: Theory and Applications*. CRC Press. doi:10.1201/b13019

Fairchild, A. L., & Bayer, R. (1999). Uses and Abuses of Tuskegee. *Science*, *284*(5416), 919–921. doi:10.1126cience.284.5416.919 PMID:10357678

Falchuk, B., Loeb, S., & Neff, R. (2018). The Social Metaverse: Battle for Privacy. *IEEE Technology and Society Magazine*, *37*(2), 52–61. doi:10.1109/MTS.2018.2826060

Falk, A., & Knell, M. (2004). Choosing the Joneses: Endogenous Goals and Reference Standards. *The Scandinavian Journal of Economics*, *106*(3), 417–435. doi:10.1111/j.0347-0520.2004.00370.x

Falvo, D., & Holland, B. E. (2017). *Medical and Psychosocial Aspects of Chronic Illness and Disability*. Jones & Bartlett Learning.

Farr, A. (1998). *Sartre's Radicalism and Oakeshott's Conservatism: The Duplicity of Freedom*. Springer. doi:10.1057/9780230380264

Feldstein, S. (2019). *The global expansion of AI surveillance* (Vol. 17). Carnegie Endowment for International Peace.

Ferguson, N. (2008). *The Ascent of Money: A Financial History of the World: 10th Anniversary Edition*. Penguin.

Féron, É. (2018). *Wartime Sexual Violence against Men: Masculinities and Power in Conflict Zones*. Rowman & Littlefield.

Ferrari, L., Hu, A. W., Rosnati, R., & Lee, R. M. (2017). Ethnic Socialization and Perceived Discrimination on Ethnic Identity Among Transracial Adoptees: A Cross-Cultural Comparison Between Italy and the United States. *Journal of Cross-Cultural Psychology*, *48*(10), 1507–1521. doi:10.1177/0022022117728338

Ferrito, M., Needs, A., & Adshead, G. (2017). Unveiling the shadows of meaning: Meaning-making for perpetrators of homicide. *Aggression and Violent Behavior*, *34*, 263–272. doi:10.1016/j.avb.2016.11.009

Festinger, L. (1962). *A Theory of Cognitive Dissonance*. Stanford University Press.

Finch, B. K., Kolody, B., & Vega, W. A. (2000). Perceived Discrimination and Depression among Mexican-Origin Adults in California. *Journal of Health and Social Behavior, 41*(3), 295–313. JSTOR. doi:10.2307/2676322

Finch, B. K., Hummer, R. A., Kol, B., & Vega, W. A. (2001). The Role of Discrimination and Acculturative Stress in the Physical Health of Mexican-Origin Adults. *Hispanic Journal of Behavioral Sciences, 23*(4), 399–429. doi:10.1177/0739986301234004

Finklea, K., Christensen, M. D., Fischer, E. A., Lawrence, S. V., & Theohary, C. A. (2015, July). *Cyber intrusion into US office of personnel management: In brief.* Library of Congress.

Flayelle, M., Brevers, D., King, D. L., Maurage, P., Perales, J. C., & Billieux, J. (2023). A taxonomy of technology design features that promote potentially addictive online behaviours. *Nature Reviews Psychology, 2*(3), 3. Advance online publication. doi:10.103844159-023-00153-4

Fleck, S., & Hachet, M. (2016). Making Tangible the Intangible: Hybridization of the Real and the Virtual to Enhance Learning of Abstract Phenomena. *Frontiers in ICT (Lausanne, Switzerland), 3*. Advance online publication. doi:10.3389/fict.2016.00030

Flegar, V. (2018). Vulnerability and the Right to Preventive Health Care for Asylum-Seeking and Undocumented Migrant Children. *Children's Rights in Health Care*, 334–356. doi:10.1163/9789004327573_015

Fletcher, J. F. (1997). *Situation Ethics: The New Morality.* Westminster John Knox Press.

Fleury, J. S. (2018). *Adolf Hitler: Trial in Absentia in Nuremberg.* Xlibris Corporation.

Flood, R. (2017, April 29). *Shock birth rates shows Greeks may be dying out over imposed austerity.* Express.Co.Uk. https://www.express.co.uk/news/world/798318/Greece-birth-rate-dying-economy-austerity-EU-debt-population-crisis

Forman, M. (Director). (1975). *One Flew Over the Cuckoo's Nest.* https://www.imdb.com/title/tt0073486/

Fornalé, E. (2017). Global–regional interaction to extend access to social protection for migrant workers: Insights from ASEAN and MERCOSUR. *International Social Security Review, 70*(3), 31–52. doi:10.1111/issr.12140

Forsberg, E.-M., Hanssen, A. B., Nielsen, H. M., & Olesen, I. (2018). Patent Ethics: The Misalignment of Views Between the Patent System and the Wider Society. *Science and Engineering Ethics, 24*(5), 1551–1576. doi:10.100711948-017-9956-5 PMID:28812235

Forsyth, R., Colver, A., Alvanides, S., Woolley, M., & Lowe, M. (2007). Participation of young severely disabled children is influenced by their intrinsic impairments and environment. *Developmental Medicine and Child Neurology, 49*(5), 345–349. doi:10.1111/j.1469-8749.2007.00345.x PMID:17489807

Fortunato, J. (2013). "Irreversibility" and the Modern Understanding of Death. *Discussions, 9*(2). http://www.inquiriesjournal.com/articles/795/irreversibility-and-the-modern-understanding-of-death

Foucault, M. (2019). *Discipline and Punish: The Birth of the Prison.* Penguin UK.

Foundation Thomson Reuters. (2017, July 24). *Turkey withdraws list of German firms suspected of backing terrorism-Germany.* News.Trust.Org. http://news.trust.org/item/20170724100356-1gk0g/

Fox, N., Ward, K., & O'Rourke, A. (2005). Pro-anorexia, weight-loss drugs and the internet: An 'anti-recovery' explanatory model of anorexia. *Sociology of Health & Illness, 27*(7), 944–971. doi:10.1111/j.1467-9566.2005.00465.x PMID:16313524

Franklin, S., & Lock, M. M. (2003). *Remaking life & death: Toward an anthropology of the biosciences.* School of American Research Press.

Frederick, K. (2019). *Hearing on "How Corporations and Big Tech Leave Our Data Exposed to Criminals, China, and Other Bad Actors"*. Academic Press.

Frege, G. (1952). The Fundamental Laws of Arithmetic II. In Translations from the Philosophical Writings of Gottlob Frege. Blackwell.

Freud, S. (2008). The Future of an Illusion. Academic Press.

Freud, S. (1939). *Moses and monotheism.* Knopf.

Friar, J. L. (1993). Three-nucleon forces and the three-nucleon systems. *Czechoslovak Journal of Physics, 43*(3), 259–287. doi:10.1007/BF01589847

Friedman, H. H., Friedman, L. W., & Adel, S. (2017). *Conscious Capitalism vs. Rapacious Capitalism: Lessons from King Leopold II* (*SSRN* Scholarly Paper No. 2988563). https://papers.ssrn.com/abstract=2988563

Friedman, T. L. (1995, February 22). Opinion | Foreign Affairs; Don't Mess With Moody's. *The New York Times.* https://www.nytimes.com/1995/02/22/opinion/foreign-affairs-don-t-mess-with-moody-s.html

Friedman, U. (2017, January 30). *Where America's Terrorists Actually Come From.* The Atlantic. https://www.theatlantic.com/international/archive/2017/01/trump-immigration-ban-terrorism/514361/

Friedman, S. H., Hall, R. C. W., & Appel, J. M. (2018). The Last Jedi Takes His Own Life: Rational Suicide and Homicide-Suicide in Star Wars. *Academic Psychiatry, 42*(4), 503–509. doi:10.100740596-018-0938-y PMID:29797244

Fromm, E. (2013). The Relation of Freud's Life and Death Instincts to Biophilia and Necrophilia. In *The Anatomy of Human Destructiveness.* Open Road Media.

Fuentes-Camacho, T. (2018). Introduction: UNESCO and the Law of Cyberspace. In The International Dimensions of Cyberspace Law. doi:10.4324/9781315181912-1

Fujieda, S. (2018). Between the Secular and the Religious: Japanese Buddhism in the Public Discourse on the Issues of Organ Transplant. *Proceedings of the XXIII World Congress of Philosophy,* 3. 10.5840/wcp232018386

Fujita, M., Slingsby, B. T., & Akabayashi, A. (2010). Transplant tourism from Japan. *The American Journal of Bioethics, 10*(2), 24–26. doi:10.1080/15265160903506384 PMID:20131168

Fukushi, T., Sakura, O., & Koizumi, H. (2007). Ethical considerations of neuroscience research: The perspectives on neuroethics in Japan. *Neuroscience Research, 57*(1), 10–16. doi:10.1016/j.neures.2006.09.004 PMID:17034890

Furness, D. (2018, July 4). *Cyborg Law: What Laws Should Govern a Society of Cyborgs.* Digital Trends. https://www.digitaltrends.com/cool-tech/cyborg-law-and-rights-of-augmented-humans/

Gaines, D. (2004). Geographical perspectives on disability: A socio-spatial analysis of the mentally disabled population in Russia. *Middle States Geographer, 37*(1), 80–89.

Galán, S., Roy, R., Solé, E., Racine, M., de la Vega, R., Jensen, M. P., & Miró, J. (2018). Committed Action, Disability and Perceived Health in Individuals with Fibromyalgia. *Behavioral Medicine (Washington, D.C.), 0*(0), 1–8. doi:10.1080/08964289.2018.1467370 PMID:29671691

Gallagher, S. (2014). The cruel and unusual phenomenology of solitary confinement. *Frontiers in Psychology, 5.* Advance online publication. doi:10.3389/fpsyg.2014.00585 PMID:24971072

Gallagher, S. (2018). Educating the Right Stuff: Lessons in Enactivist Learning. *Educational Theory, 68*(6), 625–641. doi:10.1111/edth.12337

Gallant, P. (2003, April 11). A Moriori Lesson. *National Review*. https://www.nationalreview.com/2003/04/moriori-lesson-paul-gallant/

Galliott, J. (2016). *Commercial Space Exploration: Ethics, Policy and Governance*. Routledge. doi:10.4324/9781315572857

Ganeri, J. (2017). Classical Indian philosophy. In S. Bernecker & K. Michaelian (Eds.), *The Routledge Handbook of Philosophy of Memory*., doi:10.4324/9781315687315-33

Gao, C., Ludlow, F., Matthews, J. A., Stine, A. R., Robock, A., Pan, Y., Breen, R., Nolan, B., & Sigl, M. (2021). Volcanic climate impacts can act as ultimate and proximate causes of Chinese dynastic collapse. *Communications Earth & Environment*, *2*(1), 1. Advance online publication. doi:10.103843247-021-00284-7

García-Peña, L. (2019). Immigrant Rights are Human Rights. *NACLA Report on the Americas*, *51*(1), 101–104. doi:10.1080/10714839.2019.1593700

Garrett, P. M. (2015). Constraining and confining ethnic minorities: Impoverishment and the logics of control in neoliberal Ireland. *Patterns of Prejudice*, *49*(4), 414–434. doi:10.1080/0031322X.2015.1074785

Gee, G. C., Spencer, M., Chen, J., Yip, T., & Takeuchi, D. T. (2007). The association between self-reported racial discrimination and 12-month DSM-IV mental disorders among Asian Americans nationwide. *Social Science & Medicine*, *64*(10), 1984–1996. doi:10.1016/j.socscimed.2007.02.013 PMID:17374553

Geerken, H. H. (2021). *The Gold of the Bandas: The History of the Nutmeg: The forgotten islands that once made world history*. BoD – Books on Demand.

Gelfand, E. D. (1983). *Imagination in Confinement: Women's Writings from French Prisons*. Cornell University Press.

General Assembly United Nations. (2008, July 28). Fifth quinquennial review and appraisal of the World Programme of Action concerning Disabled Persons. In *Social Development, Including Questions, Relating to the World Social Situation and to Youth, Ageing, Disabled Persons and the Family*. 63rd Session General Assembly United Nations. https://documents-dds-ny.un.org/doc/UNDOC/GEN/N08/442/48/PDF/N0844248.pdf?OpenElement

Gerrard, J. (2018, July 30). Study claims austerity has killed more in Spain than those who died in the Civil War. *Euro Weekly News Spain*. https://www.euroweeklynews.com/2018/07/30/study-claims-austerity-has-killed-more-in-spain-than-those-who-died-in-the-civil-war/

Ghadirian, P., & Bishop, I. D. (2002). Composition of augmented reality and GIS to visualize environmental changes. *Proceedings of the Joint AURISA and Institution of Surveyors Conference*, 25–30.

Giaimo, G. N. (2016). Memory, Brains, and Narratives? The Humanities as a Testing-Ground for Bioethical Scenario-Building. *Literature and Medicine*, *34*(1), 53–78. doi:10.1353/lm.2016.0003 PMID:27545476

Giger, H. R. (1994). Giger's Alien (2nd ed.). Morpheus International.

Gillett, G. (1992). Coma, Death and Moral Dues: A Response to Serafini. *Bioethics*, *6*(4), 375–377. doi:10.1111/j.1467-8519.1992.tb00214.x PMID:11652098

Gillett, G. (2015). Culture, Truth, and Science After Lacan. *Journal of Bioethical Inquiry*, *12*(4), 633–644. doi:10.100711673-015-9664-2 PMID:26615543

Giordano, J. (2012). Integrative Convergence in Neuroscience: Trajectories, Problems, and the Need for a Progressive Neurobioethics. In A. Vaseashta, E. Braman, & P. Susmann (Eds.), *Technological Innovations in Sensing and Detection of Chemical, Biological, Radiological, Nuclear Threats and Ecological Terrorism* (pp. 115–130). Springer Netherlands. doi:10.1007/978-94-007-2488-4_10

Giordano, J., & Benedikter, R. (2012). An early-and necessary-flight of the owl of Minerva: Neuroscience, neurotechnology, human socio-cultural boundaries, and the importance of neuroethics. *Journal of Evolution and Technology / WTA*, *22*(1), 14–25.

Gisler, G. R., Weaver, R. P., Mader, C. L., & Gittings, M. L. (2004). Two- and three-dimensional asteroid impact simulations. *Computing in Science & Engineering*, *6*(3), 46–55. doi:10.1109/MCISE.2004.1289308

Gladden, M. E. (2016a). *Enterprise Architecture for Neurocybernetically Augmented Organizational Systems: The Impact of Posthuman Neuroprosthetics on the Creation of Strategic, Structural, Functional, Technological, and Sociocultural Alignment* [PhD Thesis].

Gladden, M. E. (2016b). *Posthuman Management: Creating Effective Organizations in an Age of Social Robotics, Ubiquitous AI, Human Augmentation, and Virtual Worlds*. Defragmenter Media.

Glannon, W. (2013). *Brain, Body, and Mind: Neuroethics with a Human Face*. Oxford University Press.

Glaser, W. (1994). Commentary: Garry David, psychiatry, and the discourse of dangerousness. *Australian and New Zealand Journal of Criminology*, *27*(1), 46–49. doi:10.1177/000486589402700107

Glinski, S. (2020, May 4). *As adults, Afghanistan's "dancing boys" lead double lives*. News.Trust.Org. https://news.trust.org/item/20200504031922-2nvol/

Gloeckler, S., & Biller-Andorno, N. (2023). Mental health services in the metaverse: Potential and concerns. *Swiss Medical Weekly*, *153*(4), 40089. doi:10.57187mw.2023.40089 PMID:37099742

Glover, J. (1988). *I: The Philosophy and Psychology of Personal Identity*. Penguin Books.

Glowa, K. M., Egerer, M., & Jones, V. (2019). Agroecologies of displacement: A study of land access, dislocation, and migration in relation to sustainable food production in the Beach Flats Community Garden. *Agroecology and Sustainable Food Systems*, *43*(1), 92–115. doi:10.1080/21683565.2018.1515143

Gnes, D. (2013). *Maghnia: Crossing the Uncrossable Border*. https://euromedrights.org/wp-content/uploads/2013/01/Maghnia_Crossing-the-uncrossable border_EN.pdf

Goering, S., Klein, E., Dougherty, D. D., & Widge, A. S. (2017). Staying in the Loop: Relational Agency and Identity in Next-Generation DBS for Psychiatry. *AJOB Neuroscience*, *8*(2), 59–70. doi:10.1080/21507740.2017.1320320

Gofas, A., Hamati-Ataya, I., & Onuf, N. (2018). Positionality and the Pluriverse. In *The SAGE Handbook of the History, Philosophy and Sociology of International Relations*. SAGE. doi:10.4135/9781526402066

Goldston, J., Chaffer, T. J., & Martinez, G. (2022). The Metaverse as the Digital Leviathan: A Case Study of Bit. Country. *Journal of Applied Business & Economics*, *24*(2).

Gomes, S., & Duarte, V. (2018). What about ethics? Developing qualitative research in confinement settings. *European Journal of Criminology*. Advance online publication. doi:10.1177/1477370818801305

Gomez, R., & Vannini, S. (2017). Notions of Home and Sense of Belonging in the Context of Migration in a Journey through Participatory Photography. *The Electronic Journal on Information Systems in Developing Countries*, *78*(1), 1–46. doi:10.1002/j.1681-4835.2017.tb00569.x

Gong, F., Xu, J., & Takeuchi, D. T. (2017). Racial and Ethnic Differences in Perceptions of Everyday Discrimination. *Sociology of Race and Ethnicity (Thousand Oaks, Calif.)*, *3*(4), 506–521. doi:10.1177/2332649216681587

González Gómez, J. A., & Álvarez, D. G. (2022, November 17). *El balance escalofriante de la ley del "solo sí es sí": Ya son más de 1.000 los delincuentes sexuales beneficiados*. Periodista Digital. https://www.periodistadigital.com/politica/justicia/20221117/mapa-delincuentes-sexuales-beneficiados-ley-irene-montero-noticia-689404809015/

Goodman, K. K. W. (1999). Philosophy as News: Bioethics, Journalism and Public Policy. *The Journal of Medicine and Philosophy*, *24*(2), 181–200. doi:10.1076/jmep.24.2.181.2533 PMID:10344425

Goodwin, M. (2006). *Black Markets: The Supply and Demand of Body Parts*. Cambridge University Press. doi:10.1017/9781107051041

Gorayska, B., & Mey, J. L. (2004). *Cognition and Technology: Co-existence, convergence and co-evolution*. John Benjamins Publishing. doi:10.1075/z.127

Goshe, S. (2019). The lurking punitive threat: The philosophy of necessity and challenges for reform. *Theoretical Criminology*, *23*(1), 25–42. doi:10.1177/1362480617719450

Goumart, S.-I. (2019). Affective Politics and "Crisis": The Examples of HIV-Positive Women's Public Denouncement and of the Refugees' Confinement. In D. Olkowski & E. Pirovolakis (Eds.), *Deleuze and Guattari's Philosophy of Freedom: Freedom's Refrains*. Routledge.

Gramigna, R. (2022). Inside Facebook's semiosphere. How social media influence digital hate and fuel cyber-polarization. *Social Semiotics*, *32*(5), 606–633. doi:10.1080/10350330.2022.2157171

Gray, C. H. (2000). Cyborg Warriors. In *Cyborg Citizen: Politics in the Posthuman Age*. Routledge. doi:10.4324/9780203949351

Gray, C. H., Mentor, S., & Figueroa-Sarriera, H. (1995). *The cyborg handbook*. Routledge.

Green, C. R., Ndao-Brumblay, S. K., West, B., & Washington, T. (2005). Differences in Prescription Opioid Analgesic Availability: Comparing Minority and White Pharmacies Across Michigan. *The Journal of Pain*, *6*(10), 689–699. doi:10.1016/j.jpain.2005.06.002 PMID:16202962

Green, J. (2012). *The Fault in Our Stars*. Penguin.

Green, R. M., Donovan, A., & Jauss, S. A. (2008). *Global Bioethics: Issues of Conscience for the Twenty-First Century*. OUP Oxford.

Greig, D. N. (2002). *Neither Bad Nor Mad: The Competing Discourses of Psychiatry, Law, and Politics*. Jessica Kingsley Publishers.

Greig, D. N. (2006). Madness, badness, and evil. In *Forensic psychiatry* (pp. 135–152). Springer. doi:10.1007/978-1-59745-006-5_7

Gresky, J., Haelm, J., & Clare, L. (2017). Modified human crania from Göbekli Tepe provide evidence for a new form of Neolithic skull cult. *Science Advances*, *3*(6), e1700564. doi:10.1126ciadv.1700564 PMID:28782013

Grim, B. J., & Finke, R. (2010). *The Price of Freedom Denied: Religious Persecution and Conflict in the Twenty-First Century*. Cambridge University Press. doi:10.1017/CBO9780511762345

Groeneveld, E. (2017, May 15). Early Human Migration. *World History Encyclopedia*. https://www.worldhistory.org/article/1070/early-human-migration/

Grossberg, L. (1996). Identity and Cultural Studies: Is That All There Is? In S. Hall & P. du Gay (Eds.), *Questions of Cultural Identity: SAGE Publications*. SAGE.

Gross, M. L. (2014). Communitarian Bioethics: Three Case Studies. *Society*, *51*(4), 354–361. doi:10.100712115-014-9792-z

Grossman, H. (2023, April 26). *Top Swedish doctors blow whistle on trans puberty-suppressing drugs affecting children's bones: "Experimental."* Fox News. https://www.foxnews.com/media/top-swedish-doctors-blow-whistle-on-trans-puberty-suppressing-drugs-affecting-childrens-bones-experimental

Grossman, A. (2004). *Pious and Rebellious: Jewish Women in Medieval Europe*. UPNE.

Guan, S. (1987). Chartism and the first opium war. *History Workshop Journal*, *24*(1), 17–31. doi:10.1093/hwj/24.1.17

Guildford, B. J., Daly-Eichenhardt, A., Hill, B., Sanderson, K., & McCracken, L. M. (2018). Analgesic reduction during an interdisciplinary pain management programme: Treatment effects and processes of change. *British Journal of Pain*, *12*(2), 72–86. doi:10.1177/2049463717734016 PMID:29796259

Guillaume, L. S., & Finlay, E. (2018). Disabled mobility and the production of impairment: The case of Australia's migration policy framework. *Asia Pacific Viewpoint*, *59*(1), 119–131. doi:10.1111/apv.12182

Gullo, M. (2013). *Insubordinacion fundante*. Editorial Biblos.

Gupta, A. (2014). Altruism Toward Non-Humans: Lessons for Interstellar Communication. In D. A. Vakoch (Ed.), *Extraterrestrial Altruism: Evolution and Ethics in the Cosmos* (pp. 79–91). Springer Berlin Heidelberg. doi:10.1007/978-3-642-37750-1_6

Gustafsson, J. E. (2010). Did Locke defend the memory continuity criterion of personal identity. *Locke Studies*, *10*, 113–129.

Guttman, N., & Salmon, C. T. (2004). Guilt, Fear, Stigma and Knowledge Gaps: Ethical Issues in Public Health Communication Interventions. *Bioethics*, *18*(6), 531–552. doi:10.1111/j.1467-8519.2004.00415.x PMID:15580723

Hage, V. (2013, November 6). Philosophical Differences: The Falling-Out of Camus and Sartre. *Spiegel Online*. https://www.spiegel.de/international/zeitgeist/camus-and-sartre-friendship-troubled-by-ideological-feud-a-931969.html

Haldane, J. B. S. (1926). *Daedalus, Or, Science and the Future*. Kegan Paul.

Hale, B., Wright, C., & Miller, A. (2017). *A Companion to the Philosophy of Language*. John Wiley & Sons. doi:10.1002/9781118972090

Hall, C. M. (2007). *Pro-poor Tourism: Who Benefits?: Perspectives on Tourism and Poverty Reduction*. Channel View Publications. doi:10.21832/9781845410766

Hall, I. (2009). *British International Thinkers from Hobbes to Namier*. Springer. doi:10.1057/9780230101739

Hallward, P. (2006). *Out of this World: Deleuze and the Philosophy of Creation*. Verso.

Hamauzu, S. (2018). On Development from Husserl's phenomenology: Between Phenomenology of Intersubjectivity and Clinical Philosophy of Caring. 大阪大学大学院文学研究科紀要, *58*(2), 1–311.

Hancock, P., & Tyler, M. (2001). Work, Postmodernism and Organization: A Critical Introduction. *Sage (Atlanta, Ga.)*. Advance online publication. doi:10.4135/9781446218365

Han, D.-I. D., Bergs, Y., & Moorhouse, N. (2022). Virtual reality consumer experience escapes: Preparing for the metaverse. *Virtual Reality (Waltham Cross)*, *26*(4), 1443–1458. doi:10.100710055-022-00641-7

Hanushek, E. A. (2011). Valuing teachers: How much is a good teacher worth. *Education Next*, *11*(3), 40–45.

Harada, K. (1946). *The Saionji-Harada Memoirs*. Academic Press.

Harari, Y. N. (2015). *Sapiens: A Brief History of Humankind.* Harper Collins.

Hargreaves, A. P., & Shirley, D. L. (2009). *The Fourth Way: The Inspiring Future for Educational Change.* Corwin Press. doi:10.4135/9781452219523

Harrison, K., Fahy, K. M., & Thomposn, F. (1998). *Case studies in midwifery ethics: Discourses in knowledge & power.* AAHPSSS.

Harrison, R. T., & Leitch, C. M. (2019). The Dynamics of Identity, Identity Work and Identity Formation in the Family Business: Insights from Identity Process Theory and Transformative Learning. In E. Memili & C. Dibrell (Eds.), *The Palgrave Handbook of Heterogeneity among Family Firms* (pp. 673–713). Springer International Publishing. doi:10.1007/978-3-319-77676-7_25

Hart, C. L., & Curtis, D. A. (2023). *Big Liars: What Psychological Science Tells Us about Lying, and How You Can Avoid Being Duped.* American Psychological Association.

Hartman, D., & Zimberoff, D. (2002). Memory access to our earliest influences. *Journal of Heart Centered Therapies, 5*, 3–64.

Harwood, K. A. (2018). Decentering Whiteness in Feminist Bioethics: Assisted Reproductive Technologies (ART) as an Illustrative Case. In L. Campo-Engelstein & P. Burcher (Eds.), *Reproductive Ethics II* (pp. 99–112). Springer International Publishing. doi:10.1007/978-3-319-89429-4_8

Hasel, M. G. (1994). Israel in the Merneptah Stela. *Bulletin of the American Schools of Oriental Research. American Schools of Oriental Research, 296*, 45–61. doi:10.2307/1357179

Hassan, G., Brouillette-Alarie, S., Alava, S., Frau-Meigs, D., Lavoie, L., Fetiu, A., Varela, W., Borokhovski, E., Venkatesh, V., Rousseau, C., & Sieckelinck, S. (2018). Exposure to Extremist Online Content Could Lead to Violent Radicalization:A Systematic Review of Empirical Evidence. *International Journal of Developmental Science, 12*(1–2), 71–88. doi:10.3233/DEV-170233

Hastings, G. (2017). Rebels with a cause: The spiritual dimension of social marketing. *Journal of Social Marketing, 7*(2), 223–232. Advance online publication. doi:10.1108/JSOCM-02-2017-0010

Hatton, J. P. (2011). *How and why did MARS facilitate migration control? Understanding the implication of migration and refugee studies (MARS) with the restriction of human mobility by UK state agencies* [PhD Thesis]. Oxford University.

Hatzenbuehler, M. L. (2017). Structural Stigma and Health. In B. Major, J. F. Dovidio, & B. G. Link (Eds.), *The Oxford Handbook of Stigma, Discrimination, and Health.* Oxford University Press.

Hauge, Å. L. (2007). Identity and Place: A Critical Comparison of Three Identity Theories. *Architectural Science Review, 50*(1), 44–51. doi:10.3763/asre.2007.5007

Heath, R. (2018, December 13). *Brexit Britain: Small, boring and stupid.* POLITICO. https://www.politico.eu/article/brexit-britain-small-boring-and-stupid-theresa-may-eu-withdrawal-deal/

Heidegger, M. (2010). *Being and Time.* SUNY Press.

Heidelbaugh, J. J., & Yen, G. (2008). Lifestyle Risks. In J. J. Heidelbaugh (Ed.), *Clinical Men's Health* (pp. 321–337). W.B. Saunders. doi:10.1016/B978-141603000-3.10017-6

Hekman, S. J. (2014). *The Feminine Subject.* John Wiley & Sons.

Henig, R. M. (2011, May 5). The Life and Legacy of Paul de Kruif. *Alicia Patterson Foundation.* https://aliciapatterson.org/stories/life-and-legacy-paul-de-kruif

Henriksen, A.-K. (2017). Confined to Care: Girls' Gendered Vulnerabilities in Secure Institutions. *Gender & Society*, *31*(5), 677–698. doi:10.1177/0891243217726968

Herbert, B. (2010, October 12). So Utterly Inhumane. *The New York Times*. https://www.nytimes.com/2010/10/12/opinion/12herbert.html

Hersh, S. (2023, February 8). How America Took Out The Nord Stream Pipeline [Substack newsletter]. *Seymour Hersh*. https://seymourhersh.substack.com/p/how-america-took-out-the-nord-stream

Heslin, A., Deckard, N. D., Oakes, R., & Montero-Colbert, A. (2019). Displacement and Resettlement: Understanding the Role of Climate Change in Contemporary Migration. In R. Mechler, L. M. Bouwer, T. Schinko, S. Surminski, & J. Linnerooth-Bayer (Eds.), *Loss and Damage from Climate Change: Concepts, Methods and Policy Options* (pp. 237–258). Springer International Publishing. doi:10.1007/978-3-319-72026-5_10

Hessel, D. T. (2000). *Christianity and ecology: Seeking the well-being of earth and humans*. Harvard University Center for the Study of World Religions.

Heyer, K. E. (2012). Reframing Displacement and Membership: Ethics of Migration. *Theological Studies*, *73*(1), 188–206. doi:10.1177/004056391207300109

Higa, T. (1993). Weyl manifolds and Einstein-Weyl manifolds. *Rikkyo Daigaku Sugaku Zasshi*, *42*(2), 143–160.

Higonnet, E. (2011). Quiet Genocide: Guatemala 1981-1983. Transaction Publishers.

Hilberg, R. (2003). *The Destruction of the European Jews*. Yale University Press.

Hill, R. B. (1989). Economic forces, structural discrimination and black family instability. *The Review of Black Political Economy*, *17*(3), 5–23. doi:10.1007/BF02901099

Hilton, R. H. (1984). Women traders in medieval England. *Women's Studies*, *11*(1–2), 139–155. doi:10.1080/00497878.1984.9978607

Hirsch, A., Maylea, C., David, C., & Nipperess, S. (2019). The changing face of disability and refugee services and policy in Australia: Implications for social work. *Social Work and Policy Studies: Social Justice, Practice and Theory*, *2*(1). https://openjournals.library.sydney.edu.au/index.php/SWPS/article/view/12775

Hirsch, F. (2020). *Soviet Judgment at Nuremberg: A New History of the International Military Tribunal After World War II*. Oxford University Press.

Hitler, A. (1969). *Mein Kampf*. Hutchinson.

Hofmann, B., & Magelssen, M. (2018). In pursuit of goodness in bioethics: Analysis of an exemplary article. *BMC Medical Ethics*, *19*(1), 60. doi:10.118612910-018-0299-9 PMID:29902999

Hohenberg, P. M., & Lees, L. H. (2009). *The Making of Urban Europe, 1000-1994*. Harvard University Press. doi:10.2307/j.ctvjf9vvp

Hollenbach, D. (2016). Borders and Duties to the Displaced: Ethical Perspectives on the Refugee Protection System. *Journal on Migration and Human Security*, *4*(3), 148–165. doi:10.1177/233150241600400306

Holloway, K. F. (2011). *Private Bodies, Public Texts: Race, Gender, and a Cultural Bioethics*. Duke University Press.

Holmes, D., Murray, S. J., & Knack, N. (2015). Experiencing Seclusion in a Forensic Psychiatric Setting: A Phenomenological Study. *Journal of Forensic Nursing*, *11*(4), 200–213. doi:10.1097/JFN.0000000000000088 PMID:26457901

Holterman, D. (2014). The biopolitical war for life: Extractivism and the Ugandan oil state. *The Extractive Industries and Society*, *1*(1), 28–37. doi:10.1016/j.exis.2014.01.003

Holt, T. J., Freilich, J. D., & Chermak, S. M. (2017). Internet-Based Radicalization as Enculturation to Violent Deviant Subcultures. *Deviant Behavior*, *38*(8), 855–869. doi:10.1080/01639625.2016.1197704

Homer. (2011). *The Odyssey*. Collector's Library.

Hondagneu-Sotelo, P. (1999). Introduction: Gender and Contemporary U.S. Immigration. *The American Behavioral Scientist*, *42*(4), 565–576. doi:10.1177/00027649921954363

Horner, N. (2018). *What is Social Work?: Contexts and Perspectives*. Learning Matters.

Hou, F., & Balakrishnan, T. R. (1996). The Integration of Visible Minorities in Contemporary Canadian Society. *The Canadian Journal of Sociology/Cahiers Canadiens de Sociologie, 21*(3), 307–326. doi:10.2307/3341769

Houston, L. (2002). Historical and Cross-Cultural Perspectives on the Art of Medical Diagnosis. *Senior Thesis Projects, 1993-2002*, 97.

Hovil, L. (2014). *Local Integration*. The Oxford Handbook of Refugee and Forced Migration Studies., doi:10.1093/oxfordhb/9780199652433.013.0042

Hsieh, C., & Liu, B. (1983). The Pursuance of Better Quality of life: In the Long Run, Better Quality of Social Life Is the Most Important Factor in Migration. *American Journal of Economics and Sociology*, *42*(4), 431–440. doi:10.1111/j.1536-7150.1983.tb01730.x

Huang, C. J. (2018). On Being the "Right" Kind of Chronic Pain Patient. *Narrative Inquiry in Bioethics*, *8*(3), 239–245. doi:10.1353/nib.2018.0073 PMID:30595591

Huang, H.-M., Liaw, S.-S., & Lai, C.-M. (2016). Exploring learner acceptance of the use of virtual reality in medical education: A case study of desktop and projection-based display systems. *Interactive Learning Environments*, *24*(1), 3–19. doi:10.1080/10494820.2013.817436

Huang, T.-C., Chen, C.-C., & Chou, Y.-W. (2016). Animating eco-education: To see, feel, and discover in an augmented reality-based experiential learning environment. *Computers & Education*, *96*, 72–82. doi:10.1016/j.compedu.2016.02.008

Huarte, P. (2023, March 11). *Ángela "Pam" dice que en España los hombres "son bastante violadores"*. The Objective. https://theobjective.com/espana/2023-03-11/angela-pam-hombres-violadores/

Hublin, J.-J. (2015). The modern human colonization of western Eurasia: When and where? *Quaternary Science Reviews*, *118*, 194–210. doi:10.1016/j.quascirev.2014.08.011

Huddleston, T. (2022, January 31). *"This is creating more loneliness": The metaverse could be a serious problem for kids, experts say*. CNBC. https://www.cnbc.com/2022/01/31/psychologists-metaverse-could-be-a-problem-for-kids-mental-health.html

Huff, J. L., Smith, J. A., Jesiek, B. K., Zoltowski, C. B., & Oakes, W. C. (2018). Identity in Engineering Adulthood: An Interpretative Phenomenological Analysis of Early-Career Engineers in the United States as They Transition to the Workplace. *Emerging Adulthood, 2167696818780444*. Advance online publication. doi:10.1177/2167696818780444

Hughes, J. (2007). The compatibility of religious and transhumanist views of metaphysics, suffering, virtue and transcendence in an enhanced future. *Institute for Ethics and Emerging Technologies*. https://www.researchgate.net/profile/James_Hughes2/publication/237467243_The_Compatibility_of_Religious_and_Transhumanist_Views_of_Metaphysics_Suffering_Virtue_and_Transcendence_in_an_Enhanced_Future/links/02e7e5318b7ab3a5a5000000/The-Compatibility-of-Religious-and-Transhumanist-Views-of-Metaphysics-Suffering-Virtue-and-Transcendence-in-an-Enhanced-Future.pdf

Hughes, B. (2017). Impairment on the move: The disabled incomer and other invalidating intersections. *Disability & Society*, *32*(4), 467–482. doi:10.1080/09687599.2017.1298991

Hughes, J., & Churchland, P. S. (1995). My behavior made me do it: The uncaused cause of teleological behaviorism. *Behavioral and Brain Sciences*, *18*(1), 130–131. doi:10.1017/S0140525X00037705

Hughey, M. W. (2017). Bad hombres? The implicit and explicit racialization of immigration. *Humanity & Society*, *41*(1), 127–129. doi:10.1177/0160597616680016

Hume, D. (1964). *Hume on Religion*. World Publishing Company.

Humphries, S., & Gordon, P. (1992). *Out of Sight: Experiences of disability 1900–50*. Northcote.

Hurst, S. A. (2008). Vulnerability in research and health care; describing the elephant in the room? *Bioethics*, *22*(4), 191–202. doi:10.1111/j.1467-8519.2008.00631.x PMID:18405317

Hussein, S., & Manthorpe, J. (2014). Structural marginalisation among the long-term care workforce in England: Evidence from mixed-effect models of national pay data. *Ageing and Society*, *34*(1), 21–41. doi:10.1017/S0144686X12000785

Husserl, E. (1970). *The Crisis of European Sciences and Transcendental Phenomenology: An Introduction to Phenomenological Philosophy*. Northwestern University Press.

Hymes, R. (2015). Epilogue: A Hypothesis on the East Asian Beginnings of the Yersinia pestis Polytomy. *The Medieval Globe*, *1*(1), 285–308. doi:10.17302/TMG.1-1.11

Iacono, C. (2019, September 28). *No, Slavery Did Not Make America Rich | Corey Iacono*. https://fee.org/articles/no-slavery-did-not-make-america-rich/

Ibaba, S. I., & Okoye, A. C. (2015). Psychopathic leadership and the challenge of terrorism in Nigeria. *Poultry Science Reviews*, *6*(2), 149–170.

Iborra, Y. S. (2016, March 3). *Cuando la puta y el cura entraron al Walden-7*. SomAtents. http://somatents.com/es/magazine-es/cuando-la-puta-y-el-cura-entraron-al-walden-7/

Iezzoni, L. I. (2011). Eliminating Health And Health Care Disparities Among The Growing Population Of People With Disabilities. *Health Affairs*, *30*(10), 1947–1954. doi:10.1377/hlthaff.2011.0613 PMID:21976339

Iftime, O. (2006). Life sciences, apophatism and Bioethics. *European Journal of Science and Theology*, *2*(1), 21–46.

IJsselsteijn, W. A., de Kort, Y. A. W., & Haans, A. (2006). Is This My Hand I See Before Me? The Rubber Hand Illusion in Reality, Virtual Reality, and Mixed Reality. *Presence (Cambridge, Mass.)*, *15*(4), 455–464. doi:10.1162/pres.15.4.455

Imrie, R. (2004). Demystifying disability: A review of the International Classification of Functioning, Disability and Health. *Sociology of Health & Illness*, *26*(3), 287–305. doi:10.1111/j.1467-9566.2004.00391.x PMID:15043600

Inckle, K. (2018). Irrational Perspectives and Untenable Positions: Sociology, Madness and Disability. In Subcultures, Bodies and Spaces: Essays on Alternativity and Marginalization (Vol. 1–0, pp. 169–188). Emerald Publishing Limited. doi:10.1108/978-1-78756-511-120181011

Ingraham, C. (2017). The Suddener World: Photography and Ineffable Rhetoric. *Philosophy & Rhetoric, 50*(2), 129–152. doi:10.5325/philrhet.50.2.0129

Intergovernmental Bioethics Committee. (2005). *Universal Declaration on Bioethics and Human Rights.* General Conference of the United Nations Educational, Scientific and Cultural Organization.

International Association of Bioethics. (1999). *Embodying Bioethics: Recent Feminist Advances*. Rowman & Littlefield.

Iosifides, T. (2011). Theoretical and Epistemological Issues in Qualitative Research. In *Qualitative Methods in Migration Studies*. https://www.taylorfrancis.com/

Iovannone, J. J. (2019, May 30). *Leslie Feinberg: Transgender Warrior - Queer History For the People*. Medium. https://medium.com/queer-history-for-the-people/leslie-feinberg-transgender-warrior-fcb1bcaf15b2

Isaac, R. K., & Platenkamp, V. (2018). Dionysus Versus Apollo: An Uncertain Search for Identity Through Dark Tourism—Palestine as a Case Study. In P. R. Stone, R. Hartmann, T. Seaton, R. Sharpley, & L. White (Eds.), *The Palgrave Handbook of Dark Tourism Studies* (pp. 211–225). doi:10.1057/978-1-137-47566-4_9

Ivan, M. K. (2019). *Protective Obligations Owed by States to Persons with Disabilities under International Human Rights Law, Before, During and After Situations of Armed conflicts* [PhD Thesis]. Cardiff University.

Ivinson, G. M., & Taylor, C. A. (2013). Material Feminisms: New Directions for Education. *Gender and Education, 25*(6), 665–670. doi:10.1080/09540253.2013.834617

Jackson, D. Z. (2005, July 22). The Westmoreland mind-set. *The New York Times*. https://www.nytimes.com/2005/07/22/opinion/derrick-z-jackson-the-westmoreland-mindset.html

Jackson, B. A., Buehler, J. W., Cole, D., Cookson, S., Dausey, D. J., Honess-Morreale, L., Lance, S., Molander, R. C., O'Neal, P., & Lurie, N. (2006). Bioterrorism with zoonotic disease: Public health preparedness lessons from a multiagency exercise. *Biosecurity and Bioterrorism, 4*(3), 287–292. doi:10.1089/bsp.2006.4.287 PMID:16999589

Jacob, M. A., & Riles, A. (2007). The New Bureaucracies of Virtue: Introduction. *Political and Legal Anthropology Review, 30*(2), 181–191. doi:10.1525/pol.2007.30.2.181

Jacobson, J., Heard, C., & Fair, H. (2017). *Prison: Evidence of its use and over-use from around the world.* Academic Press.

Janz, U. (2010, September 3). *Conference of the States Parties to the Convention on the Rights of Persons with Disabilities*. Academic Press.

Jay Gould, S. (1978). Were Dinosaurs Dumb? In New Scientist. Reed Business Information.

Jebari, K. (2015). Sensory Enhancement. In J. Clausen & N. Levy (Eds.), *Handbook of Neuroethics* (pp. 827–838). Springer Netherlands. doi:10.1007/978-94-007-4707-4_106

Jefferson Airplane. (1966, November 3). *White Rabbit* [7-inch single]. RCA Victor.

Jefferson-Jones, J. (2013). The Exchange of Inmate Organs for Liberty: Diminishing the Yuck Factor in the Bioethics Repugnance Debate. *SSRN, 16*, 105. doi:10.2139srn.2251248

Jeffery, A. J., & Shackelford, T. K. (2018). Moral positions on publishing race differences in intelligence. *Journal of Criminal Justice, 59*, 132–135. doi:10.1016/j.jcrimjus.2017.05.008

Ji, S., Choi, Y., Lee, C.-K., & Mjelde, J. W. (2018). Comparing willingness-to-pay between residents and non-residents using a contingent valuation method: Case of the Grand Canal in China. *Asia Pacific Journal of Tourism Research, 23*(1), 79–91. doi:10.1080/10941665.2017.1399919

Johnsdotter, S. (2018). The Impact of Migration on Attitudes to Female Genital Cutting and Experiences of Sexual Dysfunction Among Migrant Women with FGC. *Current Sexual Health Reports*, *10*(1), 18–24. doi:10.100711930-018-0139-4 PMID:29541003

Johnson, D. K. (2009). *The Lavender Scare: The Cold War Persecution of Gays and Lesbians in the Federal Government*. University of Chicago Press.

Jones, S. (2023, February 16). Spain passes law allowing anyone over 16 to change registered gender. *The Guardian*. https://www.theguardian.com/world/2023/feb/16/spain-passes-law-allowing-anyone-over-16-to-change-registered-gender

Jones, R. B. (2001). Impairment, disability and handicap—Old fashioned concepts? *Journal of Medical Ethics*, *27*(6), 377–379. doi:10.1136/jme.27.6.377 PMID:11731599

Jon, G. (2017). *Understanding health and social care* (3rd ed.). Policy Press.

Jordan, A., Noel, M., Caes, L., Connell, H., & Gauntlett-Gilbert, J. (2018). A developmental arrest? Interruption and identity in adolescent chronic pain. *Pain Reports (Baltimore, Md.)*, *3*(Suppl 1), e678. Advance online publication. doi:10.1097/PR9.0000000000000678 PMID:30324170

Jørgensen, M. B. (2011). Understanding the Research–Policy Nexus in Denmark and Sweden: The Field of Migration and Integration. *British Journal of Politics and International Relations*, *13*(1), 93–109. doi:10.1111/j.1467-856X.2010.00441.x

Jörke, D., & Selk, V. (2015). Der hilflose Antipopulismus. *Leviathan (Düsseldorf)*, *43*(4), 484–500. doi:10.5771/0340-0425-2015-4-484

Jotterand, F., & Ienca, M. (2017). The Biopolitics of Neuroethics. In E. Racine & J. Aspler (Eds.), Debates About Neuroethics: Perspectives on Its Development, Focus, and Future (pp. 247–261). Springer International Publishing. doi:10.1007/978-3-319-54651-3_17

Jotterand, F., & Levin, S. B. (2019). Moral Deficits, Moral Motivation and the Feasibility of Moral Bioenhancement. *Topoi*, *38*(1), 63–71. doi:10.100711245-017-9472-x

Judge, M. (Director). (2006). *Idiocracy* [Adventure, Comedy, Sci-Fi]. Twentieth Century Fox, Ternion Pictures, Major Studio Partners.

Judge, T. A., & Cable, D. M. (2004). The Effect of Physical Height on Workplace Success and Income: Preliminary Test of a Theoretical Model. *The Journal of Applied Psychology*, *89*(3), 428–441. doi:10.1037/0021-9010.89.3.428 PMID:15161403

Juengst, E. T., Fishman, J. R., McGowan, M. L., & Settersten, R. A. Jr. (2014). Serving epigenetics before its time. *Trends in Genetics*, *30*(10), 427–429. doi:10.1016/j.tig.2014.08.001 PMID:25242336

Kade, D. (2016). Ethics of Virtual Reality Applications in Computer Game Production. *Philosophies*, *1*(1), 73–86. doi:10.3390/philosophies1010073

Kahn, S., & Lee, J. (2010). *Activism and Rhetoric: Theories and Contexts for Political Engagement*. Routledge.

Kandylis, G. (2019). Accommodation as Displacement: Notes from Refugee Camps in Greece in 2016. *Journal of Refugee Studies*, *32*(1), i12–i21. Advance online publication. doi:10.1093/jrs/fey065

Kapell, M. W., & McVeigh, S. (2011). *The Films of James Cameron: Critical Essays*. McFarland.

Kaplan, J. M. (2014). Ignorance, Lies, and Ways of Being Racist. *Critical Philosophy of Race*, *2*(2), 160–182. doi:10.5325/critphilrace.2.2.0160

Kaplan, S. A., & Tsytovich, V. N. (2013). *Plasma Astrophysics: International Series of Monographs in Natural Philosophy*. Elsevier.

Karananas, G. K., Shaposhnikov, M., Shkerin, A., & Zell, S. (2021). Scale and Weyl invariance in Einstein-Cartan gravity. *Physical Review. D*, *104*(12), 124014. doi:10.1103/PhysRevD.104.124014

Katz, B. F. (2009). *Neuroengineering The Future: Virtual Minds And The Creation Of Immortality*. Jones & Bartlett Publishers.

Kearney, P. M., & Pryor, J. (2004). The International Classification of Functioning, Disability and Health (ICF) and nursing. *Journal of Advanced Nursing*, *46*(2), 162–170. doi:10.1111/j.1365-2648.2003.02976.x PMID:15056329

Keeling, G., & Burr, C. (2022). Digital Manipulation and Mental Integrity. In *The Philosophy of Online Manipulation* (pp. 253–271). Routledge. doi:10.4324/9781003205425-15

Keidel, A. (2006). China's social unrest: The story behind the stories. *Carnegie Endowment for International Peace, 48*.

Kellert, S. R. (1985). Historical trends in perceptions and uses of animals in 20th century America. *Environmental Review*, *9*(1), 19–33. doi:10.2307/3984113

Kellmeyer, P. (2018). Neurophilosophical and Ethical Aspects of Virtual Reality Therapy in Neurology and Psychiatry. *Cambridge Quarterly of Healthcare Ethics*, *27*(4), 610–627. doi:10.1017/S0963180118000129 PMID:30198465

Kelly, M. G. E. (2015). *Biopolitical Imperialism*. John Hunt Publishing.

Kelsay, J., & Johnson, J. T. (1991). *Just War and Jihad: Historical and Theoretical Perspectives on War and Peace in Western and Islamic Traditions*. Greenwood Press.

Kemasang, A. R. T. (1985). How Dutch Colonialism Foreclosed a Domestic Bourgeoisie in Java: The 1740 Chinese Massacres Reappraised. *Review - Fernand Braudel Center*, *9*(1), 57–80.

Kemble, F. A. (2006). *Journal of a Residence on a Georgian Plantation: 1838-1839*. IndyPublish.com.

Kennedy, S. E., Shen, Y., Charlesworth, J. A., Mackie, J. D., Mahony, J. D., Kelly, J. J. P., & Pussell, B. A. (2005). Outcome of overseas commercial kidney transplantation: An Australian perspective. *The Medical Journal of Australia*, *182*(5), 224–227. doi:10.5694/j.1326-5377.2005.tb06671.x PMID:15748132

Kentikelenis, A., Karanikolos, M., Papanicolas, I., Basu, S., McKee, M., & Stuckler, D. (2011). Health effects of financial crisis: Omens of a Greek tragedy. *Lancet*, *378*(9801), 1457–1458. doi:10.1016/S0140-6736(11)61556-0 PMID:21988763

Keown, D., Prebish, C. S., Husted, W. R., & Husted, W. R. (1998). *Buddhism and Human Rights*. Psychology Press.

Kerley, K. R. (2018). *Finding Freedom in Confinement: The Role of Religion in Prison Life*. ABC-CLIO.

Kerruish, E. (2019). Arranging sensations: Smell and taste in augmented and virtual reality. *The Senses and Society*, *14*(1), 31–45. doi:10.1080/17458927.2018.1556952

Kershaw, I. (1999). *Hitler, 1889-1936: Hubris*. W.W. Norton.

Kershaw, I. (2000). *Hitler, 1936-45: Nemesis*. W.W. Norton.

Keune, J. D. (2018). Disability and the Contemporary Surgical Gestalt. *Disability Studies Quarterly*, *38*(1). Advance online publication. doi:10.18061/dsq.v38i1.5390

Keygnaert, I., Dialmy, A., Manço, A., Keygnaert, J., Vettenburg, N., Roelens, K., & Temmerman, M. (2014). Sexual violence and sub-Saharan migrants in Morocco: A community-based participatory assessment using respondent driven sampling. *Globalization and Health, 10*(1), 32. doi:10.1186/1744-8603-10-32 PMID:24885537

Khan, S. A., Fatima, Q.-U.-A., Ansari, K., Ahmad, M., Mirbahar, A. M., & Safdar, S. (2022). Advertising Mental Healthcare through Metaverse: Consequential Effects on Stigmatizing Beliefs, Psychological Safety, and Mental Health Literacy. *Pakistan Journal of Medical & Health Sciences, 16*(12), 12. Advance online publication. doi:10.53350/pjmhs20221612268

Kierkegaard, S. (1959). *A Kierkegaard anthology*. Modern Library.

Kim, H.-J., & Fredriksen-Goldsen, K. I. (2017). Disparities in Mental Health Quality of Life Between Hispanic and Non-Hispanic White LGB Midlife and Older Adults and the Influence of Lifetime Discrimination, Social Connectedness, Socioeconomic Status, and Perceived Stress. *Research on Aging, 39*(9), 991–1012. doi:10.1177/0164027516650003 PMID:27193047

Kimoto, T. (2018). The Remedy: Queer and Trans Voices on Health and Health Care edited by Zena Sharman. IJFAB: International Journal of Feminist Approaches to Bioethics. doi:10.3138/ijfab.2018.02.09.1

Kirchick, J. (2016, January 21). How Intersectionality Makes You Stupid. *Tablet Magazine*. https://www.tabletmag.com/sections/news/articles/intersectionality-makes-you-stupid

Kiss, E., & Euben, J. P. (2010). *Debating Moral Education: Rethinking the Role of the Modern University*. Duke University Press.

Kitman, J. L. (2000). The secret history of lead. *Nation, 270*(11), 11–11.

Klaas, B. (2021). *Corruptible: Who Gets Power and How It Changes Us*. Simon and Schuster.

Klee, E., Dressen, W., & Riess, V. (1991). *"The Good Old Days": The Holocaust as Seen by Its Perpetrators and Bystanders*. Konecky Konecky.

Klein, N. (2001). *No Logo: No Space, No Choice, No Jobs*. Flamingo.

Klinke, M. E., Thorsteinsson, B., & Jónsdóttir, H. (2014). Advancing Phenomenological Research: Applications of "Body Schema," "Body Image," and "Affordances" in Neglect. *Qualitative Health Research, 24*(6), 824–836. doi:10.1177/1049732314533425 PMID:24819692

Klooster, W. (2021). *Spanish American Independence Movements: A History in Documents: (From the Broadview Sources Series)*. Broadview Press.

Knight, C. (2010). Language, Lies and Lipstick: A Speculative Reconstruction of the African Middle Stone Age "Human Revolution.". In P. M. Kappeler & J. Silk (Eds.), *Mind the Gap: Tracing the Origins of Human Universals* (pp. 299–313). Springer. doi:10.1007/978-3-642-02725-3_14

Knoppers, B. M., & Thorogood, A. M. (2017). Ethics and Big Data in health. *Current Opinion in Systems Biology, 4*, 53–57. doi:10.1016/j.coisb.2017.07.001

Koch, T. (2012). *Thieves of Virtue: When Bioethics Stole Medicine*. MIT Press. doi:10.7551/mitpress/9079.001.0001

Kopelman, L. M. (2009). Bioethics as Public Discourse and Second-Order Discipline. *The Journal of Medicine and Philosophy: A Forum for Bioethics and Philosophy of Medicine, 34*(3), 261–273. 10.1093/jmp/jhp019

Korasidou, M. (2002). *Hotan hē arrōstia apeilei: Epitērēsē kai elenchos tēs hygeias tou plēthysmou stēn Hellada tou 19ou aiōna*. Topothētō Giōrgos Dardanos.

Kosfeld, M., Heinrichs, M., Zak, P. J., Fischbacher, U., & Fehr, E. (2005). Oxytocin increases trust in humans. *Nature, 435*(7042), 673–676. doi:10.1038/nature03701 PMID:15931222

Kovarik, B. (2021, December 8). *A century of tragedy: How the car and gas industry knew about the health risks of leaded fuel but sold it for 100 years anyway*. The Conversation. http://theconversation.com/a-century-of-tragedy-how-the-car-and-gas-industry-knew-about-the-health-risks-of-leaded-fuel-but-sold-it-for-100-years-anyway-173395

Kozel, S., Gibson, R., & Martelli, B. (2018). The Weird Giggle: Attending to Affect in Virtual Reality. *Transformations: Journal of Media. Culture Technique, 31*, 1–24.

Kramer, W. R. (2011). Colonizing mars—An opportunity for reconsidering bioethical standards and obligations to future generations. *Futures, 43*(5), 545–551. doi:10.1016/j.futures.2011.02.006

Krasinski, J. (Director). (2018). *A Quiet Place*. https://www.imdb.com/title/tt6644200/

Krause, E. L., & Zordo, S. D. (2012). Introduction. Ethnography and biopolitics: Tracing 'rationalities' of reproduction across the north–south divide. *Anthropology & Medicine, 19*(2), 137–151. doi:10.1080/13648470.2012.675050 PMID:22889424

Kreidler, M. (2007, March 29). Secrets and Lies. *Skeptical Inquirer*. https://skepticalinquirer.org/exclusive/secrets_and_lies/

Kreitmair, K. V., & Cho, M. K. (2017). The neuroethical future of wearable and mobile health technology. In J. Illes (Ed.), *Neuroethics: Anticipating the future*. Oxford University Press.

Kreps, D. (2019). *Understanding Digital Events: Bergson, Whitehead, and the Experience of the Digital*. Routledge. doi:10.4324/9780429032066

Krishna, G. (2017). Transgenders in Kerala-Realisation of their. *Identity*.

Krishna, L. K. R., Watkinson, D. S., & Beng, N. L. (2015). Limits to relational autonomy—The Singaporean experience. *Nursing Ethics, 22*(3), 331–340. doi:10.1177/0969733014533239 PMID:24913544

Kristeva, J. (1984). *Revolution in Poetic Language*. Columbia University Press.

Kruif, P. D. (1962). *The sweeping wind: A memoir*. Harcourt, Brace & World.

Krzyzanowski, M., & Wodak, R. (2017). The Production and Re/production of Institutional and Everyday Racism towards Migrants: The Role of Discourse 1. In *The Politics of Exclusion: Debating Migration in Austria*. Routledge. doi:10.4324/9781315133959-1

Kubrick, S. (Director). (1968). *2001: A Space Odyssey*. https://www.imdb.com/title/tt0062622/

Kuehn, B. M. (2018). Virtual and Augmented Reality Put a Twist on Medical Education. *Journal of the American Medical Association, 319*(8), 756–758. doi:10.1001/jama.2017.20800 PMID:29417140

Καλύβας, Δ. (2018). *Tools and platforms for enhanced human-computer interaction*. http://okeanis.lib2.uniwa.gr/xmlui/handle/123456789/4377

Laclau, E. (1994). *The Making of Political Identities*. Verso.

Laing, R. D. (1990). The Politics of Experience and The Bird of Paradise. Academic Press.

Lanati, M., & Thiele, R. (2018). The impact of foreign aid on migration revisited. *World Development, 111*, 59–74. doi:10.1016/j.worlddev.2018.06.021

Landolt, P., & Wei Da, W. (2005). The Spatially Ruptured Practices of Migrant Families: A Comparison of Immigrants from El Salvador and the People's Republic of China. *Current Sociology*, *53*(4), 625–653. doi:10.1177/0011392105052719

Langley, M. C., Amano, N., Wedage, O., Deraniyagala, S., Pathmalal, M. M., Perera, N., Boivin, N., Petraglia, M. D., & Roberts, P. (2020). Bows and arrows and complex symbolic displays 48,000 years ago in the South Asian tropics. *Science Advances*, *6*(24), eaba3831. Advance online publication. doi:10.1126ciadv.aba3831 PMID:32582854

Larrivee, D. (2019). *Introductory Chapter: Reconciling Neurobioethics through Nature's Lens - Metaphysical Determinants of Subjectivity*. Bioethics - Neurology, Psychiatry and Genetics., doi:10.5772/intechopen.83502

Laruelle, M. (2019). Reductio ad Hitlerum as a new frame for political and geopolitical conflicts. In *Routledge handbook of character assassination and reputation management* (pp. 307–319). Routledge. doi:10.4324/9781315150178-23

Laue, T. V. (1963). *Sergei Witte and the Industrialization of Russia*. Atheneum.

Laurent, L. (2018, December 17). *The Colossal Price of Theresa May's Immigration Obsession*. https://www.bloomberg.com/opinion/articles/2018-12-17/brexit-the-colossal-price-of-theresa-may-s-immigration-obsession

Lauritzen, P. (2008). Visual bioethics. *The American Journal of Bioethics*, *8*(12), 50–56. doi:10.1080/15265160802559146 PMID:19085479

Lavazza, A. (2016). *Frontiers in Neuroethics: Conceptual and Empirical Advancements*. Cambridge Scholars Publishing.

Lavender, N. (2019, January 12). *This is the point when cold weather becomes dangerous*. Ladders | Business News & Career Advice. https://www.theladders.com:443/career-advice/this-is-the-point-when-cold-weather-becomes-dangerous

Laws, T. (2018). Tuskegee as Sacred Rhetoric: Focal Point for the Emergent Field of African American Religion and Health. *Journal of Religion and Health*, *57*(1), 408–419. doi:10.100710943-017-0505-y PMID:29064071

Lázaro-Muñoz, G., McGuire, A. L., & Goodman, W. K. (2017). Should We Be Concerned About Preserving Agency and Personal Identity in Patients With Adaptive Deep Brain Stimulation Systems? *AJOB Neuroscience*, *8*(2), 73–75. doi:10.1080/21507740.2017.1320337 PMID:29238625

Lea, D. (1999). The Infelicities of Business Ethics in the Third World: The Melanesian Context. *Business Ethics Quarterly*, *9*(3), 421–438. doi:10.2307/3857510

Leary, M. R., & Baumeister, R. F. (2000). The nature and function of self-esteem: Sociometer theory. In Advances in experimental social psychology (Vol. 32, pp. 1–62). Academic Press. doi:10.1016/S0065-2601(00)80003-9

Leeds Revolutionary Feminist Group. (1982). Political Lesbianism: The Case Against Heterosexuality'. In The Woman Question. Fontana London.

Lee, S., Chiu, M. Y. L., Tsang, A., Chui, H., & Kleinman, A. (2006). Stigmatizing experience and structural discrimination associated with the treatment of schizophrenia in Hong Kong. *Social Science & Medicine*, *62*(7), 1685–1696. doi:10.1016/j.socscimed.2005.08.016 PMID:16174547

Lemmons, T. (2017). Justice in Migration: A Case Study for War Refugees. *CMC Senior Theses*. https://scholarship.claremont.edu/cmc_theses/1562

Lepshokova, Z., Lebedeva, N., & van de Vijver, F. J. R. (2018). The mediating role of identity incompatibility in the relationship between perceived discrimination and acculturation preferences in two generations of the ethnic Russian minority in the North Caucasus. *European Journal of Developmental Psychology*, *15*(1), 99–114. doi:10.1080/17405629.2017.1336433

Levinas, E. (1979). *Totality and Infinity: An Essay on Exteriority*. Springer Science & Business Media. doi:10.1007/978-94-009-9342-6

Levine, C., Faden, R., Grady, C., Hammerschmidt, D., Eckenwiler, L., & Sugarman, J. (2004). The limitations of "vulnerability" as a protection for human research participants. *The American Journal of Bioethics*, *4*(3), 44–49. doi:10.1080/15265160490497083 PMID:16192138

Lewis, D. K. (1983). *Philosophical Papers, I*. http://home.sandiego.edu/~baber/metaphysics/readings/Lewis.SurvivalAndIdentity.pdf

Lewis, S. (1998). *Arrowsmith*. Penguin Group.

Lewton, T. (2019, October 23). Uganda's eco-feminists are taking on mining and plantation industries. *Mongabay Environmental News*. https://news.mongabay.com/2019/10/ugandas-eco-feminists-are-taking-on-mining-and-plantation-industries/

Liang, X., Zhang, L., Wan, Y., Yu, X., Guo, Y., Chen, X., Tan, C., Huang, T., Shen, H., Chen, X., Li, H., Lv, K., Sun, F., Chen, S., & Guo, J. (2012). Changes in the Diurnal Rhythms during a 45-Day Head-Down Bed Rest. *PLoS One*, *7*(10), e47984. doi:10.1371/journal.pone.0047984 PMID:23110150

Lichterman, B. L. (2005). Soviet medical ethics (1917-1991). *Journal International De Bioethique = International Journal of Bioethics*, *16*(3–4), 33–41, 167–168.

Liebersohn, H., & Schneider, D. (2001). "My Life in Germany before and after January 30, 1933": A Guide to a Manuscript Collection at Houghton Library, Harvard University. *Transactions of the American Philosophical Society*, *91*(3), 1. doi:10.2307/3655110

Lifton, R. J. (1988). *The Nazi Doctors: Medical Killing And The Psychology Of Genocide*. Basic Books.

Lilienfeld, S. O., Latzman, R. D., Watts, A. L., Smith, S. F., & Dutton, K. (2014). Correlates of psychopathic personality traits in everyday life: Results from a large community survey. *Frontiers in Psychology*, *5*. Advance online publication. doi:10.3389/fpsyg.2014.00740 PMID:25101019

Lim, D. (2018). Rhetoric, Experimental Philosophy, and Irrelevance. *AJOB Neuroscience*, *9*(3), 160–162. doi:10.1080/21507740.2018.1496170

Lin, P., & Abney, K. (2014). Introduction to Astronaut Bioethics. *FutureTense, 8*.

Lindau, J. D. (2022). *Surveillance and the Vanishing Individual: Power and Privacy in the Digital Age*. Rowman & Littlefield.

Lin, J. Y. (1995). Why the Industrial Revolution Did Not Originate in China. *Economic Development and Cultural Change*, *43*(2), 269–292. doi:10.1086/452150

Link, B. G., & Phelan, J. C. (2001). Conceptualizing Stigma. *Annual Review of Sociology*, *27*(1), 363–385. doi:10.1146/annurev.soc.27.1.363

Lisio, D. J. (2012). *Hoover, Blacks, and Lily-Whites: A Study of Southern Strategies*. UNC Press Books.

Littlefield, M. M., & Johnson, J. (2012). *The Neuroscientific Turn: Transdisciplinarity in the Age of the Brain*. University of Michigan Press. doi:10.3998/mpub.4585194

Liu, N., Wen, S. W., Manual, D. G., Katherine, W., Bottomley, J., & Walker, M. C. (2010). Social disparity and the use of intrapartum epidural analgesia in a publicly funded health care system. *American Journal of Obstetrics and Gynecology*, *202*(3), 273.e1–273.e8. doi:10.1016/j.ajog.2009.10.871 PMID:20045506

Livingstone, F. B., & Dobzhansky, T. (1962). On the Non-Existence of Human Races. *Current Anthropology*, *3*(3), 279–281. doi:10.1086/200290

Lizardo, O. (2006). The puzzle of women's "highbrow" culture consumption: Integrating gender and work into Bourdieu's class theory of taste. *Poetics*, *34*(1), 1–23. doi:10.1016/j.poetic.2005.09.001

LoBianco, A. F., & Sheppard-Jones, K. (2007). Perceptions of Disability as Related to Medical and Social Factors. *Journal of Applied Social Psychology*, *37*(1), 1–13. doi:10.1111/j.0021-9029.2007.00143.x

Lobont, F., & Mladin, I. (2018). Philosophical-Ethical Alleviation of Perceptive Experience of Corporeality in Terminally Ill Sufferers. *Ethics of Emerging Biotechnologies*, 107–118.

Locke, J. (1836). *An Essay Concerning Human Understanding*. T. Tegg and Son.

Logan, T. (2022, February 24). Slavery Was Never an American Economic Engine. *Bloomberg.Com*. https://www.bloomberg.com/opinion/articles/2022-02-24/slavery-was-never-an-american-economic-engine

Lombard, M. (2005). Evidence of hunting and hafting during the Middle Stone Age at Sibidu Cave, KwaZulu-Natal, South Africa: A multianalytical approach. *Journal of Human Evolution*, *48*(3), 279–300. doi:10.1016/j.jhevol.2004.11.006 PMID:15737394

Long, K. (2014). *Rethinking 'Durable' Solutions*. The Oxford Handbook of Refugee and Forced Migration Studies., doi:10.1093/oxfordhb/9780199652433.013.0013

López Tovar, A. (2022, August 5). *Padres indignados por una gincana 'porno' con niños en Vilassar de Mar*. La Vanguardia. https://www.lavanguardia.com/vida/20220805/8450732/padres-indignados-gincana-porno-ninos-vilassar-mar.html

Lorge, P. A. (2012). *Chinese Martial Arts: From Antiquity to the Twenty-First Century*. Cambridge University Press.

Love, N. (Director). (2007, May). *Ross Kemp on Gangs: El Salvador*. https://www.imdb.com/title/tt1211709/

Lozano Rodríguez, M. (2019). Ethical Information Foresights: A Journey Across Uncharted Conflictology. American University of Sovereign Nations.

Lozano Rodríguez, M. (2020b). *Lost in Securitization: Displaced people in the pandemic dawn of the New World Order*. 8th International Public Health & Bioethics Ambassador Conference (IPHA8). https://scholar.google.es/citations?view_op=view_citation&hl=en&user=mQnwblkAAAAJ&citation_for_view=mQnwblkAAAAJ:35r97b3x0nAC

Lozano Rodríguez, M. (2020c, September 3). Lies, bloody lies and hiddennes on COVID-19. *The Ninth International Public Health & Bioethics Ambassador Conference (IPHA9).*

Lozano Rodríguez, M. (2021). Don't Let Kellys Go to Hell: From Welfarism to Misery in the Cleaning Ladies' Sector in Spain. Handbook of Research on the Impact of COVID-19 on Marginalized Populations and Support for the Future.

Lozano Rodríguez, M. (2021). Don't Let Kellys Go to Hell: From Welfarism to Misery in the Cleaning Ladies' Sector in Spain. In Handbook of Research on the Impact of COVID-19 on Marginalized Populations and Support for the Future (pp. 284–308). IGI Global.

Lozano Rodríguez, M. (2021, January 9). On Privileges—A reflection on the worshipping and demonization of privileges part 3. Thirteenth International Public Health Ambassador Conference.

Lozano Rodríguez, M. (2022). Shame on You! Digital Ethnography, Suicide, Stigma, and Warfare. In Practices, Challenges, and Prospects of Digital Ethnography as a Multidisciplinary Method (pp. 198-222). IGI Global.

Lozano Rodríguez, M. (2023, January 25). How COVID-19 is Changing Corruption Forever: The Crisis of the State-led Crisis Management and Why You Are Going to Pay for It. In *Conference on Disaster Management: Global Pandemic and Its Impact.* Binghamton University - State University of New York.

Lozano Rodríguez, M., & Torres, C. E. (2023). Evolving From Predictive to Liquid Maintenance in Postmodern Industry. In Encyclopedia of Data Science and Machine Learning (pp. 2182-2198). IGI Global.

Lozano Rodríguez, M., Sevimli, S., Dhastagir, S., & Jayachandran, S. (2023). COVID-19 Measures on Crime, Corruption, and Inequity of the Impacts of Control Measures Statement (World Emergency COVID19 Pandemic Ethics (WeCope) Committee). Eubios Ethics Institute.

Lozano Rodríguez, M. (2018). *Ethical Information Foresights: A journey across uncharted conflictology.* American University of Sovereign Nations.

Lozano Rodríguez, M. (2019). The Tagged Dove. In D. R. J. Macer (Ed.), *Legacies ofLove, Peace and Hope: How Education can overcome Hatred and Divide.* Eubios Ethics Institute.

Lozano Rodríguez, M. (2020a). How not to face coronavirus: The case of Spain. *Eubios Journal of Asian and International Bioethics; EJAIB, 30*(4), 152.

Lozano Rodríguez, M. (2023). *5. Are Human Biological Differences Based on Inequalities in Wealth a Threat for Planetary Health?* Planetary Health and Bioethics.

Lozano Rodríguez, M., & Choclán Roca, C. (2019). Plataforma digital para atajar el incivismo: el valor de lo intangible en la governanza metropolitana. *V Congreso de Ciudades Inteligentes*, 34-40.

Lucas, T., Wegner, R., Pierce, J., Lumley, M. A., Laurent, H. K., & Granger, D. A. (2017). Perceived Discrimination, Racial Identity, and Multisystem Stress Response to Social Evaluative Threat Among African American Men and Women. *Psychosomatic Medicine, 79*(3), 293–305. doi:10.1097/PSY.0000000000000406 PMID:27806018

Ludwig, L., & Faloyin, D. (2018, June 20). We Asked a Dinosaur Expert How Realistic the New "Jurassic World" Really Is. *Vice.* https://www.vice.com/en_us/article/d3kkzy/we-asked-a-dinosaur-expert-how-realistic-the-new-jurassic-world-really-is

Luibhéid, E. (2019). Heteronormativity: A Bridge Between Queer Migration and Critical Trafficking Studies. *Women's Studies in Communication, 0*(0), 1–5. doi:10.1080/07491409.2018.1544001

Lundahl, O. (2021). Media framing of social media addiction in the UK and the US. *International Journal of Consumer Studies, 45*(5), 1103–1116. doi:10.1111/ijcs.12636

Lu, S.-J., & Liu, Y.-C. (2015). Integrating augmented reality technology to enhance children's learning in marine education. *Environmental Education Research, 21*(4), 525–541. doi:10.1080/13504622.2014.911247

Lushetich, N. (2018). How Long? How Far? *Performance Research, 23*(4–5), 34–39. doi:10.1080/13528165.2018.1511039

Lyons, L. K., Dorsch, T. E., Bell, L. F., & Mason, L. G. (2018). Renegotiating Identity: A Phenomenological Investigation of the College Transition for Former High School Athletes No Longer Engaged in Varsity Competition. *Identity, 18*(1), 18–33. doi:10.1080/15283488.2017.1410156

Lyotard, J.-F. (1984). *The Postmodern Condition: A Report on Knowledge.* University of Minnesota Press.

Lyreskog, D. M., Pavarini, G., Jacobs, E., Bennett, V., Mawdsley, G., & Singh, I. (2023). Testing Design Bioethics Methods: Comparing a Digital Game with a Vignette Survey for Neuroethics Research with Young People. *AJOB Empirical Bioethics, 14*(1), 55–64. doi:10.1080/23294515.2022.2110964 PMID:35993968

Macer, D. (2018). *Application of the essence of bioethics as love of life to some dilemmas of our future*. American University of Sovereign Nations.

Macer, D. R. J. (1994). *Bioethics for the people by the people*. Eubios Ethics Institute.

Macer, D. R. J. (1998). *Bioethics is love of life: An alternative textbook*. Eubios Ethics Institute.

Macer, D. R. J. (2019). Is There Cross-Cultural Evidence for an Association Between Intersectionality and Bioethical Decision Making? Not Yet, but Awaiting Advances in Mental Mapping. *The American Journal of Bioethics*, *19*(2), 34–36. doi:10.1080/15265161.2018.1558311 PMID:31543012

Macias, T. (2013). 'Tortured bodies': The biopolitics of torture and truth in Chile. *International Journal of Human Rights*, *17*(1), 113–132. doi:10.1080/13642987.2012.701912

Mackelprang, R. W., Salsgiver, R. O., & Salsgiver, R. (2016). *Disability: A Diversity Model Approach in Human Service Practice*. Oxford University Press.

MacKenzie Bryers, H., & van Teijlingen, E. (2010). Risk, theory, social and medical models: A critical analysis of the concept of risk in maternity care. *Midwifery*, *26*(5), 488–496. doi:10.1016/j.midw.2010.07.003 PMID:20719418

Macklin, R. (2003). Dignity is a useless concept. *BMJ (Clinical Research Ed.)*, *327*(7429), 1419–1420. doi:10.1136/bmj.327.7429.1419 PMID:14684633

Madary, M., & Metzinger, T. K. (2016). Real virtuality: A code of ethical conduct. Recommendations for good scientific practice and the consumers of VR-technology. *Frontiers in Robotics and AI*, 3.

Madeley, J. (2000). *Hungry for Trade: How the Poor Pay for Free Trade*. Zed Books.

Maercker, A., & Guski-Leinwand, S. (2018). Psychologists' involvement in repressive "Stasi" secret police activities in former East Germany. *International Perspectives in Psychology : Research, Practice, Consultation*, *7*(2), 107–119. doi:10.1037/ipp0000085

Magli, G. (2013). *Sirius and the project of the megalithic enclosures at Gobekli Tepe*. https://doi.org//arXiv.1307.8397 doi:10.48550

Magra, I. (2018, October 28). 'They Don't Have Money': Greece's Prostitutes Hit Hard by Financial Crisis. *The New York Times*. https://www.nytimes.com/2018/10/27/world/europe/prostitutes-greece-crisis.html

Mahdavi, P. (2009). *Passionate Uprisings: Iran's Sexual Revolution*. Stanford University Press.

MahlerT.NissimN.ShalomE.GoldenbergI.HassmanG.MakoriA.KochavI.EloviciY.ShaharY. (2018). *Know Your Enemy: Characteristics of Cyber-Attacks on Medical Imaging Devices*. https://arxiv.org/abs/1801.05583

Major, B., & O'Brien, L. T. (2005). The Social Psychology of Stigma. *Annual Review of Psychology*, *56*(1), 393–421. doi:10.1146/annurev.psych.56.091103.070137 PMID:15709941

Malik, E. (2018, November 9). *Carl Sagan history timeline*. Project Management Tips and Tricks. https://www.officetimeline.com/blog/carl-sagan-timeline

Malina, R. F. (2009). Intimate science and hard humanities. *Leonardo*, *42*(3), 184–184. doi:10.1162/leon.2009.42.3.184

Malpas, J. (2018). Holism, content, and self. In *Place and Experience: A Philosophical Topography*. Routledge. doi:10.4324/9781315265445-4

Malpus, Z. (2018). Guest editorial for May issue. *British Journal of Pain*, *12*(2), 69–71. doi:10.1177/2049463718769938 PMID:29796258

Ma, M., Fallavollita, P., Seelbach, I., Von Der Heide, A. M., Euler, E., Waschke, J., & Navab, N. (2016). Personalized augmented reality for anatomy education. *Clinical Anatomy (New York, N.Y.)*, *29*(4), 446–453. doi:10.1002/ca.22675 PMID:26646315

Mandela, N. (1961, June 26). *Nelson Mandela International Day*. https://www.un.org/en/events/mandeladay/struggle.shtml

Margalit, A. (2009). *The Ethics of Memory*. Harvard University Press.

Marks, S., White, D., & Singh, M. (2017). Getting Up Your Nose: A Virtual Reality Education Tool for Nasal Cavity Anatomy. *SIGGRAPH Asia 2017 Symposium on Education*, 1:1-1:7. 10.1145/3134368.3139218

Marshall, D. (2016). Queer Pedagogies Out of Place and Time: Redrawing the Boundaries of Youth, Sexual and Gender Difference, and Education. *Journal of Homosexuality*, *63*(3), 405–415. doi:10.1080/00918369.2016.1124697 PMID:26766505

Marta, J. (1998). Whose Consent is it Anyway? A Poststructuralist framing of the Person in Medical Decision-Making. *Theoretical Medicine and Bioethics*, *19*(4), 353–370. doi:10.1023/A:1009955914018 PMID:9865139

Martial, C., Cassol, H., Charland-Verville, V., Merckelbach, H., & Laureys, S. (2018). Fantasy Proneness Correlates With the Intensity of Near-Death Experience. *Frontiers in Psychiatry*, *9*, 190. Advance online publication. doi:10.3389/fpsyt.2018.00190 PMID:29930518

Martín, L. J. V., & Francisco Peregil, M. (2021, May 17). *5.000 personas llegan a nado a Ceuta en plena escalada de tensión diplomática con Marruecos*. El País. https://elpais.com/espana/2021-05-17/un-millar-de-personas-llegan-a-nado-a-ceuta-en-plena-escalada-de-tension-diplomatica-con-marruecos.html

Martin, P., Washington, J., Sprague, J., & Purkayastha, B. (2014). *Thanks to Tracy Ore for sharing the wonderful pictures from the winter meeting, that you'll see throughout Network News*. https://socwomen.org/wp-content/uploads/2018/04/NN-spring-2014-1.pdf

Martínez-García, M., Vallejo, M., Hernández-Lemus, E., & Álvarez-Díaz, J. A. (2019). Novel methods of qualitative analysis for health policy research. *Health Research Policy and Systems*, *17*(1), 6. doi:10.118612961-018-0404-z PMID:30642358

Martín-Gutiérrez, J. (2017). Editorial: Learning Strategies in Engineering Education Using Virtual and Augmented Reality Technologies. *Eurasia Journal of Mathematics, Science and Technology Education*, *13*(2), 297–300. doi:10.12973/eurasia.2017.00630a

Martín-Gutiérrez, J., Fabiani, P., Benesova, W., Meneses, M. D., & Mora, C. E. (2015). Augmented reality to promote collaborative and autonomous learning in higher education. *Computers in Human Behavior*, *51*, 752–761. doi:10.1016/j.chb.2014.11.093

Martín-Gutiérrez, J., Mora, C. E., Añorbe-Díaz, B., & González-Marrero, A. (2017). Virtual technologies trends in education. *Eurasia Journal of Mathematics, Science and Technology Education*, *13*(2), 469–486.

Masolo, D. A. (1994). *African Philosophy in Search of Identity*. Indiana University Press.

Mason, P. H., Roy, A., & Singh, P. (2017). Reciprocity-building and the importance of interdisciplinary collaboration in tuberculosis research. *Journal of Biosocial Science*, *49*(4), 559–562. doi:10.1017/S0021932016000626 PMID:27869060

Mast, K. (2019, January 15). *State of Science: World Wakes to Perils of Plastics*. Discover Magazine. http://discovermagazine.com/2019/january/world-wakes-to-perils-of-plastics

Mathieu, C., Babiak, P., & Hare, R. D. (2020). Psychopathy in the Workplace. In *The Wiley International Handbook on Psychopathic Disorders and the Law* (pp. 607–644). John Wiley & Sons, Ltd., doi:10.1002/9781119159322.ch27

Mathieu, C., Hare, R. D., Jones, D. N., Babiak, P., & Neumann, C. S. (2013). Factor structure of the B-Scan 360: A measure of corporate psychopathy. *Psychological Assessment*, *25*(1), 288–293. doi:10.1037/a0029262 PMID:22775409

Mauthner, N. (2018). A Posthumanist Ethics of Mattering: New Materialisms and the Ethical Practice of Inquiry. In R. Iphofen & M. Tolich (Eds.), *The SAGE Handbook of Qualitative Research Ethics*. SAGE. doi:10.4135/9781526435446.n4

Mautner, M. N. (2000). *Seeding the universe with life: Securing our cosmological future*. Michael Mautner.

Mautner, M. N. (2009). Life-Centered Ethics, and the Human Future in Space. *Bioethics*, *23*(8), 433–440. doi:10.1111/j.1467-8519.2008.00688.x PMID:19077128

May, C., & Clough, C. (2017). *Transnational crime and the developing world*. Global Financial Integrity. https://www.gfintegrity.org/wp-content/uploads/2017/03/Transnational_Crime-final.pdf

May, A. T. (2003). Physician-Assisted Suicide, Euthanasia, and Christian Bioethics: Moral Controversy in Germany. *Christian Bioethics*, *9*(2–3), 273–283. doi:10.1076/chbi.9.2.273.30279 PMID:15254994

Mayes, C., Hooker, C., & Kerridge, I. (2015). Bioethics and Epistemic Scientism. *Journal of Bioethical Inquiry*, *12*(4), 565–567. doi:10.100711673-015-9667-z PMID:26646798

McCaffery, E. J. (2016). Taking wealth seriously. *Tax L. Rev.*, *70*, 305.

McCallum, D. (1997). Mental health, criminality and the human sciences. In A. R. Petersen & R. Bunton (Eds.), *Foucault, Health and Medicine*. Psychology Press.

McCormick, M., Büntgen, U., Cane, M. A., Cook, E. R., Harper, K., Huybers, P., Litt, T., Manning, S. W., Mayewski, P. A., More, A. F. M., Nicolussi, K., & Tegel, W. (2012). Climate Change during and after the Roman Empire: Reconstructing the Past from Scientific and Historical Evidence. *The Journal of Interdisciplinary History*, *43*(2), 169–220. doi:10.1162/JINH_a_00379

McCrea, J. (2014, June 9). Virtual Reality: Then and Now. *The Real McCrea*. https://therealmccrea.com/2014/06/08/virtual-reality-then-and-now/

McCurry, J. (2019, January 23). Naomi Osaka sponsor apologises for "whitewashing" tennis star in ad. *The Guardian*. https://www.theguardian.com/sport/2019/jan/23/naomi-osaka-sponsor-apologises-for-whitewashing-tennis-star-in-ad

McDonald, W. (2017). Søren Kierkegaard. In E. N. Zalta (Ed.), *The Stanford Encyclopedia of Philosophy* (Winter 2017). Metaphysics Research Lab, Stanford University. https://plato.stanford.edu/archives/win2017/entries/kierkegaard/

McGowan, M. (2019, August 26). Racism study finds one in three school students are victims of discrimination. *The Guardian*. https://www.theguardian.com/australia-news/2019/aug/27/racism-study-finds-one-in-three-school-students-are-victims-of-discrimination

McGrath, P. (2018). *A Question of Choice: Bioethical Reflections on a Spiritual Response to the Technological Imperative*. Routledge., doi:10.4324/9780429459948

McKenny, G. P. (1997). *To Relieve the Human Condition: Bioethics, Technology, and the Body*. SUNY Press.

McKinstry, L. (2018, November 16). How can mediocre Mrs May compare herself to cricket icon Geoffrey Boycott? *The Telegraph*. https://www.telegraph.co.uk/politics/2018/11/16/can-mediocrity-like-mrs-may-compare-cricket-icon-geoffrey-boycott/

McLaughlin, L. A., & Braun, K. L. (1998). Asian and Pacific Islander cultural values: Considerations for health care decision making. *Health & Social Work, 23*(2), 116–126. doi:10.1093/hsw/23.2.116 PMID:9598394

McMinn, B. G., & Hinton, L. (2000). Confined to barracks: The effects of indoor confinement on aggressive behavior among inpatients of an acute psychogeriatric unit. In American Journal of Alzheimer's Disease (Vol. 15, pp. 36–41). doi:10.1177/153331750001500106

McNeill, J. R., & McNeill, W. H. (2003). *The human web: A bird's-eye view of world history*. W.W. Norton & Company. https://archive.org/details/humanwebbirdseye00jrmc

McNeill, J. R., & McNeill, W. H. (2003). *The Human Web: A Bird's-eye View of World History*. W.W. Norton.

Meinitzer, A., Kalcher, K., Gartner, G., Halwachs-Baumann, G., März, W., & Stettin, M. (2008). Drugs and brain death diagnostics: Determination of drugs capable of inducing EEG zero line. *Clinical Chemistry and Laboratory Medicine, 46*(12), 1732–1738. doi:10.1515/CCLM.2008.335 PMID:18973460

Melber, H. (2016, April 11). *How liberators turn into oppressors: A study of southern African states*. The Conversation. http://theconversation.com/how-liberators-turn-into-oppressors-a-study-of-southern-african-states-57213

Mellifont, D. (2018). Soft affirmative action lacking traction? An early qualitative exploration of the RecruitAbility Scheme performance within the Australian Public Service. *Australian Journal of Career Development, 27*(1), 20–28. doi:10.1177/1038416217745070

Mercer, S. H., & Cunningham, M. (2003). Racial Identity in White American College Students: Issues of Conceptualization and Measurement. *Journal of College Student Development, 44*(2), 217–230. doi:10.1353/csd.2003.0021

Merleau-Ponty, M. (1973). *The Prose of the World*. Northwestern University Press.

Merleau-Ponty, M. (2013). *Phenomenology of Perception*. Routledge. doi:10.4324/9780203720714

Metz, L., Lewis, J. E., & Slimak, L. (2023). Bow-and-arrow, technology of the first modern humans in Europe 54,000 years ago at Mandrin, France. *Science Advances, 9*(8), eadd4675. Advance online publication. doi:10.1126ciadv.add4675 PMID:36812314

Meyerbröker, K., & Emmelkamp, P. M. G. (2010). Virtual reality exposure therapy in anxiety disorders: A systematic review of process-and-outcome studies. *Depression and Anxiety, 27*(10), 933–944. doi:10.1002/da.20734 PMID:20734361

Miedema, F. (2012). *Science 3.0: Real Science, Real Knowledge*. Amsterdam University Press.

Migliaccio, M. L., Zagli, G., Cianchi, G., Lazzeri, C., Bonizzoli, M., Cecchi, A., Anichini, V., Gensini, G. F., & Peris, A. (2013). Extracorporeal membrane oxygenation in brain-death organ and tissues donors: A single-centre experience. *British Journal of Anaesthesia, 111*(4), 673–674. doi:10.1093/bja/aet323 PMID:24027145

Milgram, S. (1963). Behavioral Study of obedience. *Journal of Abnormal and Social Psychology, 67*(4), 371–378. doi:10.1037/h0040525 PMID:14049516

Mindock, C. (2018, February 14). *Woman put in solitary confinement after claiming she was sexually abused by immigration guard*. The Independent. https://www.independent.co.uk/news/world/americas/laural-monterrosa-immigrant-sexual-abuse-assault-guard-texas-jail-prison-solitary-confinement-hutto-a8211206.html

Minichiello, V., & Scott, J. (2014). *Male Sex Work and Society*. Columbia University Press. doi:10.17312/harringtonparkpress/2014.09.msws

Mirza, M. (2017). *Disability-inclusive healthcare in humanitarian camps: Pushing the boundaries of disability studies and global health*. http://repositoriocdpd.net:8080/handle/123456789/1488

Mitford, M. R. (1852). *Recollections of a Literary Life: Or, Books, Places, and People*. R. Bentley.

Mlakar, J., Korva, M., Tul, N., Popović, M., Poljšak-Prijatelj, M., Mraz, J., Kolenc, M., Resman Rus, K., Vesnaver Vipotnik, T., Fabjan Vodušek, V., Vizjak, A., Pižem, J., Petrovec, M., & Avšič Županc, T. (2016). Zika Virus Associated with Microcephaly. *The New England Journal of Medicine*, *374*(10), 951–958. doi:10.1056/NEJMoa1600651 PMID:26862926

Mock, S. E., & Eibach, R. P. (2012). Stability and change in sexual orientation identity over a 10-year period in adulthood. *Archives of Sexual Behavior*, *41*(3), 641–648. doi:10.100710508-011-9761-1 PMID:21584828

Moghimi, S., Kushki, A., Guerguerian, A. M., & Chau, T. (2013). A Review of EEG-Based Brain-Computer Interfaces as Access Pathways for Individuals with Severe Disabilities. *Assistive Technology*, *25*(2), 99–110. doi:10.1080/10400435.2012.723298 PMID:23923692

Monbiot, G. (2005, December 27). *How Britain Denies its Holocausts*. George Monbiot. https://www.monbiot.com/2005/12/27/how-britain-denies-its-holocausts/

Moore, J. (2017). A Phenomenological Study of Lesbian and Gay People in Leadership Roles: How Perspectives and Priorities Shift in the Workplace as Sexual Orientation Evolves Through Social Constructs. *Doctoral Dissertations*. https://repository.usfca.edu/diss/405

Morales, H. (2013). Aristophanes' Lysistrata, the Liberian 'sex strike', and the Politics of Reception. *Greece & Rome*, *60*(2), 281–295. doi:10.1017/S0017383513000107

Moreno, J. D. (2015). Neuroethics: An Agenda for Neuroscience and Society. In H. Kuhse, U. Schüklenk, & P. Singer (Eds.), *Bioethics: An Anthology*. John Wiley & Sons.

Moro, C., Štromberga, Z., Raikos, A., & Stirling, A. (2017). The effectiveness of virtual and augmented reality in health sciences and medical anatomy. *Anatomical Sciences Education*, *10*(6), 549–559. doi:10.1002/ase.1696 PMID:28419750

Moser, K. (2018). The Sensorial, Biocentric Philosophy of Michel Serres and Michel Onfray: Rehabilitating the Human Body. *Pacific Coast Philology*, *53*(1), 92–110. doi:10.5325/pacicoasphil.53.1.0092

Moya, E. M., Loza, O., & Lusk, M. (2012). Border Health: Inequities, Social Determinants, and the Cases of Tuberculosis and HIV. In M. Lusk, K. Staudt, & E. Moya (Eds.), *Social Justice in the U.S.-Mexico Border Region* (pp. 161–178). Springer Netherlands. doi:10.1007/978-94-007-4150-8_9

Mulinari, D., & Neergaard, A. (2019). A contradiction in terms? Migrant activists in the Sweden Democrats party. *Identities (Yverdon)*, *26*(2), 222–240. doi:10.1080/1070289X.2017.1418275

Munasinghe, S., & van den Broek, N. (2005). Abortions in adolescents. *Tropical Doctor*, *35*(3), 133–136. doi:10.1258/0049475054620752 PMID:16105332

Murphy, T. F. (2015). LGBT People and the Work Ahead in Bioethics. *Bioethics*, *29*(6), ii–v. doi:10.1111/bioe.12168 PMID:26053920

Murphy, T. F. (2018). Bioethics, children, and the environment. *Bioethics*, *32*(1), 3–9. doi:10.1111/bioe.12386 PMID:28873213

Murray, S. J., & Holmes, D. (2015). Seclusive Space: Crisis Confinement and Behavior Modification in Canadian Forensic Psychiatric Settings. In K. Reiter & A. Koenig (Eds.), Extreme Punishment: Comparative Studies in Detention, Incarceration and Solitary Confinement (pp. 115–134). Palgrave Macmillan UK. doi:10.1057/9781137441157_7

Murray, S. J., & Holmes, D. (2014). Interpretive Phenomenological Analysis (IPA) and the Ethics of Body and Place: Critical Methodological Reflections. *Human Studies*, *37*(1), 15–30. doi:10.100710746-013-9282-0

Murthy, V. (2011). *The Political Philosophy of Zhang Taiyan: The Resistance of Consciousness*. BRILL. doi:10.1163/ej.9789004203877.i-268

Muschietti, A. (Director). (2017). *It*. https://www.imdb.com/title/tt1396484/

Muschietti, A. (Director). (2019). *It Chapter Two*. https://www.imdb.com/title/tt7349950/

Mutize, M. (2019, June 23). *Why credit rating agencies are still getting away with bad behaviour*. The Conversation. http://theconversation.com/why-credit-rating-agencies-are-still-getting-away-with-bad-behaviour-117549

Naimou, A. (2016). Double Vision: Refugee Crises and the Afterimages of Endless War. *College Literature*, *43*(1), 226–233. doi:10.1353/lit.2016.0004

Nash, A. (2015). An Aesthetics of Digital Virtual Environments. In *New Opportunities for Artistic Practice in Virtual Worlds* (pp. 1–22). https://www.igi-global.com/chapter/an-aesthetics-of-digital-virtual-environments/132413

Nash, M. (2011). Wittgenstein and Aesthetic Reasoning with Stories in the Bioethics Classroom. *Doctoral Dissertations*. https://trace.tennessee.edu/utk_graddiss/1105

National Research Council. (2014). Rising Incarceration Rates. In *The Growth of Incarceration in the United States*. Exploring Causes and Consequences. doi:10.17226/18613

Nemorin, S. (2017). Neuromarketing and the "poor in world" consumer: How the animalization of thinking underpins contemporary market research discourses. *Consumption Markets & Culture*, *20*(1), 59–80. doi:10.1080/10253866.2016.1160897

Neng-fu, K. (1991). Chinese Library Science in the Twelfth Century. *Libraries & Culture*, *26*(2), 357–371.

Nerozzi, T. (2022, August 20). *White people banned from off-campus UC Berkeley student housing common areas*. Fox News. https://www.foxnews.com/us/white-people-banned-off-campus-uc-berkeley-student-housing-common-areas

Nichols, S. (2017). Memory and Personal Identity. In S. Bernecker & K. Michaelian (Eds.), *The Routledge Handbook of Philosophy of Memory*. Routledge. doi:10.4324/9781315687315-14

Nikolas, K. (2012, April 2). *Greece to tackle problem of migrant communicable diseases*. Digital Journal. https://www.digitaljournal.com/article/322251

Nikolopoulos, G., Arvanitis, M., Masgala, A., & Paraskeva, D. (2005). Migration and HIV epidemic in Greece. *European Journal of Public Health*, *15*(3), 296–299. doi:10.1093/eurpub/cki064 PMID:15923211

Niño-Becerra, S. (2022). *Futuro, ¿qué futuro? Claves para sobrevivir más allá de la pandemia*. Academic Press.

Nispel, M. (Director). (2009). *Friday the 13th*. https://www.imdb.com/title/tt0758746/

Nnaemeka, O. (2015). Feminist bioethics and global responsibility: Exploring health care delivery in Kenya. *International Journal of Feminist Approaches to Bioethics*. Advance online publication. doi:10.3138/ijfab.2.1.71

Noman, A. (2016). The Tragedy of the Dancing Boys in Afghanistan: The US Silence on Bacha Bazi. *Seattle Journal for Social Justice*, *15*, 503.

Nucera, S., Tartarisco, G., Epasto, A., Smeriglio, D., Mazzeo, A., Pioggia, G., & Anastasi, A. (2018). Ubiquitous, Wearable, Mobile: Paradigm Shifts in E-Learning and Diffusion of Knowledge. Handbook of Research on Mobile Devices and Smart Gadgets in K-12 Education, 286–307. doi:10.4018/978-1-5225-2706-0.ch018

Nussbaum, M. (2003). Capabilities as Fundamental Entitlements: Sen and Social Justice. *Feminist Economics*, *9*(2–3), 33–59. doi:10.1080/1354570022000077926

Nussbaum, M. (2004). Beyond the social contract: Capabilities and global justice. an Olaf Palme lecture, delivered in Oxford on 19 June 2003. *Oxford Development Studies, 32*(1), 3–18. doi:10.1080/1360081042000184093

Nussbaum, M. C. (2013). *Creating Capabilities: The Human Development Approach*. Harvard University Press. doi:10.2307/j.ctt2jbt31

O'Brien, E. (2011). Fuelling traffic: Abolitionist claims of a causal nexus between legalised prostitution and trafficking. *Crime, Law, and Social Change, 56*(5), 547–565. doi:10.100710611-011-9334-1

Oberman, M. (2018). *Her Body, Our Laws: On the Front Lines of the Abortion War, from El Salvador to Oklahoma*. Beacon Press.

Office for National Statistics. (2015). *How has life expectancy changed over time?* Office for National Statistics. https://www.ons.gov.uk/peoplepopulationandcommunity/birthsdeathsandmarriages/lifeexpectancies/articles/howhaslifeexpectancychangedovertime/2015-09-09

Oliveira, E. C. D., Bertrand, P., Lesur, M. E. R., Palomo, P., Demarzo, M., Cebolla, A., Baños, R., & Tori, R. (2016). Virtual Body Swap: A New Feasible Tool to Be Explored in Health and Education. *2016 XVIII Symposium on Virtual and Augmented Reality (SVR)*, 81–89. 10.1109/SVR.2016.23

Ollero, D. J. (2019, January 12). *Cuando Carmena criticaba la Ley de Violencia de Género con los mismos argumentos que Vox*. ELMUNDO. https://www.elmundo.es/espana/2019/01/12/5c392b3b21efa00b688b4710.html

Ondrášek, S., & Kajanová, A. (2018). *Financial aspect of working in private sex business*. https://www.narodacek.cz/wp-content/uploads/2019/10/Ondrasek_Kajanova.pdf

Orlu, S. (2021, March 14). *Why I do not support the Black Lives Matter movement—Troy Media*. https://troymedia.com/lifestyle/why-i-do-not-support-the-black-lives-matter-movement/

Orobator, A. E. (2010). Justice for the Displace: The Challenge of a Christian understanding. In H. David (Ed.), *Driven from Home: Protecting the Rights of Forced Migrants*. Georgetown University Press.

Orwell, G. (2013). *1984*. HarperPerennial Classics.

Orwell, G. (2009). *Animal Farm: A Fairy Story*. Houghton Mifflin Harcourt.

Ostrowski, T. M., & Sikorska, I. (2015). *Health and Resilience*. Wydawnictwo UJ.

Ottosdottir, G., & Evans, R. (2016). Disability and chronic illness. In F. Thomas (Ed.), *Handbook of Migration and Health*. Edward Elgar. doi:10.4337/9781784714789.00028

Ozaki, T. (Director). (2018). *Goblin Slayer*. https://www.imdb.com/title/tt8690728/

Pachankis, J. E., Hatzenbuehler, M. L., Mirandola, M., Weatherburn, P., Berg, R. C., Marcus, U., & Schmidt, A. J. (2017). The Geography of Sexual Orientation: Structural Stigma and Sexual Attraction, Behavior, and Identity Among Men Who Have Sex with Men Across 38 European Countries. *Archives of Sexual Behavior, 46*(5), 1491–1502. doi:10.100710508-016-0819-y PMID:27620320

Padela, A. I. (2018). Conceptualizing the Human Being: Insights from the Genethics Discourse and Implications for Islamic Bioethics. *Islamic Ethics and the Genome Question*, 111–138. doi:10.1163/9789004392137_006

Padilla, B., & Miguel, J. P. (2009). Health and migration in the European Union: Building a shared vision for action. *Health and Migration in the European Union: Better Health for All in an Inclusive Society, 1a*, 15–22.

Padmanaban, D. (2017, April 12). *Where Empathy Lives in the Brain*. The Cut. https://www.thecut.com/2017/04/where-empathy-lives-in-the-brain.html

Pafford, B. (2009). The third wave—Medical tourism in the 21st century. *Southern Medical Journal*, *102*(8), 810–813. doi:10.1097/SMJ.0b013e3181aa8ce4 PMID:19593301

Palade, A. (2019). The peculiarities of effective teaching through theater. *Învăţământ, Cercetare. Creaţie*, *V*(1), 333–335.

Pałczyńska, J. (2023). *Surveillance of employees in the Metaverse: Data protection and privacy challenges*. https://ruj.uj.edu.pl/xmlui/handle/item/308702

PalharesD.dos SantosÍ. A. (2017). *Astronomic Bioethics: Terraforming X Planetary protection*. https://arxiv.org/abs/1704.08404

Palmieri, A., Calvo, V., Kleinbub, J. R., Meconi, F., Marangoni, M., Barilaro, P., Broggio, A., Sambin, M., & Sessa, P. (2014). "Reality" of near-death-experience memories: Evidence from a psychodynamic and electrophysiological integrated study. *Frontiers in Human Neuroscience*, *8*. Advance online publication. doi:10.3389/fnhum.2014.00429 PMID:24994974

Pamuk, Ş. (2007). The Black Death and the origins of the 'Great Divergence' across Europe, 1300–1600. *European Review of Economic History*, *11*(3), 289–317. doi:10.1017/S1361491607002031

Pandya, S. K. (2011). Understanding Brain, Mind and Soul: Contributions from Neurology and Neurosurgery. *Mens Sana Monographs*, *9*(1), 129–149. doi:10.4103/0973-1229.77431 PMID:21694966

Paracka, D. (2018). Year of India: Introduction to the Special Issue. *JGI, 12*(1), 1–14.

Paraskevis, D., Nikolopoulos, G., Fotiou, A., Tsiara, C., Paraskeva, D., Sypsa, V., Lazanas, M., Gargalianos, P., Psichogiou, M., Skoutelis, A., Wiessing, L., Friedman, S. R., Jarlais, D. C., Terzidou, M., Kremastinou, J., Malliori, M., & Hatzakis, A. (2013). Economic Recession and Emergence of an HIV-1 Outbreak among Drug Injectors in Athens Metropolitan Area: A Longitudinal Study. *PLoS One*, *8*(11), e78941. doi:10.1371/journal.pone.0078941 PMID:24265730

Parekh, S. (2016). *Refugees and the Ethics of Forced Displacement*. Taylor & Francis. doi:10.4324/9781315883854

Parker, N. (1991). The Garry David Case. *The Australian and New Zealand Journal of Psychiatry*, *25*(3), 371–374. doi:10.3109/00048679109062638 PMID:1958160

Parker, R., & Aggleton, P. (2003). HIV and AIDS-related stigma and discrimination: A conceptual framework and implications for action. *Social Science & Medicine*, *57*(1), 13–24. doi:10.1016/S0277-9536(02)00304-0 PMID:12753813

Parnia, S., Waller, D. G., Yeates, R., & Fenwick, P. (2001). A qualitative and quantitative study of the incidence, features and aetiology of near death experiences in cardiac arrest survivors. *Resuscitation*, *48*(2), 149–156. doi:10.1016/S0300-9572(00)00328-2 PMID:11426476

Parran, T. (1942, November 19). *Thomas Parran to Lewis H. Weed* (NARAII_0000284). PCSBI HSPI Archives.

Parr, H. (2011). A brief spatial history of community and mental illness II: Community care. In *Mental Health and Social Space: Towards Inclusionary Geographies?* John Wiley & Sons.

Pascalev, A., Pascalev, M., & Giordano, J. (2016). Head Transplants, Personal Identity and Neuroethics. *Neuroethics*, *9*(1), 15–22. doi:10.100712152-015-9245-4

Pasha-Robinson, L. (2016, October 18). *Margaret Thatcher is the worst prime minister in 100 years*. The Independent. https://www.independent.co.uk/news/uk/politics/margaret-thatcher-worst-prime-minister-pm-david-cameron-century-100-years-a7367481.html

Pateman, C. (2015). Sexual Contract. In *The Wiley Blackwell Encyclopedia of Gender and Sexuality Studies* (pp. 1–3). John Wiley & Sons, Ltd. doi:10.1002/9781118663219.wbegss468

Patterson, J. (2007). *The Angel Experiment: A Maximum Ride Novel.* Academic Press.

Paul, S. (2018). Are we doing enough? Stigma, discrimination and human rights violations of people living with schizophrenia in India: Implications for social work practice. *Social Work in Mental Health, 16*(2), 145–171. doi:10.1080/15332985.2017.1361887

Paz, S., & Semenza, J. C. (2016). El Niño and climate change—Contributing factors in the dispersal of Zika virus in the Americas? *Lancet, 387*(10020), 745. doi:10.1016/S0140-6736(16)00256-7 PMID:26850984

Pearson, A. (2018). The Debate about Wheelchair Spaces on Buses Goes Round and Round: Access to Public Transport for People with Disabilities as a Human Right. *The Northern Ireland Legal Quarterly, 69*(1), 1–17. doi:10.53386/nilq.v69i1.68

Pederson, N., Hessl, A. E., Baatarbileg, N., Anchukaitis, K. J., & Di Cosmo, N. (2014). Pluvials, droughts, the Mongol Empire, and modern Mongolia. *Proceedings of the National Academy of Sciences of the United States of America, 111*(12), 4375–4379. doi:10.1073/pnas.1318677111 PMID:24616521

Pelargos, P. E., Nagasawa, D. T., Lagman, C., Tenn, S., Demos, J. V., Lee, S. J., Bui, T. T., Barnette, N. E., Bhatt, N. S., Ung, N., Bari, A., Martin, N. A., & Yang, I. (2017). Utilizing virtual and augmented reality for educational and clinical enhancements in neurosurgery. *Journal of Clinical Neuroscience, 35*, 1–4. doi:10.1016/j.jocn.2016.09.002 PMID:28137372

Pellegrino, E. D. (2013). Medical Ethics and Moral Philosophy in an Era of Bioethics. In J. R. Garrett, F. Jotterand, & D. C. Ralston (Eds.), *The Development of Bioethics in the United States* (pp. 241–254). Springer Netherlands. doi:10.1007/978-94-007-4011-2_14

Pembrey, M. E., Bygren, L. O., Kaati, G., Edvinsson, S., Northstone, K., Sjöström, M., & Golding, J. (2006). Sex-specific, male-line transgenerational responses in humans. *European Journal of Human Genetics, 14*(2), 159–166. doi:10.1038j.ejhg.5201538 PMID:16391557

Penz, P., Drydyk, J., & Bose, P. S. (2011). *Displacement by Development: Ethics, Rights and Responsibilities.* Cambridge University Press. doi:10.1017/CBO9780511973499

Perona Calvete, C. (2023, January 7). *The Bolivar Legacy Part II: Debt and Regret.* https://europeanconservative.com/articles/essay/the-bolivar-legacy-part-ii-debt-and-regret/

Perpiñá, C., Botella, C., Baños, R., Marco, H., Alcañiz, M., & Quero, S. (1999). Body Image and Virtual Reality in Eating Disorders: Is Exposure to Virtual Reality More Effective than the Classical Body Image Treatment? *Cyberpsychology & Behavior, 2*(2), 149–155. doi:10.1089/cpb.1999.2.149 PMID:19178251

Perry, A. M., Romer, C., & Barr, A. (2021, December 20). Why is life expectancy so low in Black neighborhoods? *Brookings.* https://www.brookings.edu/blog/the-avenue/2021/12/20/why-is-life-expectancy-so-low-in-black-neighborhoods/

Peters, T. (2019). Does extraterrestrial life have intrinsic value? An exploration in responsibility ethics. *International Journal of Astrobiology, 18*(4), 304–310. doi:10.1017/S147355041700057X

Petrilli, S. (2004). *The responsibility of power and the power of responsibility: From the "semiotic" to the "semioethic" animal.* Academic Press.

Pettitt, P. B. (2002). The Neanderthal dead: Exploring mortuary variability in Middle Palaeolithic Eurasia. *Before Farming, 1*(4), 1–19. doi:10.3828/bfarm.2002.1.4

Petty, G. W., Mohr, J. P., Pedley, T. A., Tatemichi, T. K., Lennihan, L., Duterte, D. I., & Sacco, R. L. (1990). The role of transcranial Doppler in confirming brain death: Sensitivity, specificity, and suggestions for performance and interpretation. *Neurology*, *40*(2), 300–300. doi:10.1212/WNL.40.2.300 PMID:2405294

Philippi, C. L., Feinstein, J. S., Khalsa, S. S., Damasio, A., Tranel, D., Landini, G., Williford, K., & Rudrauf, D. (2012). Preserved Self-Awareness following Extensive Bilateral Brain Damage to the Insula, Anterior Cingulate, and Medial Prefrontal Cortices. *PLoS One*, *7*(8), e38413. doi:10.1371/journal.pone.0038413 PMID:22927899

Philipps, K. (2015). *Hospitality and Emerging Populations: Toward a Theology of Migration in the Context of the Catholic Church in the United States* [PhD Thesis].

Phillips, J. C. (2011). *Symbolism, complexity and intentional burial among the Neanderthals*. Academic Press.

Piché, J., & Walby, K. (2018). Abolitionist reflections on museums of confinement. *Déviance et Société*, *42*(4), 643–662.

Pillsbury, B. L. K. (1978). "Doing the month": Confinement and convalescence of Chinese women after childbirth. *Social Science & Medicine. Part B, Medical Anthropology*, *12*, 11–22. doi:10.1016/0160-7987(78)90003-0 PMID:565536

Pink Floyd. (1980, June 23). *Comfortably Numb* [7"]. Harvest (UK), Columbia (US).

Pintassilgo, S., & Carvalho, H. (2017). Trends and consequences of the technocratic paradigm of childbirth in Portugal: A population-based analysis of birth conditions and social characteristics of parents. *Sexual & Reproductive Healthcare : Official Journal of the Swedish Association of Midwives*, *13*, 58–67. doi:10.1016/j.srhc.2017.06.003 PMID:28844359

Pinto, J. M., Wroblewski, K. E., Kern, D. W., Schumm, L. P., & McClintock, M. K. (2014). Olfactory Dysfunction Predicts 5-Year Mortality in Older Adults. *PLoS One*, *9*(10), e107541. doi:10.1371/journal.pone.0107541 PMID:25271633

Pisani, M., & Grech, S. (2017). *Disability and Forced Migration: Critical Intersectionalities*. http://repositoriocdpd.net:8080/handle/123456789/1505

Pitts, L., Jr. (2010, November 22). A tragedy of 'justice'. *The Spokesman-Review*. https://www.spokesman.com/stories/2010/nov/22/a-tragedy-of-justice/

Pitts, A. J. (2018). Examining Carceral Medicine through Critical Phenomenology. *International Journal of Feminist Approaches to Bioethics*, *11*(2), 14–35. Advance online publication. doi:10.3138/ijfab.2017.08.11

Pliny the Elder. (1855). *The Natural History of Pliny*. H. G. Bohn.

Plotinus. (1952). *The six enneads* (S. Mackenna & B. S. Page, Trans.). Encyclopaedia Britannica.

Poincaré, H. (1898). On the foundations of geometry. *The Monist*, *9*(1), 1–43.

Pope, K. O., D'Hondt, S. L., & Marshall, C. R. (1998). Meteorite impact and the mass extinction of species at the Cretaceous/Tertiary boundary. *Proceedings of the National Academy of Sciences of the United States of America*, *95*(19), 11028–11029. doi:10.1073/pnas.95.19.11028 PMID:9736679

Popescu, D.-E. (2018). *Dynamic injustice: Interlocking recognition and distribution* [PhD Thesis]. The London School of Economics and Political Science (LSE).

Portes, A. (2010). Migration and Social Change: Some Conceptual Reflections. *Journal of Ethnic and Migration Studies*, *36*(10), 1537–1563. doi:10.1080/1369183X.2010.489370

Pottinger, M., Johnson, M., & Feith, D. (2022, November 30). Xi Jinping in His Own Words. *Foreign Affairs*. https://www.foreignaffairs.com/china/xi-jinping-his-own-words

Prainsack, B., & Buyx, A. (2017). *Solidarity in Biomedicine and Beyond*. Cambridge University Press. doi:10.1017/9781139696593

Presidential Commission for the Study of Bioethical Issues. (2011). *"Ethically Impossible" STD Research in Guatemala from 1946 to 1948*. BrainFeed Press. https://law.stanford.edu/wp-content/uploads/2011/09/EthicallyImpossible_PC-SBI_110913.pdf

Pressly, T. J. (2006). "The Known World" of Free Black Slaveholders: A Research Note on the Scholarship of Carter G. Woodson. *Journal of African American History*, *91*(1), 81–87. doi:10.1086/JAAHv91n1p81

Price, R. M. (2008). *Top Secret: The Truth Behind Today's Pop Mysticisms*. Prometheus Books.

Priestley, M. (1998). Childhood disability and disabled childhoods: Agendas for research. *Childhood*, *5*(2), 207–223. doi:10.1177/0907568298005002007

Prior, A. (2010, February 18). No Joke: Buffalo and Moravia Duke It Out Over Millard Fillmore. *Wall Street Journal*. https://www.wsj.com/articles/SB10001424052748703444804575071322401554694

Prokop-Prigge, K. A., Greene, K., Varallo, L., Wysocki, C. J., & Preti, G. (2016). The effect of ethnicity on human axillary odorant production. *Journal of Chemical Ecology*, *42*(1), 33–39. doi:10.100710886-015-0657-8 PMID:26634572

Prokop-Prigge, K. A., Thaler, E., Wysocki, C. J., & Preti, G. (2014). Identification of volatile organic compounds in human cerumen. *Journal of Chromatography. B, Analytical Technologies in the Biomedical and Life Sciences*, *953–954*, 48–52. doi:10.1016/j.jchromb.2014.01.043 PMID:24572763

Prusak, B. G. (2005). Rethinking "Liberal Eugenics": Reflections and Questions on Habermas on Bioethics. *The Hastings Center Report*, *35*(6), 31–42. doi:10.2307/3528563 PMID:16396203

Pye, L. W., & Bix, H. P. (2000). Hirohito and the Making of Modern Japan. *Foreign Affairs*, *79*(5), 128. doi:10.2307/20049894

Qiu, J. (2016). Nanotechnology development in China: Challenges and opportunities. *National Science Review*, *3*(1), 148–152. doi:10.1093/nsr/nww007

Quesada, J., Hart, L. K., & Bourgois, P. (2011). Structural Vulnerability and Health: Latino Migrant Laborers in the United States. *Medical Anthropology*, *30*(4), 339–362. doi:10.1080/01459740.2011.576725 PMID:21777121

Quine, W. V. (1950). Identity, Ostension, and Hypostasis. *The Journal of Philosophy*, *47*(22), 621–633. doi:10.2307/2021795

Quine, W. V. (1951). Main Trends in Recent Philosophy: Two Dogmas of Empiricism. *The Philosophical Review*, *60*(1), 20–43. doi:10.2307/2181906

Racine, E. (2010). *Pragmatic Neuroethics: Improving Treatment and Understanding of the Mind-brain*. MIT Press. doi:10.7551/mitpress/8027.001.0001

Racine, E., & Sample, M. (2018). Two Problematic Foundations of Neuroethics and Pragmatist Reconstructions. *Cambridge Quarterly of Healthcare Ethics*, *27*(4), 566–577. doi:10.1017/S0963180118000099 PMID:30720412

Radu, S. (2019, February 14). *China Overtook the U.S. and Will Hold the Largest Share of World's Data at Least by 2025*. https://www.usnews.com/news/best-countries/articles/2019-02-14/china-overtook-the-us-and-will-hold-the-largest-share-of-worlds-data-at-least-by-2025

Rahman, A. U. (2020). *Reasons behind the Great Bengal Famine in 1770: British claim vs Reality*. Academic Press.

Ramos, S. R., Warren, R., Shedlin, M., Melkus, G., Kershaw, T., & Vorderstrasse, A. (2019). A Framework for Using eHealth Interventions to Overcome Medical Mistrust Among Sexual Minority Men of Color Living with Chronic Conditions. *Behavioral Medicine (Washington, D.C.), 45*(2), 166–176. doi:10.1080/08964289.2019.1570074 PMID:31343963

Ramsey, P. (1967). *Deeds and rules in Christian ethics*. Scribner.

Rask, S. (2018). *Diversity and Health in the Population: Findings on Russian, Somali and Kurdish origin populations in Finland.* https://helda.helsinki.fi/handle/10138/233970

Rathunde, K., & Isabella, R. (2017). Play, Flow, and Tailoring Identity in Middle Adulthood. In J. D. Sinnott (Ed.), *Identity Flexibility During Adulthood: Perspectives in Adult Development* (pp. 211–232). Springer International Publishing. doi:10.1007/978-3-319-55658-1_14

Rawlinson, M. C. (2001). The Concept of a Feminist Bioethics. *The Journal of Medicine and Philosophy: A Forum for Bioethics and Philosophy of Medicine, 26*(4), 405–416. 10.1076/jmep.26.4.405.3010

Rawlinson, M. C. (2017, March 21). *The Concept of a Feminist Bioethics: IJFAB at Ten (world)* [Letter]. doi:10.3138/ijfab.10.1.1

Rawlinson, M. C. (2016). *Just Life: Bioethics and the Future of Sexual Difference*. Columbia University Press. doi:10.7312/rawl17174

Ray, M. L. (2022). *Nurturing with Nature: Exploring Anorexia Nervosa through an Ecotherapeutic Lens* [PhD Thesis]. Pacifica Graduate Institute.

Raymond, C., Matthews, T., & Horton, R. M. (2020). The emergence of heat and humidity too severe for human tolerance. *Science Advances, 6*(19). doi:10.1126/sciadv.aaw1838

Redding, P. (2007). *Analytic Philosophy and the Return of Hegelian Thought*. Cambridge University Press. doi:10.1017/CBO9780511487620

Reed, R. D., & Antonsen, E. L. (2018). Should NASA Collect Astronauts' Genetic Information for Occupational Surveillance and Research? *AMA Journal of Ethics, 20*(9), 849–856. doi:10.1001/amajethics.2018.849 PMID:30242816

Reer, F., & Quandt, T. (2021). Games Addiction: A Comprehensive Overview. In Oxford Research Encyclopedia of Communication. doi:10.1093/acrefore/9780190228613.013.1154

Rees, N., & Fuller, R. (2020). *The toxic truth: Children's exposure to lead pollution undermines a generation of future potential.* UNICEF.

Reynolds, J. M. (2018). Dangerous Discourses of Disability, Subjectivity and Sexuality by Margrit Shildrick. *International Journal of Feminist Approaches to Bioethics, 11*(1), 162–167. Advance online publication. doi:10.3138/ijfab.11.1.162

Rhodes, P. J., Small, N. A., Ismail, H., & Wright, J. P. (2008). 'What really annoys me is people take it like it's a disability', epilepsy, disability and identity among people of Pakistani origin living in the UK. *Ethnicity & Health, 13*(1), 1–21. doi:10.1080/13557850701803031 PMID:18066735

Richards, H. (2012, February 21). *How Goldman Sachs helped mask Greece's debt.* The Bureau of Investigative Journalism. https://www.thebureauinvestigates.com/stories/2012-02-21/how-goldman-sachs-helped-mask-greeces-debt

Richardson, H. S. (2000). Specifying, Balancing, and Interpreting Bioethical Principles. *The Journal of Medicine and Philosophy: A Forum for Bioethics and Philosophy of Medicine, 25*(3), 285–307. 10.1076/0360-5310(200006)25:3;1-H;FT285

Richie, C. (2016). Lessons from Queer Bioethics: A Response to Timothy F. Murphy. *Bioethics*, *30*(5), 365–371. doi:10.1111/bioe.12246 PMID:26833492

Ridenour, A. A. (2017). The Meaning of Death and the Goal of Medicine: An Augustinian and Barthian Reassessment. *Christian Bioethics*, *23*(1), 60–76. doi:10.1093/cb/cbw021

Rieder, T. N., Hutchinson, L. A., & Kahn, J. P. (2022). Why the World Needs Bioethics Communication. *Perspectives in Biology and Medicine*, *1*. Advance online publication. doi:10.1353/pbm.0.0174 PMID:36468391

Riedweg, C. (2005). *Pythagoras: His Life, Teaching, and Influence*. Cornell University Press.

Riehl, J. (2012, August 22). *The roots of human self-awareness*. Iowa Now. https://now.uiowa.edu/2012/08/roots-human-self-awareness

Riva, G. (2005). Virtual Reality in Psychotherapy [Review]. *Cyberpsychology & Behavior*, *8*(3), 220–230. doi:10.1089/cpb.2005.8.220 PMID:15971972

Riva, G., & Mantovani, F. (2014). 1 Extending the Self through the Tools and the Others: A General Framework for Presence and Social Presence in Mediated Interactions. In J. Waterworth & D. Murray (Eds.), *Interacting with Presence* (pp. 9–31). Sciendo Migration. doi:10.2478/9783110409697.1

Rivera-Amarillo, C., & Camargo, A. (2019). Zika assemblages: Women, populationism, and the geographies of epidemiological surveillance. *Gender, Place and Culture*, *0*(0), 1–17. doi:10.1080/0966369X.2018.1555518

Roberts, S. (2005, February 21). More Africans Enter U.S. Than in Days of Slavery. *The New York Times*. https://www.nytimes.com/2005/02/21/nyregion/more-africans-enter-us-than-in-days-of-slavery.html

Roberts, M. L. (2013). *What Soldiers Do: Sex and the American GI in World War II France*. University of Chicago Press. doi:10.7208/chicago/9780226923123.001.0001

Robertson, C. L., & Hoffman, S. J. (2014). Conflict and Forced Displacement: Human Migration, Human Rights, and the Science of Health. *Nursing Research*, *63*(5), 307–308. doi:10.1097/NNR.0000000000000058 PMID:25171556

Robertson, M. (2011). Symposium: Neuroethics and Mental Health—Old Wine in New Bottles or a Legitimate New Field of Bioethical Inquiry. *Journal of Bioethical Inquiry*, *8*(1), 13–14. doi:10.100711673-010-9267-x

Robotham, M. (2020, July 30). Lying Is Fundamental to Who We Are as Humans. *CrimeReads*. https://crimereads.com/lying-is-fundamental-to-who-we-are-as-humans/

Rodolfo, K. S. (2003). *Tsunamites and seismites*. Access Science.

Rodríguez Wallenius, C. A., & Cruz Arenas, R. A. (2013). *El México bárbaro del siglo XXI*. Universidad Autónoma Metropolitana.

Rodríguez, I. O. (2017). Animal Citizenship, Phenomenology, and Ontology: Some reflections on Donaldsons & Kymlickas Zoopolis. *Bangladesh Journal of Bioethics*, *8*(1), 21–32. doi:10.3329/bioethics.v8i1.31080

Rodriguez, M. A., & García, R. (2013). First, Do No Harm: The US Sexually Transmitted Disease Experiments in Guatemala. *American Journal of Public Health*, *103*(12), 2122–2126. doi:10.2105/AJPH.2013.301520 PMID:24134370

Roehrs, T., & Roth, T. (2011). Medication and Substance Abuse. In M. H. Kryger, T. Roth, & W. C. Dement (Eds.), *Principles and Practice of Sleep Medicine* (5th ed., pp. 1512–1523). W.B. Saunders., doi:10.1016/B978-1-4160-6645-3.00132-8

Rogatchevski, A. (2014). Non-Totalitarian Imprisonment under Western and Eastern Eyes: Lord Archer, Eduard Limonov, and Theories of Human Motivation. *Punishment as a Crime?*, 163.

Rollin, B. E. (2007). Animal Mind: Science, Philosophy, and Ethics. *The Journal of Ethics*, *11*(3), 253–274. doi:10.100710892-007-9018-3

Romani, L., Holck, L., & Risberg, A. (2018). Benevolent discrimination: Explaining how human resources professionals can be blind to the harm of diversity initiatives. *Organization*. Advance online publication. doi:10.1177/1350508418812585

Rosenberg (Director). (1980). *Brubaker*. https://www.imdb.com/title/tt0080474/

Rosenberg, L. (2023). The Metaverse and Conversational AI as a Threat Vector for Targeted Influence. *2023 IEEE 13th Annual Computing and Communication Workshop and Conference (CCWC)*, 504–510. 10.1109/CCWC57344.2023.10099167

Rosenberg, A. (2004). *The Myth of the Twentieth Century: An Evaluation of the Spiritual-intellectual Confrontations of Our Age*. Historical Review Press.

Rosenfeld, A. (1999). The Journalist's Role in Bioethics. *The Journal of Medicine and Philosophy: A Forum for Bioethics and Philosophy of Medicine, 24*(2), 108–129. 10.1076/jmep.24.2.108.2532

Roser, T. (2017). Simon Peng-Keller: Sinnereignisse in Todesnähe. Traum- und Wachvisionen Sterbender und Nahtoderfahrungen im Horizont von Spiritual Care (Studies in Spiritual Care 1), Berlin / Boston (de Gruyter) 2017, 188 S., ISBN-13: 978–3110473162, Eur (D) 79,95. *International Journal of Practical Theology*, *21*(2), 323–325. doi:10.1515/ijpt-2017-0034

Roskies, A. (2002). Neuroethics for the new millennium. *Neuron*, *35*(1), 21–23. doi:10.1016/S0896-6273(02)00763-8 PMID:12123605

Ross, M. (2018). Virtual Reality's New Synesthetic Possibilities. *Television & New Media, 1527476418805240*. Advance online publication. doi:10.1177/1527476418805240

Rothblatt, M. (2001). Key Legal Issues: Determining the Rise of Geoethics as the Optimum Analytical Paradigm. *Graft*, *4*(2), 143–145. doi:10.1177/152216280100400218

Roudometof, V. (2018). *Glocal Religions: An Introduction*. Multidisciplinary Digital Publishing Institute.

Rowan, A., & Kartal, T. (2018). Dog Population & Dog Sheltering Trends in the United States of America. *Animals (Basel)*, *8*(5), 68. Advance online publication. doi:10.3390/ani8050068 PMID:29710771

Rowe, E. A. (2018). *Life-saving and life-changing: The decision-making processes of people seeking asylum* [PhD]. Queensland University of Technology. https://eprints.qut.edu.au/116152/

Roy, D. (2012). Neuroethics, Gender and the Response to Difference. *Neuroethics*, *5*(3), 217–230. doi:10.100712152-011-9130-8

Rumbaut, R. G. (1991). *Migration, Adaptation, and Mental Health: The Experience of Southeast Asian Refugees in the United States* (*SSRN* Scholarly Paper ID 1886652). Social Science Research Network. https://papers.ssrn.com/abstract=1886652

Rupp, T., & Delaney, K. A. (2004). Inadequate analgesia in emergency medicine. *Annals of Emergency Medicine*, *43*(4), 494–503. doi:10.1016/j.annemergmed.2003.11.019 PMID:15039693

Russell, B. (1924). *Icarus: Or, The Future of Science*. Dutton.

Russo, A., & Russo, J. (Directors). (2019). *Avengers: Endgame*. https://www.imdb.com/title/tt4154796/

Rust, G., Nembhard, W. N., Nichols, M., Omole, F., Minor, P., Barosso, G., & Mayberry, R. (2004). Racial and ethnic disparities in the provision of epidural analgesia to Georgia Medicaid beneficiaries during labor and delivery. *American Journal of Obstetrics and Gynecology, 191*(2), 456–462. doi:10.1016/j.ajog.2004.03.005 PMID:15343221

Ryan, C. P. (2008). Zoonoses likely to be used in bioterrorism. *Public Health Reports, 123*(3), 276–281. doi:10.1177/003335490812300308 PMID:19006970

Ryder, R. D. (1983). *Victims of Science: The Use of Animals in Research.* National Anti-Vivisection Society Limited.

Sadler, J. Z. (2010). Dignity, Arête, and Hubris in the Transhumanist Debate. *The American Journal of Bioethics, 10*(7), 67–68. doi:10.1080/15265161003728845 PMID:20582838

Sager, A. (2018). Political Philosophy, Migration, and Methodological Nationalism. In A. Sager (Ed.), *Toward a Cosmopolitan Ethics of Mobility: The Migrant's-Eye View of the World* (pp. 17–36). Springer International Publishing. doi:10.1007/978-3-319-65759-2_2

Sager, A. (2018). Toward a Political Philosophy of Mobility. In A. Sager (Ed.), *Toward a Cosmopolitan Ethics of Mobility: The Migrant's-Eye View of the World* (pp. 91–95). Springer International Publishing. doi:10.1007/978-3-319-65759-2_6

Said, E. (1990). In the shadow of the West: Edward Said. *Discourses: Conversations in Postmodern Art and Culture*, 93–104.

Said, E. W. (2014). *Culture And Imperialism.* Random House.

Saint-Paul, R. (1955). *0112683—Jean-Paul Sartre (1905-1980).* Granger. https://www.granger.com/results.asp?image=0112683

Saint-Upéry, M. (2010, October 21). *Left at the Crossroads: Ogling the poor.* https://sputniknews.com/analysis/20101021161035393/

Sakamoto, H. (2002). Chapter 2. A Challenge to the Concept of Human Rights and Human Dignity. *Journal International de Bioethique, 13*(2), 23–30. doi:10.3917/jib.132.0023

Sakellaropoulos, S. (2016). Winners and Losers in the Greek Financial Crisis. *Global Dialogue, 6*(4). https://globaldialogue.isa-sociology.org/winners-and-losers-in-the-greek-financial-crisis/

Salkeld, D. (1994). *Madness and Drama in the Age of Shakespeare.* Manchester University Press.

Sallan, L. C., & Coates, M. I. (2014). The long-rostrumed elasmobranch Bandringa Zangerl, 1969, and taphonomy within a Carboniferous shark nursery. *Journal of Vertebrate Paleontology, 34*(1), 22–33. doi:10.1080/02724634.2013.782875

Samanta, J., Samanta, A., & Madhloom, O. (2018). A rights-based proposal for managing faith-based values and expectations of migrants at end-of-life illustrated by an empirical study involving South Asians in the UK. *Bioethics, 32*(6), 368–377. doi:10.1111/bioe.12447 PMID:29882999

Samuels, G. (2016, June 7). *Refugee men are selling sex for less than price of a sandwich to survive.* The Independent. https://www.independent.co.uk/news/world/europe/dozens-of-refugee-men-sell-sex-for-as-little-as-2-to-survive-in-athens-a7068741.html

Samuels, A. (2016). *I Rebel, Therefore We Are.* New Political Thinking on Individual Responsibility for Group, Society, Culture, and Planet.

Sánchez-Quinto, F., Malmström, H., Fraser, M., Girdland-Flink, L., Svensson, E. M., Simões, L. G., George, R., Hollfelder, N., Burenhult, G., Noble, G., Britton, K., Talamo, S., Curtis, N., Brzobohata, H., Sumberova, R., Götherström, A., Storå, J., & Jakobsson, M. (2019). Megalithic tombs in western and northern Neolithic Europe were linked to a kindred society. *Proceedings of the National Academy of Sciences of the United States of America*, *116*(19), 9469–9474. doi:10.1073/pnas.1818037116 PMID:30988179

Sandakova, L. B., & Tolstobrova, L. I. (2019, July). *Proliferation of Neuroscientific Knowledge in Pedagogy and Education*. "Humanities and Social Sciences: Novations, Problems, Prospects" (HSSNPP 2019). https://www.atlantis-press.com/proceedings/hssnpp-19/125913454

Sander, R. (Director). (2016, June 15). *Human Rights and the Cyrus Cylinder: Freedom: A Shared Dream*. https://www.farhang.org/freedom

Sandler, R. (2016). *Ethics and Emerging Technologies*. Springer.

Sandu, A. (2019). Towards a Phenomenology of the Digitalization of Consciousness. The Virtualization of the Social Space. *Postmodern Openings*, *10*(2), 155–161. doi:10.18662/po/77

Sanghi, S., & Smaldone, A. (2022, January 27). *The Evolution of the Racial Gap in U.S. Life Expectancy*. https://www.stlouisfed.org/on-the-economy/2022/january/evolution-racial-gap-us-life-expectancy

Sapphire. (1996). *Push*. Vintage Contemporaries/Vintage Books.

Saramo, S. (2017). The meta-violence of Trumpism. *European Journal of American Studies*, *12*(2), 12–2. doi:10.4000/ejas.12129

Sarango, C. (2020). Las Leyes de Burgos y Valladolid un antecedente temprano de una igualdad humana. *Ciencia e Interculturalidad*, *26*(1), 1. Advance online publication. doi:10.5377/rci.v26i01.9887

Sardelis, C. (2018, June). Christoforos Sardelis: The role of the insurance sector in the post-program era. *Economy Watch*. http://www.economywatch.gr/christoforos-sardelis-the-role-of-the-insurance-sector-in-the-post-program-era/

Sartre, J.-P. (1992). *Being and Nothingness*. Simon and Schuster.

Sawin, C. (2017). Bioethics in Space Exploration. Encyclopedia of Bioastronautics, 1–6.

Sax, L. (2002). How common is intersex? A response to Anne Fausto-Sterling. *Journal of Sex Research*, *39*(3), 174–178. doi:10.1080/00224490209552139 PMID:12476264

Saycell, J., Lomax, M., Massey, H., & Tipton, M. (2019). How cold is too cold? Establishing the minimum water temperature limits for marathon swim racing. *British Journal of Sports Medicine*, *53*(17), 1078–1084. doi:10.1136/bjsports-2018-099978 PMID:30824457

Scheepers, D., Spears, R., Doosje, B., & Manstead, A. S. R. (2003). Two Functions of Verbal Intergroup Discrimination: Identity and Instrumental Motives as a Result of Group Identification and Threat. *Personality and Social Psychology Bulletin*, *29*(5), 568–577. doi:10.1177/0146167203029005002 PMID:15272991

Schenker, J. G. (2005). Assisted reproduction practice: Religious perspectives. *Reproductive Biomedicine Online*, *10*(3), 310–319. doi:10.1016/S1472-6483(10)61789-0 PMID:15820035

Schermer, M., & Keulartz, J. (2002). How Pragmatic is Bioethics? In J. Keulartz, M. Korthals, M. Schermer, & T. Swierstra (Eds.), *Pragmatist Ethics for a Technological Culture* (pp. 41–68). Springer Netherlands., doi:10.1007/978-94-010-0301-8_4

Schiffer, J. P. (2010). Implications of old physics simplify the understanding of nuclei. *Physics (College Park, Md.)*, *3*, 2. doi:10.1103/Physics.3.2

Schiff, N. D., & Fins, J. J. (2016). Brain death and disorders of consciousness. *Current Biology*, *26*(13), R572–R576. doi:10.1016/j.cub.2016.02.027 PMID:27404252

Schilder, P. (1938). The Social Neurosis. *Psychoanalytic Review*, 1–19.

Schlegel, L. (2020). Jumanji Extremism? How games and gamification could facilitate radicalization processes. *Journal for Deradicalization*, *23*, 1–44.

Schmeink, L. (2017). New Ways to Think about Zombies. *Science Fiction Studies*, *44*(3), 623–625. doi:10.5621ciefictstud.44.3.0623

Schmidt, U. (2011). Church, Public And Bioethics: Religion's Construction Of Public Significance Through The Bioethical Discourse. *The Public Significance of Religion*, 191–213.

Schneider, J. (2013). Philosophical issues in the search for extraterrestrial life and intelligence1. *International Journal of Astrobiology*, *12*(3), 259–262. doi:10.1017/S1473550413000025

Scholl, L. (2018). *Medicine, Health and Being Human*. Routledge. doi:10.4324/9780203732700

Scott, R. (Director). (1982, June 25). *Blade Runner* [Action, Drama, Sci-Fi]. The Ladd Company, Shaw Brothers, Warner Bros.

Scott, J., Onyango, M. A., Burkhardt, G., Mullen, C., Rouhani, S., Haider, S., Albutt, K., Greiner, A., VanRooyen, M., & Bartels, S. (2018). A qualitative analysis of decision-making among women with sexual violence-related pregnancies in conflict-affected eastern Democratic Republic of the Congo. *BMC Pregnancy and Childbirth*, *18*(1), 322. doi:10.118612884-018-1942-7 PMID:30089448

Scott, K. M., & Collings, S. C. D. (2010). Gender and the association between mental disorders and disability. *Journal of Affective Disorders*, *125*(1), 207–212. doi:10.1016/j.jad.2010.06.022 PMID:20637507

Scott, W., McCracken, L. M., & Trost, Z. (2014). A psychological flexibility conceptualisation of the experience of injustice among individuals with chronic pain. *British Journal of Pain*, *8*(2), 62–71. doi:10.1177/2049463713514736 PMID:26516537

Seedhouse, E. (2010). *Prepare for Launch: The Astronaut Training Process*. Springer Science & Business Media. doi:10.1007/978-1-4419-1350-0

Sellars, W. (1974). Meaning as functional classification. *Synthese*, *27*(3), 417–437. doi:10.1007/BF00484606

Sen, A. (2005). Human Rights and Capabilities. *Journal of Human Development*, *6*(2), 151–166. doi:10.1080/14649880500120491

Senor, T. D. (2005). *Epistemological problems of memory*. https://stanford.library.sydney.edu.au/entries/memory-episprob/

Sententia, W. (2004). Neuroethical considerations: Cognitive liberty and converging technologies for improving human cognition. *Annals of the New York Academy of Sciences*, *1013*(1), 221–228. doi:10.1196/annals.1305.014 PMID:15194617

Sereny, G. (1983). *Into That Darkness: An Examination of Conscience*. Knopf Doubleday Publishing Group.

Sextus Empiricus. (1933). *Sextus Empiricus: Outlines of Pyrrhonism*. Heinemann.

Shafaeddin, M. (1998). How did developed countries industrialize. In *The History of Trade and Industrial Policy: The Cases of Great Britain and the USA*. Academic Press.

Shakespeare, T., & Watson, N. (1998). Theoretical Perspectives on Research. *Growing Up With Disability, 34*, 13.

Shakur, A. (2001). *Assata: An Autobiography*. Zed Books.

Shamim, A., Banu, S., Blassingame, J., Guilford, K. L., & Williams, J. (2023). Delusions and Fixed Beliefs in the Age of the Internet: When Do Conspiracy Theories Become Illness? *Psychiatric Annals*, *53*(4), 171–174. doi:10.3928/00485713-20230313-03

Sharkey, A. (2019). Autonomous weapons systems, killer robots and human dignity. *Ethics and Information Technology*, *21*(2), 75–87. doi:10.100710676-018-9494-0

Shaw, D. (2014). Neuroenhancing public health. *Journal of Medical Ethics*, *40*(6), 389–391. doi:10.1136/medethics-2012-101300 PMID:23793059

Shelley, M. W. (1869). Frankenstein, or, The Modern Prometheus. *Dent, Contemporary Dentistry*.

Shemie, S. D. (2014). Rethinking Mortality: Exploring the Boundaries between Life and Death. *Annals of the New York Academy of Sciences*, *1330*(1), 101–104. doi:10.1111/nyas.12564 PMID:25351389

Shergold, A. (2016, June 9). France's World Cup win in 98 was not appreciated until 12 years later. *Mail Online*. https://www.dailymail.co.uk/sport/football/article-3632753/France-s-World-Cup-win-1998-not-properly-appreciated-12-years-later-squad-went-strike-claims-Frank-Leboeuf.html

Shi, D.-P., & Yu, L. (2004). Euthanasia should be Legalized in China: Personal Perspective. In R.-Z. Qiu (Ed.), *Bioethics: Asian Perspectives: A Quest for Moral Diversity* (pp. 157–169). Springer Netherlands. doi:10.1007/978-94-017-0419-9_15

Shildrick, M. (1997). Feminist theory and postmodernism. In *Leaky Bodies and Boundaries*. https://www.taylorfrancis.com/

Shildrick, M. (2000). Becoming Vulnerable: Contagious Encounters and the Ethics of Risk. *The Journal of Medical Humanities*, *21*(4), 215–227. doi:10.1023/A:1009025125203

Shildrick, M. (2009). *Dangerous discourses of disability, subjectivity and sexuality*. Palgrave Macmillan. doi:10.1057/9780230244641

Shildrick, M. (2015). *Leaky Bodies and Boundaries: Feminism, Postmodernism and (Bio)ethics*. Routledge. doi:10.4324/9781315004952

Shildrick, M., Carnie, A., Wright, A., McKeever, P., Jan, E. H.-C., Luca, E. D., Bachmann, I., Abbey, S., Bo, D. D., Poole, J., El-Sheikh, T., & Ross, H. (2018). Messy entanglements: Research assemblages in heart transplantation discourses and practices. *Medical Humanities*, *44*(1), 46–54. doi:10.1136/medhum-2017-011212 PMID:28972037

Shimazono, Y. (2007). The state of the international organ trade: A provisional picture based on integration of available information. *Bulletin of the World Health Organization*, *85*(12), 955–962. doi:10.2471/BLT.06.039370 PMID:18278256

Shoemaker, D. (2010). The Insignificance of Personal Identity for Bioethics. *Bioethics*, *24*(9), 481–489. doi:10.1111/j.1467-8519.2009.01719.x PMID:19438443

Shoemaker, S. (1963). *Self-knowledge and Self-identity*. Cornell University Press.

Shook, J. R. (2012). Neuroethics and the possible types of moral enhancement. *AJOB Neuroscience*, *3*(4), 3–14. doi:10.1080/21507740.2012.712602

Shook, J. R., & Giordano, J. (2014). A principled and cosmopolitan neuroethics: Considerations for international relevance. *Philosophy, Ethics, and Humanities in Medicine; PEHM*, *9*(1), 1. doi:10.1186/1747-5341-9-1 PMID:24387102

Shotter, J., Fleming, S., & Hall, B. (2021, December 5). How migration became a weapon in a 'hybrid war.' *Financial Times*. https://www.ft.com/content/83ece7e4-cc71-45b5-8db7-766066215612

Shrader-Frechette, K. S. (1991). *Risk and Rationality: Philosophical Foundations for Populist Reforms*. University of California Press. doi:10.1525/9780520320789

Shuman, A., & Bohmer, C. (2016). The Uncomfortable Meeting Grounds of Different Vulnerabilities: Disability and the Political Asylum Process. *Feminist Formations*, *28*(1), 121–145. doi:10.1353/ff.2016.0018

Sieferle, R. P. (2017). *Das Migrationsproblem: Über die Unvereinbarkeit von Sozialstaat und Masseneinwanderung*. Manŭscriptum.

Silani, G., Lamm, C., Ruff, C. C., & Singer, T. (2013). Right supramarginal gyrus is crucial to overcome emotional egocentricity bias in social judgments. *The Journal of Neuroscience : The Official Journal of the Society for Neuroscience*, *33*(39), 15466–15476. doi:10.1523/JNEUROSCI.1488-13.2013 PMID:24068815

Silva, W. H. S., Lopes, G. L. B., Yano, K. M., Tavares, N. S. A., Rego, I. A. O., & Cavalcanti, F. A. da C. (2015). Efecto de un programa de rehabilitación utilizando la realidad virtual para el equilibrio y la funcionalidad de los pacientes con accidente cerebrovascular crónico. *Motriz*, *21*(3), 237–243. doi:10.1590/S1980-65742015000300003

Silver, C., & Hyman, D. A. (2018). *Overcharged: Why Americans Pay Too Much for Health Care*. Cato Institute.

Singer, I. (2004). *Three Philosophical Filmmakers: Hitchcock, Welles*. Renoir.

Siricharoen, W. (2019). The Effect of Virtual Reality as a form of Escapism. *CONF-IRM 2019 Proceedings*. https://aisel.aisnet.org/confirm2019/36

Skarlatos, D., Agrafiotis, P., Balogh, T., Bruno, F., Castro, F., Petriaggi, B. D., Demesticha, S., Doulamis, A., Drap, P., Georgopoulos, A., Kikillos, F., Kyriakidis, P., Liarokapis, F., Poullis, C., & Rizvic, S. (2016). Project iMARECULTURE: Advanced VR, iMmersive Serious Games and Augmented REality as Tools to Raise Awareness and Access to European Underwater CULTURal heritagE. In M. Ioannides, E. Fink, A. Moropoulou, M. Hagedorn-Saupe, A. Fresa, G. Liestøl, V. Rajcic, & P. Grussenmeyer (Eds.), *Digital Heritage. Progress in Cultural Heritage: Documentation, Preservation, and Protection* (pp. 805–813). Springer International Publishing. doi:10.1007/978-3-319-48496-9_64

Skidelsky, E. (2018). Moral Enhancement and the Human Condition. *Royal Institute of Philosophy*, *83*(Supplements, 83), 109–120. doi:10.1017/S1358246118000310

Skidmore, G. (2011). *Haley Barbour speaking at CPAC in Washington D.C. on February 12, 2011*. https://commons.wikimedia.org/wiki/File:Haley_Barbour_by_Gage_Skidmore.jpg

Skinner, B. F. (1966). *Walden Two*. Hackett Publishing.

Skinner, J. (1961). Ritual Matricide: A Study of the Origins of Sacrifice. *The American Imago*, *18*(1), 71–102.

Skinner, R. T. (2019). Walking, talking, remembering: An Afro-Swedish critique of being-in-the-world. *African and Black Diaspora: An International Journal*, *12*(1), 1–19. doi:10.1080/17528631.2018.1467747

Skotnicki, A. (2019). *Conversion and the Rehabilitation of the Penal System: A Theological Rereading of Criminal Justice*. Oxford University Press. doi:10.1093/oso/9780190880835.001.0001

Slater, M. E., Godette, D., Huang, B., Ruan, W. J., & Kerridge, B. T. (2017). Sexual Orientation-Based Discrimination, Excessive Alcohol Use, and Substance Use Disorders Among Sexual Minority Adults. *LGBT Health*, *4*(5), 337–344. doi:10.1089/lgbt.2016.0117 PMID:28876167

Slattery, D. R. (2008). Virtual reality and hallucination: A technoetic perspective. *The Engineering Reality of Virtual Reality, 2008*(6804), 680407. doi:10.1117/12.772650

Smagurauskaitė, A., & Šarauskaitė, G. (2023). Buy, borrow, die: How rich live off their paper wealth and avoid paying taxes and why we let them do it? *Vilnius University Open Series*, 6–14.

Smith, J., & Ashford, B. (2018, November 29). *Rainbow lightens up migrants' view of US at the Mexican border*. Mail Online. https://www.dailymail.co.uk/news/article-6444247/Rainbow-lightens-sky-migrant-camp-border.html

Smith, M. (2022, August 31). Women Who Stay Single and Don't Have Kids Are Getting Richer. *Bloomberg.Com*. https://www.bloomberg.com/news/articles/2022-08-31/women-not-having-kids-get-richer-than-men

Smith, R. C. (2017, September 29). *Transnational Localities: Community, Technology and the Politics of Membership within the Context of Mexico and U.S. Migration 1*. Transnationalism from Below. doi:10.4324/9781351301244-7

Smith, S. (2018). *(De)pathologizing Discourse: The Problematization of Trans and Gender-Nonconforming Mental Health in Ontario* [Thesis]. Université d'Ottawa/University of Ottawa. doi:10.20381/ruor-22327

Smith, T. (2008). *Bioethics for the masses: The negotiation of bioethics in film and fiction*. https://search.proquest.com/openview/4f3f42eca10d14c80f751711fbbb8250/1?pq-origsite=gscholar&cbl=18750&diss=y

Smith, A. (2007). *Wealth of Nations*. Cosimo, Inc.

Smith, J. A. (1999). Identity development during the transition to motherhood: An interpretative phenomenological analysis. *Journal of Reproductive and Infant Psychology, 17*(3), 281–299. doi:10.1080/02646839908404595

Smith, R., Martínez Álvarez, M., & Chanda, R. (2011). Medical tourism: A review of the literature and analysis of a role for bi-lateral trade. *Health Policy (Amsterdam), 103*(2–3), 276–282. doi:10.1016/j.healthpol.2011.06.009 PMID:21767892

Snopkowski, K., Towner, M. C., Shenk, M. K., & Colleran, H. (2016). Pathways from education to fertility decline: A multi-site comparative study. *Philosophical Transactions of the Royal Society B: Biological Sciences, 371*(1692). doi:10.1098/rstb.2015.0156

Socialstyrelsen. (2020, February 12). *Vanligt med flera psykiatriska diagnoser hos personer med könsdysfori*. Socialstyrelsen. https://www.socialstyrelsen.se/om-socialstyrelsen/pressrum/press/vanligt-med-flera-psykiatriska-diagnoser-hos-personer-med-konsdysfori/

Socialstyrelsen. (2022, February 22). *Uppdaterade rekommendationer för hormonbehandling vid könsdysfori hos unga*. Socialstyrelsen. https://www.socialstyrelsen.se/om-socialstyrelsen/pressrum/press/uppdaterade-rekommendationer-for-hormonbehandling-vid-konsdysfori-hos-unga/

Society of Women Engineers. (2023). *Earnings Gap*. https://swe.org/research/2023/earning-gap/

Soldatic, K., Meekosha, H., & Somers, K. (2012). Finding Ernesto: Temporary Labour Migration and Disabled Children's Health. *International Journal of Population Research, 2012*, 1–9. Advance online publication. doi:10.1155/2012/696753

Solomon, O., & Lawlor, M. C. (2018). Beyond V40.31: Narrative Phenomenology of Wandering in Autism and Dementia. *Culture, Medicine and Psychiatry, 42*(2), 206–243. doi:10.100711013-017-9562-7 PMID:29368117

Sosa Wagner, F. (2004). *El mito de la autonomia universitaria*. Academic Press.

Southern, R. W. (1970). *Western Society and the Church in the Middle Ages*. Penguin Books.

Spielberg, S. (Director). (1986, February 7). *The Color Purple* [Drama]. Warner Bros., Amblin Entertainment, The Guber-Peters Company.

Spielberg, S. (Director). (1993). *Jurassic Park*. https://www.imdb.com/title/tt0107290/

Spinney, L. (2018, February 27). *Did Human Sacrifice Help People Form Complex Societies?* The Atlantic. https://www.theatlantic.com/health/archive/2018/02/did-human-sacrifice-help-people-form-complex-societies/554327/

Spitz, R., Pereira, C., Jr., Queiroz, F., Leite, L. C., Dam, P., Ferranti, M. P., Kogut, R., & Oliveira, W. (2017). Gamification, citizen science and civic engagement: In search of the common good. *6th International Conference Balance-Unbalance (BunB)*. UN-Habitat. Https://Unhabitat. Org/Acessado Em, 2.

Spivey, R. J. (2015, June). *A cosmological hypothesis potentially resolving the mystery of extraterrestrial silence with falsifiable implications for neutrinos*. https://www.ingentaconnect.com/content/pe/pe/2015/00000028/00000002/art00019

Stahl, R. Y. (2013, March 28). *Stop Rape: A WWII Chaplain's Advice*. Nursing Clio. https://nursingclio.org/2013/03/28/stop-rape-a-wwii-chaplains-advice/

Stanley, E., & Mihaere, R. (2018). Challenging Māori Imprisonment and Human Rights Ritualism. In E. Stanley (Ed.), *Human Rights and Incarceration: Critical Explorations* (pp. 79–102). Springer International Publishing. doi:10.1007/978-3-319-95399-1_4

Star, S. L. (1990). Power, technology and the phenomenology of conventions: On being allergic to onions. *The Sociological Review, 38*(S1), 26–56. doi:10.1111/j.1467-954X.1990.tb03347.x

Staub, E. (1989). *The Roots of Evil: The Origins of Genocide and Other Group Violence*. Cambridge University Press.

Steadman, H. J., Monahan, J., Duffee, B., Hartstone, E., & Robbins, P. C. (1984). The impact of state mental hospital deinstitutionalization on United States prison populations, 1968-1978. *The Journal of Criminal Law & Criminology, 75*(2), 474. doi:10.2307/1143164

Stearns, H. (1922). *Civilization in the United States: An Inquiry by Thirty Americans*. Academic Press.

Steele, N. M. (2013). *Trolls under the bridge: Anonymous online comments and gatekeeping in the digital realm*. University of Missouri-Columbia.

Steiner, R. (1988). "Paths to Xanadu …" Some Notes on the Development of Dream Displacement and Condensation in Sigmund Freud's "Interpretation of Dreams." *Int. R. Psycho-Anal., 15*, 415–454.

Stein, M. A., & Lord, J. E. (2011). Enabling Refugee and IDP Law and Policy: Implications of the U.N. Convention on the Rights of Persons with Disabilities. *Arizona Journal of International and Comparative Law, 28*, 401.

Stelarc. (1997). From psycho to cyber strategies: Prosthetics, robotics and remote existence. *Cultural Values, 1*(2), 241–249. doi:10.1080/14797589709367146

Stenslie, S. (2015). Stelarc: On the Body as an Artistic Material. *The Journal of Somaesthetics, 1*(1).

Stephens, N. (2010). In Vitro Meat: Zombies on the Menu. *SCRIPTed: A Journal of Law. Technology in Society, 7*, 394.

Stewart, J. L., Spivey, L. A., Widman, L., Choukas-Bradley, S., & Prinstein, M. J. (2019). Developmental patterns of sexual identity, romantic attraction, and sexual behavior among adolescents over three years. *Journal of Adolescence, 77*(1), 90–97. doi:10.1016/j.adolescence.2019.10.006 PMID:31693971

Stonington, S., & Ratanakul, P. (2006). Is There a Global Bioethics? End-of-Life in Thailand and the Case for Local Difference. *PLoS Medicine, 3*(10), e439. doi:10.1371/journal.pmed.0030439 PMID:17076564

Story, L., Thomas, L., Jr., & Schwartz, N. D. (2010, February 13). *Wall St. Helped Greece to Mask Debt Fueling Europe's Crisis—NYTimes.com*. https://www.nytimes.com/2010/02/14/business/global/14debt.html

Strathdee, S. A., Magis-Rodriguez, C., Mays, V. M., Jimenez, R., & Patterson, T. L. (2012). The Emerging HIV Epidemic on the Mexico-US Border: An International Case Study Characterizing the Role of Epidemiology in Surveillance and Response. *Annals of Epidemiology, 22*(6), 426–438. doi:10.1016/j.annepidem.2012.04.002 PMID:22626001

Strathern, M. (2006). Don't eat unwashed lettuce. *American Ethnologist, 33*(4), 532–534. doi:10.1525/ae.2006.33.4.532

Straube, T. (2018, July 11). *Trump Agency Is Using Federal HIV Funds to Separate Immigrant Families*. POZ. https://www.poz.com/article/trump-agency-using-federal-hiv-funds-separate-immigrant-families

Strollo, F., Macchi, C., Eberini, I., Masini, M. A., Botta, M., Vassilieva, G., Nichiporuk, I., Monici, M., Santucci, D., Celotti, F., Magni, P., & Ruscica, M. (2018). Body composition and metabolic changes during a 520-day mission simulation to Mars. *Journal of Endocrinological Investigation, 41*(11), 1267–1273. doi:10.100740618-018-0861-9 PMID:29532423

Sturm, S. (2001). Second Generation Employment Discrimination: A Structural Approach. *Columbia Law Review, 101*(3), 458. doi:10.2307/1123737

Subcommittee on Nanoscale Science, Engineering, and Technology. (2022). *The national nanotechnology initiative supplement to the president's 2022 budget*. National Science and Technology Council.

Subramanian, S. (2017, February 26). Worse than Tuskegee. *Slate Magazine*. https://www.slate.com/articles/health_and_science/cover_story/2017/02/guatemala_syphilis_experiments_worse_than_tuskegee.html

Sullivan, W. F., Heng, J., De Bono, C., Gleeson, G., Goulding, G., & Jamieson, C. (2018). Healing Relationships and Transformations in Health Care: IACB Consensus Statement on Ethical Discernment and Practical Recommendations. *The National Catholic Bioethics Quarterly, 18*(2), 319–327. doi:10.5840/ncbq201818229

Sun, L. (2023, April 8). *What Animals Can Teach Us About Truth and Lies*. Time. https://time.com/6269499/animals-truth-lies-lessons/

Suslov, M. (2016). Of Planets and Trenches: Imperial Science Fiction in Contemporary Russia. *The Russian Review, 75*(4), 562–578. doi:10.1111/russ.12102

Suzuki, K., Garfinkel, S. N., Critchley, H. D., & Seth, A. K. (2013). Multisensory integration across exteroceptive and interoceptive domains modulates self-experience in the rubber-hand illusion. *Neuropsychologia, 51*(13), 2909–2917. doi:10.1016/j.neuropsychologia.2013.08.014 PMID:23993906

Swanson, R. N. (1995). *Religion and Devotion in Europe, C.1215- C.1515*. Cambridge University Press.

Sweden's National Public Television Broadcaster (Director). (2019, April 2). The trans train. In Uppdrag granskning. Author.

Swedish Agency for Health Technology Assessment and Assessment of Social Services. (2019, December 20). *Gender dysphoria in children and adolescents: An inventory of the literature*. https://www.sbu.se/en/publications/sbu-bereder/gender-dysphoria-in-children-and-adolescents-an-inventory-of-the-literature/

Swedish Agency for Health Technology Assessment and Assessment of Social Services. (2022, February 22). *Hormonbehandling vid könsdysfori – barn och unga* [Systematisk litteraturöversikt; HTA; Systematic review]. https://www.sbu.se/sv/publikationer/SBU-utvarderar/hormonbehandling-vid-konsdysfori/

Szreter, S., & Fisher, K. (2010). *Sex Before the Sexual Revolution: Intimate Life in England 1918–1963*. Cambridge University Press. doi:10.1017/CBO9780511778353

Tabor, A. S., & Milfont, T. L. (2011). Migration change model: Exploring the process of migration on a psychological level. *International Journal of Intercultural Relations, 35*(6), 818–832. doi:10.1016/j.ijintrel.2010.11.013

Tackett, K. (2018). The Composition of Worldviews: The Relationships between Conservatism, Religiosity, Empathy, Dogmatism, and Psychological Flexibility. *Morehead State Theses and Dissertations*. https://scholarworks.moreheadstate.edu/msu_theses_dissertations/151

Takahashi, R. (2019). *The Science, Philosophy and Bioethics of Gerontology: An Individual and Community Journey from Japan*. LAP LAMBERT Academic Publishing.

Talbot, M. (2008, February 25). *The Lost Children*. https://www.newyorker.com/magazine/2008/03/03/the-lost-children

Taleb, N. N. (2016). *The most intolerant wins: The dictatorship of the small minority*. Incerto.

Talwar, S. K. (2018). Identity Matters: Questioning Trauma and Violence through Art, Performance and Social Practice. In *Art Therapy for Social Justice: Radical Intersections*. Routledge. doi:10.4324/9781315694184-3

Tan, A. C., Kerridge, I. H., & Pavlakis, N. (2018). Refusal of treatment in a young patient with recurrent medulloblastoma. *Journal of Case Reports and Images in Oncology*, *4*, 1–4. doi:10.5348/Z10-2018-48-CR

Tan, S., Zhao, Y., Fan, F., Zou, Y., Jin, Z., Zen, Y., Zhu, X., Yang, F., Tan, Y., & Zhou, D. (2015). Brain Correlates of Self-Evaluation Deficits in Schizophrenia: A Combined Functional and Structural MRI Study. *PLoS One*, *10*(9), e0138737. doi:10.1371/journal.pone.0138737 PMID:26406464

Tattersall, I. (2007). Homo ergaster and its contemporaries. Handbook of Paleoanthropology, 3, 1633–1653.

Taylor, P. (2012, December 26). *The Growing Electoral Clout of Blacks Is Driven by Turnout, Not Demographics | Pew Research Center*. https://www.pewsocialtrends.org/2012/12/26/the-growing-electoral-clout-of-blacks-is-driven-by-turnout-not-demographics/

Taylor, C. (2007). *A Secular Age*. Harvard University Press.

Taylor, L. E. (2006). The Enneagram: A tool for self-reflection, critical appraisal and introducing spirituality in the classroom. Presented at the *First North American Conference on Spirituality and Social Work*.

Taylor, P. C., Alcoff, L. M., & Anderson, L. (2017). *The Routledge Companion to the Philosophy of Race*. Routledge. doi:10.4324/9781315884424

Telloglou, T., & Kakaounaki, M. (Directors). (2011). Three weeks and one night on duty. In *Folders*. Greece: Skai TV. https://scholar.google.com/scholar_lookup?hl=en&author=Telloglou+T&author=Kakaounaki+M&title=Three+weeks+and+one+night+on+duty

ten Have, H. (2016). *Global Bioethics: An introduction*. Routledge. doi:10.4324/9781315648378

ten Have, H. (2019). *Wounded Planet: How Declining Biodiversity Endangers Health and How Bioethics Can Help*. JHU Press. doi:10.1353/book.66171

ten Have, H., & Purtilo, R. B. (2004). *Ethical Foundations of Palliative Care for Alzheimer Disease*. JHU Press.

Tepperman, J. (2022, December 19). China's Dangerous Decline. *Foreign Affairs*. https://www.foreignaffairs.com/china/chinas-dangerous-decline

Terry, D. (2013, August 21). *DHS Employee Promotes Race War in Spare Time, Advocates Mass Murder of Whites*. Southern Poverty Law Center. https://www.splcenter.org/hatewatch/2013/08/21/dhs-employee-promotes-race-war-spare-time-advocates-mass-murder-whites

Teunissen, T., Lindhout, P., Schipper, K., & Abma, T. (2019). Living a Meaningful Life and Taking Good Care of Oneself in Times of Illness: Highlighting a Dilemma. *International Journal of Feminist Approaches to Bioethics, 12*(1), 44–60. doi:10.3138/ijfab.12.1.03

Tham, J., Durante, C., & Gómez, A. G. (2018). *Religious Perspectives on Social Responsibility in Health: Towards a Dialogical Approach*. Springer. doi:10.1007/978-3-319-71849-1

Tharoor, I. (2016, June 8). Europe's top soccer teams reflect the power of multiculturalism. *Washington Post*. https://www.washingtonpost.com/news/worldviews/wp/2016/06/08/europes-top-soccer-teams-reflect-the-power-of-multiculturalism/

The Crickets. (2016, July 8). *Please Don't Ever Change 1961-1962 Original Recordings Remastered* [Audio CD]. Jasmine Music.

The Economist. (2020). *China wants to put itself back at the centre of the world* (China's Belt and Road). https://www.economist.com/special-report/2020/02/06/china-wants-to-put-itself-back-at-the-centre-of-the-world

The Meaning of the Holy Qur'an: Complete Translation with Selected Notes. (2015). Kube Publishing Ltd.

The Mercury News. (2016, May 13). Cartoons: Playing politics with Donald Trump. *The Mercury News*. https://www.mercurynews.com/2016/05/13/cartoons-playing-politics-with-donald-trump-2/

The State of Obesity. (2018, September). Adult Obesity in the United States. *The State of Obesity*. https://www.stateofobesity.org/adult-obesity/

Themis, A. (2017, May). Brothels. *Athens Social Atlas*. https://www.athenssocialatlas.gr/en/article/brothels/

Theodoridis, A. G., & Nelson, A. J. (2012). Of BOLD Claims and Excessive Fears: A Call for Caution and Patience Regarding Political Neuroscience. *Political Psychology, 33*(1), 27–43. doi:10.1111/j.1467-9221.2011.00860.x

Thomas, B. (2017, April 20). *Eating People Is Wrong—But It's Also Widespread and Sacred*. Sapiens. https://www.sapiens.org/biology/cannibalism-ritualized-sacred/

Thomas, A., & Benjamin, L. (2017). Policies and mechanisms to address climate-induced migration and displacement in Pacific and Caribbean small island developing states. *International Journal of Climate Change Strategies and Management, 10*(1), 86–104. doi:10.1108/IJCCSM-03-2017-0055

Thompson, A. L., & Thompson, A. S. (2015). *But If a Zombie Apocalypse Did Occur: Essays on Medical, Military, Governmental, Ethical, Economic and Other Implications*. McFarland.

Thomson, H. (2019, November 20). *Why the line between life and death is now more blurred than ever*. New Scientist. https://www.newscientist.com/article/mg24432570-300-why-the-line-between-life-and-death-is-now-more-blurred-than-ever/

Thonnard, M., Charland-Verville, V., Brédart, S., Dehon, H., Ledoux, D., Laureys, S., & Vanhaudenhuyse, A. (2013). Characteristics of Near-Death Experiences Memories as Compared to Real and Imagined Events Memories. *PLoS One, 8*(3), e57620. doi:10.1371/journal.pone.0057620 PMID:23544039

Thurer, S. L. (1988). Rebel with a cause: The disabled beggar. *The Journal of Medical Humanities and Bioethics, 9*(2), 153–163. doi:10.1007/BF01139240

Tichtchenko, P. D., & Yudin, B. G. (1992). Towards a Bioethics in Post-Communist Russia. *Cambridge Quarterly of Healthcare Ethics, 1*(4), 295–303. doi:10.1017/S0963180100006472 PMID:1342350

Timmermans, S., & Tietbohl, C. (2018). Fifty years of sociological leadership at Social Science and Medicine. *Social Science & Medicine, 196*, 209–215. doi:10.1016/j.socscimed.2017.11.007 PMID:29254656

TO BHMA. (2012, April 1). *Λοβέρδος - Χρυσοχοΐδης: Υποχρεωτικό πιστοποιητικό υγείας για μετανάστες*. TO BHMA. https://www.tovima.gr/2012/04/01/society/loberdos-xrysoxoidis-ypoxrewtiko-pistopoiitiko-ygeias-gia-metanastes/

Todes, D. (2000). *Ivan Pavlov: Exploring the Animal Machine*. Oxford University Press.

Tognini, G., & Porterfield, C. (2023, May 4). *How Rich Is King Charles III? Inside The New Monarch's Outrageous Fortune*. Forbes. https://www.forbes.com/sites/giacomotognini/2023/05/04/how-rich-is-king-charles-iii-inside-the-new-monarchs-outrageous-fortune/

Toledano, P., & Myers, D. (2011, February 21). *TIME Magazine Cover: 2045: The Year Man Becomes Immortal - Feb. 21, 2011*. https://content.time.com/time/covers/0,16641,20110221,00.html

Tong, R. (2001). Towards a Feminist Global Bioethics: Addressing Women's Health Concerns Worldwide. *Health Care Analysis*, *9*(2), 229–246. doi:10.1023/A:1011390521518 PMID:11561998

Tong, R. (2004). Feminist Approaches to Bioethics. In G. Khushf (Ed.), *Handbook of Bioethics: Taking Stock of the Field from a Philosophical Perspective* (pp. 143–161). Springer Netherlands., doi:10.1007/1-4020-2127-5_7

Tong, R. (2018). Feminist Reflections on the Role of Theories in a Global Bioethics. In S. Sherwin (Ed.), *Globalizing Feminist Bioethics*., doi:10.4324/9780429500176

Tonry, M. (2019). *Proportionality Theory in Punishment Philosophy: Fated for the Dustbin of Otiosity?* (*SSRN* Scholarly Paper ID 3331136). Social Science Research Network. https://papers.ssrn.com/abstract=3331136

Torres, D. (2018, April 17). *Spain's degree scandal hits at heart of government*. POLITICO. https://www.politico.eu/article/cristina-cifuentes-spain-corruption-mariano-rajoy-pp-degree-scandal-hits-at-heart-of-government/

Torres, S. (1999). A culturally-relevant theoretical framework for the study of successful ageing. *Ageing and Society*, *19*(1), 33–51. doi:10.1017/S0144686X99007242

Torres, S. (2003). A Preliminary Empirical Test of a Culturally-Relevant Theoretical Framework for the Study of Successful Aging. *Journal of Cross-Cultural Gerontology*, *18*(1), 79–100. doi:10.1023/A:1024830629385 PMID:14617959

Torrey, E. F. (1998). *Out of the shadows: Confronting America's mental illness crisis*. Academic Press.

Trachtenberg, A. J., & Manns, B. (2017). Cost analysis of medical assistance in dying in Canada. *Canadian Medical Association Journal*, *189*(3), E101–E105. doi:10.1503/cmaj.160650 PMID:28246154

Trelease, J. (2006). *All in the Family: The Bushes and the McGraws*. http://www.trelease-on-reading.com/whatsnu_bush-mcgraw.html

Trothen, T. J., & Mercer, C. (2017). *Religion and Human Enhancement: Death, Values, and Morality*. Springer. doi:10.1007/978-3-319-62488-4

Tsuda, T. (1998). The Stigma of Ethnic Difference: The Structure of Prejudice and "Discrimination" toward Japan's New Immigrant Minority. *Journal of Japanese Studies*, *24*(2), 317–359. doi:10.2307/133237

Tunstall, K. E. (2006). *Displacement, Asylum, Migration: The Oxford Amnesty Lectures 2004*. OUP Oxford.

Turim, M. C. (1985). *Abstraction in avant-garde films*. UMI Research Press.

Turmusani, M. (2018). The Rise of Disability Politics. In *Disabled People and Economic Needs in the Developing World: A Political Perspective from Jordan*. Routledge., doi:10.4324/9781315257730

Turner, L. (2004). Bioethics needs to rethink its agenda. *BMJ (Clinical Research Ed.)*, *328*(7432), 175. doi:10.1136/bmj.328.7432.175 PMID:15202475

Turse, N. (2013, October 9). For America, Life Was Cheap in Vietnam. *The New York Times*. https://www.nytimes.com/2013/10/10/opinion/for-america-life-was-cheap-in-vietnam.html

Tuzzeo, J. (2008). Changing our Minds: Dystopian Psychological Conditioning in Nineteen Eighty-Four, Brave New World, and Walden Two. *English Master's Theses*. https://digitalcommons.brockport.edu/eng_theses/88

Twenge, J. M. (2000). The age of anxiety? The birth cohort change in anxiety and neuroticism, 1952–1993. *Journal of Personality and Social Psychology*, *79*(6), 1007–1021. doi:10.1037/0022-3514.79.6.1007 PMID:11138751

Twiss, S. B. (2018). Explorations. In *Global Ethics: Comparative Religious Ethics And Interreligious Dialogue*. Routledge. doi:10.4324/9780429500626

Tyrovolas, S., Kassebaum, N. J., Stergachis, A., Abraha, H. N., Alla, F., Androudi, S., Car, M., Chrepa, V., Fullman, N., Fürst, T., Haro, J. M., Hay, S. I., Jakovljevic, M. B., Jonas, J. B., Khalil, I. A., Kopec, J. A., Manguerra, H., Martopullo, I., Mokdad, A., ... Kotsakis, G. A. (2018). The burden of disease in Greece, health loss, risk factors, and health financing, 2000–16: An analysis of the Global Burden of Disease Study 2016. *The Lancet. Public Health*, *3*(8), e395–e406. doi:10.1016/S2468-2667(18)30130-0 PMID:30055996

Ubukata, T. (Director). (2016, September 15). *Psycho-Pass* [Animation, Action, Crime]. Dentsu, Fuji Television Network, Index.

Uğur, S., & Kurubacak, G. (2019). Artificial Intelligence to Super Artificial Intelligence, Cyber Culture to Transhumanist Culture: Change of the Age and Human. Handbook of Research on Learning in the Age of Transhumanism, 1–16. doi:10.4018/978-1-5225-8431-5.ch001

Umbrello, S., & Lombard, J. (2018). Silence of the Idols. *Appropriating the Myth of Sisyphus for Posthumanist Discourses.*, *9*(4), 98–121. doi:10.18662/po/47

UN General Assembly. (1979). *Convention on the elimination of all forms of discrimination against women*. Author.

UN Human Rights Office. (2015, September 25). *A step forward for intersex visibility and human rights*. OHCHR. https://www.ohchr.org/en/stories/2015/09/step-forward-intersex-visibility-and-human-rights

United Nations High Commissioner for Refugees. (1984, November 22). *Cartagena Declaration on Refugees, adopted by the Colloquium on the International Protection of Refugees in Central America, Mexico and Panama, Cartagena de Indias, Colombia, 22 November 1984*. UNHCR. https://www.unhcr.org/about-us/background/45dc19084/cartagena-declaration-refugees-adopted-colloquium-international-protection.html

United Nations. (1948, December 10). *Universal Declaration of Human Rights*. https://www.un.org/en/universal-declaration-human-rights/

United Nations. (2006). *Convention on the Rights of Persons with Disabilities*. https://www.un.org/disabilities/documents/convention/convoptprot-e.pdf

US Department of Health and Human Services. (1976). *The Belmont Report*. Office of the Secretary.

US Department of Labor. (2021). *Occupations with the smallest gender earnings gap*. https://www.dol.gov/agencies/wb/data/earnings/smallest-gender-earnings-gap

US Department of Labor. (2022). *Earnings ratios by race, ethnicity, and educational attainment (annual)*. https://www.dol.gov/agencies/wb/data/earnings/earnings-ratio-race-ethnicity-education-annual

Usmani, S. S., Sharath, M., & Mehendale, M. (2022). Future of mental health in the metaverse. *General Psychiatry*, *35*(4), e100825. doi:10.1136/gpsych-2022-100825 PMID:36189180 van den Ameele, S., Keygnaert, I., Rachidi, A., Roelens, K., & Temmerman, M. (2013). The role of the healthcare sector in the prevention of sexual violence against sub-Saharan transmigrants in Morocco: A study of knowledge, attitudes and practices of healthcare workers. *BMC Health Services Research*, *13*(1), 77. doi:10.1186/1472-6963-13-77 PMID:23442386

Van den Branden, S., & Broeckaert, B. (2010). Necessary Interventions: Muslim Views on Pain and Symptom Control in English Sunni E-Fatwas. *Ethical Perspectives*, *17*(4), 626–651.

Van den Branden, S., & Broeckaert, B. (2011). Living in the hands of God. English Sunni e-fatwas on (non-)voluntary euthanasia and assisted suicide. *Medicine, Health Care, and Philosophy*, *14*(1), 29–41. doi:10.100711019-010-9280-1 PMID:20809196

Van Dijk, R. M. (2013). The use of bucrania in the architecture of First Dynasty Egypt. *Journal for Semitics*, *22*(2), 449–463. doi:10.10520/EJC147422

Van Gordon, W., Shonin, E., Dunn, T. J., Sheffield, D., Garcia-Campayo, J., & Griffiths, M. D. (2018). Meditation-Induced Near-Death Experiences: A 3-Year Longitudinal Study. *Mindfulness*, *9*(6), 1794–1806. doi:10.100712671-018-0922-3 PMID:30524512

Vanderstraeten, G. (2009). Dreaming of virtual reality and rehabilitation technology in the future: Virtual reality, robotics and telerehabilitation. *Rehabilitation in the Future*, *8*, 120–123.

Vargas, M. (2019, March 13). *The globalisation of border control and peoples' resistance* (K. Lang, Trans.). Globalisation of Border Control and the Peoples' Resistance. https://www.tni.org/en/article/the-globalisation-of-border-control-and-peoples-resistance

Veatch, R. M. (2000). *Cross-cultural Perspectives in Medical Ethics*. Jones & Bartlett Learning.

Ventres, W., & Gusoff, G. (2014). Poverty Blindness: Exploring the Diagnosis and Treatment of an Epidemic Condition. *Journal of Health Care for the Poor and Underserved*, *25*(1), 52–62. doi:10.1353/hpu.2014.0025 PMID:24509012

Verhoeven, P. (Director). (1997). *Starship Troopers*. https://www.imdb.com/title/tt0120201/

Verkuyten, M. (2008). Life Satisfaction Among Ethnic Minorities: The Role of Discrimination and Group Identification. *Social Indicators Research*, *89*(3), 391–404. doi:10.100711205-008-9239-2

Vesely, P. (2018). Conservative Colonization: An Ethical Analysis of Galactic Expansion. *UReCA: The NCHC Journal of Undergraduate Research & Creative Activity*. https://digitalcommons.unl.edu/ureca/51

Vesely, C. K., Letiecq, B. L., & Goodman, R. D. (2017). Immigrant Family Resilience in Context: Using a Community-Based Approach to Build a New Conceptual Model. *Journal of Family Theory & Review*, *9*(1), 93–110. doi:10.1111/jftr.12177

Victoria Wilches, P. (2019). *El terror bolivariano: Guerra y genocidio contra España durante la independencia de Colombia y Venezuela en el siglo XIX*. Academic Press.

Vieira de Faria, J. W., Teixeira, M. J., de Moura Sousa, L., Pinhata Otoch, J., & Gadelha Figueiredo, E. (2016). Virtual and stereoscopic anatomy: When virtual reality meets medical education. *Journal of Neurosurgery*, *125*(5), 1105–1111. doi:10.3171/2015.8.JNS141563 PMID:26871375

Villaverde, L. E., & Carter, R. A. (2013). Singularity, Cyborgs, Drones, Replicants and Avatars: Coming to Terms with the Digital Self. *Science Fiction and Speculative Fiction*, 119–131. https://doi.org/ doi:10.1007/9789462093805_008

Vlachadis, N., Vrachnis, N., Ktenas, E., Vlachadi, M., & Kornarou, E. (2014). Mortality and the economic crisis in Greece. *Lancet*, *383*(9918), 691. doi:10.1016/S0140-6736(14)60250-6 PMID:24560049

Vollmann, J., & Winau, R. (1996). Informed consent in human experimentation before the Nuremberg code. *BMJ (Clinical Research Ed.)*, *313*(7070), 1445–1447. doi:10.1136/bmj.313.7070.1445 PMID:8973233

Voltaire. (1765). *Dictionnaire philosophique, portatif.* Academic Press.

von Beyme, K. (2019). Definitions of Populism and Its Stages of Development. In K. von Beyme (Ed.), *Rightwing Populism: An Element of Neodemocracy* (pp. 27–34). Springer International Publishing. doi:10.1007/978-3-030-03177-0_2

Vopson, M. M. (2021, May 4). *The world's data explained: How much we're producing and where it's all stored.* The Conversation. http://theconversation.com/the-worlds-data-explained-how-much-were-producing-and-where-its-all-stored-159964

Wade, M. (2016). *Intensive Labours, Expansive Visions: Emerging ideals of the ethical subject amidst the rise of cognitive neuroscience*. https://openresearch-repository.anu.edu.au/handle/1885/105355

Wade, M. (2019). Risky disciplining: On interdisciplinarity between sociology and cognitive neuroscience in the governing of morality. *European Journal of Social Theory*. Advance online publication. doi:10.1177/1368431018810330

Waldby, C. (2000). *The Visible Human Project: Informatic Bodies and Posthuman Medicine*. Psychology Press. doi:10.4324/9780203360637_chapter_1

Walker, A. (1982). *The Color Purple*. Harcourt.

Walker, A. K., & Fox, E. L. (2018). Why Marginalization, Not Vulnerability, Can Best Identify People in Need of Special Medical and Nutrition Care. *AMA Journal of Ethics*, *20*(10), 941–947. doi:10.1001/amajethics.2018.941 PMID:30346922

Walker, C. (2018). What Is "Sharp Power"? *Journal of Democracy*, *29*(3), 9–23. doi:10.1353/jod.2018.0041

Walker, M. (2003). *Genetic virtue*. Institute for Ethics and Emerging Technologies.

Walker, M. J. (2018). On Replacement Body Parts. *Journal of Bioethical Inquiry*. Advance online publication. doi:10.100711673-018-9889-y PMID:30565032

Walter, E. V. (1969). *Terror and resistance; a study of political violence, with case studies of some primitive African communities*. Oxford University Press. https://archive.org/details/terrorresistance0000walt

Walter, M. (2012). Human experiments: First, do harm. *NATNews*, *482*(7384), 148–152. doi:10.1038/482148a PMID:22318585

Wang, Q., & Sun, X. (2016). The Role of Socio-political and Economic Factors in Fertility Decline: A Cross-country Analysis. *World Development*, *87*, 360–370. doi:10.1016/j.worlddev.2016.07.004

Warmington, J. (2021, September 10). *WARMINGTON: School board says it got burned in Indigenous book burning project*. Torontosun. https://torontosun.com/news/national/warmington-school-board-says-it-got-burned-in-indigenous-book-burning-project

Warnecke-Berger, H. (2019). The Cultural Political Economy of Violence: A Conclusion. In H. Warnecke-Berger (Ed.), *Politics and Violence in Central America and the Caribbean* (pp. 261–277). Springer International Publishing. doi:10.1007/978-3-319-89782-0_8

Waters, M. W. (2022). *King of the World: The Life of Cyrus the Great*. Oxford University Press. doi:10.1093/oso/9780190927172.001.0001

Watson, K. (2018, October 24). *Synesthesia: Definition, Examples, Causes, Symptoms, and Treatment*. Healthline. https://www.healthline.com/health/synesthesia

Watson, N., Shakespeare, T., Cunningham-Burley, S., Barnes, C., Corker, M., Davis, J., & Priestley, M. (2001). *Life as a disabled child: A qualitative study of young people's experiences and perspectives*. University of Edinburgh.

Watson, T. J. (2008). Managing Identity: Identity Work, Personal Predicaments and Structural Circumstances. *Organization, 15*(1), 121–143. doi:10.1177/1350508407084488

Watts, J., Sheehan, O., Atkinson, Q. D., Bulbulia, J., & Gray, R. D. (2016). Ritual human sacrifice promoted and sustained the evolution of stratified societies. *Nature, 532*(7598), 7598. Advance online publication. doi:10.1038/nature17159 PMID:27042932

Webb, S. (2002). *If the Universe Is Teeming with Aliens ... WHERE IS EVERYBODY? Fifty Solutions to the Fermi Paradox and the Problem of Extraterrestrial Life*. Springer Science & Business Media.

Weber, B. (2014, November 24). Leslie Feinberg, Writer and Transgender Activist, Dies at 65. *The New York Times*. https://www.nytimes.com/2014/11/25/nyregion/leslie-feinberg-writer-and-transgender-activist-dies-at-65.html

Weber, S. (2018). Participatory Visual Research with Displaced Persons: 'Listening' to Post-conflict Experiences through the Visual. *Journal of Refugee Studies*. Advance online publication. doi:10.1093/jrs/fey038

Weiner, E. (2008, March 9). Heads Up: Poverty Tours - Slum Visits: Tourism or Voyeurism? Travel. *The New York Times*. https://www.nytimes.com/2008/03/09/travel/09heads.html

Weinstein, J., & Colebrook, C. (2017). Preface: Postscript on the Posthuman. In *Posthumous Life: Theorizing Beyond the Posthuman*. Columbia University Press. doi:10.7312/wein17214

Weng, N. G., Bee, O. Y., Yew, L. H., & Hsia, T. E. (2016). An Augmented Reality System for Biology Science Education in Malaysia. *International Journal of Innovative Computing, 6*(2). http://kp.fsksm.utm.my/ijic/index.php/ijic/article/view/128

White, A. (2010, October 7). *EU Stats Office: Greek Economy Figures Unreliable*. ABC News. https://web.archive.org/web/20101007143658/http://abcnews.go.com/Business/wireStory?id=9541636

White, M. (2009, October 14). *The Industrial Revolution*. https://www.bl.uk/georgian-britain/articles/the-industrial-revolution

Whitehouse, P. J. (2006). Quality of life: The bridge from the cholinergic basal forebrain to cognitive science and bioethics. *Journal of Alzheimer's Disease, 9*(s3), 447–453. doi:10.3233/JAD-2006-9S351 PMID:16914884

White, S. E. (1980). A Philosophical Dichotomy in Migration Research. *The Professional Geographer, 32*(1), 6–13. doi:10.1111/j.0033-0124.1980.00006.x

Wiederhold, B. K. (2023). (Mental) Healthcare Consumerism in the Metaverse: Is There a Benefit? [editorial]. *Cyberpsychology, Behavior, and Social Networking, 26*(3), 145–146. doi:10.1089/cyber.2023.29269.editorial PMID:36880891

Wilde, O. (2007). *The Collected Works of Oscar Wilde: The Complete Plays Poems and Stories Including the Picture of Dorian Gray and De Profundis*. Wordsworth Editions.

Williams, B. G. (2010). The CIA's Covert Predator Drone War in Pakistan, 2004–2010: The History of an Assassination Campaign. *Studies in Conflict and Terrorism, 33*(10), 871–892. doi:10.1080/1057610X.2010.508483

Williamson, B. (2017). Moulding student emotions through computational psychology: Affective learning technologies and algorithmic governance. *Educational Media International, 54*(4), 267–288. doi:10.1080/09523987.2017.1407080

Williamson, T. (2005). I *—Armchair Philosophy, Metaphysical Modality and Counterfactual Thinking. *Proceedings of the Aristotelian Society, 105*(1), 1–23. doi:10.1111/j.0066-7373.2004.00100.x

Williams, S. J. (1997). Modern medicine and the "uncertain body": From corporeality to hyperreality? *Social Science & Medicine*, *45*(7), 1041–1049. doi:10.1016/S0277-9536(97)00031-2 PMID:9257396

Wilson, B. (2010, February 4). What is 'Poverty Porn' and are we guilty of indulging in it? *Trespass Magazine*. https://www.trespassmag.com/what-is-%e2%80%98poverty-porn%e2%80%99-and-are-we-guilty-of-indulging-in-it/

Wilson, E. G. (2006). Matter and Spirit in the Age of Animal Magnetism. *Philosophy and Literature*, *30*(2), 329–345. doi:10.1353/phl.2006.0042

Wilson, E., Kenny, A., & Dickson-Swift, V. (2018). Ethical Challenges in Community-Based Participatory Research: A Scoping Review. *Qualitative Health Research*, *28*(2), 189–199. doi:10.1177/1049732317690721 PMID:29235941

Wilson, J. (2014). *Hitler's Alpine Headquarters*. Pen and Sword.

Winter, C. J. (2019, February). *Decolonising Dignity for Inclusive Democracy*. https://doi.org/info:doi/10.3197/096327119X15445433913550

Witmer, B. G., & Singer, M. J. (1998). Measuring Presence in Virtual Environments: A Presence Questionnaire. *Presence (Cambridge, Mass.)*, *7*(3), 225–240. doi:10.1162/105474698565686

Wittgenstein, L. (2007). *Wittgenstein: Lectures and Conversations on Aesthetics, Psychology and Religious Belief*. University of California Press.

Wittgenstein, L. (2009). *Philosophical Investigations*. John Wiley & Sons.

Wolbring, G. (2003). Disability Rights Approach Toward Bioethics? *Journal of Disability Policy Studies*, *14*(3), 174–180. doi:10.1177/10442073030140030701

Wolffe, R. (2018, July 1). Trump's only fixed position on abortion is his disdain for women | Richard Wolffe. *The Guardian*. https://www.theguardian.com/commentisfree/2018/jul/01/donald-trump-abortion-supreme-court

Wolf, S. (1999). Morality and the View From Here. *The Journal of Ethics*, *3*(3), 203–223. doi:10.1023/A:1009833100856

Wolpe, H. (2018). Industrialism and Race in South Africa. In S. Zubaida (Ed.), *Race and Racialism*. doi:10.4324/9781351039345-8

Wolpoff, M. H., & Lee, S.-H. (2001). The Late Pleistocene Human species of Israel. *Bulletins et Mémoires de La Société d'Anthropologie de Paris. Bulletins et Memoires de la Société d'Anthropologie de Paris*, *13*(3–4), 3–4. doi:10.4000/bmsap.6220

Wong, C. C. Y., Paulus, D. J., Lemaire, C., Leonard, A., Sharp, C., Neighbors, C., Brandt, C. P., Lu, Q., & Zvolensky, M. J. (2018). Examining HIV-Related stigma in relation to pain interference and psychological inflexibility among persons living with HIV/AIDS: The role of anxiety sensitivity. *Journal of HIV/AIDS & Social Services*, *17*(1), 1–15. doi:10.1080/15381501.2017.1370680 PMID:30034300

Wood, B. (2011). Did early Homo migrate "out of" or "in to" Africa? *Proceedings of the National Academy of Sciences of the United States of America*, *108*(26), 10375–10376. doi:10.1073/pnas.1107724108 PMID:21677194

World Health Organization. (2022, October 23). *Almost 1 million people die every year due to lead poisoning, with more children suffering long-term health effects*. https://www.who.int/news/item/23-10-2022-almost-1-million-people-die-every-year-due-to-lead-poisoning--with-more-children-suffering-long-term-health-effects

Würth, K., Langewitz, W., Reiter-Theil, S., & Schuster, S. (2018). Their view: Difficulties and challenges of patients and physicians in cross-cultural encounters and a medical ethics perspective. *BMC Medical Ethics*, *19*(1), 70. doi:10.118612910-018-0311-4 PMID:29973220

Xu, J., Jiang, C., Wei, W., & Li, Y. (2022). The Research of Medical Metaverse Application Under the Background of the Normalization of the New Crown Epidemic. In L.-J. Zhang (Ed.), *Metaverse – METAVERSE 2022* (pp. 68–80). Springer Nature Switzerland. doi:10.1007/978-3-031-23518-4_6

X-Wealth. (2021). *Preservation And Succession: Family Wealth Transfer 2021*. https://go.wealthx.com/download-preservation-and-succession-family-wealth-transfer-2021

Yang, L. H., Kleinman, A., Link, B. G., Phelan, J. C., Lee, S., & Good, B. (2007). Culture and stigma: Adding moral experience to stigma theory. *Social Science & Medicine*, *64*(7), 1524–1535. doi:10.1016/j.socscimed.2006.11.013 PMID:17188411

Yilmaz, R. M. (2016). Educational magic toys developed with augmented reality technology for early childhood education. *Computers in Human Behavior*, *54*, 240–248. doi:10.1016/j.chb.2015.07.040

Yu, R. (Director). (2003). *Freddy contra Jason*. https://www.imdb.com/title/tt0329101/

Zachary, P. (1990, January 23). Artificial Reality: Computer Simulations One Day May Provide Surreal Experiences. *Wall Street Journal*.

Zanini, A., Montalti, M., Caola, B., Leadbetter, A., & Martini, M. (2017). Pain during illusory own arm movement: A study in immersive virtual reality. *European Medical Journal*, *2*(2), 90–97.

Zhao, Y. (1999). Labor Migration and Earnings Differences: The Case of Rural China. *Economic Development and Cultural Change*, *47*(4), 767–782. doi:10.1086/452431

Ziegler, M. (2018). Some Form of Punishment: Penalizing Women for Abortion. *The William and Mary Bill of Rights Journal*, *26*(3), 735.

Zimbardo, P. (2011). *The Lucifer Effect: How Good People Turn Evil*. Random House.

Zimmerman, Q. (2023, March 29). *Winds of Change: SWCA Assists Standing Rock Sioux Tribe with Environmental Services for Anpetu Wi Wind Farm*. SWCA. https://www.swca.com/news/2023/03/winds-of-change-swca-assists-standing-rock-sioux-tribe-with-environmental-services-for

Zornoza, M. G. (2021, November 14). *Cuáles son las empresas europeas "cómplices" de los "crímenes de guerra" de Marruecos en el Sáhara Occidental*. https://www.publico.es/internacional/son-empresas-europeas-complices-crimenes-guerra-marruecos-sahara-occidental.html

Zwart, H. (2017a). Conceptual and Methodological Framework: Lacanian Psychoanalysis. In H. Zwart (Ed.), *Tales of Research Misconduct: A Lacanian Diagnostics of Integrity Challenges in Science Novels* (pp. 25–55). Springer International Publishing., doi:10.1007/978-3-319-65554-3_2

Zwart, H. (2017b). Introduction: An Oblique Perspective on Research Misconduct. In H. Zwart (Ed.), *Tales of Research Misconduct: A Lacanian Diagnostics of Integrity Challenges in Science Novels* (pp. 1–23). Springer International Publishing., doi:10.1007/978-3-319-65554-3_1

About the Author

Manuel Lozano Rodríguez, aka Manel, is an author born in 1978 Barcelona whose passion for knowledge and unique perspective has shaped his exceptional academic career. With a self-taught beginnings in many areas of expertise, his academic journey as a polymath has been nothing short of extraordinary. His insatiable curiosity and dedication to exploring uncharted territories have led him to become a disruptive voice in the field of bioethics. Despite encountering various challenges along the way, including severe impairments, Manel's unique combination of talents and disabilities has shaped his perspective and contributed to the distinct character of his work. In 2018, at the age of 40, encouraged by Professor Macer, Manel embarked on a Master's program in Sustainability Sciences at his esteemed American University of Sovereign Nations, where he not only started but also successfully completed his studies. Manel holds a Ph.D. in Bioethics, Sustainability, and Global Public Health, as well as a Master of Science in Sustainability, Peace, and Development, both from the American University of Sovereign Nations and a postgraduate course in Organisational Sustainability in Harvard. Throughout his career, Manel has produced a diverse body of work that reflects his deep understanding of ethical dilemmas and their implications for society. His scholarly papers, while remaining faithful to the principles of scientific rigour, showcase an unorthodox approach. Within the boundaries of academia, Manel weaves humour, lyricism, and moments of raw honesty, providing a refreshing and thought-provoking experience. As he eloquently describes it, "I shape my thoughts with the hammer that I get from Nietzsche's corpse." Manel's contributions include groundbreaking research on topics such as the impact of socioeconomic inequalities on planetary health, the intersection of digital ethnography with issues of suicide, stigma, and warfare, and the effects of the COVID-19 pandemic on marginalized populations. Manel's work has been widely recognized and published in prestigious journals, conferences, and research compilations, solidifying his reputation as a disruptive and visionary academic. In addition to his scholarly pursuits, Manuel has authored several thought-provoking works that explore the multifaceted complexities of our modern world. His writings touch on subjects ranging from the challenges faced by displaced populations in the midst of a changing global landscape to the power of education in fostering love, peace, and hope. As readers delve into "Bioethics of Displacement," they will be guided by Manuel's extensive knowledge, his unique perspective, and his unwavering commitment to addressing the ethical implications of displacement. With each page, they will witness the power of an author who fearlessly challenges conventional norms, offering new insights and perspectives that will inspire and enlighten. In Manuel Lozano Rodríguez, readers will discover an author whose unyielding curiosity, and innovative approach combine to cook an uncommon food for thought. To sum up, through his work, Manel invites readers to engage with complex issues and embrace the possibility of transformative change.

Index

A

Accountability 25, 27-28, 42, 152, 210, 229, 238, 242
Activism 11, 45, 48, 110-111, 130
Agency 11-12, 14, 27, 29, 38, 61, 66, 110, 113, 115, 121, 123, 125, 128, 131-132, 139, 148, 161, 210, 222, 225, 227, 233-234, 260, 262, 272, 276-277
Alien Life 274, 281
Alienation 184, 197, 210
Artificial Intelligence (AI) 22
Asylum 61, 152, 154-156, 161-163, 188, 218, 221, 226, 237, 265
Augmented Reality (AR) 260, 270

B

Belonging 1, 15, 30, 33, 38, 41, 53-56, 58, 61-62, 66-67, 75, 122-124, 128-129, 132, 144, 152, 158, 163-164, 167, 170, 185, 187, 202, 206, 229, 242, 245-247, 257, 276-278, 281-283, 285, 287
Biocitizenship 113, 122
Bioethics 1-9, 11, 14-62, 66-67, 69, 72-73, 82, 87, 92-93, 99-100, 112-114, 117-118, 120, 123-133, 135-143, 147, 151, 161-162, 165-170, 174-175, 177-178, 181, 183, 185-186, 192, 196-197, 199-201, 203-204, 206-208, 210, 212, 223-224, 227, 229, 231, 234, 236-240, 242-243, 246-247, 250-251, 254, 256-257, 259-260, 262, 264, 266-268, 271-274, 276-285, 287-288, 290, 292
Biopolitics 1-2, 7, 16, 41, 43, 92-93, 99, 103, 128, 134, 144, 150-153, 166, 181, 187, 191, 206-207, 213, 221-222, 230, 234, 257, 266, 285
Brain Dynamics 257, 270
Burden 15, 58-62, 71-72, 107, 111, 115, 127, 150-151, 153, 165, 179, 222, 227-229, 260

C

Capitalism 58, 62, 67, 83, 85, 96, 103, 107, 151, 170, 285
Citizenship 148, 151, 156, 163, 187, 196, 229-230, 259
Climate Change 52, 87, 110, 116, 122, 134, 141, 164, 166, 176, 178
Clinical Practice 2, 98, 103, 157
Cognitive Dissonance 239, 246, 250, 254
Cognitive Freedom 258, 270
Colonial Institution 217-218, 238
Communication 4, 6, 35-37, 45, 48-49, 51-53, 55, 57-58, 71, 97-98, 100, 116, 124, 160, 165, 175, 214, 225, 227, 231-232, 248, 257, 265, 268, 278-279, 281, 284, 287
Community 15-16, 38, 40-41, 60, 66, 68-69, 95, 105, 112, 115-116, 118, 122, 130, 135, 157, 162, 165, 170, 176, 186-187, 196, 206, 208, 236, 251, 286, 293
Computing Era 82, 90
Concupiscence 39, 48
Confinement 137, 140, 157, 202, 217-224, 229-236, 238-239, 272
Conscious Machines 239, 244, 254
Consent 59, 71, 91, 107, 114, 134, 136, 221, 241-242, 248, 251, 254-255, 279
Corporeality 168, 179, 257, 260-262, 266, 270, 289
Corruption 42, 64, 79, 83, 87, 108, 111, 113, 122, 163, 174, 222-223, 230, 234
Cultural Dynamics 66, 199, 201, 203, 216
Cultural Humility 165, 179
Cultural Traditions 181-182, 191, 197
Cyber-Physical 22, 27
Cyborg 134-135, 164, 167-169, 174, 176, 285, 293

D

Data 10, 12, 22, 28, 31-33, 43, 51, 68, 70, 72, 82-84, 88, 90, 94-95, 115, 134, 136, 138, 141, 145, 149, 153, 156, 166, 168, 186, 202, 208-210, 214, 223, 225, 243, 263, 282-284, 290
Data Analytics 22
Death 6, 8, 11, 16-17, 20, 46, 59, 67-68, 75, 77, 80-81,

88, 93, 95, 108, 115, 117-118, 134-135, 138, 149, 152, 168-170, 179, 220, 222, 239-240, 242-246, 248-250, 252-255, 273, 276-277, 284, 289
Decision-Making 2, 11, 15-16, 22, 24-38, 45, 48-53, 55-56, 66, 71, 95, 101, 122, 128-129, 132, 161, 180, 200, 208, 210, 223, 242, 248-249, 251, 276-277, 279, 290
Disability 15, 60-61, 67-70, 99, 126-127, 130, 164, 166, 168, 170, 176, 186, 189, 193, 195-196, 202, 204, 214, 217-218, 224, 226-238, 248, 273, 288
Discrimination 11-12, 18, 23, 25, 61, 64-67, 93, 126-127, 131, 144, 149-151, 155-156, 158-159, 162, 165, 182, 186-187, 189, 192-196, 199-204, 206, 210, 212-216, 222, 225-227, 230-231, 235, 288
Displacement 1-2, 5, 24-30, 32-33, 35-38, 42, 49-50, 53-58, 60-62, 66-68, 72-73, 82, 84, 92-94, 98-99, 112-114, 123-124, 126-130, 132, 136-139, 143-144, 148-152, 154, 156-157, 159-162, 164-166, 168, 170, 173-176, 178, 181-184, 186-187, 189, 199-204, 206, 208, 210, 220, 223-224, 232, 239, 242-243, 246-249, 255, 257, 262, 273-274, 276, 282-288, 290-291
Diversity 2, 4, 6-7, 14, 17, 21, 26-27, 37, 110, 128, 144, 148, 163, 177, 195-196, 203, 226, 228, 231, 242, 248, 260

E

Early Globalisation 77-78, 90
Ecopoiesis 274, 281
Energy 49, 53-56, 62, 72, 79
Engagement in Bioethics of Displacement 143
Epigenetics 204, 206, 216, 290-291
Epistemology 39, 202, 287, 290, 293
Equality 11, 14, 42, 44, 58, 60, 71, 107, 110, 127, 156, 174, 185, 207, 211, 229
Ethical Decision-Making 24-30, 32-33, 35-38, 45, 48-53, 55, 128-129, 132, 248-249, 276-277
Ethical Dilemmas 2, 4, 32-33, 35, 38, 117, 137, 139, 143, 181, 183, 191, 197, 278, 284, 289
Ethical Frameworks 52-53, 57
Ethics 3-6, 9, 11, 15-23, 26, 29, 33-36, 40, 47, 52, 56-57, 59, 68-69, 87, 116-118, 120-121, 126-129, 131, 137, 140-142, 148, 157, 159-162, 165, 178, 183, 192, 195, 212, 215, 226, 231, 233-235, 240, 246, 249, 256, 258, 263, 266, 268-269, 272, 274, 276-280, 289-291
Extraterrestrial Life 271, 274, 276, 280-281

F

Feminism 9-11, 14, 21, 48, 64, 71, 107, 109-110, 127, 184, 215
Force 49, 53-55, 57, 66-67, 79, 152, 170, 204, 218, 249, 257
Forced Migration 144-145, 151-152, 154, 159-160, 163, 182, 236
Fourth Industrial Revolution (4IR) 22
Freedom 12, 20, 23, 46, 55, 58, 62, 67, 71, 75, 79, 105, 111, 114, 128, 131, 133, 148, 153, 167, 187, 202, 211-212, 221, 223-224, 229, 232-234, 258, 270

G

Good Bioethics 49-50, 52-53, 55, 57-58, 279
Governance 58, 60, 67, 104, 113, 116-117, 122, 136, 156, 208, 210, 215, 223, 230, 279, 290
Group Dignity 134-135, 143-144

H

History 11-12, 18, 55, 63, 69, 72-75, 77-78, 81-89, 91, 100, 102, 108-109, 120, 125, 140, 158, 182, 187, 195, 219, 231, 236, 238-239, 244, 246-247, 251, 257, 274, 282-283, 290-291
Human Body 93-94, 97, 99, 101, 167-168, 183, 185, 197, 282
Human Rights 5, 9, 11, 18, 20, 23, 32, 37, 47, 75, 93, 95, 110, 121, 127-128, 130-131, 143, 147, 152, 158, 161, 165, 174, 190, 214, 234, 237, 243, 269
Humanitarian Crises 147, 226-227, 238

I

Identity 2, 14, 16-17, 45, 58, 60-61, 63, 70, 95, 108, 110, 113, 119, 121, 126-127, 131, 153, 155, 157, 166-167, 169, 177, 181-182, 185-197, 199, 202-206, 211-212, 214, 216-217, 220, 225, 227-228, 233-234, 245-246, 249, 268, 288, 291
Ideology 48, 105, 109, 127-128, 139, 188, 248-249
Inclusivity 62, 71, 210
Indigenous Populations 182-184, 197
Individual 26-27, 29, 38-39, 42, 51, 53-55, 62, 64-66, 76, 90, 98, 114, 123-125, 127-129, 131, 133, 136, 139, 143-144, 147, 162, 180-182, 184, 187, 199, 201, 203, 206, 208, 210, 213, 220, 222, 224, 242, 247, 271, 274, 284, 288, 290, 293
Individual Dignity 133, 143
Inflexibility 35, 38, 45, 83, 92, 94-99, 102-105, 107, 112, 116, 123, 125, 203, 239, 246, 248

Injustice 37, 102, 123, 125, 142, 150, 163, 179, 208, 229, 285
Institution 71, 84, 122, 157, 217-218, 221, 232, 238, 263, 265
Intercorporeal Stigma 204, 206-207, 216
Interpersonal Stigma 207, 216
Intersectionality 58, 62-69, 71, 82, 104-105, 108, 191, 226-227, 229

J

Justice 2, 4-5, 11, 14, 20, 46, 58, 60, 62, 69-71, 82, 111, 126-128, 130, 142-143, 148-149, 160, 164-165, 167, 174-175, 194, 197, 218-220, 222, 227, 229, 231, 237, 240, 242, 247, 267

L

Life 1-4, 6, 8, 11-13, 15-21, 25, 33, 40-42, 44-48, 54-55, 60, 63-65, 69-71, 73-74, 81, 83-84, 90, 93-94, 96, 98, 103, 105, 107-113, 124-129, 131, 133, 135, 137, 140, 142-143, 150, 154-155, 158-162, 164-165, 168, 170, 172, 177, 179, 182-185, 187, 190-191, 196-197, 200, 203-204, 206, 212, 218, 220, 223, 225, 227, 229, 234, 237, 239-240, 242-245, 250-251, 253-256, 259, 261, 270-277, 280-281, 283-288, 290, 292-293
Liquid Maintenance 22, 24, 27-28, 31, 33
Local Invariance 28, 30-33, 38, 57

M

Macrocosm 282-283, 290, 293
Marginalisation 15, 60, 66, 68, 71, 126, 186, 197, 200, 209
Meta-Ethics 57, 66
Microcosm 283, 287, 290, 293
Migration 61, 68, 73, 86, 144-145, 147-167, 174-178, 181-182, 194, 196, 226, 233, 236-237, 268, 273, 287
Mindset 29, 38, 44, 103, 108, 123, 127-129, 131-133, 135, 155, 182-183, 219, 227, 246, 286
Moral Imagination 164-165, 180, 285
Moral Philosophy 2, 20, 25, 27, 33, 36, 38

N

Natural World 73, 91, 94, 284
Neoliberalism 93, 104, 107-108, 111-112, 116, 122, 124, 151, 220, 248
Neurobioethics 256, 259, 265-266, 270
Neuroethics 114, 118, 120, 232, 234, 249, 251, 256, 258-259, 263-270
Nihilism 76, 239, 255

O

Off-Earth Bioethics 271-274, 279, 281-282
Oppression 8, 62, 64-67, 71, 108-109, 182, 188

P

Pain 9, 11, 22, 26, 60, 84, 92-93, 95-105, 107, 112, 123, 126, 173-174, 190, 194, 202-204, 206, 212, 222, 246, 248, 260, 264, 270, 285-288, 290
Panspermia 274-276, 281
Person 5, 28-30, 38, 42, 45, 53-54, 62-63, 75, 98, 108, 110-111, 113-115, 124-125, 131, 133, 135, 143, 147, 151-152, 167, 182, 185, 187, 189-191, 203, 206, 221, 224, 227, 229, 238, 242, 247-248, 251, 257, 261-262, 273, 275, 279, 281, 288-290
Phenomenologically Transparent 257, 270
Phenomenology 16, 56, 140, 144, 154-155, 165, 176, 178, 189, 196, 223, 230, 232, 236, 238, 269
Philosophical Discourse 36, 45, 48
Philosophy 2, 5, 15-18, 20, 22, 25, 27, 33, 36-38, 42, 47, 53, 60, 68, 94, 101, 112, 120, 125, 128, 137-140, 142, 145, 147-149, 156, 158, 161-162, 175-176, 178, 180, 187, 192-197, 213, 220, 223, 227-235, 237-239, 244, 250, 253-256, 258, 263-264, 269, 285, 290-291
Philosophy of Necessity 220, 233, 238
Political Dignity 132, 135, 143
Politics 1, 7, 9, 19, 38, 40, 42-43, 45, 47, 56, 60, 64, 70, 82, 84, 92-93, 95-97, 99, 103-105, 107, 110, 112, 114, 116, 118, 120, 123, 126, 128-129, 153-154, 158-160, 170, 173-174, 176, 182-184, 187-188, 193-194, 196, 202-204, 222-223, 226, 228-229, 232-233, 239, 241, 258, 276, 285-287, 289-290, 292
Post-Structuralism 255
Poverty 64, 79-80, 97, 100, 102, 108, 121, 135, 142-143, 152, 164, 167, 174, 228, 232, 286
Power 1-2, 8, 11, 18, 32, 35, 38-39, 42, 44, 62, 71-72, 74-78, 81-84, 93-94, 99, 103, 105, 107-109, 111, 114, 118, 125, 127-129, 131, 133-134, 137, 139, 143, 149, 151-152, 162, 164, 166, 170, 178, 182, 184, 199-201, 203-206, 208, 210-211, 213, 216, 226, 239, 244, 246-247, 251, 255, 258, 268, 272, 283, 288, 290, 292
Power Dynamics 32, 42, 72, 84, 99, 182, 184, 200-201, 204, 206, 210, 216, 239

Predictive Maintenance (PdM) 22, 27
Preemptive Bioethics 24-25, 27, 29, 32, 34-36, 38, 45, 128, 200
Presence 75, 78, 104, 118, 207, 260-265, 268, 270, 275
Public Health 60-61, 93, 97-98, 100, 114, 119-120, 136, 144-145, 149-150, 164, 170, 172, 177, 179-181, 186, 191, 199, 201, 211, 221, 223, 230, 234, 241, 243, 253, 283-284, 291-292

R

Rebellion 123, 125, 127, 132, 163
Refugee 66, 130, 142, 144-145, 159-160, 173, 178, 180, 222, 226-227, 233, 237, 285, 287
Resilience 69, 116, 122, 162, 200-201, 236
Resistance 12, 90, 107, 112, 123-124, 131, 138, 162, 195, 201, 272
Respect 27, 29-30, 52, 71, 126, 135, 143, 165, 170, 241, 248, 276, 289

S

Silk Road 72, 76-79, 91, 108
Slavery 73, 79, 81, 91-92, 109, 111, 119-120, 138, 218, 229-230
Social Justice 2, 46, 58, 69-70, 111, 126-128, 130, 143, 148, 160, 197, 227, 229, 267
Social Stakeholders 242, 255
Social Structure 123-124, 129, 197, 231
Society 1, 17-18, 25, 27-28, 40, 42-43, 45-46, 56, 60-61, 65-66, 68, 70-71, 75, 83, 89, 92-93, 98-99, 101, 103-105, 112-117, 120-124, 128, 131, 138, 140, 142, 147-149, 156-157, 160, 162, 176, 178, 189, 193-194, 196-197, 204, 206, 210-212, 220, 222, 224, 229, 231-233, 239, 241, 244, 253, 255, 257, 266-267, 270, 273, 287, 290, 293
Solidarity 58, 60, 62, 67, 164-165, 175, 180, 195, 282-283
Spiritual Machine 244, 255
Stigma 97, 100, 102, 188, 199-200, 202-204, 206-208, 211-217, 231
Structural Injustice 150, 163, 285
Structural Stigma 203-204, 207, 212, 214, 216
Structural Theories 58, 60, 71
Surrogate Realities 125, 136, 208, 256-260, 262-263, 270-271, 289
Syncretic Ethical Decision-Making 30, 35-36, 45, 48
Syncretist Bioethics 34

T

Taxonomies 23
Terraforming 274, 280-281
Transcendence 118, 151, 168, 207, 289, 293
Transcendental 49, 53-55, 57, 127-128, 140, 274, 279, 284
Transhumanism: Using 293

V

Values 2, 4, 9, 11, 19, 27, 29, 33, 36-38, 40, 48, 51-52, 61-62, 105, 112, 117, 126-127, 129, 131, 143, 165, 178-180, 190, 197, 201, 210, 216-217, 247-249, 255, 262, 277, 282-283, 290, 292
Virtual Reality (VR) 270
Vulnerability 35, 37, 58, 61, 64, 127-128, 130-131, 134, 138, 150, 158, 161, 182, 218, 223, 225-226, 263, 293

W

War Crime 95, 103
Wholeness 242, 255

X

Xenobioethics 271, 274-275, 279, 281-282
Xenophobia 147, 163

www.ingramcontent.com/pod-product-compliance
Lightning Source LLC
LaVergne TN
LVHW080534300926
849162LV00007B/1604
* 9 7 8 1 6 6 8 4 4 8 0 8 3 *